DIELECTRIC RESONATORS

The Artech House Microwave Library

Introduction to Microwaves by Fred E. Gardiol
Microwaves Made Simple by W. Stephen Cheung and Frederic Levien
Microwave Tubes by A. Scott Gilmour, Jr.
Electric Filters by Martin Hasler and Jacques Neirynck
Microwave Technology by Erich Pehl
Receiving Systems Design by Stephen J. Erst
Applications of GaAs MESFETs by R.A. Soares, et al.
GaAs Processing Techniques by R.E. Williams
GaAs FET Principles and Technology by J.V. DiLorenzo and D. Khandelwal
Dielectric Resonators, Darko Kajfez and Pierre Guillon, eds.
Modern Spectrum Analyzer Theory and Applications by Morris Engelson
Microwave Materials and Fabrication Techniques by Thomas S. Laverghetta
Handbook of Microwave Testing by Thomas S. Laverghetta
Microwave Measurements and Techniques by Thomas S. Laverghetta
Principles of Electromagnetic Compatibilty by Bernhard E. Keiser
Design and Analysis of Linear Active Circuits by William Rynone, Jr.
The Design of Impedance-Matching Networks for Radio-Frequency and Microwave Amplifiers by P.L.D. Abrie
Microwave Filters, Impedance Matching Networks and Coupling Structures by G.L. Matthaei, Leo Young, and E.M.T. Jones
Microwave Engineer's Handbook, 2 vol., Theodore Saad, ed.
Handbook of Microwave Integrated Circuits by R.K. Hoffmann
Microwave Integrated Circuits, Jeffrey Frey and Kul Bhasin, eds.
Computer-Aided Design of Microwave Circuits by K.C. Gupta, R. Garg, and R. Chadha
Microstrip Lines and Slotlines by K.C. Gupta, R. Garg, and I.J. Bahl
Microstrip Antennas by I.J. Bahl and P. Bhartia
Antenna Design Using Personal Computers by David M. Pozar
Microwave Circuit Design Using Programmable Calculators by J. Lamar Allen and Max Medley, Jr.
Stripline Circuit Design by Harlan Howe, Jr.
Microwave Transmission Line Filters by J.A.G. Malherbe
Electrical Characteristics of Transmission Lines by W. Hilberg
Multiconductor Transmission Line Analysis by Sidney Frankel
Microwave Diode Control Devices by Robert V. Garver
Tables for Active Filter Design by Mario Biey and Amedeo Premoli
Active Filter Design by A.B. Williams
Laser Applications by W.V. Smith
Ferrite Control Components, 2 vol., Lawrence Whicker, ed.
Microwave Remote Sensing, 3 vol., by F.T. Ulaby, R.K. Moore, and A.K. Fung

DIELECTRIC RESONATORS

Darko Kajfez
and Pierre Guillon,
Editors

ARTECH HOUSE, INC.
610 Washington Street
Dedham, MA 02026

International Standard Book Number: 0-89006-201-3
Library of Congress Catalog Card Number: 86-70447

86 87 10 9 8 7 6 5 4 3 2 1

The Authors

Hesham A. Auda, Assistant Professor, University of Mississippi, University, MS.

Allen W. Glisson, Jr., Associate Professor, University of Mississippi, University, MS.

Pierre Guillon, Professor, University of Limoges, France.

Donald F. Hanson, Associate Professor, University of Mississippi, University, MS.

Darko Kajfez, Professor, University of Mississippi, University, MS.

Amarpal S. Khanna, Senior Member of Technical Staff, Avantek, Santa Clara, CA.

Krzysztof A. Michalski, Assistant Professor, University of Mississippi, University, MS.

Contents

The Authors

Hesham A. Auda, Assistant Professor, University of Mississippi, University, MS.

Allen W. Glisson, Jr., Associate Professor, University of Mississippi, University, MS.

Pierre Guillon, Professor, University of Limoges, France.

Donald F. Hanson, Associate Professor, University of Mississippi, University, MS.

Darko Kajfez, Professor, University of Mississippi, University, MS.

Amarpal S. Khanna, Senior Member of Technical Staff, Avantek, Santa Clara, CA.

Krzysztof A. Michalski, Assistant Professor, University of Mississippi, University, MS.

Contents

Chapter 1
INTRODUCTION
Darko Kajfez

Reducing the cost of microwave circuits goes hand in hand with reducing their size. In this respect, microstrip and stripline have been essential in eliminating bulky waveguides and rigid coaxial lines in a great majority of microwave systems. Only in a few and very demanding applications, such as high power transmission, or low-loss filtering, are waveguides still being used. A more recent advance in miniaturization of microwave circuits has been the appearance of the low-loss temperature-stable dielectric resonators. These dielectric resonators are used to replace waveguide filters in such demanding applications as satellite communications where microstrip and stripline resonators cannot be used because of their inherently high losses. Furthermore, carefully designed microwave oscillators which utilize dielectric resonators can equal the temperature stability of conventional microwave resonant cavities machined from invar.

Handbooks and textbooks on microwave devices often devote many pages to the theory of hollow resonant cavities, but they provide little information on cavities containing dielectric resonators. On the other hand, the analysis and design of dielectric resonators are discussed extensively in many professional journals and conferences. The present book attempts to bring this knowledge together in an organized manner.

The size of a dielectric resonator is considerably smaller than the size of an empty resonant cavity operating at the same frequency, provided the relative dielectric constant of the material is a number substantially larger than unity. Only recently, materials having a dielectric constant between 30 and 40 with good temperature stability and low dielectric losses have become available.

The shape of a dielectric resonator is usually a short, solid cylinder, but one can also find tubular, spherical, and parallelepiped shapes. A commonly used resonant mode in cylindrical dielectric resonators is denoted $TE_{01\delta}$. The magnetic field intensity for this mode is sketched in Fig. 1.1. For a distant observer this mode appears as a magnetic dipole, and for this reason some authors call it a "magnetic dipole mode," instead of using the term $TE_{01\delta}$ introduced by Cohn. The electric field lines are simple circles concentric with the axis of the

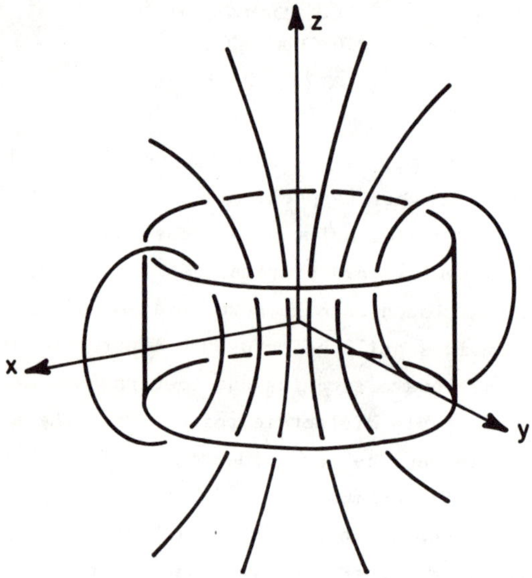

Fig. 1.1 Magnetic field lines of the resonant mode $TE_{01\delta}$ in an isolated dielectric resonator

cylinder. When the relative dielectric constant is around 40, more than 95 % of the stored electric energy of $TE_{01\delta}$ mode, as well as a great part of the stored magnetic energy (typically over 60 %), are located within the cylinder. The remaining energy is distributed in the air around the resonator, decaying rapidly with distance away from the resonator surface.

Although the geometrical form of a dielectric resonator is extremely simple, an exact solution of the Maxwell equations is considerably more difficult than for hollow metal cavities. This difficulty holds true for an isolated dielectric resonator such as shown in Fig. 1.1, and even more so for a dielectric resonator mounted on a microstrip, or placed within a shielding metal cavity. For this reason, the exact resonant frequency of a certain resonant mode, such as $TE_{01\delta}$, can only be computed by rather complicated numerical procedures. For an approximate estimation of the resonant frequency of the isolated dielectric resonator, the following simple formula can be used:

$$f_{GHz} = \frac{34}{a_{mm}\sqrt{\epsilon_r}}\left(\frac{a}{L} + 3.45\right) \qquad (1.1)$$

The radius of the resonator is denoted by a, and its length by L. The lengths are expressed in millimeters, and the frequency in gigahertz. The relative dielectric constant of the material is ϵ_r. The above formula is accurate to about 2 % in the range

$$0.5 < a/L < 2 \text{ and } 30 < \epsilon_r < 50$$

The formula has been obtained by fitting a straight line to the results of the numerical solution which is described in detail in Ch. 6.

The simplest way to incorporate the dielectric resonator into a microwave network is placing it on top of a microstrip substrate as shown in Fig. 1.2. The lateral distance between the resonator and the microstrip conductor determines the amount of coupling between the resonator and the microstrip transmission line. In order to prevent losses due to radiation, the entire device is usually enclosed in a shielding box made of a metal, most often aluminum.

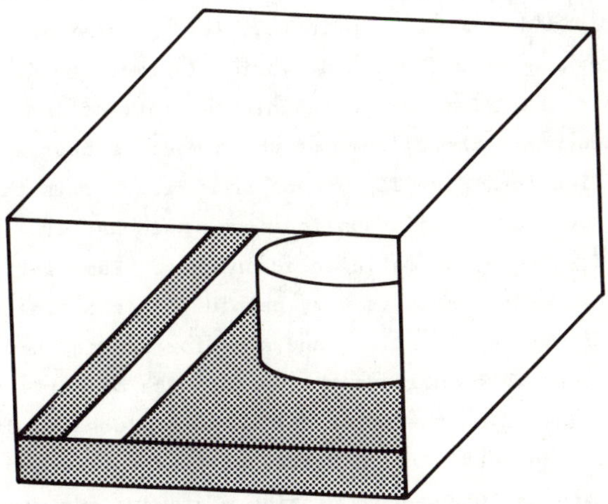

Fig. 1.2 Dielectric resonator mounted on microstrip

By bringing the metal enclosure close to the dielectric resonator, the resonant frequency of the $TE_{01\delta}$ mode is modified from the value given by (1.1) to a new, increased value. The reason for such behavior of the resonant frequency can be found in the cavity perturbation theory. Namely, when a metal wall of a resonant cavity is moved inward, the resonant frequency will decrease if the stored energy of the displaced field is predominantly electric. Otherwise, when the stored energy close to the metal wall is mostly magnetic, as is the case for the shielded $TE_{01\delta}$ dielectric resonator considered here, the resonant frequency will increase when the wall moves inward.

The magnetic field of a dielectric resonator located inside a cylindrical metal cavity is shown in Fig. 1.3. The resonant mode is the same $TE_{01\delta}$ as in Fig. 1.1, but the field is obviously modified due to the presence of metal boundaries. The dielectric resonator shown here is made of material having a relative dielectric constant of 38, resting on a substrate having a relative dielectric constant of 10. The size of the resonator (radius 5.25 mm and height 4.6 mm) is such that, in free space, its resonant frequency would be 4.83 GHz. For the cavity shown in Fig. 1.3, the resonant frequency becomes 5.36 GHz. The computational procedure which has been used to obtain these results is discussed in Ch. 5 of this book.

As is the case with all resonant cavities, there are many possible resonant modes which can be excited in dielectric resonators. These modes can be divided into three main families: transverse electric (TE), transverse magnetic (TM), and hybrid electromagnetic (HEM) modes. Each of the three families has an infinite variety of individual modes so that one encounters the dilemma of which mode is best suited for a particular application. The $TE_{01\delta}$ mode from Fig. 1.1 is the one used traditionally, but for certain applications, such as for a dual-mode filter, the $HEM_{11\delta}$ mode has definite advantages. Familiarity with individual lowest-order modes is very helpful in selecting the proper mode for a particular application, and also for knowing how to eliminate the undesired modes which degrade the performance of certain devices.

This book is based on the notes for a short course on dielectric resonators at the University of Mississippi first presented in the spring of 1985 and is intended to provide microwave engineers with information on the design procedures of devices containing dielectric resonators. Such a presentation also requires an explanation of

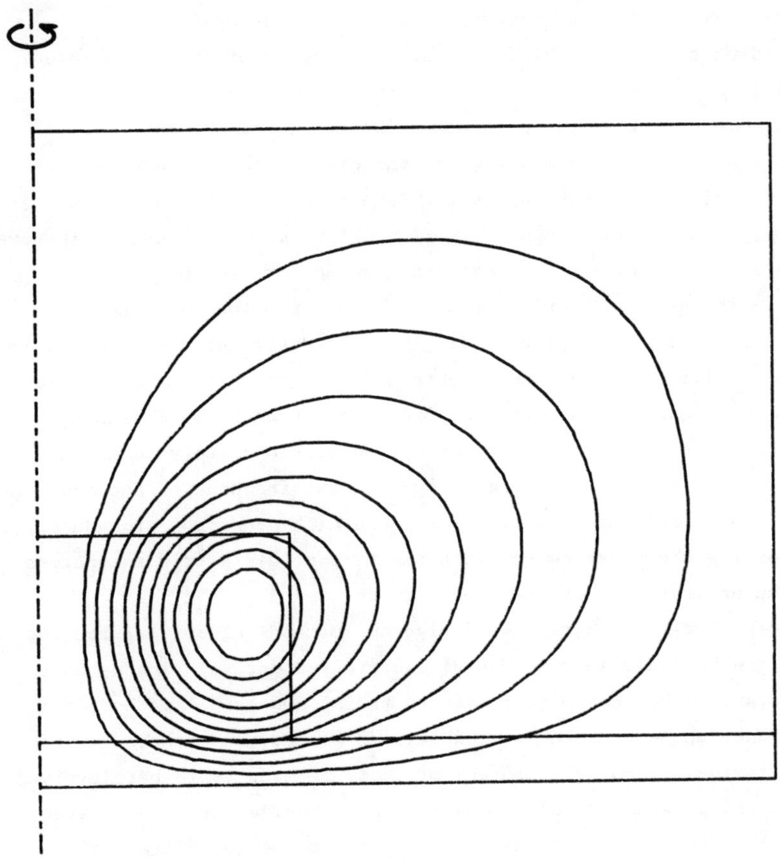

Fig. 1.3 Dielectric resonator within a cylindrical metal cavity

theoretical principles which have been used in deriving the practical results. A brief description of the book's contents is given in the following.

Chapter 2 reviews the properties of microwave resonators and gives some examples of the most commonly used cavities. Computation and measurement of the Q factor are discussed in detail.

Chapter 3 develops the description of the electromagnetic field in a dielectric rod waveguide. The solution of the Maxwell equations in this infinitely long structure is very similar to the well-known solutions for hollow metal waveguides. An important consequence of the field solution is the classification of modes which can propagate in this structure. Field plots of several modes provide a physical understanding that is useful for later description of fields in actual dielectric resonators.

Chapter 4 describes two simple approximate models of the electromagnetic field distribution in the dielectric resonator mounted on a microstrip. One is known as the Cohn model, and the other is called the Itoh and Rudokas model. With slight modifications to improve the accuracy, both models have been programmed for use on a personal computer. The program listings are given in Appendices 4.A and 4.B.

For some practical applications, like the design of a dielectric resonator oscillator, the results obtained with approximate models may be entirely adequate. For other applications, such as in designing filters for satellite communications, it becomes necessary to use more sophisticated procedures for computing resonant frequencies and coupling parameters with sufficient accuracy. These high-precision procedures for evaluating the electromagnetic parameters of dielectric resonators in various environments are described in Ch. 5 and 6.

Chapter 5 presents a review of rigorous methods of solving for the electromagnetic field in shielded dielectric resonators. The methods described are: the mode-matching (both radial and axial) method, the Green's function method, the perturbational method, and others. While the mathematical formulation in any of these methods is rather involved, once they are programmed on a computer, they provide rather accurate results. Chapter 5 presents sample results obtained by different authors in the form of diagrams which may be useful in design work.

Numerical solution for the resonant modes in isolated dielectric resonators is presented in Ch. 6. The integral equation method, which is used in this case, is generally applicable because it handles radiation effects, applies to a dominant mode as well as any of the higher-order modes, and, finally, because the shape of the resonator does not have to be cylindrical, but instead may be an arbitrary body of revolution. The mode chart and the field plots for several lowest-order modes, generated by the computer, are also included in the chapter.

Ceramic materials used for manufacturing dielectric resonators must have a very low loss tangent and a controllable temperature stability. Mathematical descriptions of these two properties, together with the procedures for their measurement, are given in Ch. 7. A table of data for commercially available materials can be found in Appendix 7.A.

The computation of coupling between the dielectric resonator and an external circuit is the topic of Ch. 8. The formulas and diagrams provide practical data for coupling of dielectric resonators to microstrips, inductive loops, to waveguides below cutoff, and finlines. Coupling through narrow irises and coupling between dual modes are also included.

The application of dielectric resonators in designing microwave filters is the subject of Ch. 9. Band-pass and band-stop filters in microstrip and in rectangular waveguides are discussed. Design information is provided on special filters, like elliptic type and dual-mode filters.

Chapter 10 provides comprehensive coverage of the design procedure for stable microwave transistor oscillators using dielectric resonators. The procedure utilizes measured three-port S-parameters of the transistor. Description of a procedure for measurement of oscillator performance with the network analyzer is also included in the chapter.

It may be useful to explain briefly some mathematical symbols used in this book. Scalar quantities are described by the ordinary Roman and Greek letters, like E_0, β, γ, Γ, S_{11}. No distinction in the notation is made between real and complex numbers. The asterisk * is used to denote a complex conjugate value. Three-dimensional field vectors are underlined, and the three-dimensional unit vectors wear a hat ^ as in

$$\underline{E} = \hat{x}E_x + \hat{y}E_y + \hat{z}E_z \tag{1.2}$$

Matrices carry a wavy underline symbol. For instance

$$\underset{\sim}{S} = \begin{pmatrix} S_{11} & S_{12} & S_{13} \\ S_{11} & S_{12} & S_{13} \\ S_{31} & S_{32} & S_{33} \end{pmatrix} \tag{1.3}$$

Column matrices (also called column vectors) are denoted by the Dirac symbol:

$$|a\rangle = \begin{pmatrix} a_3 \\ a_2 \\ a_3 \end{pmatrix} \tag{1.4}$$

The conjugate transpose of $|a\rangle$ is denoted by $\langle a|$, and it represents a row matrix (also called row vector):

$$\langle a| = (a_1^* \quad a_2^* \quad a_3^*) \tag{1.5}$$

The preparation of notes for the short course on dielectric resonators, as well as their transformation into a manuscript of this book, was made possible through the support and encouragement of Dr. Charles E. Smith, Chairman of the Department of Electrical Engineering at the University of Mississippi.

Some of the numerical and experimental results presented in Ch. 2, 3, 4, 6, and 7 were obtained by Joseph James, Mihailo Crnadak, W. Perry Wheless, R. Todd Ward, and Man-Chun Yu. The majority of illustrations were drawn by M. Kris Aune and Michael G. Metcalfe. The difficult job of setting the text and formulas on a relatively inexpensive word processing system was accomplished by Priscilla Ragsdale.

The author of this introductory chapter feels indebted to Dr. Ferdo Ivanek of Harris Corporation, Farinon Division, for instigating an interest in dielectric resonators and for his advice on problems of special importance for practical applications.

Material presented in Ch. 4 and 6 is based upon work supported by the National Science Foundation under Grants ECS-8304442 and ECS-8443558 and performed at the Department of Electrical Engineering, University of Mississippi. The results in Ch. 8, 9, and parts of Ch. 10 originated from the work supported by the French government institutions CNET, CNES, CNRS, and DRET. Corresponding research was performed in the Laboratory of Optical and Microwave Communications at the University of Limoges.

Chapter 2
MICROWAVE RESONATORS
Donald F. Hanson

2.1 Introduction

Resonators are important components in microwave communication circuits. They create, filter, and select frequencies in oscillators, amplifiers, and tuners. Fields inside a resonator store energy at the resonant frequency where equal storage of electric and magnetic energies occurs. The input impedance at resonance is purely real since the reactance is proportional to the difference between electric and magnetic energy storage.

The Q factor is an important figure of merit for a resonant circuit. The Q factor relates a resonant circuit's capacity for electromagnetic energy storage with its energy dissipation through heat. Microwave Q factors can be as high as 10,000. At lower frequencies, the Q factor is usually between 50 and 500. Resonator bandwidth is inversely proportional to Q factor. Thus, high Q factor resonators have narrow bandwidths. Resonators and the Q factor and its measurement are discussed in this chapter.

2.2 Q Factor

The figure of merit for assessing the performance or quality of a resonator is the quality factor, Q, which is a measure of energy loss or dissipation per cycle as compared to the energy stored in the fields inside the resonator. Q factor is defined by

$$Q = 2\pi \frac{\text{maximum energy storage during a cycle}}{\text{average energy dissipated per cycle}} \tag{2.1}$$

$$= \frac{2\pi W_0}{PT} = \frac{\omega_0 W_0}{P}$$

where W_0 is stored energy, P is power dissipation, ω_0 is resonant radian frequency, and T is period $= 2\pi/\omega_0$.

Some properties of resonators can be examined by starting with the differential equation for a simple resonator circuit:

$$\frac{d^2 v}{dt^2} + 2\sigma \frac{dv}{dt} + \omega_0^2 v = f(t) \tag{2.2}$$

This can be obtained by proper manipulation of Maxwell's equations. When $\sigma = 0$ in this equation, the homogeneous solution is

$$v(t) = A \sin \omega_0 t + B \cos \omega_0 t \tag{2.3}$$

The presence of $\sigma > 0$ corresponds to a resonator with losses. Laplace transforming the equation and solving for the transfer function, one obtains

$$T(s) = V(s)/F(s) = \frac{1}{s^2 + 2\sigma s + \omega_0^2} \tag{2.4}$$

The denominator can be factored to give

$$T(s) = \frac{j}{2\omega_L} \left\{ \frac{1}{s + \sigma + j\omega_L} - \frac{1}{s + \sigma - j\omega_L} \right\} \tag{2.5}$$

where the loaded natural resonant frequency is

$$\omega_L = \sqrt{\omega_0^2 - \sigma^2} \tag{2.6}$$

Thus, the presence of loss ($\sigma > 0$) results in a change in resonant frequency. This is called frequency pulling due to loss. Note that $\omega_L = \omega_0$ only when $\sigma = 0$.

The natural response of the differential equation is

$$v(t) = V\, e^{-\sigma t} \sin \omega_L t \tag{2.7}$$

The stored energy W is proportional to the average value of $v^2(t)$, which for small σ is

$$W = \frac{1}{2} V^2\, e^{-2\sigma t} \tag{2.8}$$

The average power P in the system is

$$P = -\frac{dW}{dt} = 2\sigma W \tag{2.9}$$

Therefore,

$$\sigma = \frac{P}{2W} \tag{2.10}$$

Since $Q = \omega_0 W/P$, one obtains

$$Q = \omega_0/2\sigma \tag{2.11}$$

The loaded natural resonant frequency now becomes

$$\omega_L = \omega_0 \sqrt{1 - \frac{1}{4Q^2}} \tag{2.12}$$

Substituting $\sigma = \omega_0/2Q$ in the original differential equation, one obtains

$$\frac{d^2v}{dt^2} + \frac{\omega_0}{Q}\frac{dv}{dt} + \omega_0^2\, v = f(t) \tag{2.13}$$

For a perfect resonator, $Q \to \infty$, and the first derivative drops out. For a practical resonator, Q is finite and the first derivative must be retained.

For $s = j\omega$ and $\sigma = \omega_0/2Q$, the transfer function T(s) becomes

$$T(\omega) = V(j\omega)/F(j\omega) = \frac{1}{\omega_0^2 - \omega^2 + j\omega\omega_0/Q} \tag{2.14}$$

The denominator in this expression can also be written as

$$\omega_0^2 - \omega^2 + j\omega\frac{\omega_0}{Q} = j\frac{\omega\omega_0}{Q}\left[1 + jQ\left(\frac{\omega}{\omega_0} - \frac{\omega_0}{\omega}\right)\right] \tag{2.15}$$

The ω-dependence can be factored to give

$$\frac{\omega}{\omega_0} - \frac{\omega_0}{\omega} = \left(\frac{\omega - \omega_0}{\omega_0}\right)\left(\frac{\omega_0}{\omega} + 1\right) \tag{2.16}$$

Since for high Q, ω is very close to ω_0, $\omega_0/\omega + 1 \simeq 2$. Therefore,

$$\frac{\omega}{\omega_0} - \frac{\omega_0}{\omega} \simeq 2\left(\frac{\omega - \omega_0}{\omega_0}\right) = 2\delta \tag{2.17}$$

where $\delta = (\omega - \omega_0)/\omega_0$ is the frequency tuning parameter [1]. The approximate transfer function then becomes

$$T(\omega) = \frac{-jQ/(\omega\omega_0)}{1 + j2Q\delta} \tag{2.18}$$

The magnitude of the transfer function $T(\omega)$ is a bell-shaped curve whose form is highly Q-dependent.

The half-power bandwidth B is defined to be the frequency spread $\Delta\omega$, where $\Delta\omega$ is defined by the difference in half-power frequencies ω_1 and ω_2, where

$$|T(\omega)| = \frac{1}{\sqrt{2}}|T(\omega_0)| \tag{2.19}$$

for $\omega = \omega_1$ and $\omega = \omega_2$. Using the approximate transfer function, the above equation becomes

$$\frac{Q/(\omega\omega_0)}{\sqrt{1 + 4Q^2\delta^2}} = \frac{1}{\sqrt{2}}\frac{Q}{\omega_0^2} \tag{2.20}$$

This is true when $4Q^2\delta^2 = 1$, or when

$$\delta = \pm \frac{1}{2Q} \tag{2.21}$$

The solution for approximate half-power frequencies then becomes

$$\omega_i = \omega_0 \pm \frac{\omega_0}{2Q}, \quad i = 1,2 \tag{2.22}$$

The approximate bandwidth B is

$$B = \Delta\omega = |\omega_1 - \omega_2| \simeq \omega_0/Q = 2\sigma \tag{2.23}$$

To a very good approximation, then, quality factor is given by

$$Q = \frac{\omega_0}{\Delta\omega} = \frac{f_0}{\Delta f} \tag{2.24}$$

In summary, knowledge of the Q factor allows for rapid determination of resonator bandwidth and loaded natural resonant frequency ω_L.

When a resonant circuit or cavity is used as a load in a microwave circuit, several different Q factors can be defined. The first Q factor accounts for internal losses. It is the unloaded Q factor, Q_0. Next, the external Q factor, Q_e, accounts for external losses. It is present because, in order to be useful, a resonator must be attached to some external circuit. Lastly, the loaded Q factor, Q_L, is the overall Q factor, and includes both internal and external losses.

The unloaded Q is the Q factor due to the losses in the cavity or resonator itself:

$$Q_0 = \frac{\omega_0 W}{P_0} \tag{2.25}$$

where P_0 is the internal power dissipation. For cavity resonators, power loss by conductors, dielectric fills, and radiation can contribute to unloaded Q. These losses can be individually accounted for by defining conductor quality factor Q_c, dielectric quality factor Q_d, and radiation quality factor Q_r. If the conductor power loss is P_c, then Q_c is given by

$$Q_c = \frac{\omega_0 W}{P_c} \tag{2.26}$$

where ω_0 is the resonant radian frequency and W is the maximum stored energy. The loss tangent for a dielectric material is defined by

$$\tan\delta = \sigma/(\omega\epsilon_0\epsilon_r) \tag{2.27}$$

where $\epsilon_r\epsilon_0$ is the dielectric constant, σ is conductivity of the medium, and ω is radian frequency. The dielectric quality factor Q_d for homogeneous dielectrics is

$$Q_d = \frac{\omega_0 W}{P_d} = \frac{\omega_0\epsilon\int|E|^2 dV}{\sigma\int|E|^2 dV} = \frac{\omega_0\epsilon}{\sigma} = \frac{1}{\tan\delta} \tag{2.28}$$

The loss tangent is the reciprocal of Q_d. The radiation quality factor Q_r is related to the power radiated from the cavity. For a closed cavity with highly conducting walls, radiation is essentially non-existant and $Q_r \to \infty$. If, on the other hand, the cavity contains an aperture, then it is possible for radiation to occur. In this case, the radiation quality factor becomes

$$Q_r = \frac{\omega_0 W}{P_r} \tag{2.29}$$

where P_r is the radiated power. The total power loss present in the cavity itself is the sum of these three individual internal components

$$P_0 = P_c + P_d + P_r \tag{2.30}$$

Substituting this in (2.25), the unloaded Q becomes

$$Q_0 = \frac{\omega W}{P_c + P_d + P_r} \tag{2.31}$$

The more power loss, the lower the Q. The relationship between Q_0 and Q_c, Q_d, and Q_r is found by rearranging the above formula:

$$\frac{1}{Q_0} = \frac{P_c}{\omega W} + \frac{P_d}{\omega W} + \frac{P_r}{\omega W} = \frac{1}{Q_c} + \frac{1}{Q_d} + \frac{1}{Q_r} \tag{2.32}$$

One manifestation of this is that the lowest Q of the three dominates, and is approximately equal to Q_0. This is true if the other two Q's are more than 10 times the smallest Q.

In order to be useful, a cavity or resonator must deliver power to an external load. The power loss due to the presence of an external load in a cavity system results in the external quality factor Q_e. It is defined by

$$Q_e = \frac{\omega W}{P_e} \tag{2.33}$$

The stored energy W in the numerator is still the energy stored inside the cavity, but the power loss in the denominator is an external drain on the internal energy reserves.

The loaded Q is the total Q for the system including power losses both internal and external to the resonator system. The loaded Q is

$$Q_L = \frac{\omega W}{P_T} \tag{2.34}$$

Since $P_T = P_e + P_0$, one obtains

$$\frac{1}{Q_L} = \frac{1}{Q_e} + \frac{1}{Q_0} \tag{2.35}$$

Again, the smallest quality factor dominates. The loaded Q, Q_L, and the unloaded Q, Q_0, are related by the coupling coefficient $\kappa = P_e/P_0$. The relationship is [2]

$$Q_L = \frac{Q_0}{1 + \kappa} \tag{2.36}$$

This is useful in describing the results of measurements.

2.3 Lumped Element Resonant Circuits

A microwave circuit is a circuit designed for operation in the frequency range of 1 GHz to 100 GHz. The free-space wavelength is 30 cm at 1 GHz. Above 1 GHz, coaxial line, microstrip lines, or hollow waveguide are usually used for the construction of resonators. To give a compact circuit below 1 GHz, resonators are usually designed using discrete components, such as a series or a parallel combination of an inductor and a capacitor.

A typical discrete component one port resonator is shown in Fig. 2.1. The input admittance Y for this is

$$Y = G + j\,\frac{1}{R_0}\left[\frac{\omega}{\omega_0} - \frac{\omega_0}{\omega}\right] = G + jB \qquad (2.37)$$

where $R_0 = \sqrt{L/C}$. The angular resonant frequency ω_0 is

$$\omega_0 = \frac{1}{\sqrt{LC}} \qquad (2.38)$$

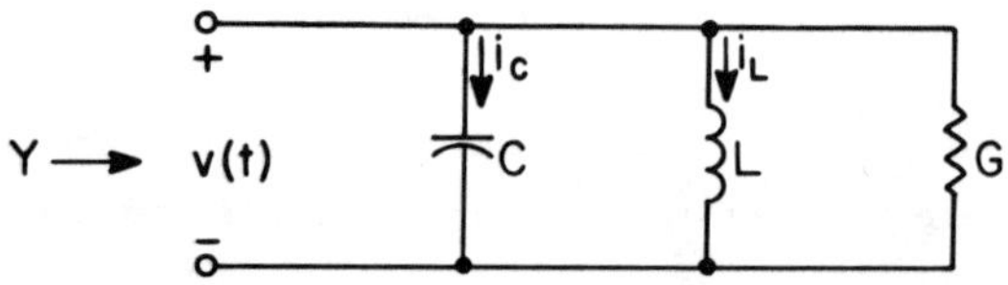

Fig. 2.1 Discrete component resonator

This parallel resonant circuit's susceptance B goes to zero when $\omega = \omega_0$, where the admittance Y = G. Under the condition of resonance notice that the susceptance B is zero because the inductor's susceptance is the negative of that due to the capacitor. The general admittance equation can also be written in the form:

$$Y = G\left\{1 + j\,\frac{1}{GR_0}\left[\frac{\omega}{\omega_0} - \frac{\omega_0}{\omega}\right]\right\}$$

$$= G\left\{1 + j\,Q_0\left[\frac{\omega}{\omega_0} - \frac{\omega_0}{\omega}\right]\right\} \qquad (2.39)$$

For this parallel resonant circuit, the unloaded Q factor, Q_0, is

$$Q_0 = \frac{\omega_0 C}{G} = \frac{1}{\omega_0 LG} = \frac{1}{R_0 G} \tag{2.40}$$

It is the ratio of the magnitude of the susceptance of either L or C at resonance to the conductance G. For a series resonant circuit, the unloaded Q factor is

$$Q_0 = \frac{\omega_0 L}{R} = \frac{1}{\omega_0 CR} \tag{2.41}$$

Consider Fig. 2.1 with $v(t) = V \cos \omega t$. Then $i_L = (V/\omega L) \sin \omega t$. The instantaneous electric and magnetic energies are

$$w_e(t) = \frac{1}{2} Cv^2(t) = \frac{1}{2} CV^2 \cos^2 \omega t \tag{2.42a}$$

$$w_m(t) = \frac{1}{2} Li_L^2 = \frac{1}{2} \frac{V^2}{\omega^2 L} \sin^2 \omega t \tag{2.42b}$$

At the resonant frequency $\omega_0 = 1/\sqrt{LC}$, thus,

$$w_e(t) = W_0 \cos^2 \omega_0 t \tag{2.43a}$$

$$w_m(t) = W_0 \sin^2 \omega_0 t \tag{2.43b}$$

where $W_0 = (CV^2)/2 = (V^2/2\omega_0^2 L)$ and is the maximum energy storage. The stored energy is constant since

$$w_e(t) + w_m(t) = W_0 \tag{2.44}$$

The instantaneous energy oscillates between the inductor and the capacitor. For the parallel resonant circuit in Fig. 2.1, the unloaded Q factor is

$$Q_0 = \frac{\omega_0 W_0}{P_G} = \frac{\frac{1}{2} \omega_0 CV^2}{\frac{1}{2} GV^2} = \frac{\omega_0 C}{G} \tag{2.45}$$

where P_G is the average dissipated power in G. A graph of the susceptance of a parallel resonant circuit versus ω is given in Fig. 2.2. It shows that the susceptance varies between $-\infty$ at $\omega = 0$ and ∞ as $\omega \to \infty$. At $\omega = \omega_0$, $B(\omega) = 0$, where $Y = G$.

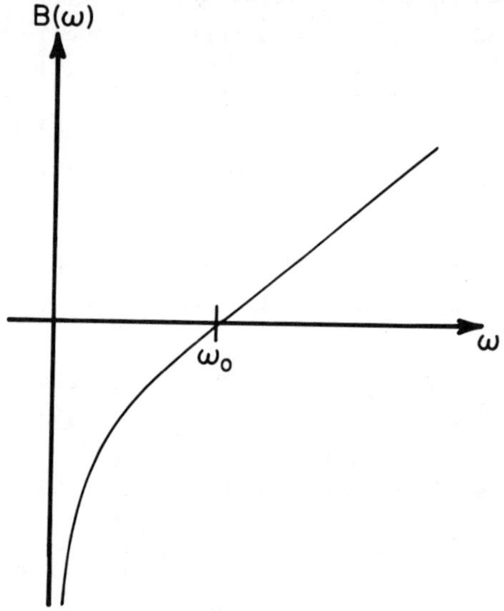

Fig. 2.2 Susceptance of parallel resonant circuit

To obtain Q_e, the external Q factor, the parallel resonant circuit must be externally loaded or used as a load. For example, consider a transmission line with a lumped parallel resonant circuit load as shown in Fig. 2.3. The parallel resonant circuit is matched or critically

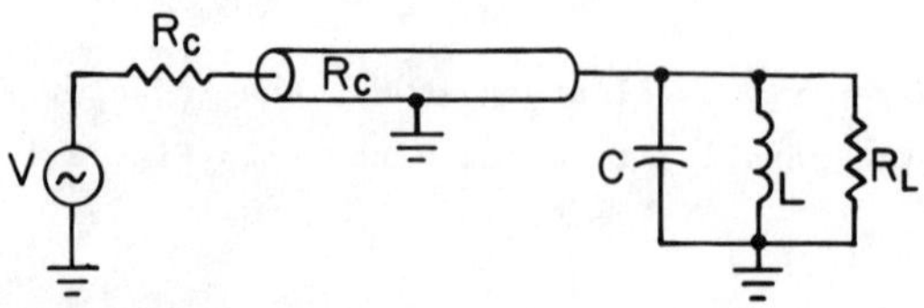

Fig. 2.3 Resonant circuit with an external source

coupled to the line only if $\omega = \omega_0$ and $R_L = R_c$, the line's characteristic impedance. Figure 2.4 shows an equivalent situation.

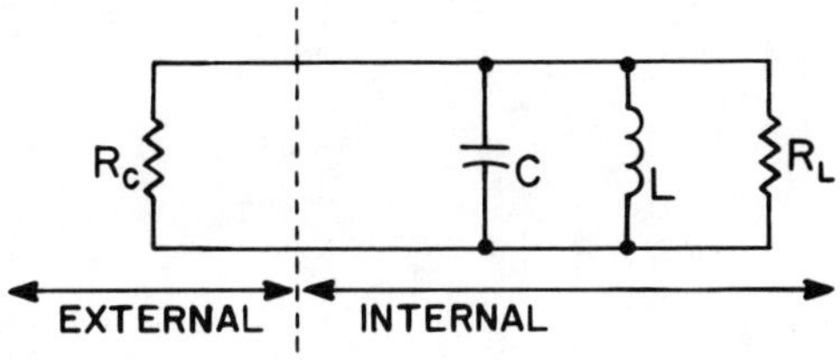

Fig. 2.4 Lumped resonant circuit

For this circuit, the unloaded and external Q factors are

$$\begin{aligned} Q_0 &= \omega_0 C\, R_L \\ Q_e &= \omega_0 C\, R_c \end{aligned} \tag{2.46}$$

The total or loaded Q factor is then

$$Q_L = \omega_0 C \left(\frac{R_c R_L}{R_c + R_L} \right) \tag{2.47}$$

The coupling coefficient κ [1] for this case is

$$\kappa = \frac{Q_0}{Q_L} - 1 = \frac{R_L}{R_c} \tag{2.48}$$

When $\kappa = 1$, the external resistor losses and the cavity losses are equal. The cavity is then critically coupled. For $\kappa < 1$, the cavity is undercoupled to the external component. Otherwise, when $\kappa > 1$, the cavity and the external component are overcoupled. The coupling coefficient can easily be measured. This provides a figure of merit indicating the quality of match.

Many concepts of the discrete component resonator are useful at microwave frequencies. The difference is that at microwave frequencies the susceptance versus ω graph contains infinitely many zeros and poles. These zeros and poles correspond to resonant frequencies for the microwave circuit. The cases $B = 0$ and $B \to \infty$ can be interpreted as open circuits and short circuits, respectively.

2.4 Transmission Line Resonators

Consider a short-circuited length of lossless coaxial line or microstrip as illustrated in Fig. 2.5. The L and C of the line are

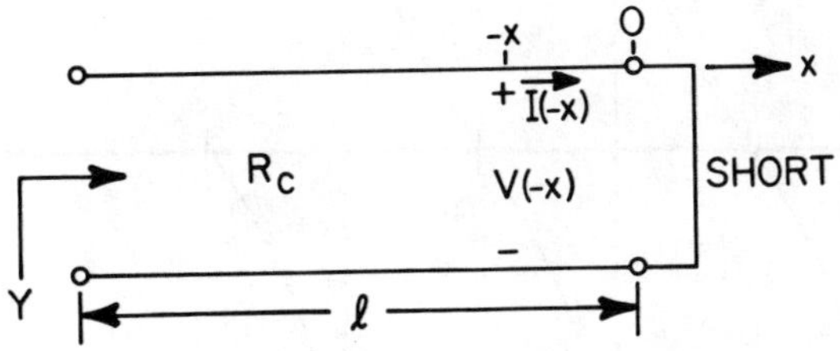

Fig. 2.5 Lossless shorted line

distributed along the length of the line and their ratio is equal to the characteristic impedance squared:

$$R_c^2 = \frac{L}{C} \tag{2.49}$$

where L and C are inductance and capacitance per meter along the line. Assuming negligible attenuation along the line, the voltage and current distributions along the line as a function of x are

$$V(-x) = jV^+ \sin \beta x$$

$$I(-x) = \frac{V^+}{R_c} \cos \beta x \tag{2.50}$$

V^+ is the peak value of voltage. Taking $Y(-\ell) = I(-\ell)/V(-\ell)$, it is easy to see that the input admittance Y of a lossless short-circuited transmission line is

$$Y = \frac{-j}{R_c} \cot \beta\ell \tag{2.51}$$

As a function of ℓ, this admittance can take on any value of susceptance from minus infinity through zero to infinity. This property is periodic in ℓ every $\lambda/2$ as shown in Fig. 2.6.

To design a resonator using microstrip or coaxial line at microwave

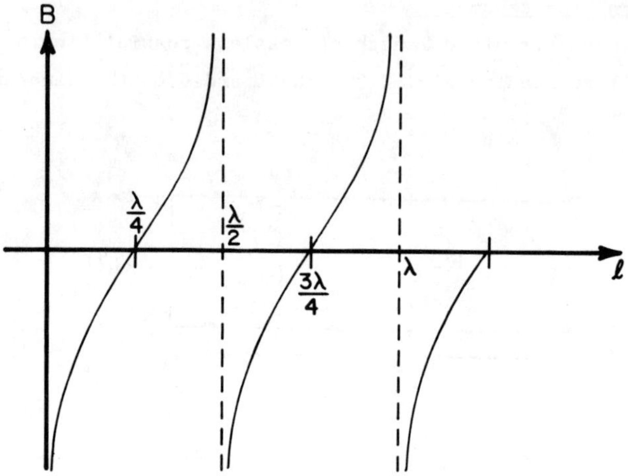

Fig. 2.6 Susceptance of shorted line

frequencies, one must choose the correct length ℓ such that the susceptance is zero or infinity. This corresponds to resonance or anti-resonance, respectively. The first resonance occurs when $\beta\ell = \pi/2$ for $\ell = \lambda/4$. A line of length $\lambda/4$ gives a circuit at microwave frequencies that is essentially equivalent to the lumped element parallel resonant circuit at lower frequencies. If ℓ is increased to $\lambda/2$, the line susceptance goes to plus or minus infinity and the line behaves very much like a short circuit which is the first anti-resonance. This is a behavior of series resonant circuits. This behavior repeats (in the ideal case) every $\lambda/2$ meters.

In order to compute the Q factor of the line, we must take into account the distributed nature of energy storage and dissipation. From (2.50), the instantaneous values of voltage and current along the line are

$$v(t,-x) = V^+ \sin \beta x \sin \omega t$$

$$i(t,-x) = \frac{V^+}{R_c} \cos \beta x \cos \omega t \tag{2.52}$$

The stored energies along the line in a length dx are

$$dw_e(t) = \frac{1}{2} C\, dx\, v^2(t) = \frac{1}{2} C\, V^{+2} \sin^2 \beta x \sin^2 \omega t\, dx \tag{2.53a}$$

$$dw_m(t) = \frac{1}{2} L\, dx\, i^2(t) = \frac{1}{2} \frac{L\, V^{+2}}{R_c^2} \cos^2 \beta x \cos^2 \omega t\, dx \tag{2.53b}$$

The stored energies along the line of length ℓ are

$$\begin{aligned} w_e(t) &= \frac{1}{2} CV^{+2} \sin^2\omega t \int_{-\ell}^{0} \sin^2 \beta x\, dx \\ &= CV^{+2} \frac{1}{4} \left(\ell - \frac{\sin 2\beta\ell}{2\beta}\right) \sin^2 \omega t \end{aligned} \tag{2.54a}$$

$$\begin{aligned} w_m(t) &= \frac{1}{2} \frac{LV^{+2}}{R_c^2} \cos^2 \omega t \int_{-\ell}^{0} \cos^2 \beta x\, dx \\ &= \frac{1}{4} \frac{LV^{+2}}{R_c^2} \left(\ell + \frac{\sin 2\beta\ell}{2\beta}\right) \cos^2 \omega t \end{aligned} \tag{2.54b}$$

$$w_e + w_m = \frac{1}{4} C\ell V^{+2} \left[1 + \left(\frac{\sin 2\beta\ell}{2\beta\ell}\right) \cos 2\omega t\right] \tag{2.54c}$$

The maximum stored energy on the line occurs when $\omega t = 0$ so

$$W_0 = \frac{1}{4} C\ell V^{+2} \left(1 + \frac{\sin 2\beta\ell}{2\beta\ell}\right) \tag{2.55}$$

Two special cases are of interest. These are the resonance at $\ell = \lambda/4$ and the anti-resonance at $\ell = \lambda/2$. For these resonant line lengths,

$$W_0 = \frac{1}{4} C\ell V^{+2} \tag{2.56}$$

The energy dissipation per period $W_d = P_d T$ of the line is found from the series resistance per unit length R and from the shunt conductance per unit length G. R and G account for conductor heating and insulation dielectric heating, respectively. The power use in a length dx due to these two effects are

$$dp_c = \frac{1}{2}\, R dx\, i^2 \tag{2.57}$$

$$dp_d = \frac{1}{2}\, G dx\, v^2 \tag{2.58}$$

The average power use becomes

$$P_c = \frac{1}{4}\, \frac{R \ell V^{+2}}{R_c^2} \left(1 + \frac{\sin 2\beta\ell}{2\beta\ell}\right) \tag{2.59}$$

$$P_d = \frac{1}{4}\, G \ell V^{+2} \left(1 - \frac{\sin 2\beta\ell}{2\beta\ell}\right) \tag{2.60}$$

For $\ell = \lambda/4$ or $\ell = \lambda/2$, the second term in parenthesis vanishes. The average power equations become

$$P_c = \frac{1}{4}\, \frac{R \ell V^{+2}}{R_c^2} \tag{2.61}$$

$$P_d = \frac{1}{4}\, G \ell V^{+2} \tag{2.62}$$

These average power dissipations have to be multiplied by the period T to obtain the energy use per period.

The quality factor Q_0 can now be found from (2.1):

$$Q_0\left(\ell = \frac{\lambda}{4} \text{ or } \frac{\lambda}{2}\right) = \frac{2\pi}{T}\, \frac{W_0}{P_c + P_d} = \frac{\omega_0 C}{\frac{R}{R_c^2} + G} = \frac{\omega_0 LC}{RC + GL} \tag{2.63}$$

This assumes line losses R and G are small. Note that Q_d and Q_c of Sec. 2.2 are related to Q_0 by (2.32) and that $Q_r \to \infty$. The quality factor Q_0 can be quite large. Notice that if R and G are zero, Q_0 is infinite. The propagation constant for the line is [3]

$$\gamma = \alpha + j\beta = \sqrt{(R + j\omega L)(G + j\omega C)} \tag{2.64}$$

If $R \ll R_c = \sqrt{L/C}$, $GR_c \ll 1$, and the frequency is large enough, then an approximate formula can be obtained. This is

$$\alpha \simeq \frac{1}{2}\left(\frac{R}{R_c} + GR_c\right) \tag{2.65a}$$

$$\beta \simeq \omega\sqrt{LC} \tag{2.65b}$$

For a quarter- or half-wavelength long transmission line, the Q factor can be written

$$Q_0 \simeq \frac{\beta}{2\alpha} \tag{2.66}$$

2.5 Cavity Resonators

An important resonator circuit at microwave frequencies is the metal cylindrical hollow waveguide resonator. Very high Q factors and accompanying narrow bandwidths can be obtained with this component. External circuits are coupled to the cavity through transmission line probes. As in the resonant transmission line case where resonance occurred at many frequencies, the hollow cylindrical waveguide resonator has many resonance frequencies and accompanying field distributions or modes. Electromagnetic fields cannot be sustained within a lossless cavity except at a resonant frequency. The field of the mode with the lowest or dominant resonant frequency is called the dominant mode. Such resonant cavities are useful for oscillator, filter, and frequency meter design. The input excitation provides only the amount of energy necessary to match the cavity losses. The cavity fields can be very large compared to the input because the constant stored energy oscillates between the electric and magnetic fields. Figure 2.7 shows a coaxial line coupled to a hollow circular cylindrical cavity and an equivalent circuit model in the neighborhood of one resonance which is valid when the coaxial line length ℓ approaches zero. The cavity

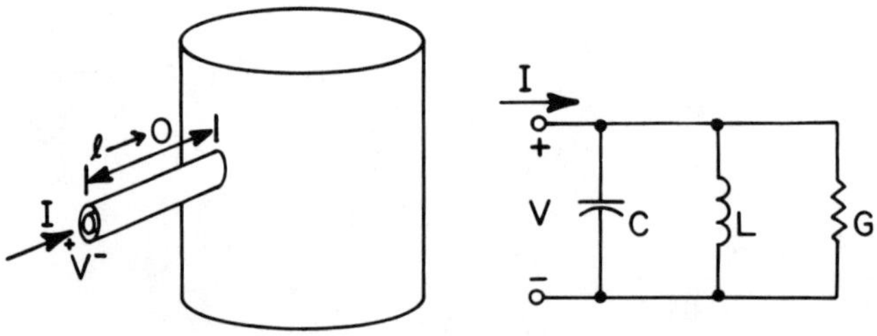

Fig. 2.7 Cavity and equivalent circuit

contains a linear, homogeneous medium modeled by parameters μ, ϵ, and σ, the medium's permeability, permittivity, and conductivity, respectively. The cavity walls are assumed to be perfectly conducting. This means $Q_0 = Q_d$.

To determine the equivalent circuit for the cavity of Fig. 2.7 at a resonant frequency, the internal fields must be determined. Once the fields have been found, the internal power dissipation, stored energies, and energy flow out of the cavity can be found by applying their respective definitions. On a pointwise basis, the fields are related to the power and energies by the following equations [4]:

$p_d = \frac{1}{2}\sigma|\underline{E}|^2$ (dissipated power density) (2.67a)

$w_m = (1/4)\mu|\underline{H}|^2$ (average stored magnetic energy density) (2.67b)

$w_e = (1/4)\epsilon|\underline{E}|^2$ (average stored electric energy density) (2.67c)

$\frac{1}{2}\underline{E} \times \underline{H}^*$ (power flux density) (2.67d)

$\frac{1}{2}(\nabla \cdot \underline{E} \times \underline{H}^*)$ (volume density of power leaving a point) (2.67e)

Manipulation of Maxwell's equations yields an equation for conservation of energy at any point inside the cavity. Maxwell's first and second equations are

$$\begin{aligned} \nabla \times \underline{H} &= j\omega\epsilon\,\underline{E} + \sigma\underline{E} + \underline{J}^i \\ -\nabla \times \underline{E} &= j\omega\mu\,\underline{H} + \underline{M}^i \end{aligned} \tag{2.68}$$

where $\underline{J}^i$ and $\underline{M}^i$ are impressed source currents. The complex power density leaving a point can be suitably written using the vector identity:

$$\nabla \cdot (\underline{E} \times \underline{H}^*) = \underline{H}^* \cdot \nabla \times \underline{E} - \underline{E} \cdot \nabla \times \underline{H}^* \tag{2.69}$$

Applying Maxwell's equations to this, one obtains

$$\begin{aligned} &\nabla \cdot (\underline{E} \times H^*) + \sigma|E|^2 + j\omega(\mu|H|^2 - \epsilon|E|^2) \\ &= -(\underline{H}^* \cdot {}^i\underline{M} + \underline{E} \cdot {}^i\underline{J}^*) \end{aligned} \tag{2.70}$$

The fields throughout a closed volume V whose surface S is the internal surface of the cavity-coaxial line combination (out to the point where I is shown) need to be accounted for in finding the total power dissipation and energy storage. Assuming sources $\underline{J}^i$ and $\underline{M}^i$ do not exist within V, so that the right-hand side of (2.70) is zero everywhere in V, one can obtain the following equation for total complex power leaving V [4]:

$$\frac{1}{2}\oiint_S \underline{E}\times\underline{H}^*\cdot d\underline{s} + \frac{1}{2}\iiint_V \sigma\,|\underline{E}|^2 dv$$

$$+ j2\omega \iiint_V \left(\frac{1}{4}\mu\,|\underline{H}|^2 - \frac{1}{4}\epsilon\,|\underline{E}|^2\right) dv = 0 \qquad (2.71)$$

If the walls of the cavity-coaxial line combination are taken to be perfectly conducting, and since $\underline{E}\times\underline{H}^*\cdot\hat{n} = \underline{H}^*\cdot(\hat{n}\times\underline{E})$ by vector identity, then

$$\frac{1}{2}\oiint_S \underline{E}\times\underline{H}^*\cdot d\underline{s} = \frac{1}{2}\iint_{S_0} \underline{H}^*\cdot(\hat{n}\times\underline{E})ds$$

$$= +\frac{1}{2}\int_{\rho=a}^{\rho=b} E_\rho \int_0^{2\pi} H_\phi^*\,\rho d\phi\, d\rho = \frac{-VI^*}{2} \qquad (2.72)$$

The surface S_0 is the cross-sectional surface of the coaxial line. The integral over S reduces to the integral over S_0 because $\hat{n}\times\underline{E}$ is zero over a perfectly conducting surface. The above assumes that the dominant TEM mode exists in the coaxial line. The relationship between V and I due to the presence of the cavity becomes

$$\frac{VI^*}{2} = \frac{1}{2}\iiint_V \sigma\,|\underline{E}|^2\,dv + j2\omega\iiint_V \left(\frac{1}{4}\mu|\underline{H}|^2 - \frac{1}{4}\epsilon|\underline{E}|^2\right)dv \qquad (2.73)$$

The admittance Y of the cavity-coaxial structure at the point illustrated in Fig. 2.7 is given by

$$I = VY \qquad (2.74)$$

$$Y_{in}^* = G-jB = |V|^{-2}\left(\iiint_V \sigma|\underline{E}|^2 dv + j4\omega\iiint_V \left(\frac{1}{4}\mu|\underline{H}|^2 - \frac{1}{4}\epsilon|\underline{E}|^2\right)dv\right) \qquad (2.75)$$

The complete fields within the resonant cavity must be determined before the admittance function Y_{in} can be found. This will be done next for the case of a circular cylindrical cavity.

Initially, recall that the cavity of the circular cylinder of Fig. 2.8 has perfectly conducting walls, and so $Q_0 = Q_d$. The internal

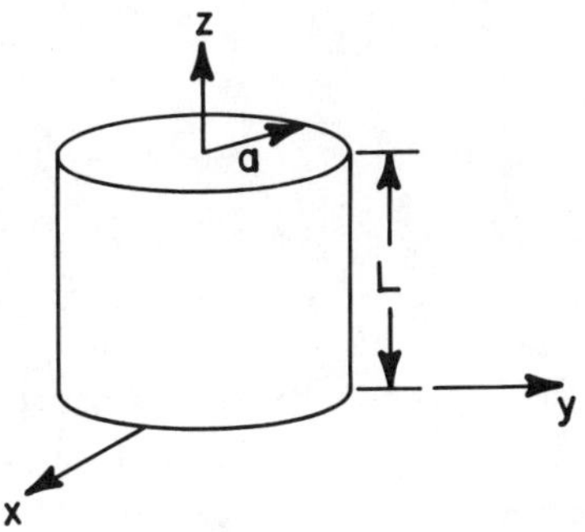

Fig. 2.8 Circular cylindrical hollow cavity

dimensions of the cavity are radius a and height L. The fields inside the cavity must obey Maxwell's equations for source free regions. Thus,

$$\begin{aligned} \nabla \times \underline{H} &= (\sigma + j\omega\epsilon)\underline{E} \\ -\nabla \times \underline{E} &= j\omega\mu\underline{H} \\ \nabla \cdot \underline{E} &= 0 \\ \nabla \cdot \underline{H} &= 0 \end{aligned} \tag{2.76}$$

Applying the curl operator and substituting the first equation in the second equation, one obtains

$$-\nabla \times \nabla \times \underline{E} + (\omega^2\mu\epsilon - j\omega\mu\sigma)\underline{E} = 0 \tag{2.77}$$

Using the identity

$$\nabla^2\underline{A} = \nabla(\nabla \cdot \underline{A}) - \nabla \times \nabla \times \underline{A} \tag{2.78}$$

and the fact that $\nabla \cdot \underline{E} = 0$, one obtains

$$\nabla^2\underline{E} + k^2\underline{E} = 0 \tag{2.79}$$

where $k^2 = \omega^2\mu\epsilon - j\omega\mu\sigma$. This is the Helmholtz wave equation that must be solved. The real and imaginary parts of $k = k_r - jk_i$ are

$$k_r^2 = k_0^2 \frac{\sqrt{1 + \left(\frac{\sigma}{\omega\epsilon}\right)^2} + 1}{2} \tag{2.80a}$$

$$k_i^2 = k_0^2 \frac{\sqrt{1 + \left(\frac{\sigma}{\omega\epsilon}\right)^2} - 1}{2} \tag{2.80b}$$

For $\sigma = 0$, $k_r = k_0$ and $k_i = 0$. Maxwell's first and second equations can be used to obtain the radial and azimuthal components from the z-components of the E and H fields. Specifically, if the z components are known, then the ρ and ϕ components are given by (3.18). Because each radial and azimuthal field component can be written in terms of E_z and H_z, one usually divides field solutions into transverse magnetic (TM) modes, where $H_z = 0$, and transverse electric (TE) modes, where $E_z = 0$. Assuming the medium is linear and homogeneous, superposition can be used. Therefore, each mode can be studied separately and, under proper excitation, can exist separately.

The modes can be solved for by separation of variables as shown in Sec. 3.3. The solution for $\psi(TM) = E_z$ or $\psi(TE) = H_z$ is [4]

$$\psi(\rho,\phi,z) = B_m(k_\rho \rho)h(m\phi)h(\beta z) \tag{2.81}$$

$B_m(k_\rho \rho)$ is a Bessel or Hankel function of the first or second kind of integer order m. Depending on boundary conditions, $h(m\phi)$ and $h(\beta z)$ are harmonic functions sine, cosine, $e^{-j\theta}$ or $e^{+j\theta}$. The characteristic equation is

$$k_\rho^2 + \beta^2 = k^2 \tag{2.82}$$

This equation must be satisfied in order for the scalar Helmholtz wave equation to have a solution; a fact that can be shown by substitution. The allowed values of k_ρ and β are referred to as eigenvalues and can have only certain fixed values derived from the boundary conditions.

The E_z and H_z fields for the TM and TE cases can be determined from the boundary conditions $E_z(\rho=a) = E_\phi(\rho=a) = 0$ and $E_\rho(z=0) = E_\rho(z=L) = E_\phi(z=0) = E_\phi(z=L)=0$. For the TM case, $H_z = 0$ and

$$E_z = \psi(TM_{mnp}) = J_m(k_\rho \rho)(A \cos m\phi + B \sin m\phi) \cos \beta z \qquad (2.83)$$

satisfies the boundary conditions at $\rho = 0$ and $z = 0$. A and B are constants. By rotating a resonator with cos $m\phi$ behavior through $\phi_0 = \pi/2m$ radians, a $-\sin m\phi$ variation results. A $\sin m\phi$ variation results if the guide is rotated by $\phi_1 = 3\pi/2m$. To enforce $E_z(\rho{=}a) = 0$, one must choose k_ρ such that

$$J_m(k_\rho a) = 0 \qquad (2.84)$$

If the zeros of the mth-order Bessel function are labeled x_{mn}, where n is the zero crossing number, $n = 1,2,3,\ldots$, k_ρ must be chosen to have certain discrete values

$$k_\rho = \frac{x_{mn}}{a} \qquad (2.85)$$

where $m = 0,1,2,3,\ldots$ The boundary condition on E_ρ or E_ϕ applied at $z = L$ yields the allowable values for β. Taking the $\partial E_z/\partial z$ at $z = L$, one obtains

$$\sin \beta L = 0 \qquad (2.86)$$

Therefore, β is required to have certain fixed values:

$$\beta = \frac{p\pi}{L} \qquad (2.87)$$

where $p = 0,1,2,3,\ldots$ The final result for the TM electric field is

$$\psi(TM_{mnp}) = J_m\left(\frac{x_{mn}\rho}{a}\right)(A \cos m\phi + B \sin m\phi) \cos \left(\frac{p\pi}{L} z\right) \qquad (2.88)$$

Since $k_\rho^2 + \beta^2 = k^2$, one easily obtains the resonant frequency. If $\sigma = 0$ and $k^2 = \omega^2\mu\epsilon$, it is

$$f_r(TM_{mnp}) = \frac{1}{2\pi\sqrt{\mu\epsilon}} \left(\left(\frac{x_{mn}}{a}\right)^2 + \left(\frac{p\pi}{L}\right)^2\right)^{1/2} \qquad (2.89)$$

One sees that the resonant frequency is dependent on m, n, and p, and the dimensions a and L of the resonator only. The significance of this is that in the ideal case Maxwell's equations do not have a solution in the cavity except when $f = f_r$. At this frequency, the solution is given by $\psi(TM_{mnp})$.

The boundary conditions applied to the TE case yield

$$H_z = \psi(TE_{mnp}) = J_m(k_\rho \rho)(A \cos m\phi + B \sin m\phi) \sin \beta z \qquad (2.90)$$

where A and B are constants. To enforce $E_\phi(\rho=a) = 0$, one must choose

$$J'_m(k_\rho a) = 0 \qquad (2.91)$$

If the zeros of the derivative of the mth-order Bessel function are labeled x'_{mn}, where n is the zero crossing number such that

$$J'_m(x'_{mn}) = 0 \qquad (2.92)$$

for n = 1,2,3,..., k_ρ must be chosen to have certain discrete values:

$$k_\rho = \frac{x'_{mn}}{a} \qquad (2.93)$$

The boundary condition on E_ρ or E_ϕ applied at z = L results in the allowable values for β in the TE case. They are

$$\beta = \frac{p\pi}{L} \qquad (2.94)$$

where p = 1,2,3,... as before. The resulting TE longitudinal field component is

$$H_z = \psi(TE_{mnp}) = J_m\left(\frac{x'_{mn}}{a}\rho\right)(A \cos m\phi + B \sin m\phi) \sin\left(\frac{p\pi}{L} z\right) \qquad (2.95)$$

The characteristic equation is used to find the resonant frequency. For $\sigma = 0$ and $k^2 = \omega^2\mu\epsilon$, the resonant frequency is

$$f_r(TE_{mnp}) = \frac{1}{2\pi\sqrt{\mu\epsilon}} \left[\left(\frac{x'_{mn}}{a}\right)^2 + \left(\frac{p\pi}{L}\right)^2\right]^{1/2} \qquad (2.96)$$

The TE mode solutions to Maxwell's equations appear when $f = f_r(TE_{mnp})$.

The zeros x_{mn} and x'_{mn} can be determined from Tables 4.3 and 4.4.

The smallest roots x_{01} and x'_{11} generate the dominant mode depending on the value of L/a. The ratio of $f_r(TM_{010})$ to $f_r(TE_{111})$ is

$$f_r(TM_{010})/f_r(TE_{111}) = x_{01}/[x'^2_{11} + (\pi a/L)^2]^{1/2} \tag{2.97}$$

When $f_r(TM_{010})/f_r(TE_{111}) < 1$, the TM_{010} mode is dominant. Solving for L/a under this condition, one obtains

$$\frac{L}{a} < \pi/\sqrt{x^2_{01} - x'^2_{11}} = 2.03 \tag{2.98}$$

for TM_{010} dominant. If $L/a > 2.03$, the TE_{111} mode is dominant instead.

For instructive purposes, it will be beneficial to examine three modes in detail. First, if the cylinder is short and $L < 2a$, then the TM_{010} mode appears first, or is dominant. From (2.83) and (3.18), the field components for the TM_{010} mode are

$$E_z = E_0 J_0(x_{01}\rho/a) \tag{2.99}$$

$$H_\phi = \frac{x_{01}}{j\omega\mu a} E_0\, J'_0(x_{01}\rho/a) = \frac{-1}{j\omega\mu}\,\frac{x_{01}}{a}\, E_0\, J_1\left(\frac{x_{01}\rho}{a}\right) \tag{2.100}$$

where E_0 is a constant. All other components are zero. The fields with time variation are

$$E_z = E_0 J_0(x_{01}\rho/a)\cos\omega t \tag{2.101a}$$

$$H_\phi = \frac{x_{01}}{\omega\mu a} E_0 J'_0(x_{01}\rho/a)\sin\omega t \tag{2.101b}$$

The z-component of E field oscillates with $\cos\omega t$ and the peak value is in the center at $\rho = 0$, gradually decreasing to zero at $\rho = a$. The H field, on the other hand, is in the ϕ direction and oscillates with $\sin\omega t$. The peak value is at $\rho = a$, while the minimum is at $\rho = 0$. This behavior is shown in Fig. 2.9. This mode is similar to the dielectric resonator's $TM_{01\delta}$ mode. The double arrows indicate fields from 0 dB to -3 dB. The long lines indicate fields from -3 dB to -6 dB and the short lines represent fields from -6 dB to -20 dB. Fields below -20 dB are not drawn.

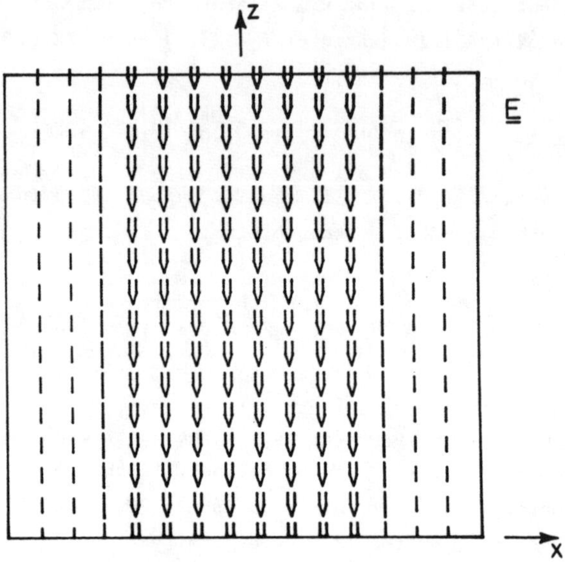

Fig. 2.9a Electric field of the TM_{010} mode

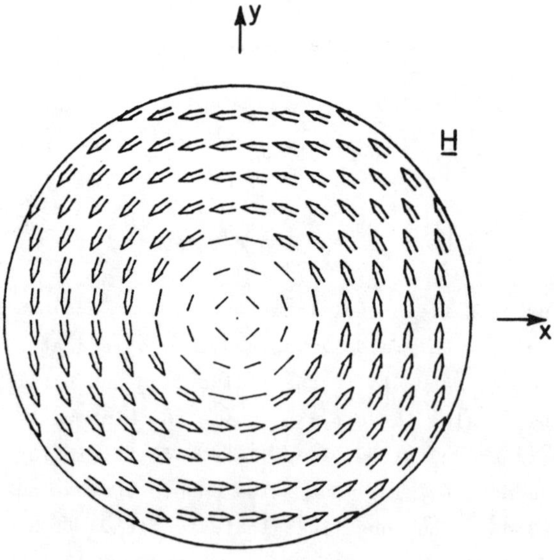

Fig. 2.9b Magnetic field of the TM_{010} mode

The fields shown in Fig. 2.9 occur for maximum energy storage in the electric field (Fig. 2.9a) and magnetic field (Fig. 2.9b). To show this, the pointwise stored electric energy is given by

$$w_e(\rho,t) = (1/2)\ \epsilon\ E_z^2 = \frac{1}{2}\ \epsilon\ E_0^2\ J_0^2(x_{01}\rho/a)\ \cos^2\omega t \tag{2.102}$$

To find the total energy storage, the volume integral of $w_e(\rho,t)$ over the volume of the resonator must be performed. The total stored electric energy is

$$\begin{aligned} W_e(t) &= \int_{\text{cavity}} w_e dv = \pi L\epsilon\ E_0^2\ \cos^2\omega t \int_0^a \rho\ J_0^2\left(\frac{x_{01}\rho}{a}\right) d\rho \\ &= \frac{\pi L\epsilon}{2}\ E_0^2 a^2\ \cos^2\omega t\ J_1^2(x_{01}) \end{aligned} \tag{2.103}$$

where the integral was evaluated using the Appendix 2.A. The time-average stored electric energy is

$$W_e = \frac{\pi L\epsilon}{4}\ E_0^2 a^2 J_1^2(x_{01}) \tag{2.104}$$

This can also be obtained by integrating the phasor field squared $(\epsilon/4)|\underline{E}|^2$ over the cavity.

The pointwise stored magnetic energy is given by [4]

$$w_m(\rho,t) = \frac{1}{2}\ \mu H^2 = \frac{1}{2}\frac{1}{\omega^2\mu}\left(\frac{x_{01}}{a}\right)^2 E_0^2\ J_1^2(x_{01}\rho/a)\ \sin^2\ \omega t \tag{2.105}$$

The total stored magnetic energy is obtained by integrating this over the cavity:

$$\begin{aligned} W_m(t) &= \int_{\text{cavity}} w_m dv = \frac{\pi L}{\omega^2\mu}\left(\frac{x_{01}}{a}\right)^2 E_0^2\ \sin^2\ \omega t \int_0^a \rho\ J_1^2\left(\frac{x_{01}}{a}\rho\right) d\rho \\ &= \frac{1}{2}\frac{\pi L}{\omega^2\mu}\ E_0^2\ x_{01}^2\ J_1^2(x_{01}) \sin^2\ \omega t \end{aligned} \tag{2.106}$$

The time-average magnetic energy then becomes

$$W_m = \frac{\pi L}{4\omega^2 \mu} E_0^2 \, x_{01}^2 \, J_1^2(x_{01}) \tag{2.107}$$

This can also be obtained from the phasor field by integrating $(\mu/4)|\underline{H}|^2$ over the volume of the cavity.

The total stored energy for the TM_{010} mode is

$$W(t) = W_m(t) + W_e(t) = \frac{\epsilon \pi L}{2} E_0^2 \, J_1^2(x_{01}) \left[a^2 \cos^2 \omega t + \left(\frac{x_{01}}{k} \right)^2 \sin^2 \omega t \right] \tag{2.108}$$

Since $\beta = 0$ for this mode, $k^2 = (x_{01}/a)^2$, and one obtains

$$W(t) = \frac{1}{2} \epsilon \pi L \, E_0^2 \, a^2 \, J_1^2(x_{01}) \tag{2.109}$$

Under the condition of resonance, the total energy is constant as a function of time. The energy alternates between being stored in the electric and the magnetic field. In the phasor approach, the average values of $W_m(t)$ and $W_e(t)$ are equal. Therefore,

$$W_m = W_e = \frac{\epsilon \pi L}{4} E_0^2 \, a^2 \, J_1^2 \, (x_{01}) \tag{2.110}$$

The total stored energy in the cavity is twice the average value:

$$W(t) = 2W_e = \frac{1}{2} \epsilon \pi L \, E_0^2 \, a^2 \, J_1^2 \, (x_{01}) \tag{2.111}$$

This is the same result as obtained before. The fact that $W_m = W_e$ results in a zero input susceptance. From (2.75), one obtains

$$|V|^2 B = -4\omega \iiint \left(\frac{1}{4} \mu |H|^2 - \frac{1}{4} \epsilon \, |E|^2 \right) dv = -4\omega \, (W_m - W_e) = 0 \tag{2.112}$$

at resonance. Therefore, the input admittance is real at resonance. See Sec. 2.8 for further discussion.

If the cylinder is tall and $L > 2a$, then the TE_{111} mode is dominant. From (2.95) and (3.18) the field components of this mode are

$$E_z = 0 \tag{2.113a}$$

$$H_z = H_0 \, J_1(x'_{11} \rho/a) \cos\phi \, \sin \frac{\pi}{L} z \tag{2.113b}$$

$$H_\rho = \frac{\pi a}{x'_{11}L} H_0 J'_1(x'_{11}\rho/a)\cos\phi \cos\frac{\pi}{L} z \tag{2.113c}$$

$$E_\rho = \frac{j\omega\mu}{\rho} H_0 \left(\frac{a}{x'_{11}}\right)^2 J_1(x'_{11}\rho/a)\sin\phi \sin\frac{\pi}{L} z \tag{2.113d}$$

$$H_\phi = \frac{-\pi}{L\rho} H_0 \left(\frac{a}{x'_{11}}\right)^2 J_1(x'_{11}\rho/a)\sin\phi \cos\frac{\pi}{L} z \tag{2.113e}$$

$$E_\phi = j\omega\mu\, H_0 \left(\frac{a}{x'_{11}}\right) J'_1(x'_{11}\rho/a)\cos\phi \sin\frac{\pi}{L} z \tag{2.113f}$$

The Bessel function of order one, $J_1(x'_{11}\rho/a)$ is zero at $\rho = 0$ and reaches 0.58 at $\rho = a$. The derivative $J_1'(x'_{11}\rho/a)$ is 0.5 at $\rho = 0$ and decreases to zero at $\rho = a$. It can be written as

$$J'_1\,(x'_{11}\rho/a) = \frac{1}{2}\,\{J_0(x'_{11}\rho/a) - J_2(x'_{11}\rho/a)\} \tag{2.114}$$

The field distributions can be simplified by using

$$\frac{J_1(x'_{11}\rho/a)}{x'_{11}\rho/a} = \frac{1}{2}\,\{J_0(x'_{11}\rho/a) + J_2(x'_{11}\rho/a)\} \tag{2.115}$$

Defining $C_H = H_0\,(\pi a/2x'_{11}L)\cos(\pi z/L)$, the transverse-to-z field $\underline{H}_t$ becomes

$$\underline{H}_t = C_H \cos\phi[J_0(x'_{11}\rho/a) - J_2(x'_{11}\rho/a)]\hat{\rho}$$
$$- C_H \sin\phi[J_0(x'_{11}\rho/a) + J_2(x'_{11}\rho/a)]\hat{\phi} \tag{2.116}$$

The magnitude of this function can be written as

$$|\underline{H}_t|^2 = C_H^2\,(J_0^2 + J_2^2 - 2\,J_0\,J_2\cos 2\phi) \tag{2.117}$$

The H field is strongest at $\rho = 0$ and generally decreases with increasing ρ. It is symmetrical about both the $\phi = 0$ and $\phi = \pi/2$ axes. Figure 2.10 shows the transverse H field $\underline{H}_t$ and Fig. 2.11 shows the field in an azimuthal plane $\phi = 0,\pi$.

At resonance, the stored energy alternates between the H field and the E field. The E field has only a transverse-to-z component.

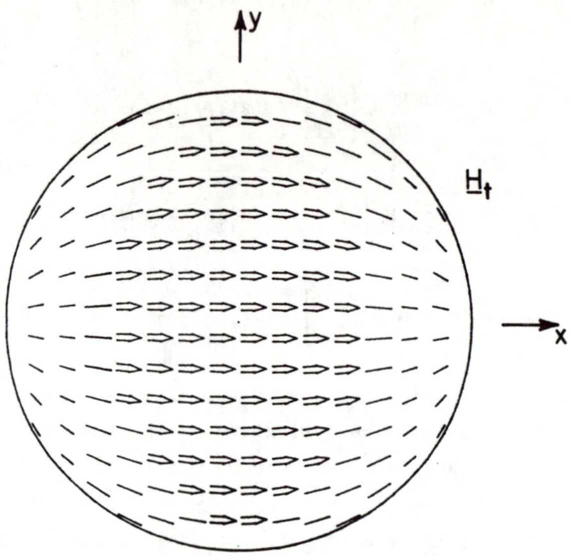

Fig. 2.10 Transverse magnetic field of the mode TE_{111}

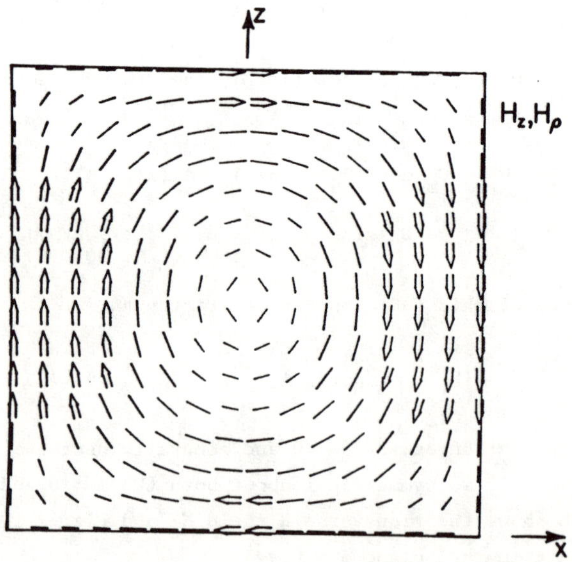

Fig. 2.11 Magnetic field in the meridian plane, mode TE_{111}

Defining $C_E = j\omega\mu\, H_0\ (a/2x'_{11})\ \sin\ (\pi z/L)$, it can be written as

$$\underline{E}_t = C_E\ [J_0(x'_{11}\rho/a) + J_2(x'_{11}\rho/a)]\ \sin\phi\ \hat{\rho}$$

$$+ C_E[J_0(x'_{11}\rho/a) - J_2(x'_{11}\rho/a)]\ \cos\phi\ \hat{\phi} \qquad (2.118)$$

Note that $\underline{E}_t \cdot \underline{H}_t = 0$. The transverse E and H field vectors are normal to one another. The magnitude of the electric field has the same ρ behavior as that of the magnetic field. Their ratio is

$$\frac{|\underline{E}_t|}{|\underline{H}_t|} = \frac{|C_E|}{|C_H|} = \frac{k\eta L}{\pi}\ \left|\tan \frac{\pi z}{L}\right| \qquad (2.119)$$

Figure 2.12 presents an illustration of the transverse electric field.

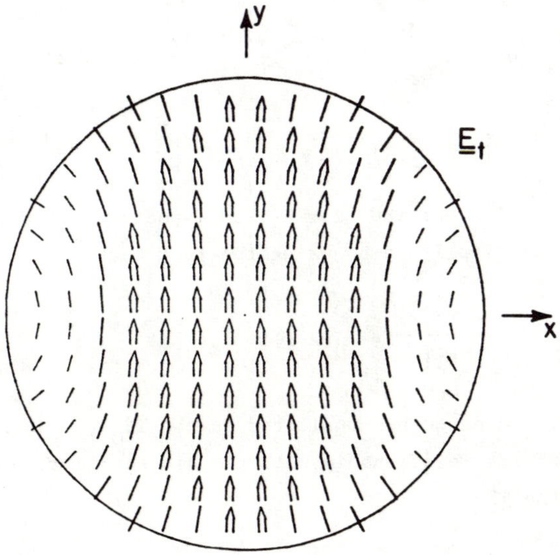

Fig. 2.12 Transverse electric field, mode TE_{111}

The average stored electric energy is

$$W_e = \frac{\epsilon}{4} \int_{\text{cavity}} E^2 dv$$

$$= \frac{\epsilon}{4} (\omega\mu)^2 H_0^2 \left(\frac{a}{2x'_{11}}\right)^2 \int_{\rho=0}^{a} \int_{\phi=0}^{2\pi} [(J_0^2 + J_2^2) - 2\, J_0 J_2 \cos 2\phi]\ \rho d\rho d\phi \int_0^L \sin^2 \frac{\pi}{L} z\, dz \tag{2.120}$$

Performing the integration with respect to ϕ and z, one obtains

$$W_e = \frac{L\epsilon\pi}{4} (\omega\mu)^2 H_0^2 \left(\frac{a}{2x'_{11}}\right)^2 \left[\int_0^a \rho\, J_0^2(x'_{11}\rho/a) d\rho + \int_0^a \rho\, J_2^2(x'_{11}\rho/a) d\rho\right] \tag{2.121}$$

The integrals in (2.121) are given in Appendix 2.A. The final result for the average stored electric energy is

$$W_e = \frac{\mu\pi L}{32} H_0^2 (ka)^2 \left(\frac{a}{x'_{11}}\right)^2 \Big[J_0^2(x'_{11}) + J_1^2(x'_{11}) + J_2^2(x'_{11})$$

$$- J_1(x'_{11})\, J_3(x'_{11})\Big] \tag{2.122}$$

This same result is also obtained for W_m, as expected. If desired, this equation can be simplified by using the fact that

$$\frac{J_1(x'_{11})}{x'_{11}} = J_0(x'_{11}) = J_2(x'_{11}) \tag{2.123}$$

The final cavity mode to be considered is the TE_{011}. This mode is included because of its relationship with the dielectric resonator's $TE_{01\delta}$ mode. For the TE_{011} mode, $E_z = E_\rho = H_\phi = 0$. The other components are

$$H_z = H_0\, J_0(x'_{01}\rho/a) \sin \frac{\pi}{L} z \tag{2.124a}$$

$$H_\rho = H_0 \frac{\pi a}{x'_{01} L} J'_0(x'_{01}\rho/a) \cos \frac{\pi}{L} z \tag{2.124b}$$

$$E_\phi = j\omega\mu\, H_0 \frac{a}{x'_{01}} J'_0\, (x'_{01}\rho/a) \sin \frac{\pi}{L} z \tag{2.124c}$$

The most obvious feature of this mode is its ϕ-independence. The radial variation of E_ϕ varies from zero at $\rho = 0$ and $\rho = a$ to a maximum at $\rho/a = x'_{11}/x'_{01}$, or $\rho = a/2.08$. The ρ component of H, H_ρ, has the same radial variation. The radial variation of H_z varies between one and -0.4. In a transverse plane, $\underline{E}_t$ is ϕ directed and $\underline{H}_t$ is ρ-directed. The TE_{011} mode is shown in Fig. 2.13. The average stored electric energy is computed as for the previous cases. The result is

$$W_e = \frac{\mu \pi L}{8} H_0^2 a^2 \left[1 + \left(\frac{\pi a}{L x'_{01}}\right)^2\right] J_0^2(x'_{01}) \qquad (2.125)$$

It can be shown that increased energy storage results if either ϵ_r or L is increased.

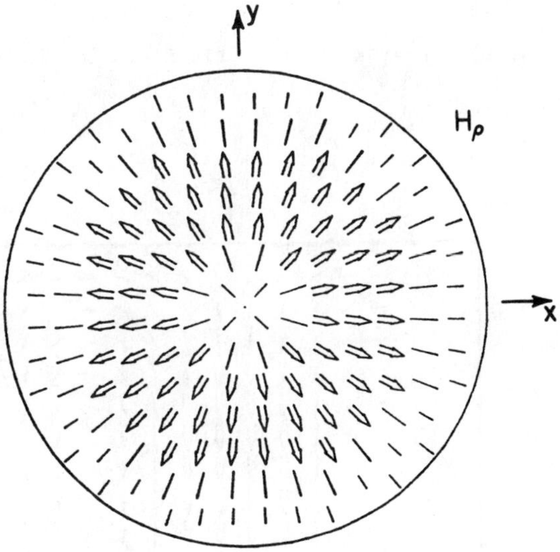

Fig. 2.13(a) TE_{011} mode, magnetic field in the equatorial plane

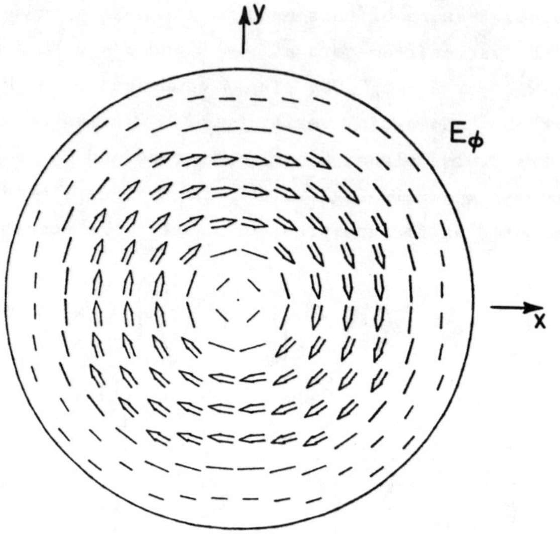

Fig. 2.13(b) TE_{011} mode, electric field in the equatorial plane

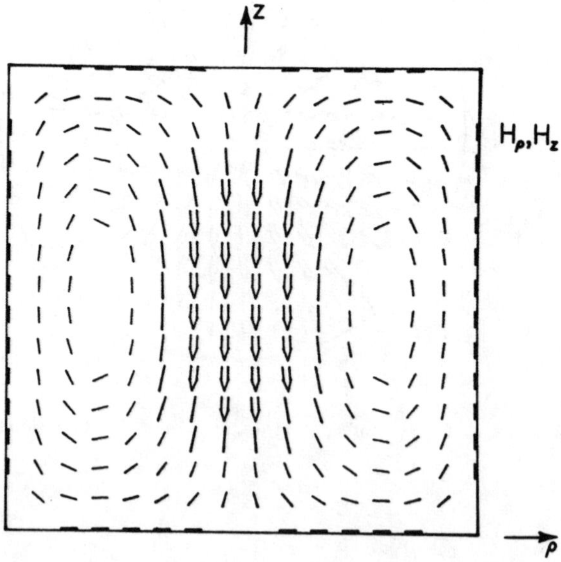

Fig. 2.13(c) TE_{011} mode, magnetic field in the meridian plane

2.6 Conductor and Dielectric Losses [4,5]

The conductor and dielectric losses P_c and P_d, respectively, are needed to find the conductor quality factor, Q_c, and dielectric quality factor, Q_d. For conductors the constitutive parameters are σ, ϵ, and μ. In dielectrics, they can be either ϵ and μ (complex ϵ), or $\sigma(\omega)$, ϵ, and μ. In the first case, σ is taken to be zero, but ϵ is taken to be complex:

$$\epsilon = \epsilon' - j\epsilon'' \tag{2.126}$$

This makes

$$j\omega\epsilon = j\omega\epsilon' + \omega\epsilon'' \tag{2.127}$$

The second term is equivalent to a dielectric loss. In the second case, the constitutive parameters are σ, ϵ, and μ. In this case, σ is allowed to be a function of frequency $\sigma(\omega)$. The two cases have the same properties if $\sigma(\omega) = \omega\epsilon''$. For purposes of our discussion here, the constitutive parameters for either a conductor or a dielectric will be taken to be σ, ϵ, and μ, which are all real, but possibly functions of frequency.

The field behavior and power dissipation at an air-metal interface is needed in order to determine the conductor losses in a cavity. The skin depth $\delta = \sqrt{2/(\omega\mu\sigma)}$, the depth at which fields have decayed to 37 % of their surface values, is very small at microwave frequencies. Therefore, conductor loss, P_c, can be determined approximately using an equivalent surface impedance, Z_s, and surface resistivity, R_s. Assuming that $\sigma \gg \omega\epsilon$ in (2.80), $\nabla \cdot \underline{E} = 0$, and $J = \sigma E$, the solution to Maxwell's equations at a dielectric-conductor boundary at x = 0 becomes

$$J_v = \sigma(\omega)e^{-(1+j)x/\delta}, \quad x > 0 \tag{2.128}$$

where $\delta = \sqrt{2/(\omega\mu\sigma)}$ is the skin depth. The equivalent surface current is obtained by integrating the volume current from zero to infinity in x:

$$J = \int_0^\infty \sigma\, e^{-(1+j)x/\delta}\, dx = \sigma \frac{\delta}{1 + j} \tag{2.129}$$

The surface impedance becomes

$$Z_s = \frac{1}{\sigma\delta}(1 + j) \tag{2.130}$$

Then, $E_{tan} = Z_s J_s$, where J_s is the surface current density. The surface resistivity is the real part of this, or

$$R_s = \frac{1}{\delta\sigma} \tag{2.131}$$

The power loss in conducting cavity walls can be approximated by

$$P_c = \frac{1}{2} R_s |\underline{J}_s|^2 = \frac{1}{2} R_s |\underline{H}|^2 \tag{2.132}$$

Here, $\underline{J}_s$ and $\underline{H}$ are the current and field, respectively, for perfectly conducting walls. This approximation is valid for good conductors at microwave frequencies.

It is of interest to apply this approximation to find the conductor quality factor Q_c for the TE_{011} mode of the cylindrical cavity. The currents are given by $\underline{J}_s = \hat{n} \times \underline{H}$, where $\hat{n}$ is the normal to the metal and $\underline{H}$ is evaluated at the boundary. The currents for the top, side, and bottom of the cavity are, respectively,

$$\underline{J}_{st} = H_0 \frac{\pi a}{x'_{01} L} J'_0 (x'_{01}\, \rho/a)\, \hat{\phi} \tag{2.133a}$$

$$\underline{J}_{ss} = H_0\, J_0(x'_{01}) \sin \frac{\pi}{L} z\, \hat{\phi} \tag{2.133b}$$

$$\underline{J}_{sb} = H_0 \frac{\pi a}{x'_{01} L} J'_0(x'_{01}\, \rho/a)\, \hat{\phi} \tag{2.133c}$$

Note that $\underline{J}_{sb} = \underline{J}_{st}$. The total top and bottom power dissipation is

$$P_1 = R_s\, H_0^2 \left(\frac{\pi a}{x'_{01} L}\right)^2 2\pi \int_0^a \rho\, J_0'^2\, (x'_{01}\rho/a)\, d\rho \tag{2.134}$$

The integral in (2.134) is

$$\int_0^a \rho\, J_0'^2\, (x'_{01}\, \rho/a)\, d\rho = \frac{a^2}{2} J_0^2\, (x'_{01}) \tag{2.135}$$

The side power dissipation is given by

$$P_2 = \frac{1}{2}\int_0^L R_s \, |\underline{J}_{ss}|^2 \, 2\pi a \, dz$$

The total conductor power dissipation in the cavity becomes

$$P_c = P_1 + P_2 = R_s \, H_0^2 \, \pi a \, J_0^2(x'_{01}) \left\{ \frac{L}{2} + a\left(\frac{\pi a}{x'_{01} L}\right)^2 \right\} \tag{2.136}$$

Using the previous result for W_e, the conductor quality factor Q_c becomes

$$Q_c = \frac{\eta}{2R_s} \frac{\left\{\left(\frac{x'_{01}}{a}\right)^2 + \left(\frac{\pi}{L}\right)^2\right\}^{3/2}}{\frac{1}{a}\left(\frac{x'_{01}}{a}\right)^2 + \frac{2}{L}\left(\frac{\pi}{L}\right)^2} \tag{2.137}$$

where $\eta = \sqrt{\mu/\epsilon}$ and $R_s = \sqrt{(\omega\mu/2\sigma)}$.

For a cavity field with a homogeneous dielectric, the dielectric quality factor Q_d is given by

$$Q_d = \frac{\omega\epsilon}{\sigma} = \frac{1}{\tan\delta} \tag{2.138}$$

It is the reciprocal of the loss tangent, $\tan\delta$, for the material. Note that δ is not the skin depth here, but is

$$\delta = \tan^{-1}\frac{\sigma}{\omega\epsilon} \tag{2.139}$$

For inhomogeneously filled cavities, one should use electric filling factors, as explained later in Ch. 7.

2.7 Incremental Rules

Incremental rules allow for the calculation of conductor quality factor, Q_c, for certain special cases. Using incremental rules, the calculation of wall currents is avoided. The incremental inductance rule of Wheeler [6,7] is valid for determining the Q_c and the attenuation constant α of TEM transmission lines. It is also useful for microstrip lines as shown by Pucel, et al. [8]. The Q and α are calculated from values of characteristic impedance R_c and R_c'. R_c is the characteristic impedance of the transmission line made of perfect conductors. R_c' is the characteristic impedance of the same line with all of its conducting walls receded by $\delta/2$, where $\delta = \sqrt{(2/\omega\mu\sigma)}$ is the skin depth. The Q factor due to conductor losses in the transmission line is then obtained from

$$Q_c = \frac{R_c}{R_c' - R_c} \tag{2.140}$$

The term "incremental inductance rule" originated from the fact that under appropriate assumptions the incremental inductance ΔL is proportional to the change in the stored magnetic energy within the conductor, which, in turn, is proportional to the dissipated power within the conductor [7].

The attenuation constant α of the transmission line can be obtained from Q_c by using (2.66) as follows:

$$\alpha = \frac{\beta}{2Q_c} \tag{2.141}$$

where β is the phase constant of the transmission line, evaluated under the assumption that the conductors are perfect.

An incremental rule has also been derived for general hollow waveguides of arbitrary cross section. The rule [9] allows for the computation of the attenuation constant, α, by computing a change in the propagation constant, β, when the metal walls of the waveguide are moved a small distance.

The incremental frequency rule of Kajfez [10] applies to rotationally symmetric modes. For the circular cylindrical cavity studied previously, it applies to the TE_{0np} modes only. Another condition used during the derivation is that the local radius of curvature must be much larger than δ, the skin depth. To apply this rule, the resonant frequency f_0 of the resonant structure with perfectly conducting walls is computed. Then the walls are moved into the cavity by a full skin depth δ. The resulting change in resonant frequency $\Delta f_0(\delta)$ due to this perturbation of cavity size is computed. The conductor quality factor Q_c is then determined from

$$Q_c = \frac{f_0}{\Delta f_0(\delta)} \tag{2.142}$$

In Sec. 4.9, this rule is applied to a dielectric resonator on a microstrip substrate. The degradation of Q due to the presence of the conductor is examined.

An alternative way of applying the incremental frequency rule is through the use of differentials. For example, consider the circular cylindrical resonator's TE_{011} mode shown in Fig. 2.13. Its resonant frequency is given by

$$f_0 = \frac{1}{2\pi\sqrt{\mu\epsilon}} \sqrt{\left(\frac{x'_{01}}{a}\right)^2 + \left(\frac{p\pi}{L}\right)^2} \tag{2.143}$$

The differential of f_0, due to an inward movement of the cavity walls by one skin depth δ, is computed as follows:

$$\Delta f_0(\delta) = -\frac{\partial f_0}{\partial a}\,\delta - \frac{\partial f_0}{\partial L}\,2\delta \tag{2.144}$$

From (2.143), one obtains

$$\Delta f_0(\delta) = \frac{\delta}{2\pi\sqrt{\mu\epsilon}} \left\{ \frac{\left(\frac{x'_{01}}{a}\right)^2 \frac{1}{a} + \frac{2}{L}\left(\frac{p\pi}{L}\right)^2}{\left[\left(\frac{x'_{01}}{a}\right)^2 + \left(\frac{p\pi}{L}\right)^2\right]^{1/2}} \right\} \tag{2.145}$$

Using this value, one obtains exactly the same formula (2.137) for Q_c as was obtained previously (for p = 1):

$$Q_c = \frac{\eta}{2R_s} \frac{\left(\left(\frac{x'_{01}}{a}\right)^2 + \left(\frac{\pi}{L}\right)^2\right)^{3/2}}{\frac{1}{a}\left(\frac{x'_{01}}{a}\right)^2 + \frac{2}{L}\left(\frac{\pi}{L}\right)^2} \qquad (2.146)$$

These techniques are useful for computing Q_c for enclosed dielectric resonators.

2.8 Circuit Modeling of Resonators

For a resonator to have predictable performance in a circuit, it must have an accurate model. For example, consider a coaxial feed terminated in a loop inside a cavity as shown in Fig. 2.14. The direction normal to the plane of the loop is shown in the ϕ-direction, but it could be chosen to be in any direction. The TM_{010} mode would be easily excited by the loop placement shown, since the H field is in the ϕ-direction. The orientation of the plane of the loop can be changed to excite other modes.

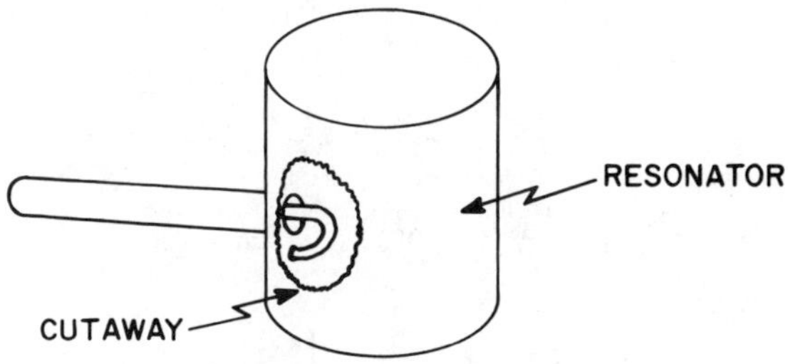

Fig. 2.14 Coaxial feed terminated in a loop

The voltage at the loop input to the resonator is found by applying Maxwell's equation:

$$-\oint_{\ell} \underline{E} \cdot d\underline{\ell} = j\omega\mu \int_{S} \underline{H} \cdot d\underline{S} \tag{2.147}$$

where S is the surface of the loop and ℓ is the boundary of S. If one assumes the magnetic field H_0 threading the loop is the same as if the loop were not present, then the right-hand side is equal to $j\omega\mu\, H_0 A$, where A is the area of S. Since the tangential E field on the surface of the loop wire is zero, the line integral reduces to the integral between the inner and outer conductors of the coaxial line. This is the voltage V at that point on the line. Therefore,

$$V = j\omega\mu\, H_0 A \tag{2.148}$$

The input admittance is found from (2.75). Thus, for $\sigma = 0$, we have

$$Y_{in} = G + jB = jB$$

$$= \frac{-1}{(\omega\mu)^2 H_0^2 A^2} j4\omega \iiint_V \left(\frac{1}{4} \mu|H|^2 - \frac{1}{4} \epsilon |E|^2\right) dV \quad (2.149)$$

$$= \frac{-j\omega 4(W_m - W_e)}{(\omega\mu)^2 H_0^2 A^2}$$

For the TM_{010} mode,

$$W_m - W_e = \frac{\pi L E_0^2}{4\omega^2 \mu} x_{01}^2 J_1^2 (x_{01})$$

$$- \frac{\pi L \epsilon}{4} E_0^2 a^2 J_1^2(x_{01}) \quad (2.150)$$

The value of H_0 is given by (2.100), evaluated for $\rho = a$:

$$|H_\phi|^2 = H_0^2 \simeq \frac{1}{\omega^2 \mu^2} \left(\frac{x_{01}}{a}\right)^2 E_0^2 J_1^2(x_{01}) \quad (2.151)$$

Therefore, from (2.149), one obtains

$$Y_{in} = \frac{-j\omega\pi L a^2}{\mu A^2} \left(\frac{1}{\omega^2} - \frac{a^2 \mu\epsilon}{x_{01}^2}\right)$$

$$(2.152a)$$

$$= \frac{1}{j\omega L_p} + j\omega\, C_p$$

where

$$L_p = \frac{\mu A^2}{\pi L a^2} \quad \text{and} \quad C_p = \frac{\pi L a^4 \epsilon}{A^2 x_{01}^2} \quad (2.152b)$$

Therefore, including the $\sigma \neq 0$, case, one finds that the resonator can be modeled by the parallel lumped element resonant circuit as shown in Fig. 2.7. The reference plane for this is the inside wall of the resonator. If a quarter wavelength of line of characteristic impedance R_c is attached, the input impedance is given by the quarter-wave transformer equation:

$$Z_{in}/R_c = \frac{1}{Z_L/R_c} \tag{2.153}$$

Since Z_L is the impedance of a parallel resonant circuit,

$$Z_L = \frac{1}{G_p + j\omega\, C_p + \dfrac{1}{j\omega L_p}} \tag{2.154}$$

The input impedance is

$$Z_{in} = R_c^2\left(G_p + j\omega\, C_p + \frac{1}{j\omega L_p}\right) \tag{2.155}$$

This is a series resonant circuit with elements $R_s = R_c^2 G_p$, $L_s = R_c^2 C_p$, $C_s = L_p/R_c^2$. Therefore, the particular model chosen depends on the reference plane for the transmission line.

Another configuration that converts a parallel resonator into a series resonator is the linear transformer. The input of the linear transformer of Fig. 2.15 can be shown to be

$$Z_{in} = j\omega L_1 + \frac{1}{Y'_{in}} \tag{2.156}$$

where

$$Y'_{in} = \frac{j\omega L_2 + Z_L}{\omega^2 M} \tag{2.157}$$

If Z_L is a series resonant circuit, the equivalent is a parallel resonant circuit.

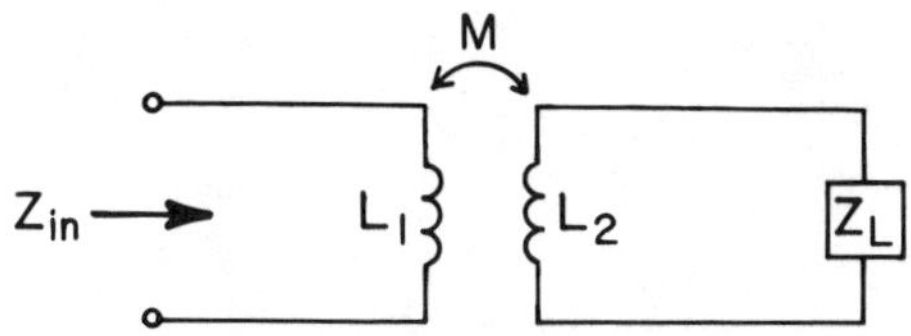

Fig. 2.15 Equivalent circuit using a linear transformer

The equivalent circuit of a cavity resonator at the cavity wall reference plane can be drawn [1] as shown in Fig. 2.16(a). One parallel resonant circuit is present for each resonator mode. The circuit of Fig. 2.16(b) is equivalent according to the previous discussion. Foster's reactance theorem [5] is also useful for analyzing resonators.

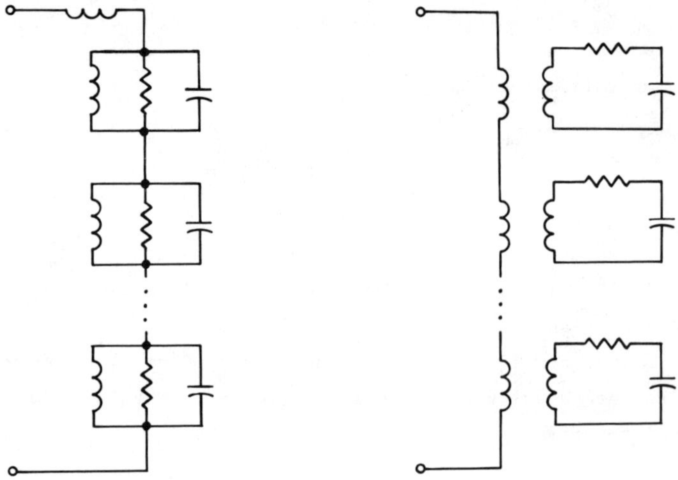

(a) (b)

Fig. 2.16 Equivalent circuits for cavity with multiple resonant modes

2.9 Q Measurement Techniques

Techniques for Q measurement are either frequency domain or time domain methods. Three frequency domain methods are useful in Q measurement. These are the reflection method, the reaction method and the transmission method. Figure 2.17 shows block diagrams for each of these three methods.

The network analyzer [11] displays the magnitude and phase of reflection coefficient S_{11} or transmission coefficient S_{21}. The reflection coefficient can be conveniently displayed in polar coordinates on the network analyzer. A Smith chart overlay for the polar reflection coefficient display allows for the impedance to be read directly. Swept-frequency or point-by-point measurements can be performed and displayed with the network analyzer.

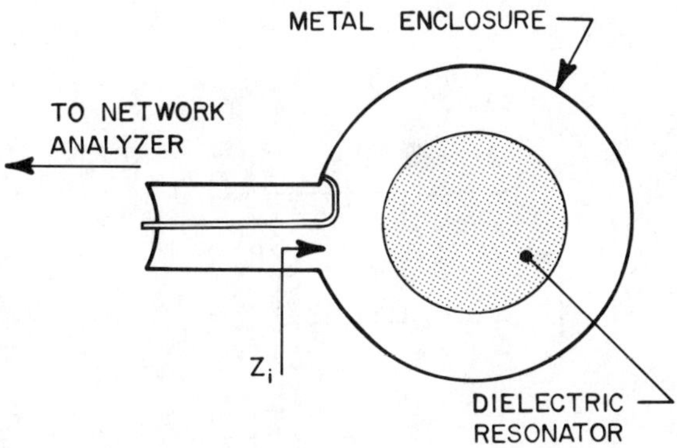

Fig. 2.17(a) Reflection method of Q measurement

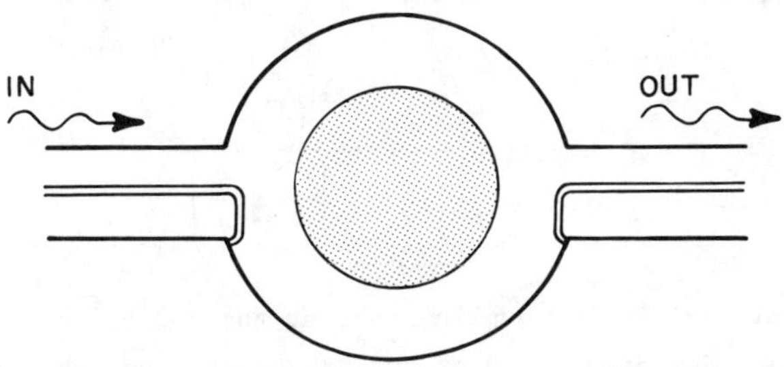

Fig. 2.17(b) Transmission method of Q measurement

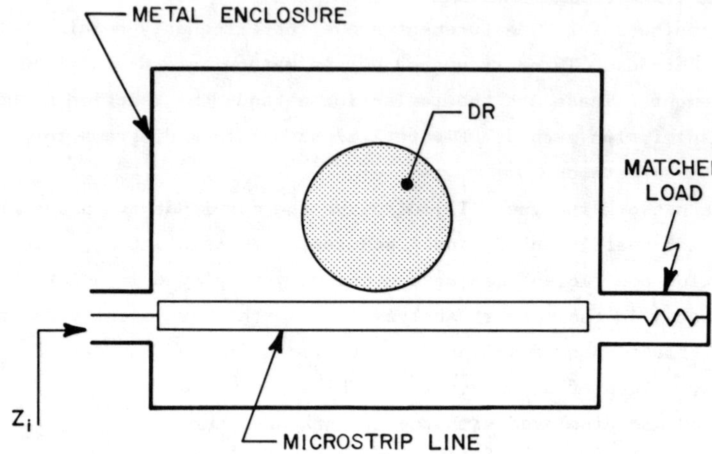

Fig. 2.17(c) Reaction method of Q measurement

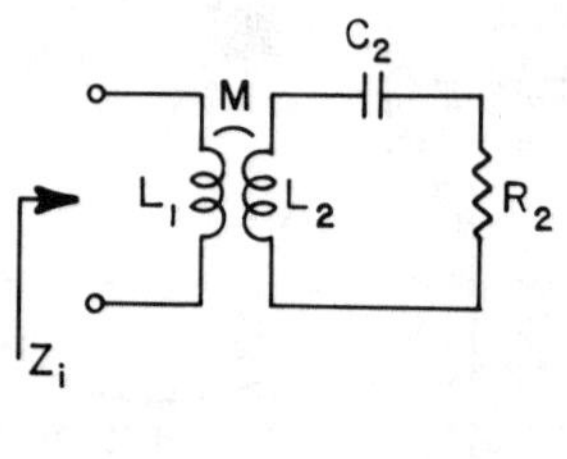

Fig. 2.18 Equivalent circuit for the reflection method of Q measurement

Figure 2.18 is a lumped-element equivalent circuit of the reflection type of measurement from Fig. 2.17(a). The input impedance is given by

$$Z_i = j\omega L_1 + \frac{\dfrac{(\omega M)^2}{R_2}}{1 + jQ_0\left(\dfrac{\omega}{\omega_0} - \dfrac{\omega_0}{\omega}\right)} \qquad (2.158)$$

L_1 is the inductance of the coupling loop at the end of the coaxial line (see Fig. 2.14). For simplicity of further discussion, the reactance ωL_1 will be neglected. The term

$$R_i = \frac{(\omega M)^2}{R_2} \tag{2.159}$$

is the induced input resistance, and in the narrow frequency band around the resonant frequency ω_0, the value of R_i shows little change with frequency. It is further convenient to simplify the frequency dependence using (2.17), thus,

$$\frac{\omega}{\omega_0} - \frac{\omega_0}{\omega} \simeq 2\,\frac{\Delta\omega}{\omega_0} \tag{2.160}$$

where $\Delta\omega = \omega - \omega_0$. The simplified expression for the input impedance becomes

$$Z_i \simeq \frac{R_i}{1 + jQ_0\, 2\,\frac{\Delta\omega}{\omega_0}} \tag{2.161}$$

The corresponding input reflection coefficient is

$$\Gamma_i = \frac{Z_i - R_c}{Z_i + R_c} \tag{2.162}$$

When the resonator is detuned, we may choose $\Delta\omega \to \infty$, so that $Z_i \to 0$, and the input reflection coefficient becomes $\Gamma_i = \Gamma_D = -1$. At frequencies close to the resonant frequency ω_0, the input reflection coefficient describes a circle, as shown in Fig. 2.19. The complex number $\Gamma_i - \Gamma_D$ is given by an expression which is easy to interpret:

$$\Gamma_i - \Gamma_D = \frac{\dfrac{R_i}{R_c}}{1 + \dfrac{R_i}{R_c} + j\, Q_0 2\,\dfrac{\Delta\omega}{\omega_0}} \tag{2.163}$$

The ratio of the input resistance R_i to the characteristic impedance R_c is defined to be the coupling coefficient κ:

$$\kappa = \frac{R_i}{R_c} \tag{2.164}$$

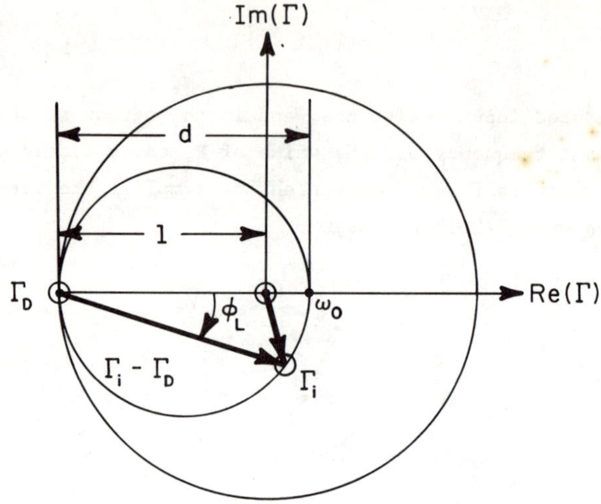

Fig. 2.19 Reflection coefficient versus ω

The loaded coupling coefficient is defined as in (2.36) and repeated below:

$$Q_L = \frac{Q_0}{1 + \kappa} \tag{2.165}$$

From (2.163), we obtain

$$\Gamma_i - \Gamma_D = \frac{2}{\left(1 + \frac{1}{\kappa}\right)\left(1 + j\ Q_L 2\ \frac{\Delta\omega}{\omega}\right)} \tag{2.166}$$

At resonance, $\Delta\omega = 0$ and the circle $\Gamma_i - \Gamma_D$ intersects the real axis. From (2.166), it follows that the diameter of the circle is

$$d = |\Gamma_i - \Gamma_D|_{max} = \frac{2\kappa}{1 + \kappa} \tag{2.167}$$

Therefore, if we measure the diameter of the circle on the polar display of the network analyzer, we may compute the coupling coefficient with the aid of (2.167), to be

$$\kappa = \frac{d}{2 - d} \tag{2.168}$$

If one observes the angle ϕ_L in Fig. 2.19 as a function of frequency, one may determine the value of the loaded Q factor. Namely, from (2.166), it follows that

$$\tan \phi_L = - Q_L \, 2 \frac{\Delta\omega}{\omega_0} \tag{2.169}$$

To measure Q_L, one selects two frequencies, denoted by f_3 and f_4, where $\phi_L = -45°$, and +45° respectively. This gives

$$Q_L = \frac{f_0}{f_3 - f_4} \tag{2.170}$$

The unloaded Q can now be found from (2.165):

$$Q_0 = Q_L(1 + \kappa) \tag{2.171}$$

The above expressions can be combined to find Q_0 directly. The result is

$$Q_0 = \frac{- \tan \phi_L}{(2 - d) \; 2\Delta\omega/\omega_0} \tag{2.172}$$

To find Q_0, one must have $\tan \phi_L = 1 - d/2$. This is a circle of radius $\sqrt{2}$ passing through Γ_D and $-\Gamma_D$. Figure 2.20 gives a template for overlaying on the reflection coefficient polar display. The straight lines correspond to $\phi_L = 45°$ and $\phi_L = 26.6°$ for Q_L measurement. The arcs are circles of radius $\sqrt{2}$ for Q_0 measurement, and the scale from .1 to 10 gives the coupling coefficient κ.

This procedure is illustrated for an inductance-coupled dielectric resonator [12]. The frequencies for $\phi_L = \pm 45°$, measured with the microwave counter, were f_3 = 7.1730 GHz, f_4 = 7.1439 GHz, and f_0 = 7.1575 GHz. The measured diameter is d = 1.45 from which the coupling coefficient is computed to be κ = 2.64. Thus, from (2.170), Q_L = 246; and, from (2.171), Q_0 = 895.

Because of the finite value of the inductance L_1 in Fig. 2.18, the measured circle on the polar display of the network analyzer is not centered on the real axis, but rather above the real axis, as shown in Fig. 2.21. The loaded resonant frequency f_L is also slightly different from the unloaded resonant frequency f_0, as pointed out in [13]. When the coupling to the input transmission line is achieved by a capacitive probe, the circle appears below the real axis of the Smith chart [7].

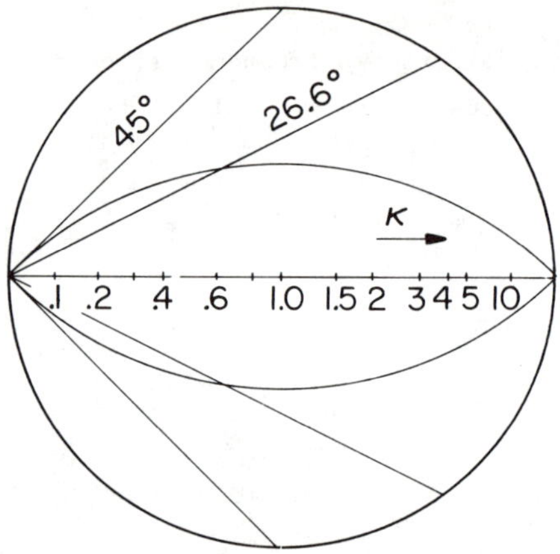

Fig. 2.20 Overlay for Q measurement (reference [12], ©1984 IEEE)

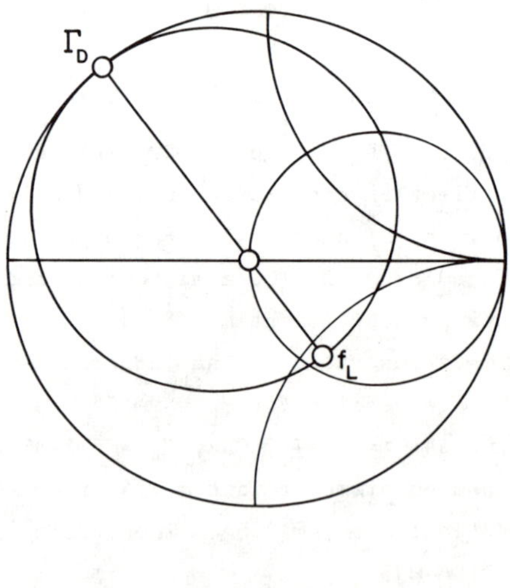

Fig. 2.21 Input reflection coefficient for inductively coupled resonator

Figure 2.17b shows the set-up for a transmission measurement of Q_L. The loaded and unloaded Q's are related by

$$Q_L = \frac{Q_0}{1 + \kappa_1 + \kappa_2} \qquad (2.173)$$

where κ_1 and κ_2 are the coupling coefficients for ports one and two, respectively. Figure 2.22 shows a typical transmission curve. The loaded Q is

$$Q_L = \frac{f_0}{\Delta f} \qquad (2.174)$$

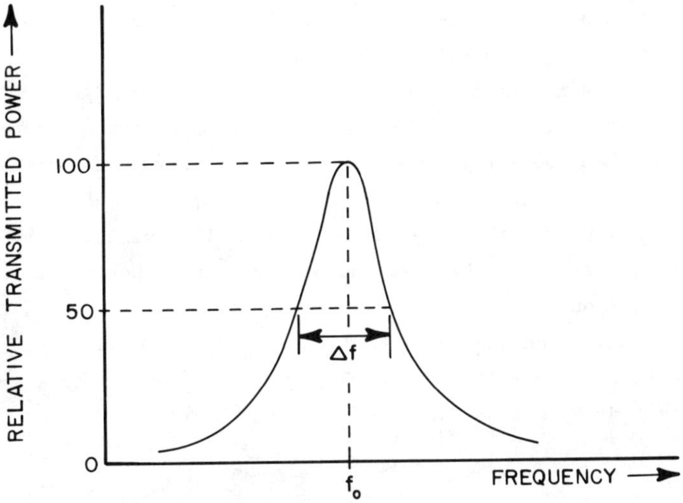

Fig. 2.22 Transmission curve

When the magnitudes of κ_1 and κ_2 are estimated by a separate experiment, the transmission method yields approximately the same accuracy for unloaded Q as does the reflection method. Error analysis [14] shows that by using a microwave counter for the frequency measurement, the main source of error is in the amplitude resolution of the network analyzer. Assuming the amplitude resolution to be ± 0.1 dB, the accuracy of the measurement of Q_0 is estimated to be ± 1.5 %.

Figure 2.17(c) illustrates a reaction type of measurement of the Q factor, such as described by Podcameni, et al. [15]. The dielectric resonator is placed next to the microstrip line. The coaxial port at the left-hand side in the figure is used for measurement of the reflection coefficient with the network analyzer, and the coaxial port on the right-hand side is terminated in a matched resistance.

For the reaction type of measurement, the loaded and unloaded Q's are related by

$$Q_L = \frac{Q_0}{1 + \kappa/2} \tag{2.175}$$

The polar display of S_{11} in the reaction-type measurement is shown in Fig. 10.3. The circles are very similar to the reflection-type measurement circles in Fig. 2.19, except that they are one-half the size. It is also possible to measure the transmission coefficient S_{12} and determine the Q factor from the observed magnitude of S_{12}. The details of the measurement procedure can be found in [16] and [17].

Error analysis of the reaction measurement of the Q factor has not been published, but it is believed to be inferior to either the reflection or the transmission method. On the other hand, the method is very convenient for those resonators which will eventually be mounted on a microstrip structure because the measurement is performed in an actual working environment.

The final technique to be mentioned here for measurement of Q is the time domain method [2]. From (2.8), the stored energy in a cavity exhibits a decay of the form

$$W = W_0 \, e^{-\omega_0 t/Q} \tag{2.176}$$

The time constant is $\tau = Q/\omega_0$. This time constant is measured to determine Q, thus,

$$Q_L = \omega_0 \, \tau \tag{2.177}$$

Appendix 2.A INTEGRALS OF BESSEL FUNCTIONS

Two integrals required to evaluate (2.103) and (2.106) are

$$\int_0^a \rho\, J_0^2\left(\frac{x_{0n}}{a}\rho\right) d\rho = \frac{a^2}{2} J_1^2(x_{0n}) \qquad (2.A1)$$

$$\int_0^a \rho\, J_1^2\left(\frac{x_{0n}}{a}\rho\right) d\rho = \frac{a^2}{2} J_1^2(x_{0n}) \qquad (2.A2)$$

where x_{0n} is a zero of $J_0(x)$ such that $J_0(x_{0n}) = 0$. These are special cases of (4.41) and (4.42). For higher-order Bessel functions, the integral to be utilized is

$$\int_0^a \rho\, J_m^2(k_\rho \rho) d\rho = \frac{a^2}{2}\left[J_m'^{\,2}(k_\rho a) + \left[1 - \frac{m^2}{(k_\rho a)^2}\right] J_m^2(k_\rho a)\right] \qquad (2.A3)$$

Substituting m = 0 and $k_\rho = x'_{11}/a$, one obtains (4.41) and for m = 2, and after some manipulation, one derives:

$$\int_0^a \rho\, J_2^2(x'_{11}\rho/a) d\rho = \frac{a^2}{2}\left[J_2^2(x'_{11}) - J_1(x'_{11}) J_3(x'_{11})\right] \qquad (2.A4)$$

References

[1] E.L. Ginzton, Microwave Measurements. New York: McGraw-Hill, 1957.

[2] M. Sucher and J. Fox, Handbook of Microwave Measurements, vol. II. Brooklyn, NY: Polytechnic Press, 1963.

[3] W.C. Johnson, Transmission Lines and Networks. New York: McGraw-Hill, 1950.

[4] R.F. Harrington, Time-Harmonic Electromagnetic Fields. New York: McGraw-Hill, 1961.

[5] S. Ramo, J.R. Whinnery, and T. Van Duzer, Fields and Waves in Communication Electronics. New York: John Wiley & Sons, 1965.

[6] H.A. Wheeler, "Formulas for the skin effect," Proc. IRE, vol. 30, pp. 412-424, Sept. 1942.

[7] D. Kajfez, Notes on Microwave Circuits, vol. 1, Oxford, MS: Kajfez Consulting, 1984.

[8] R.A. Pucel, D.J. Masse, and C.P. Hartwig, "Losses in microstrip," IEEE Trans. Microwave Theory Tech., vol. MTT-16, pp. 342-350, June 1968.

[9] X. Yansheng, "General formula for calculation of attenuation constants of EM waves in waveguides," Microwave J., vol. 28, pp. 202-203, Sept. 1985.

[10] D. Kajfez, "Incremental frequency rule for computing the Q-factor of a shielded TE_{omp} Dielectric Resonator," IEEE Trans. Microwave Theory Tech., vol. MTT-32, pp. 941-943, Aug. 1984.

[11] S.F. Adam, Microwave Theory and Applications. Englewood Cliffs, NJ: Prentice-Hall, 1969.

[12] D. Kajfez and E.J. Hwan, "Q-factor measurement with network analyzer," IEEE Trans. Microwave Theory Tech., vol. MTT-32, pp. 666-670, July 1984.

[13] D. Kajfez, "Correction for measured resonant frequency of unloaded cavity," Electron. Lett., vol. 20, no. 2, pp. 81-82, Jan. 1984.

[14] D. Kajfez and M. Crnadak, "Precision measurement of the unloaded Q factor of shielded dielectric resonators," Proc. of the IEEE Southeastcon Conference, pp. 83-87, March 3, 1985.

[15] A. Podcameni, L.F.M. Conrado, and M.M. Musso, "Unloaded quality factor measurement for MIC dielectric resonator applications," Electron. Lett., vol. 17, pp. 656-658, 1981.

[16] A.P.S. Khanna, "Q measurement of microstrip-coupled dielectric resonators," Microwaves & RF, pp. 81-86, Jan. 1984.

[17] A.P.S. Khanna and Y. Garault, "Determination of loaded, unloaded and external quality factors of a dielectric resonator coupled to a microstrip line," IEEE Trans. Microwave Theory Tech., vol. MTT-31, pp. 261-264, March 1983.

Chapter 3
DIELECTRIC ROD WAVEGUIDES
Hesham A. Auda and Darko Kajfez

3.1 Introduction

This chapter is concerned with the field analysis of the dielectric rod waveguide shown in Fig. 3.1. Such a structure has found its applications in the areas of dielectric rod antennas and optical fiber

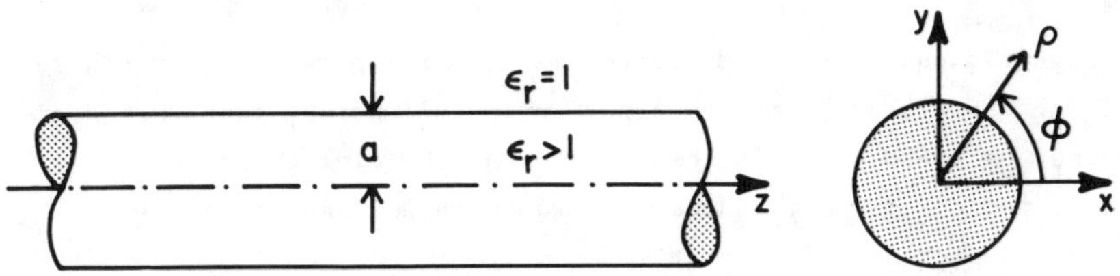

Fig. 3.1 The dielectric rod waveguide

waveguides, to name a few. The analysis is later specialized to the parallel-plate dielectric resonator shown in Fig. 3.2. The use of the resonator in measuring dielectric constants is also discussed.

A familiarity with electromagnetic waves in dielectric rod waveguides is very helpful in understanding the operation of dielectric resonators. Just as a truncated hollow waveguide becomes a resonant cavity, so the dielectric rod waveguide when truncated becomes a dielectric resonator.

The topics considered are classical electromagnetic field problems. Many of the concepts and formulas established in this chapter form a basis for some of the simple as well as rigorous models of dielectric resonators discussed in later chapters. The presentation here is, therefore, rather detailed. In the first two sections, the basic

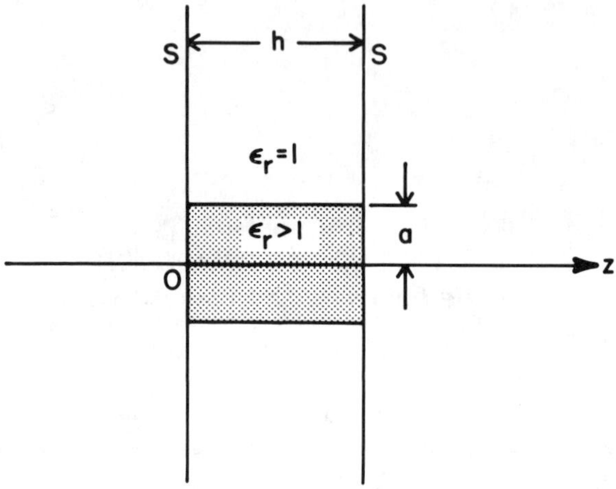

Fig. 3.2 The parallel-plate dielectric resonator

electromagnetic equations used and the method of separation of variables for solving the scalar Helmholtz equation in cylindrical coordinates are introduced. A field analysis then shows that the dielectric rod waveguide, for any given frequency and rod material and radius, is capable of supporting different field configurations, called the modes of the rod, at certain discrete values called the characteristic values, or eigenvalues, of the rod. An equation whose zeros are the eigenvalues is derived and solved numerically. Diagrams for the eigenvalues and field pattern plots of the modes are given. The use of the eigenvalue diagram for computing different mode wavelengths is illustrated. The classification of the modes of the rod, in terms of their field configurations and different power ratios, is also discussed. The parallel-plate dielectric resonator is then considered. A graphical method for determining the resonant frequencies of such a resonator is presented and used to compute the resonant frequencies for some resonators. Finally, the Courtney method for measuring dielectric constants is discussed.

3.2 Basic Equations

This section deals with the basic time-harmonic electromagnetic field equations in a linear, homogeneous, and isotropic medium. These equations are presented to the extent that they are used. For a more detailed discussion of the equations and their origin, reference should be made to one of the many texts on electromagnetic theory, for instance, the texts by Johnk [1] and Cheng [2].

The electric and magnetic fields satisfy Maxwell's two curl equations:

$$\nabla \times \underline{H} = j\omega\epsilon\underline{E} + \underline{J} \tag{3.1}$$

$$\nabla \times \underline{E} = -j\omega\mu\underline{H} \tag{3.2}$$

together with Gauss's law:

$$\nabla \cdot \epsilon\underline{E} = q \tag{3.3}$$

The phasors $\underline{E}$, $\underline{H}$, and $\underline{J}$ in (3.1), (3.2), and (3.3) are, respectively, the electric field, magnetic field, and current density vectors. The constant scalars ϵ and μ are the permittivity and permeability of the medium, respectively, while q is the charge density. Taking the divergence of (3.1), and using (3.3), there results the continuity equation:

$$\nabla \cdot \underline{J} + j\omega q = 0 \tag{3.4}$$

Similarly, taking the divergence of (3.2), the magnetic field is readily seen to be solenoid, thus,

$$\nabla \cdot \underline{H} = 0 \tag{3.5}$$

At source points, $\underline{J}$ denotes impressed sources. In the case of a lossy medium with a non-vanishing conductivity σ, $\underline{J}$ accounts for the conduction current generated by the electric field according to Ohm's law:

$$\underline{J} = \sigma\underline{E} \tag{3.6}$$

At source-free regions, (3.1) and (3.3), with the help of (3.4) and (3.6), can be rewritten as

$$\nabla \times \underline{H} = j\omega\epsilon\underline{E} \tag{3.7}$$

$$\nabla \cdot \underline{E} = 0 \tag{3.8}$$

In (3.7),

$$\epsilon = \epsilon_0\left(\epsilon_r - j\,\frac{\sigma}{\omega\epsilon_0}\right) \tag{3.9}$$

where $\epsilon_0 \simeq 8.854 \times 10^{-12}$ Farad/m is the permittivity of free space, and ϵ_r is the relative permittivity or relative dielectric constant of the medium. The loss-free case can then be recovered by setting $\sigma = 0$ in (3.9).

The main concern of this chapter is the solution of the set of equations (3.2), (3.5), (3.7), and (3.8), collectively referred to as Maxwell's equations, for $\underline{E}$ and $\underline{H}$. However, rather than solving the coupled set of equations, both $\underline{E}$ and $\underline{H}$ are indpendently solved. Taking the curl of (3.2), then using (3.7), we have

$$\nabla \times (\nabla \times \underline{E}) - k^2\underline{E} = 0 \tag{3.10}$$

where

$$k = \omega\sqrt{\mu\epsilon} = \frac{2\pi}{\lambda} \tag{3.11}$$

is the wave number of the medium and λ is its wavelength. Similarly, taking the curl of (3.7), then using (3.2), we have

$$\nabla \times (\nabla \times \underline{H}) - k^2\underline{H} = 0 \tag{3.12}$$

The duo (3.10) and (3.12) are usually called the vector Helmholtz, or wave, equations. Furthermore, using the identity:

$$\nabla \times (\nabla \times \underline{A}) = \nabla(\nabla \cdot \underline{A}) - \nabla^2\underline{A} \tag{3.13}$$

for any vector field $\underline{A}$, together with (3.5) and (3.8), (3.10) and (3.12) become

$$\left(\nabla^2 + k^2\right)\underline{E} = 0 \tag{3.14}$$

$$\left(\nabla^2 + k^2\right)\underline{H} = 0 \tag{3.15}$$

Equations (3.14) and (3.15) can each be resolved into three scalar Helmholtz, or wave, equations. Two more scalar equations are obtained from (3.8) and (3.5). A total of eight scalar partial differential equations are, therefore, involved in the solution of electromagnetic field problems.

The amount of difficulty can be greatly reduced by realizing that only two field components directed along a constant direction are needed to derive all other components. This is shown below, arbitrarily chosing E_z and H_z as the generating fields. In anticipation of the applications to follow, the cylindrical coordinate system shown in Fig. 3.3 is used.

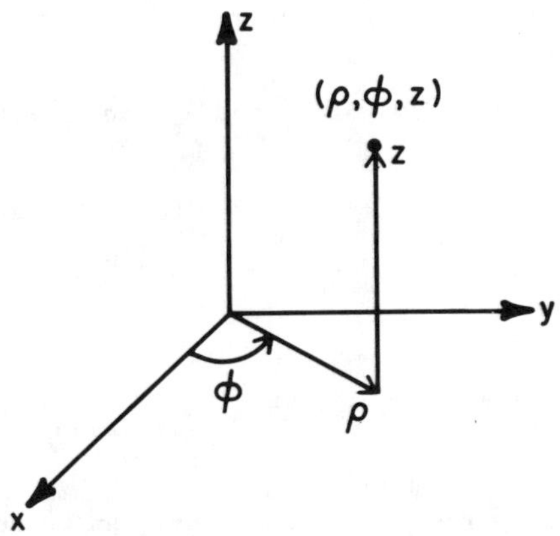

Fig 3.3 The cylindrical coordinate system

Expanding (3.2) and (3.7) in cylindrical coordinates, we have

$$\frac{1}{\rho}\frac{\partial}{\partial\phi}E_z - \frac{\partial}{\partial z}E_\phi = -j\omega\mu H_\rho \tag{3.16a}$$

$$\frac{\partial}{\partial z}E_\rho - \frac{\partial}{\partial\rho}E_z = -j\omega\mu H_\phi \tag{3.16b}$$

$$\frac{1}{\rho}\left[\frac{\partial}{\partial\rho}(\rho E_\phi) - \frac{\partial}{\partial\phi}E_\rho\right] = -j\omega\mu H_z \tag{3.16c}$$

and

$$\frac{1}{\rho}\frac{\partial}{\partial\phi}H_z - \frac{\partial}{\partial z}H_\phi = j\omega\epsilon E_\rho \tag{3.17a}$$

$$\frac{\partial}{\partial z} H_\rho - \frac{\partial}{\partial \rho} H_z = j\omega\epsilon E_\phi \tag{3.17b}$$

$$\frac{1}{\rho}\left[\frac{\partial}{\partial \rho}(\rho H_\phi) - \frac{\partial}{\partial \phi} H_\rho\right] = j\omega\epsilon E_z \tag{3.17c}$$

respectively. Substituting (3.17b) for E_ϕ into (3.16a), it becomes

$$\left[\frac{\partial^2}{\partial z^2} + k^2\right] H_\rho = j\omega\epsilon \frac{1}{\rho}\frac{\partial}{\partial \phi} E_z + \frac{\partial^2}{\partial z \partial \rho} H_z \tag{3.18a}$$

Similarly, substituting (3.16a) for H_ρ into (3.17b), we have

$$\left[\frac{\partial^2}{\partial z^2} + k^2\right] E_\phi = \frac{1}{\rho}\frac{\partial^2}{\partial z \partial \phi} E_z + j\omega\mu \frac{\partial}{\partial \rho} H_z \tag{3.18b}$$

Expressions for E_ρ and H_ϕ in terms of E_z and H_z can be similary obtained, yielding

$$\left[\frac{\partial^2}{\partial z^2} + k^2\right] E_\rho = \frac{\partial^2}{\partial z \partial \rho} E_z - j\omega\mu \frac{1}{\rho}\frac{\partial}{\partial \phi} H_z \tag{3.18c}$$

$$\left[\frac{\partial^2}{\partial z^2} + k^2\right] H_\phi = -j\omega\epsilon \frac{\partial}{\partial \rho} E_z + \frac{1}{\rho}\frac{\partial^2}{\partial z \partial \phi} H_z \tag{3.18d}$$

As affirmed, all the transverse field components can be obtained once E_z and H_z are determined. The longitudinal field components themselves satisfy

$$\left[\nabla^2 + k^2\right] E_z = 0 \tag{3.19}$$

$$\left[\nabla^2 + k^2\right] H_z = 0 \tag{3.20}$$

In an inhomogeneous medium of regions with different permittivities, (3.18), (3.19), and (3.20) are satisfied within each region, evidently with different medium parameters and wave numbers. Across the boundaries, the tangential field components on the opposite sides are continuous, a consequence of Maxwell's equations. Furthermore, in any unbounded region, the Sommerfeld condition:

$$\lim_{\rho \to \infty} \sqrt{\rho} \left(\frac{\partial}{\partial \rho} + jk \right) \begin{pmatrix} E_z \\ H_z \end{pmatrix} = 0 \tag{3.21}$$

uniformly for all directions $\underline{\rho}$, must be satisfied. There is, at most, one complex valued function which assumes given values on the boundaries, and satisfies (3.19), or (3.20), within each region, as well as the Sommerfeld condition (3.21) for the infinite region [3].

There remains the problem of solving (3.19) and (3.20), respectively, for E_z and H_z. This is readily accomplished using the method of separation of variables.

3.3 Solution of the Wave Equation in Cylindrical Coordinates

In this section, the method of separation of variables is used to obtain solutions for the wave equation in the cylindrical coordinate system. The wave equation (3.19), or (3.20), with the Laplacian ∇^2 expressed in cylindrical coordinates, is given by

$$\frac{1}{\rho}\frac{\partial}{\partial\rho}\left(\rho\frac{\partial}{\partial\rho}\psi\right) + \frac{1}{\rho^2}\frac{\partial^2}{\partial\phi^2}\psi + \frac{\partial^2}{\partial z^2}\psi + k^2\psi = 0 \tag{3.22}$$

where ψ stands for either longitudinal field component.

In essence, the method of separation of variables seeks a solution of (3.22) of the form

$$\psi(\rho,\phi,z) = P(\rho)F(\phi)Z(z) \tag{3.23}$$

Substituting (3.23) into (3.22), then dividing by ψ, it becomes

$$\frac{1}{\rho P}\frac{d}{d\rho}\left(\rho\frac{d}{d\rho}P\right) + \frac{1}{\rho^2 F}\frac{d^2F}{d\phi^2} + \frac{1}{Z}\frac{d^2Z}{dz^2} + k^2 = 0 \tag{3.24}$$

The third term is explicitly independent of ρ and ϕ. It is also necessarily independent of z, if (3.24) is to sum to zero for all (ρ,ϕ,z). Thus,

$$\frac{1}{Z}\frac{d^2Z}{dz^2} = -\beta^2 \tag{3.25}$$

where β is a constant. Substituting (3.25) into (3.24), then multiplying throughout by ρ^2, it becomes

$$\rho\frac{1}{P}\frac{d}{d\rho}\left[\rho\frac{d}{d\rho}P\right] + \frac{1}{F}\frac{d^2F}{d\phi^2} + \left[k^2 - \beta^2\right]\rho^2 = 0 \tag{3.26}$$

However, the second term in (3.26) is a function of ϕ only, whereas the rest of the equation is a function of ρ only. By the same argument, then,

$$\frac{1}{F}\frac{d^2F}{d\phi^2} = -m^2 \tag{3.27}$$

where m is a constant. Substituting (3.27) into (3.26), then multiplying throughout by P, there results

$$\rho \frac{d}{d\rho}\left(\rho \frac{d}{d\rho} P\right) + \left((k_\rho \rho)^2 - m^2\right) P = 0 \tag{3.28}$$

where

$$k_\rho^2 = k^2 - \beta^2 \tag{3.29}$$

By (3.28), the wave equation is separated into three equations, each of which determines only one of the functions P, F, or Z. The first two equations, (3.25) and (3.27), are harmonic equations, whose solutions are harmonic functions. The last equation (3.28) is a Bessel equation of the mth order, whose solutions are Bessel functions. The choice of the constants β and m, as well as the solutions for (3.25), (3.27), and (3.28), depends on the physical geometry for the structure considered, the conditions at the boundaries, and the type of field to be supported by the structure. Equation (3.29) is an important relationship between the radial wave number k_ρ, the medium wave number k, and the waveguide propagation constant β. In guided wave propagation, k_ρ can take only on certain discrete values, each of them corresponding to a different mode of propagation. In the next section, the dielectric rod is seen capable of supporting different fields -- the so-called modes of the rod.

3.4 <u>The Eigenvalue Equation</u>

Consider the dielectric rod waveguide shown in Fig. 3.1. The rod material is assumed to be a perfect dielectric characterized by the real scalar permittivity $\epsilon = \epsilon_0 \epsilon_r$ and the real scalar permeability μ_0.

Inside the rod, E_z and H_z are finite at the origin, and are periodic with a period of 2π with respect to ϕ. Furthermore, they are to represent waves traveling in the positive z-direction. These requirements are met by chosing E_z and H_z in the form:

$$E_{z1} = A\, J_m(k_{\rho 1}\rho) \begin{Bmatrix} \cos(m\phi) \\ \sin(m\phi) \end{Bmatrix} e^{-j\beta z} \tag{3.30}$$

$$H_{z1} = B\, J_m(k_{\rho 1}\rho) \begin{Bmatrix} \sin(m\phi) \\ \cos(m\phi) \end{Bmatrix} e^{-j\beta z} \tag{3.31}$$

where m is an integer, J_m is the Bessel function of the first kind and mth order, and

$$k_{\rho 1}^2 = k^2 - \beta^2 \tag{3.32}$$

In (3.32), $k = k_0 \sqrt{\epsilon_r}$ is the wave number of the dielectric, where k_0 is the wave number of free space. Outside the rod, E_z and H_z are similarly periodic with a period of 2π with respect to ϕ, and represent traveling waves along the z-axis. However, unlike the components inside, they are to be exponentially decaying in the ρ direction, if the rod is to be a waveguide. Thus,

$$E_{z2} = C\, K_m(k_{\rho 2}\rho) \begin{Bmatrix} \cos(m\phi) \\ \sin(m\phi) \end{Bmatrix} e^{-j\beta z} \tag{3.33}$$

$$H_{z2} = D\, K_m(k_{\rho 2}\rho) \begin{Bmatrix} \sin(m\phi) \\ \cos(m\phi) \end{Bmatrix} e^{-j\beta z} \tag{3.34}$$

where K_m is the modified Bessel function of the second kind and mth order, and

$$k_{\rho 2}^2 = \beta^2 - k_0^2 \tag{3.35}$$

Note that k_0 and β have been swaped in (3.35), so that the argument of K_m is real.

The field components, whether inside or outside the rod, are expressed as the product of factors, each of which satisfies only one of the equations (3.25), (3.27), or (3.28), as well as the requirements imposed by the physical geometry, the analyticity of the field at the origin, or the utilization of the rod as a waveguide transmitting energy in the z direction. Furthermore, since E_z and H_z must be linearly independent over $0 \le \phi \le 2\pi$, E_z and H_z are chosen to vary as cosine and sine functions, respectively. For m = 0, the fields are split into transverse electric to z (TE to z) and transverse magnetic to z (TM to z) fields with no ϕ-variation, completely independent of each other. For all m > 0, the fields are hybrid, or HEM, having both TE and TM parts.

Substituting (3.30) through (3.35) into (3.18), the transverse field components inside the rod are found to be

$$H_{\rho 1} = \frac{1}{k_{\rho 1}^2}\left[-j\omega\epsilon m \frac{A}{\rho} J_m(k_{\rho 1}\rho) - j\beta k_{\rho 1} B\, J_m'(k_{\rho 1}\rho)\right]\sin(m\phi)\, e^{-j\beta z} \tag{3.36a}$$

$$E_{\phi 1} = \frac{1}{k_{\rho 1}^2}\left[j\beta m \frac{A}{\rho} J_m(k_{\rho 1}\rho) + j\omega\mu_0 k_{\rho 1} B\, J_m'(k_{\rho 1}\rho)\right]\sin(m\phi)\, e^{-j\beta z} \tag{3.36b}$$

$$E_{\rho 1} = \frac{1}{k_{\rho 1}^2}\left[-j\beta k_{\rho 1} A\, J_m'(k_{\rho 1}\rho) - j\omega\mu_0 m \frac{B}{\rho} J_m(k_{\rho 1}\rho)\right]\cos(m\phi)\, e^{-j\beta z} \tag{3.36c}$$

$$H_{\phi 1} = \frac{1}{k_{\rho 1}^2}\left[-j\omega\epsilon k_{\rho 1} A\, J_m'(k_{\rho 1}\rho) - j\beta m \frac{B}{\rho} J_m(k_{\rho 1}\rho)\right]\cos(m\phi)\, e^{-j\beta z} \tag{3.36d}$$

whereas those outside the rod are given by

$$H_{\rho 2} = \frac{-1}{k_{\rho 2}^2}\left[-j\omega\epsilon_0 m \frac{C}{\rho} K_m(k_{\rho 2}\rho) - j\beta k_{\rho 2} D\, K_m'(k_{\rho 2}\rho)\right]\sin(m\phi)\, e^{-j\beta z} \tag{3.37a}$$

$$E_{\phi 2} = \frac{-1}{k_{\rho 2}^2}\left[j\beta m \frac{C}{\rho} K_m(k_{\rho 2}\rho) + j\omega\mu_0 k_{\rho 2} D\, K_m'(k_{\rho 2}\rho)\right]\sin(m\phi)\, e^{-j\beta z} \tag{3.37b}$$

$$E_{\rho 2} = \frac{-1}{k_{\rho 2}^2} \left[-j\beta k_{\rho 2} C\, K_m'(k_{\rho 2}\rho) - j\omega\mu_0 m \frac{D}{\rho} K_m(k_{\rho 2}\rho) \right] \cos(m\phi)\, e^{-j\beta z} \quad (3.37c)$$

$$H_{\phi 2} = \frac{-1}{k_{\rho 2}^2} \left[-j\omega\epsilon_0 k_{\rho 2} C\, K_m'(k_{\rho 2}\rho) - j\beta m \frac{D}{\rho} K_m(k_{\rho 2}\rho) \right] \cos(m\phi)\, e^{-j\beta z} \quad (3.37d)$$

In (3.36) and (3.37), primes denote differentiation with respect to the argument of the function.

The particular choice of solution in the ρ-direction outside the rod has automatically satisfied the Sommerfeld condition. The continuity of the tangential field components across the surface of the rod, however, requires that

$$E_{z1} = E_{z2} \quad (3.38a)$$

$$H_{z1} = H_{z2} \quad (3.38b)$$

$$E_{\phi 1} = E_{\phi 2} \quad (3.38c)$$

$$H_{\phi 1} = H_{\phi 2} \quad (3.38d)$$

at $\rho = a$. Substituting (3.36) and (3.37) into (3.38), we obtain the boundary conditions:

$$A\, J_m(k_{\rho 1}a) - C\, K_m(k_{\rho 2}a) = 0 \quad (3.39a)$$

$$B\, J_m(k_{\rho 1}a) - D\, K_m(k_{\rho 2}a) = 0 \quad (3.39b)$$

$$\frac{\beta}{k_{\rho 1}^2 a} m\, A\, J_m(k_{\rho 1}a) + \frac{j\omega\mu_0}{k_{\rho 1}} B\, J_m'(k_{\rho 1}a)$$

$$+ \frac{\beta}{k_{\rho 2}^2 a} m\, C\, K_m(k_{\rho 2}a) + \frac{j\omega\mu_0}{k_{\rho 2}} D\, K_m'(k_{\rho 2}a) = 0 \quad (3.39c)$$

$$\frac{\omega\epsilon}{k_{\rho 1}} A J'_m(k_{\rho 1}a) + \frac{\beta}{k^2_{\rho 1}a} m B J_m(k_{\rho 1}a)$$

$$+ \frac{\omega\epsilon_0}{k_{\rho 2}} C K'_m(k_{\rho 2}a) + \frac{\beta}{k^2_{\rho 2}a} m D K_m(k_{\rho 2}a) = 0 \qquad (3.39d)$$

These equations can be simplified by introducing the new variables:

$$x = k_{\rho 1}a \qquad (3.40)$$

and

$$y = k_{\rho 2}a = \sqrt{(k_0a)^2(\epsilon_r - 1) - x^2} \qquad (3.41)$$

Hence,

$$\beta a = \sqrt{(k_0a)^2\epsilon_r - x^2} \qquad (3.42)$$

The equations (3.39) then become

$$A J_m(x) - C K_m(y) = 0 \qquad (3.43a)$$

$$B J_m(x) - D K_m(y) = 0 \qquad (3.43b)$$

$$\frac{\beta am}{x^2} A J_m(x) + \frac{\omega\mu_0 a}{x} B J'_m(x) + \frac{\beta am}{y^2} C K_m(y) + \frac{\omega\mu_0 a}{y} D K'_m(y) = 0 \qquad (3.43c)$$

$$\frac{\omega\epsilon a}{x} A J'_m(x) + \frac{\beta am}{x^2} B J_m(x) + \frac{\omega\epsilon_0 a}{y} C K'_m(y) + \frac{\beta am}{y^2} D K_m(y) = 0 \qquad (3.43d)$$

In matrix form, (3.43) becomes

$$\underset{\sim}{F}|U\rangle = 0 \qquad (3.44)$$

where $\underset{\sim}{F}$ is the 4 × 4 matrix:

$$\underset{\sim}{F} = \begin{bmatrix} J_m(x) & 0 & -K_m(y) & 0 \\ 0 & J_m(x) & 0 & -K_m(y) \\ \frac{\beta am}{x^2} J_m(x) & \frac{\omega\mu_0 a}{x} J'_m(x) & \frac{\beta am}{y^2} K_m(y) & \frac{\omega\mu_0 a}{y} K'_m(y) \\ \frac{\omega\epsilon a}{x} J'_m(x) & \frac{\beta am}{x^2} J_m(x) & \frac{\omega\epsilon_0 a}{y} K'_m(y) & \frac{\beta am}{y^2} K_m(y) \end{bmatrix} \tag{3.45}$$

and |U> is the 4 × 1 column vector of the amplitudes A, B, C, and D:

$$|U\rangle = \begin{bmatrix} A \\ B \\ C \\ D \end{bmatrix} \tag{3.46}$$

A non-trivial solution of (3.44) exists only at those x's for which

$$\det(\underset{\sim}{F}) = 0 \tag{3.47}$$

where det(•) denotes the determinant of the matrix. Expanding the determinant of $\underset{\sim}{F}$ in terms of the elements of the first column, (3.47) becomes

$$\frac{\omega\epsilon a}{x} J'_m(x)\, K_m(y) \left[\frac{\omega\mu_0 a}{y} J_m(x)\, K'_m(y) + \frac{\omega\mu_0 a}{x} K_m(y)\, J'_m(x)\right]$$

$$- \frac{\beta am}{x^2} J_m(x)\, K_m(y) \left[\frac{\beta am}{y^2} J_m(x)\, k_m(y) + \frac{\beta am}{x^2} J_m(x)\, K_m(y)\right]$$

$$+ J_m^2(x) \left[- \frac{(\beta am)^2}{y^4} K_m^2(y) + \frac{\omega^2 \mu_0 \epsilon_0 a^2}{y^2} K_m'^2(y)\right]$$

$$+ J_m(x)\, K_m(y) \left[\frac{\omega^2 \mu_0 \epsilon_0 a^2}{xy} J'_m(x)\, K'_m(y) - \frac{(\beta am)^2}{x^2 y^2} J_m(x)\, K_m(y)\right] = 0 \tag{3.48}$$

Dividing throughout (3.48) by $(K_m^2(y)/\omega^2\mu_0\epsilon a^2)$ and regrouping, (3.48) can then be put in the form [6]:

$$F_1(x)F_2(x) - F_3^2(x) = 0 \tag{3.49}$$

where

$$F_1(x) = \frac{J'_m(x)}{x} + \frac{K'_m(y)\ J_m(x)}{\epsilon_r y K_m(y)} \tag{3.50}$$

$$F_2(x) = \frac{J'_m(x)}{x} + \frac{K'_m(y)\ J_m(x)}{y K_m(y)} \tag{3.51}$$

$$F_3(x) = \frac{\beta a m}{k_0 a \sqrt{\epsilon_r}}\ J_m(x) \left(\frac{1}{x^2} + \frac{1}{y^2} \right) \tag{3.52}$$

The solution of (3.49) determines the x's for which a non-trivial solution of (3.44) exists. Equation (3.49) is called the eigenvalue equation for the dielectric rod waveguide, the zeros x of which are called the eigenvalues of the dielectric rod waveguide.

In (3.41), x should not exceed a certain value x_{max} given by

$$x_{max} = k_0 a \sqrt{\epsilon_r - 1} \tag{3.53}$$

because y then becomes purely imaginary, thereby changing the modified Bessel functions K_m into Hankel functions $H_m^{(2)}$ representing outwardly traveling waves. There are, therefore, only a finite number of eigenvalues for any specified m. Another subscript n is, therefore, needed to enumerate the eigenvalues. It then becomes convenient to denote the eigenvalues by x_{mn}. The fields $(\underline{E}_{mn}, \underline{H}_{mn})$ corresponding to x_{mn} are called the modes of the dielectric rod waveguide. For m = 0, F_3 vanishes identically, and (3.49) splits into the two equations:

$$\frac{J_1(x)}{x} + \frac{K_1(y)\ J_0(x)}{\epsilon_r y K_0(y)} = 0 \tag{3.54}$$

and

$$\frac{J_1(x)}{x} + \frac{K_1(y)\ J_0(x)}{y K_0(y)} = 0 \tag{3.55}$$

corresponding, respectively, to TM and TE fields. This can be readily established by noting that, for m = 0, $\underset{\sim}{F}$ transforms into a block diagonal matrix with one block corresponding to TE fields and the other to TM fields.

3.5 The Modes of the Rod

The eigenvalues for the hybrid modes, as well as for TM and TE modes, can be determined by solving the transcendental equation (3.49). The evaluation of the eigenvalues has to be carried out numerically. This can be accomplished by first locating the zeros of the equation within small intervals, then employing a suitable iterative procedure for the solution of nonlinear equations for actually determining these zeros to the accuracy desired. The localization can be done by increasing x gradually and noting the changes of sign of the equation. The forms of F_1, F_2, and F_3 are, therefore, particularly helpful, since none of them is singular over the range of x. The first few eigenvalues so computed are plotted versus k_0a in Figs. 3.4 and 3.5 for $\epsilon_r = 20.0$ and $\epsilon_r = 38.0$, respectively.

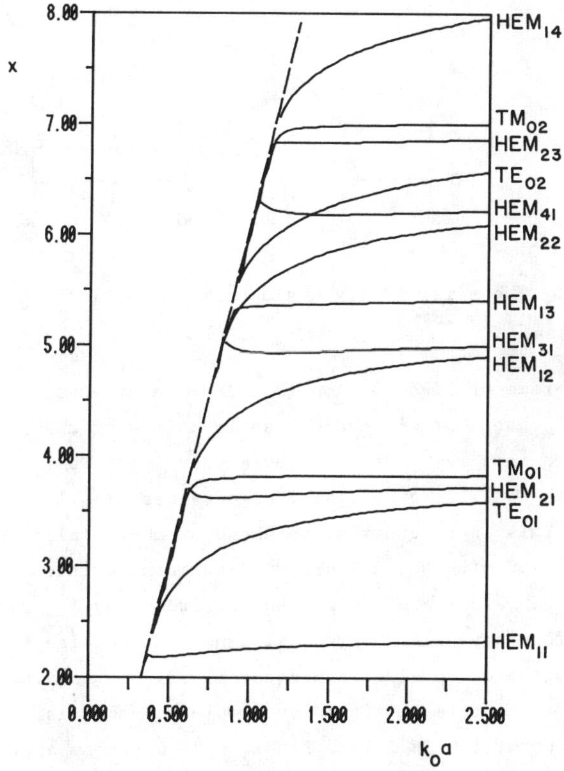

Fig. 3.4 The first 13 eigenvalues of the dielectric rod waveguide ($\epsilon_r = 20.0$)

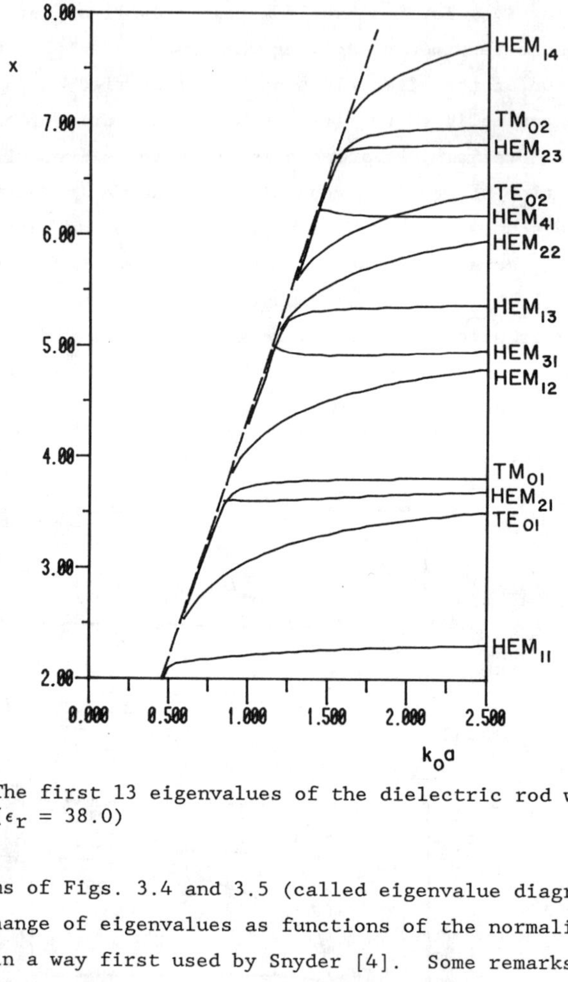

Fig. 3.5 The first 13 eigenvalues of the dielectric rod waveguide (ϵ_r = 38.0)

The diagrams of Figs. 3.4 and 3.5 (called eigenvalue diagrams) describe the change of eigenvalues as functions of the normalized frequency k_0a in a way first used by Snyder [4]. Some remarks can be drawn from these two figures. The eigenvalues clearly vary with frequency. This is in contrast to those of the metallic cylindrical waveguide filled with the uniform dielectric material. In the latter case, the eigenvalues are independent of frequency [5, sec. 5-2], and would therefore appear as horizontal lines on the eigenvalue diagram. Furthermore, the eigenvalues are generally slowly varying functions of k_0a, and can therefore be fitted by simple polynomials. For instance, the eigenvalue of the TE_{01} mode for ϵ_r = 38.0 can be approximated by the polynomial [6]:

$$x_{01}^{TE} \simeq 3.832 - \frac{0.6258}{k_0 a} + \frac{0.08285}{(k_0 a)^2} + \frac{0.02222}{(k_0 a)^3} \tag{3.56}$$

which has a relative error of magnitude less than 2.3×10^{-4} for $0.45 \le k_0 a \le 2.0$. Other eigenvalues, like that of the TM_{01} mode, change even more slowly, and can be approximated by even simpler polynomials. It is interesting to note that the first term in (3.56) is equal to the eigenvalue of the TE_{01} mode of the metallic waveguide. Although (3.56) is valid only in the specified range of $k_0 a$, it can be concluded that the horizontal line corresponding to the eigenvalue of the TE_{01} metallic waveguide mode is the asymptote ($k_0 a \to \infty$) to the eigenvalue curve for the same dielectric rod waveguide mode. This is actually true for all the TE modes, a fact that is readily established by letting $k_0 a \to \infty$ in (3.55).

To illustrate the use of the eigenvalue diagram in Fig. 3.5, the wavelength of the mode HEM_{21} dielectric rod waveguide is computed below. The waveguide wavelength λ_g is defined as

$$\lambda_g = \frac{2\pi}{\beta} \tag{3.57}$$

The diagram is valid for the material having $\epsilon_r = 38.0$, and the rod is of radius a = 1 cm and operates at frequency 5.0 GHz. From (3.11), one finds $k_0 a = 1.047$ radians. The corresponding eigenvalue is then read from Fig. 3.5 as x = 3.62. The propagation constant multiplied by the radius can now be computed from (3.42), the result being $(\beta a)_{21} = 5.34$ radians. The wavelength is then obtained from (3.57) to be $\lambda_{g21} =$ 1.18 cm. This value is considerably less than the wavelength in free space, which is $\lambda_0 = 6.0$ cm. On the other hand, if the whole space was filled with a dielectric of $\epsilon_r = 38.0$, the wavelength would be only $6.0/\sqrt{38.0} = 0.973$ cm. The wavelength of the HEM_{21} mode thus falls between these two extremes. In a similar manner, Figs. 3.4 and 3.5 can be used to determine the wavelengths of any other mode among the thirteen modes shown.

All the modal fields are evanescent in the radial direction outside the rod by virtue of the modified Bessel functions in (3.37). Those modes corresponding to an $x > k_0 a \sqrt{\epsilon_r}$ are evanescent in the z direction as well, since β then becomes purely imaginary. The modal fields corresponding to x_{mn} can be determined from (3.30), (3.31), (3.33),

(3.34), (3.36), and (3.37), with y_{mn} determined from (3.41) and the phase constant β_{mn} from (3.42). Simple relations for the amplitude coefficients A_{mn}, B_{mn}, C_{mn}, and D_{mn} are derived in the next section. Since only one of these amplitudes can be arbitrarily chosen, these The field lines represent the direction of the field in space. In any transverse-to-z plane, then, the electric field lines satisfy the differential equation:

$$\rho \frac{d\phi}{d\rho} = \frac{E_\phi}{E_\rho} \tag{3.58}$$

The magnetic field lines satisfy (3.58) with magnetic field components replacing their electric counterparts. Solving (3.58) is a very simple matter in the case of TE and TM modes. As can be seen from (3.36) and (3.37), one field component in (3.58), whether E_ρ and H_ϕ for the TE modes or E_ϕ and H_ρ for the TM modes, is always zero. The electric field lines are, therefore, given by

$$\rho = \text{constant}, \qquad \text{for TE modes} \tag{3.59a}$$

$$\phi = \text{constant}, \qquad \text{for TM modes} \tag{3.59b}$$

The equations for the magnetic field lines are similar to those for the electric field lines except for the modes reversing roles in (3.59). The exact integration of (3.58) for the hybrid modes, however, is rather difficult, and computer-generated plots of field lines must then be sought.

Computer-generated plots for the transverse-to-z field distributions of the HEM_{11}, TE_{01}, HEM_{21}, TM_{01}, and HEM_{12} are given in Figs. 3.6 through 3.11 for $\epsilon_r = 38.0$ at different values of x and k_0a. Each mode has separate illustrations for the electric field and separate for the magnetic field. The key to the arrow maps is as follows:

1. Fields with amplitudes within 1.0 dB of the maximum amplitude are called strong fields and shown as double arrows;
2. Fields with amplitudes 1.0 to 3.0 dB of the maximum amplitude are called medium fields and shown as long single lines;

3. Fields with amplitudes 3.0 to 10.0 dB of the maximum amplitude are called weak fields and shown as short single lines;
4. Fields with amplitudes of more than 10.0 dB of the maximum amplitude are neglected. The points with such fields are then left blank in the plots.

It is worth remembering that the arrow or line drawn at a point is the tangent to the field line at that point, and the blank portions do not necessarily imply zero field, since z-components of the fields may exist there. This form of plotting is very useful in identifying the locations of strong fields; a much needed piece of information when selecting the coupling device best suited to a particular mode. For instance, a short electric probe would be most efficient when oriented along an electric field line and placed at a location of a strong electric field.

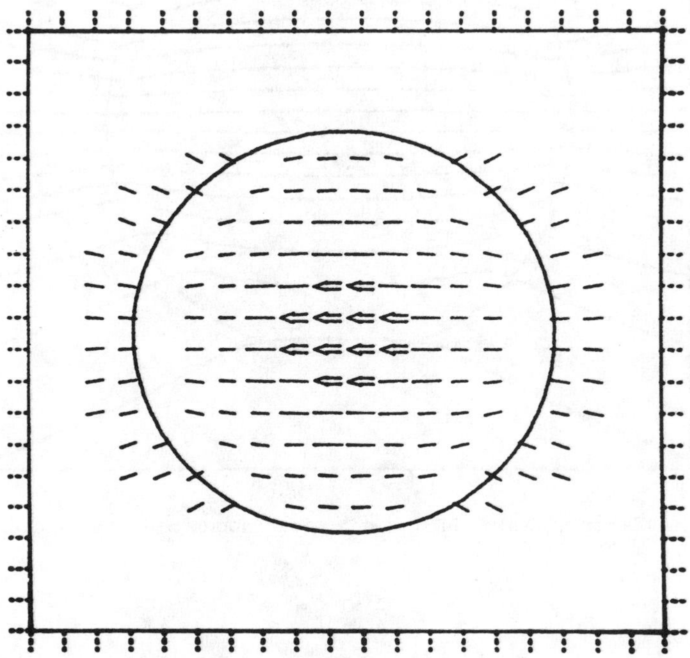

Fig. 3.6(a) Electric field, HEM_{11} mode, discretized plot (ϵ_r = 38.0, k_0a = 0.5, x = 2.1864)

In passing, note that the field lines corresponding to the TE and TM modes follow the pattern indicated in (3.59). Generally speaking, these field patterns bear some similarity to those of the corresponding modes of the metallic cylindrical waveguide filled with the same dielectric material. Furthermore, as k_0a increases, the field intensity for the HEM_{11} increases inside the rod indicating better field containment characteristics at higher frequencies. Indeed, this is shown in the next section to be the case for all the dielectric rod waveguide modes, not only the first hybrid mode.

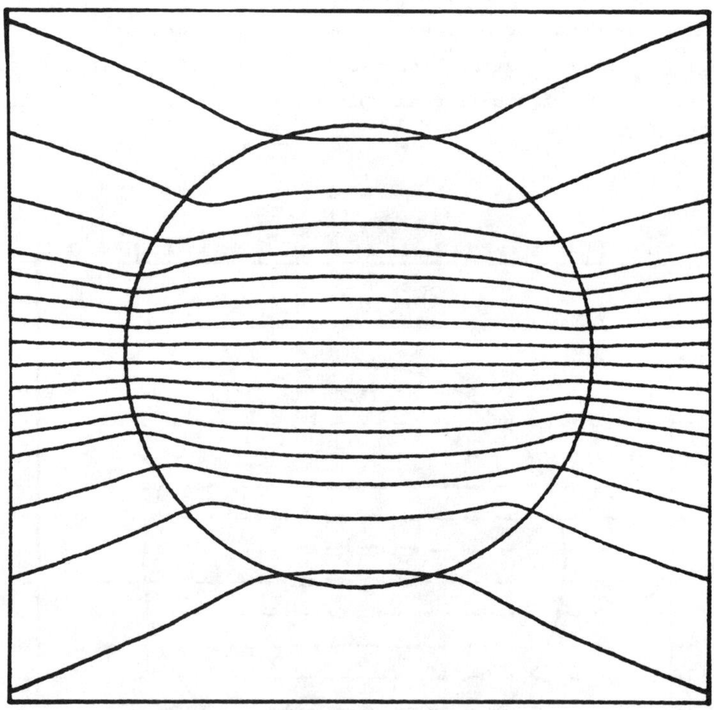

Fig. 3.6(b) Electric field, HEM_{11} mode, continuous plot (same data as in Fig. 3.6(a))

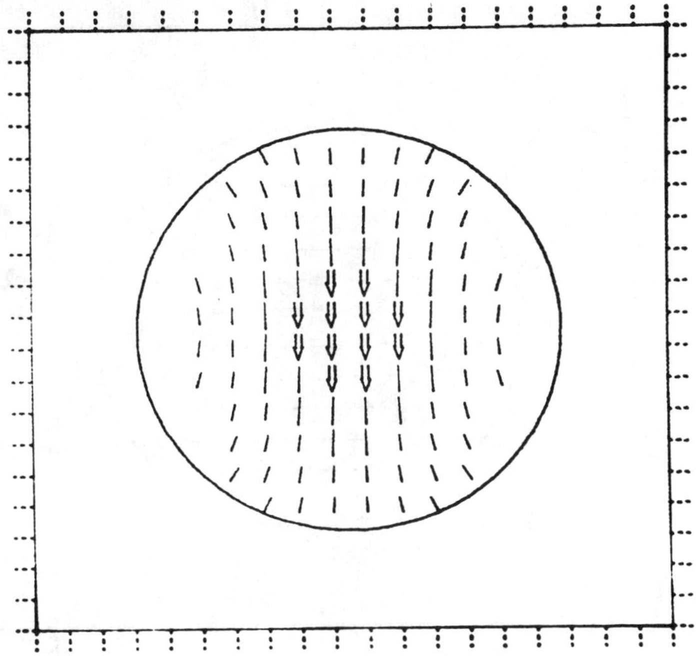

Fig. 3.6(c) Magnetic field, HEM_{11} mode, discretized plot (same data as in Fig. 3.6(a))

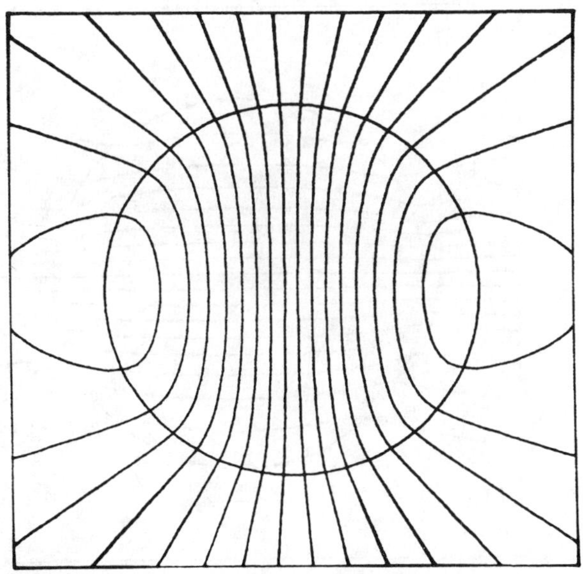

Fig. 3.6(d) Magnetic field, HEM_{11} mode, continuous plot (same data as in Fig. 3.6(a))

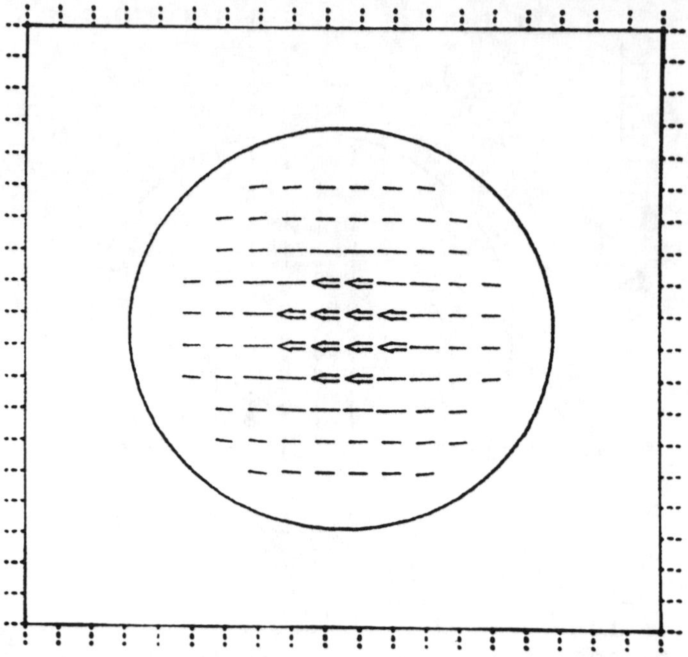

Fig. 3.7(a) Electric field, HEM_{11} mode, discretized plot (ϵ_r = 38.0, k_0a = 1.0, x = 2.2583)

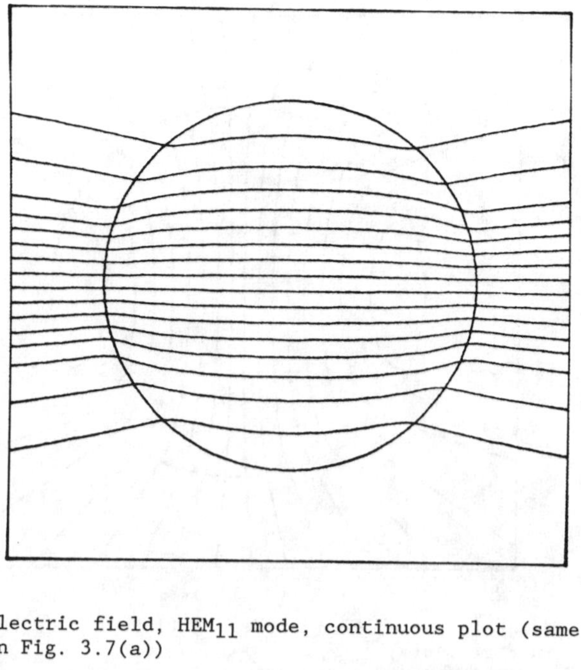

Fig. 3.7(b) Electric field, HEM_{11} mode, continuous plot (same data as in Fig. 3.7(a))

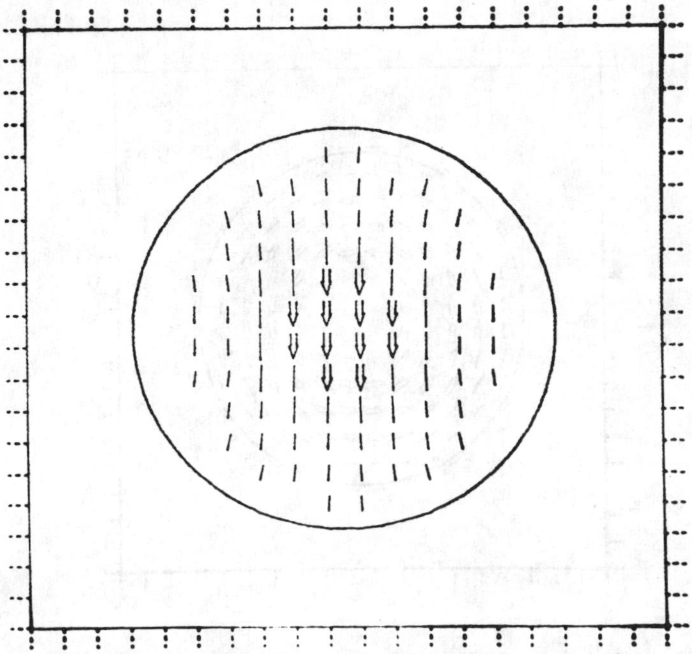

Fig. 3.7(c) Magnetic field, HEM_{11} mode, discretized plot (same data as in Fig. 3.7(a))

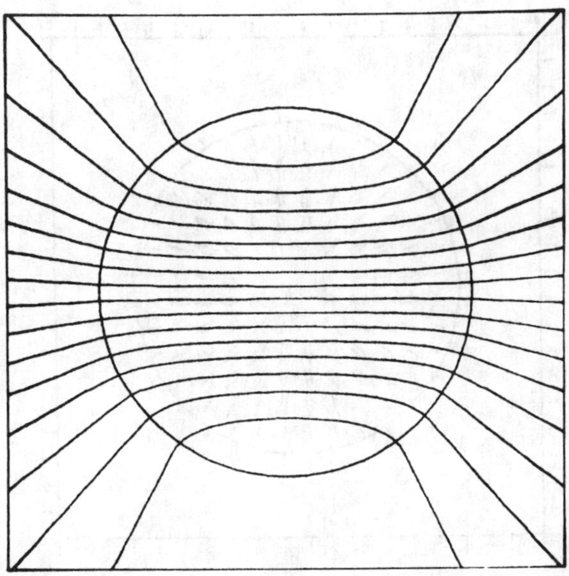

Fig. 3.7(d) Magnetic field, HEM_{11} mode, continuous plot (same data as in Fig. 3.7(a))

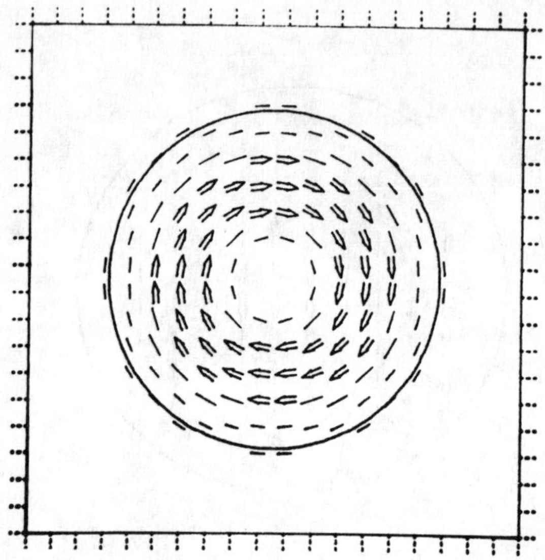

Fig. 3.8(a) Electric field, TE_{01} mode, discretized plot (ϵ_r = 38.0, k_0a, x = 3.2672)

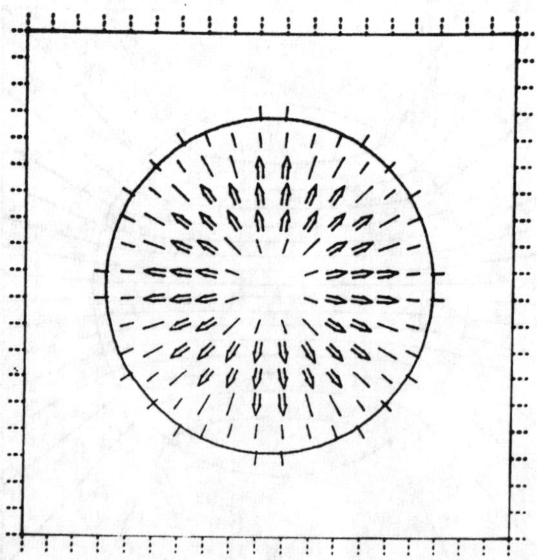

Fig. 3.8(b) Magnetic field, TE_{01} mode, discretized plot (same data as in Fig. 3.8(a))

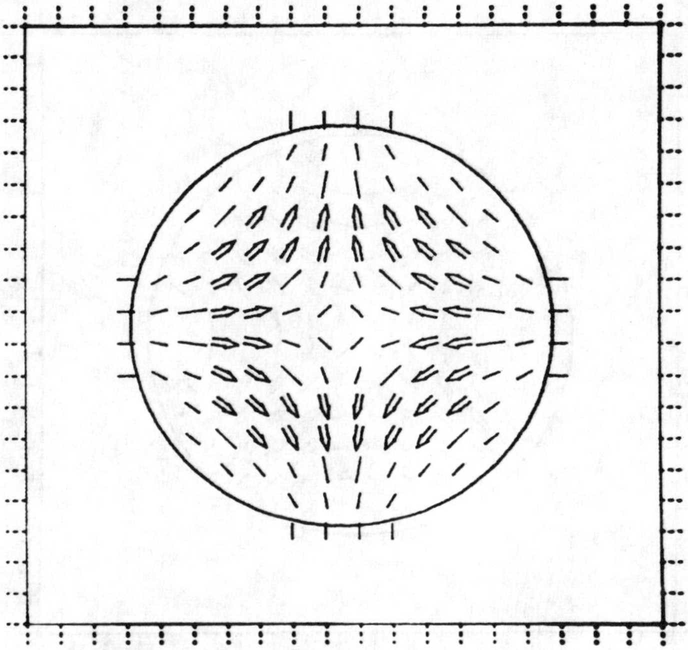

Fig. 3.9(a) Electric field, HEM_{21} mode, discretized plot ($\epsilon_r = 38.0$, $k_0a = 1.0$, $x = 3.6339$)

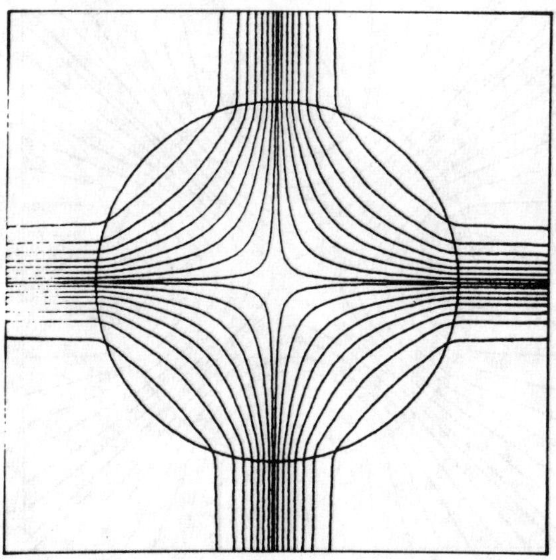

Fig. 3.9(b) Electric field, HEM_{21} mode, continuous plot (same data as in Fig. 3.9(a))

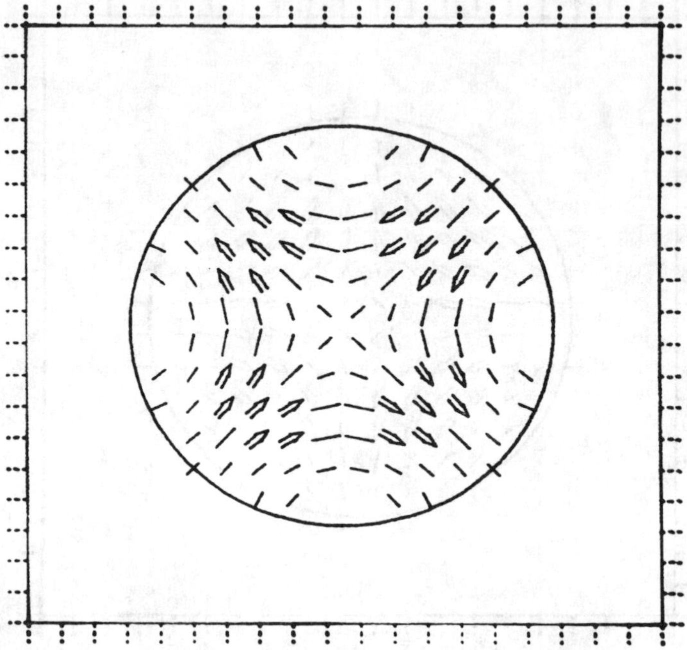

Fig. 3.9(c) Magnetic field, HEM_{21} mode, discretized plot (same data as in Fig. 3.9(a))

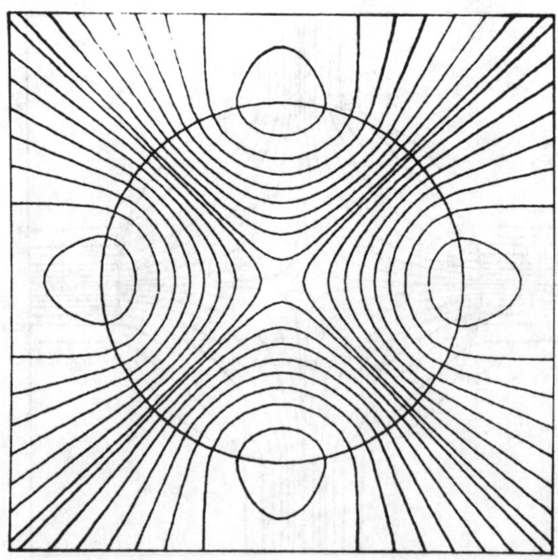

Fig. 3.9(d) Magnetic field, HEM_{21} mode, continuous plot (same data as in Fig. 3.9(a))

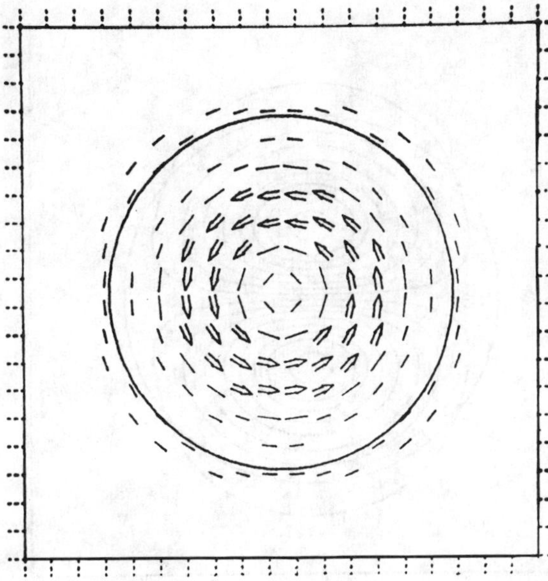

Fig. 3.10 Magnetic field TM_{01} mode, discretized plot ($\epsilon_r = 38.0$, $k_0a = 1.0$, $x = 3.8088$)

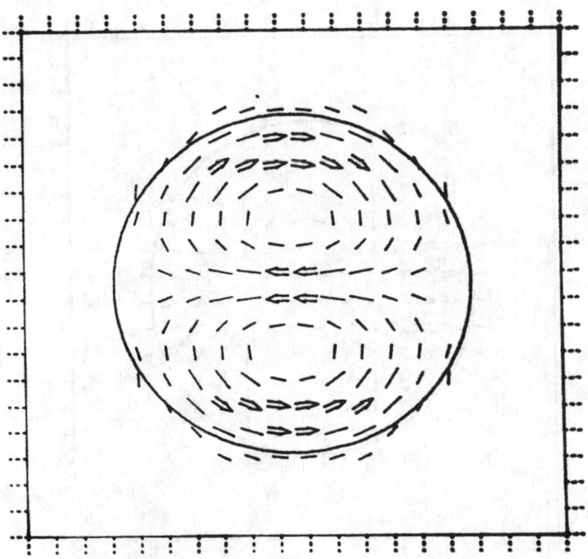

Fig. 3.11(a) Electric field, HEM_{12} mode, discretized plot ($\epsilon_r = 38.0$, $k_0a = 1.0$, $x = 4.4509$)

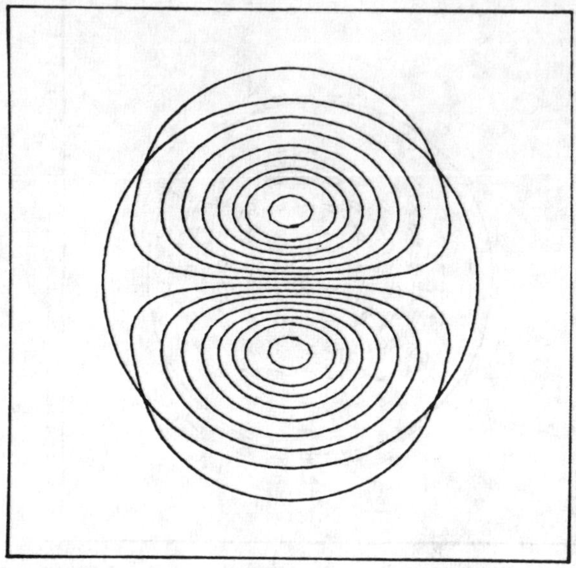

Fig. 3.11(b) Electric field, HEM_{12}, continuous plot (same data as in Fig. 3.11(a))

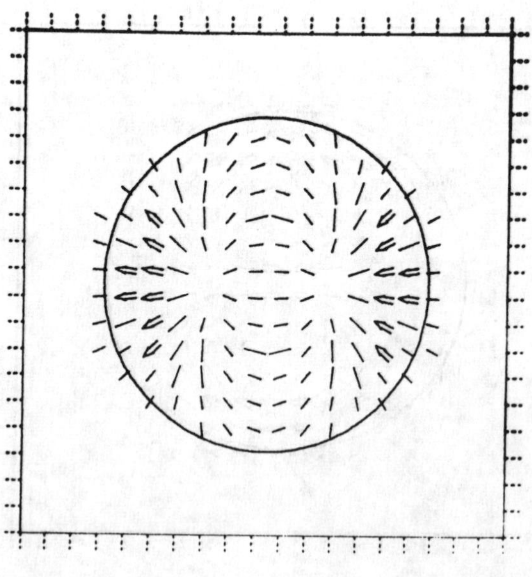

Fig. 3.11(c) Magnetic field, HEM_{12} mode, discretized plot (same data as in Fig. 3.11(a))

3.6 Power Considerations

An important characterization of the propagating modes for the dielectric rod waveguide is in terms of the ratio of the total power transmitted by the mode outside the rod to that transmitted inside it, which gives a measure of field containment in the rod. The modes can also be classified by the portions of the total power transmitted inside the rod by the different components of the field. These ratios are derived in this section, but first some relations between the amplitude coefficients A_{mn}, B_{mn}, C_{mn}, and D_{mn} are established.

The first relation:

$$\frac{C_{mn}}{A_{mn}} = \frac{D_{mn}}{B_{mn}} = \frac{J_m(x_{mn})}{K_m(y_{mn})} \tag{3.60}$$

follows immediately from (3.43a) and (3.43b). Substituting (3.59) into (3.43c) and (3.43d) and regrouping, we have

$$A_{mn}\, F_1(x_{mn}) + \eta\, B_{mn}\, F_3(x_{mn}) = 0 \tag{3.61a}$$

$$A_{mn}\, \frac{1}{\eta}\, F_3(x_{mn}) + B_{mn}\, F_2(x_{mn}) = 0 \tag{3.61b}$$

upon using (3.50), (3.51), and (3.52). In (3.61), $\eta = \sqrt{\mu_0/\epsilon}$ is the wave impedance of the dielectric medium. Consequently,

$$\frac{B_{mn}}{A_{mn}} = -\frac{1}{\eta}\frac{F_1(x_{mn})}{F_3(x_{mn})} = -\frac{1}{\eta}\frac{F_3(x_{mn})}{F_2(x_{mn})} \tag{3.62}$$

or

$$\frac{B_{mn}^2}{A_{mn}^2} = \frac{D_{mn}^2}{C_{mn}^2} = \frac{1}{\eta^2}\frac{F_1(x_{mn})}{F_2(x_{mn})} \tag{3.63}$$

The time-average power transmitted by the mnth mode is basically

$$P_{mn} = \frac{1}{2}\,\mathrm{Re}\int_0^{2\pi}\!\int_0^{\infty}\left(\underline{E}_{mn} \times \underline{H}_{mn}^*\right)\cdot \hat{z}\;\rho\, d\rho\, d\phi \tag{3.64}$$

where Re denotes the real part , and * denotes complex conjugate.

Substituting from (3.36) and (3.37), (3.64) can be written as

$$P_{mn} = P_{mn1} + P_{mn2} \tag{3.65}$$

where

$$P_{mn1} = \frac{1}{2}\,\mathrm{Re}\int_0^{2\pi}\int_0^{a}\left(\underline{E}_{mn1} \times \underline{H}^*_{mn1}\right) \cdot \hat{\underline{z}}\ \rho d\rho d\phi$$

$$= \frac{\pi a^2 \beta_{mn}\omega}{2\,x_{mn}^2}\left(\epsilon A_{mn}^2 + \mu_0 B_{mn}^2\right)\int_0^a \left[J'^2_m\left(\frac{x_{mn}}{a}\rho\right) + \frac{m^2 a^2}{x_{mn}^2 \rho^2} J_m^2\left(\frac{x_{mn}}{a}\rho\right)\right]\rho d\rho$$

$$+ \frac{\pi a^4}{2\,x_{mn}^4}\, m\, A_{mn} B_{mn}\left(\beta_{mn}^2 + k^2\right) J_m^2(x_{mn}) \tag{3.66}$$

is the time-average power transmitted inside the rod, and

$$P_{mn2} = \frac{1}{2}\,\mathrm{Re}\int_0^{2\pi}\int_a^{\infty} \underline{E}_{mn2} \times \underline{H}^*_{mn2} \cdot \hat{\underline{z}}\ \rho d\rho d\phi$$

$$= \frac{\pi a^2 \beta_{mn}\omega}{2\,y_{mn}^2}\left(\epsilon_0 C_{mn}^2 + \mu_0 D_{mn}^2\right)\int_a^\infty \left[K'^2_m\left(\frac{y_{mn}}{a}\rho\right) + \frac{a^2 m^2}{y_{mn}^2 \rho^2} K_m^2\left(\frac{y_{mn}}{a}\rho\right)\right]\rho d\rho$$

$$- \frac{\pi a^4}{2\,y_{mn}^4}\, m\, C_{mn} D_{mn}\left(\beta_{mn}^2 + k_0^2\right) K_m^2(y_{mn}) \tag{3.67}$$

is the time-average power transmitted outside the rod. As can be seen from (3.66), the transmitted power in the rod consists of three distinct parts. The first part is due to the TM part of the field, and can be recognized by the fact that its amplitude is proportional to A_{mn}^2. It can be recalled from (3.30) that A is the amplitude of the z-component of the electric field. The second part is due to the TE part of the field, which is proportional to B_{mn}^2. The third part of the transmitted power is of mixed nature, its amplitude being proportional to the product $A_{mn}B_{mn}$. A similar situation exists in region 2, where C_{mn}^2

signifies the TM part, D_{mn}^2 the TE part, and $C_{mn}D_{mn}$ the mixed part. The transmitted power in region q, q = 1,2, can then be written as

$$P_{mnq} = P_{mnq}^{TE} + P_{mnq}^{TM} + P_{mnq}^{MIX} \tag{3.68}$$

For m = 0, $P_{mnq}^{MIX} = 0$, and the transmitted power decomposes into TE and TM powers, in the same way as the eigenvalue equation (3.49) splits into two equations for TE and TM fields.

Two of the desired power ratios can now be immdediately written. The ratio of the power transmitted by the TE and TM parts inside the rod, denoted R_{mn}, is given by

$$R_{mn} = \frac{P_{mn1}^{TE}}{P_{mn1}^{TM}} = \frac{F_1(x_{mn})}{F_2(x_{mn})} \tag{3.69}$$

The ratio of the TE to the TM part of the transmitted power outside the rod is then

$$\frac{P_{mn2}^{TE}}{P_{mn2}^{TM}} = \epsilon_r R_{mn} \tag{3.70}$$

The different components of the transmitted power both inside and outside the rod can then be determined once the ratio:

$$\frac{P_{mnq}^{TE} + P_{mnq}^{TM}}{P_{mnq}} = \frac{1}{1 + \dfrac{P_{mnq}^{MIX}}{P_{mnq}^{TE} + P_{mnq}^{TM}}}, \quad q = 1,2 \tag{3.71}$$

is determined. It is convenient to introduce auxiliary constants U_{mn} and V_{mn} as follows:

$$U_{mn} = \frac{J_m'^2(x_{mn})}{J_m^2(x_{mn})} + \frac{2J_m'(x_{mn})}{x_{mn}J_m(x_{mn})} + \left(1 - \frac{m^2}{x_{mn}^2}\right) \tag{3.72a}$$

$$V_{mn} = -\frac{K'^2_m(y_{mn})}{K^2_m(y_{mn})} - \frac{2K'_m(y_{mn})}{y_{mn}K_m(y_{mn})} + \left(1 + \frac{m^2}{y^2_{mn}}\right) \tag{3.72b}$$

The ratio which can be used to compute the mixed power inside the rod is then

$$D_1 = \frac{P^{MIX}_{mn1}}{P^{TE}_{mn1} + P^{TM}_{mn1}} = -\frac{2mk_0\sqrt{\epsilon_r}R_{mn}\,(1 + \beta^2_{mn}/k^2_0\epsilon_r)}{\beta_{mn}x^2_{mn}(1 + R_{mn})U_{mn}} \tag{3.73}$$

Outside the rod, the corresepnding ratio becomes

$$D_2 = \frac{P^{MIX}_{mn2}}{P^{TE}_{mn2} + P^{TM}_{mn2}} = \frac{2mk_0\sqrt{\epsilon_r}R_{mn}\,(1 + \beta^2_{mn}/k^2_0)}{\beta_{mn}y^2_{mn}(1 + \epsilon_r R_{mn})V_{mn}} \tag{3.74}$$

Finally, the ratio of the total power transmitted in the air to the total power transmitted in the rod is given by

$$\frac{P_{mn2}}{P_{mn1}} = \frac{x^2_{mn}\left(\frac{1}{\epsilon_r} + R_{mn}\right)V_{mn}}{y^2_{mn}\left(1 + R_{mn}\right)U_{mn}}\left[\frac{1 + D_2}{1 + D_1}\right] \tag{3.75}$$

It should be noted that in reference [6], the mixed parts of the transmitted power (both D_1 and D_2) are mistakenly taken to be zero.

Figure 3.12, which shows the ratio R_{mn} from (3.69), can be used to classify the hybrid modes into two separate categories, depending on whether the larger part of the power transmitted within the rod is carried by the TE part of the field or by the TM part. The modes with $P^{TE}_{mn1} > P^{TM}_{mn1}$ are called quasi-TE modes, and those with $P^{TE}_{mn1} < P^{TM}_{mn1}$ are called quasi-TM modes. It is interesting to note that HEM_{mn} modes are quasi-TM for an odd value of n, and quasi-TE for an even value of n. In the literature on optical fibers, the quasi-TE modes are commonly called HE modes, in accordance with the notation introduced by Snitzer [10]. Similarly, quasi-TM modes are called EH modes. The correspondence between various notations is as follows:

quasi-$TM_{m,n}$ (or $EH_{m,n}$) equivalent to $HEM_{m,2n-1}$

quasi-$TE_{m,n}$ (or $HE_{m,n}$) equivalent to $HEM_{m,2n}$

It should be mentioned that HEM notation was standardized by the IRE (nowadays known as IEEE) in [11].

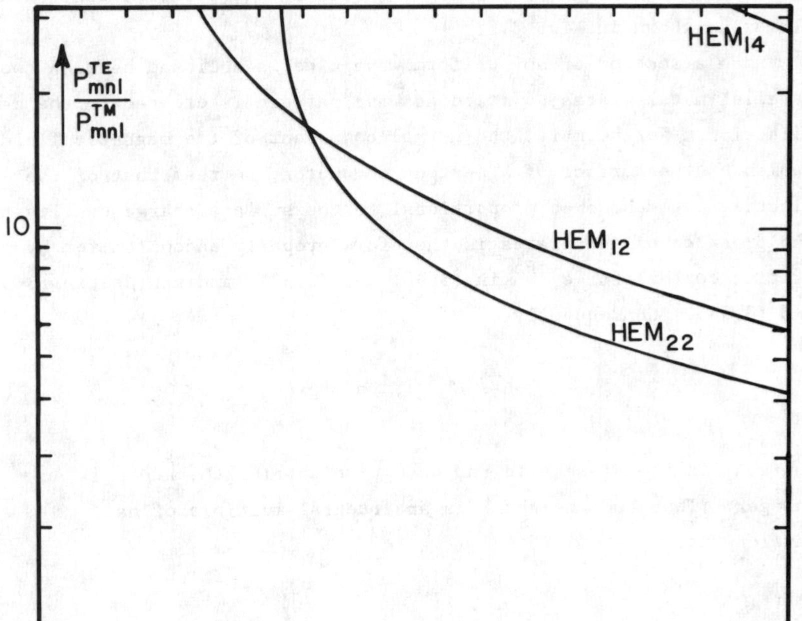

Fig. 3.12 The ratio of the TE and TM parts of the power transmitted inside the dielectric rod waveguide (ϵ_r = 38.0) (reference [6], ©1983 Microwave Systems News)

3.7 The Parallel-Plate Dielectric Resonator

Dielectric resonators can take on various shapes. The simplest, perhaps, can be constructed by placing a section of the cylindrical dielectric rod waveguide of length h between two parallel conducting plates as shown in Fig. 3.2.

When a section of any uniform waveguide is enclosed between two parallel metal plates, a standing wave pattern is created in the z-direction. Furthermore, the normal component of the magnetic field must vanish at the surface of a perfect conductor, whereas that of the electric field becomes proportional to the surface charge density there. The presence of the plates is therefore properly accounted for by writing $\cos(\beta z)$ for $e^{-j\beta z}$ in (3.30) and (3.33), and $\sin(\beta z)$ in (3.31) and (3.34). Consequently,

$$\beta h = \frac{2\pi}{\lambda_g} h = p\pi \tag{3.76}$$

where λ_g is the dielectric rod waveguide wavelength, and p is an integer. Thus, the height h is an integral multiple of half wavelengths,

$$h = p\,\frac{\lambda_g}{2}\,, \qquad p = 1,2,\ldots \tag{3.77}$$

There is only one discrete set of frequencies that satisfies both (3.42) and (3.76). Such frequencies, called the resonant frequencies of the parallel-plate dielectric resonator, can be determined graphically with the help of the eigenvalue diagram. Using (3.42) and (3.76), the value of x at resonance is determined by

$$x^2 = (k_0a)^2\epsilon_r - \left(\frac{a}{h}\,p\pi\right)^2 \tag{3.78}$$

or

$$\frac{(k_0a)^2}{u_p^2} - \frac{x^2}{v_p^2} = 1 \tag{3.79}$$

where

$$u_p = p\,\frac{a\pi}{h\sqrt{\epsilon_r}} = pu_1 \tag{3.80}$$

and

$$v_p = \sqrt{\epsilon_r}\, u_p \tag{3.81}$$

Relationship (3.79) is readily recognized as a family of hyberbolas with foci at $\pm\, pu_1 \sqrt{\epsilon_r+1}$ and an eccentricity of $\sqrt{\epsilon_r+1}$ [8, Ch. 8]. The eigenvalue diagram is of universal nature, valid for all frequencies f and all radii a. For given resonator dimensions a and h, the family of hyberbolas can be plotted for different p's. Superimposing the curves, the two families of curves intersect at a certain set of points (x_{mnp}, $(k_0a)_{mnp}$). The resonant frequencies are then given by

$$f_{mnp} = \frac{150}{\pi a} (k_0 a)_{mnp} \qquad \text{(GHz)} \tag{3.82}$$

where a is measured in millimeters. Figures 3.13 and 3.14 illustrate the graphical procedure for parallel-plate dielectric resonators with $\epsilon_r = 20.0$ and $\epsilon_r = 38.0$, respectively.

The graphical method is a good tool for predicting and identifying the various modes of the dielectric resonator. A more accurate evaluation of the resonant frequencies has to be carried out numerically. The advantage of the graphical procedure, such as in Fig. 3.13 or 3.14, is that one can easily recognize the order in which resonant frequencies of various modes will appear. For fixed radius a and fixed length h, the abscissa k_0a grows proportionally to frequency. Thus, one can see that the lowest resonant frequency belongs to the HEM_{11} resonance, the next higher is TE_{011}, and then HEM_{211}, TM_{011}, etc., follow. These resonant modes are simply found by moving along the appropriate hyperbola for p = 1 from left to right, and observing the intersections with the eigenvalues of various modes. For example, for $\epsilon_r = 38.0$ and a = h = 4.25 mm, the hyberbola p = 1 intersects the TM_{01} mode of the dielectric rod waveguide at $k_0a \simeq 0.82$. It then follows from (3.82) that $f_{011}^{TM} \simeq 9.21$ GHz. The mode corresponding to f_{011}^{TM} is evidently the TM_{011} mode, whose field distribution in any transverse-to-z plane is that of the TM_{01} rod waveguide mode. Table 3.1 shows some of the resonant frequencies obtained graphically for a = h = 4.25 mm.

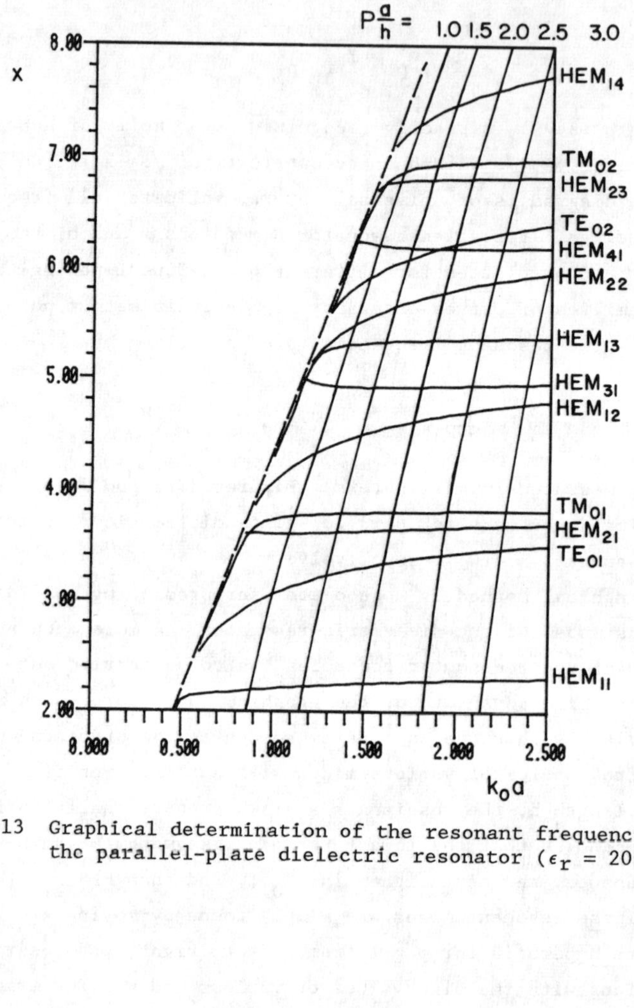

Fig. 3.13 Graphical determination of the resonant frequencies of the parallel-plate dielectric resonator (ϵ_r = 20.0)

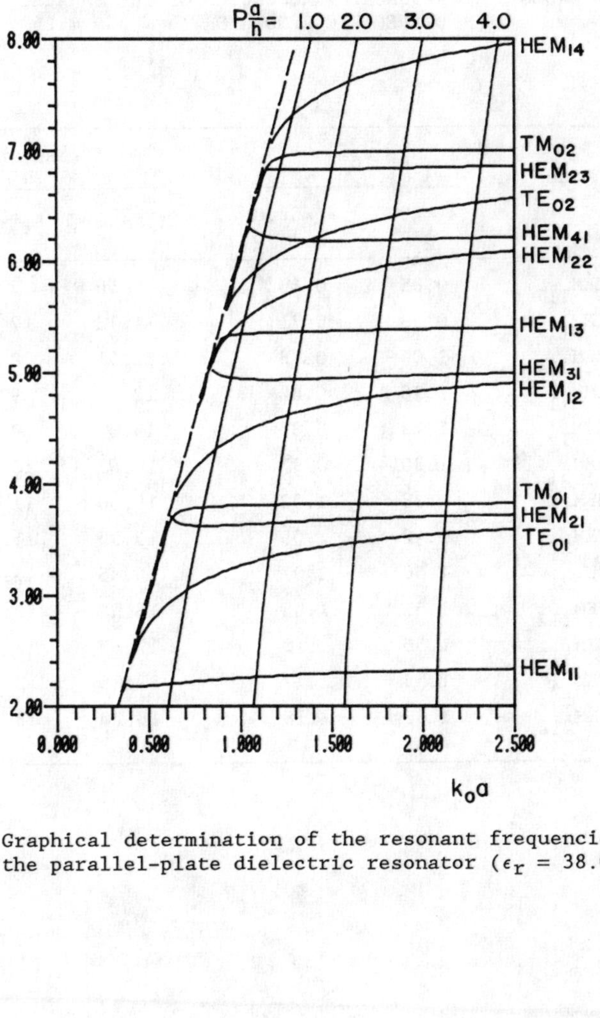

Fig. 3.14 Graphical determination of the resonant frequencies of the parallel-plate dielectric resonator ($\epsilon_r = 38.0$)

TABLE 3.1 RESONANT FREQUENCIES OF THE PARALLEL-PLATE DIELECTRIC RESONATOR (a = h = 4.25 mm)

Mode	k_0a		f (GHz)	
	ϵ_r=20.0	ϵ_r=38.0	ϵ_r=20.0	ϵ_r=38.0
HEM_{111}	0.85	0.63	9.55	7.07
TE_{011}	0.99	0.71	11.12	7.98
HEM_{211}	1.06	0.78	11.91	8.76
TM_{011}	1.10	0.82	12.36	9.21
HEM_{121}	1.19	0.87	13.37	9.77
HEM_{311}	1.30	0.97	14.61	10.90
HEM_{131}	1.38	1.02	15.50	11.46
HEM_{221}	1.42	1.05	15.95	11.80
TE_{021}	1.50	1.07	16.85	12.02
HEM_{112}	1.51	1.08	16.96	12.13
HEM_{411}	1.56	1.13	17.53	12.69
TE_{012}	1.59	1.17	17.86	13.14
HEM_{212}	1.62	1.18	18.20	13.26

3.8 Measurement of the Dielectric Constant

The parallel-plate dielectric resonator can be used to measure the dielectric constant of a dielectric material. The principles involved in the measurement are rather simple. The resonant frequency of a given mode, preferably TE_{011}, is measured for a resonator of known radius a and length h. Afterwards, the dielectric constant is computed from (3.40), (3.49), and (3.76). The procedure was first introduced by Hakki and Coleman [12]. The error analysis and the investigation of temperature effects were made by Courtney [9]. The method is now commonly known as the Courtney method.

If the distance h between the two parallel plates is smaller than one-half wavelength, the parallel-plate dielectric resonator operating in the TE_{011} mode does not radiate. In this case, the sides of the resonator can be left open for easy inspection, or for inserting semirigid coaxial cables which serve for input and output coupling. The sample holder used by Courtney is shown in Fig. 3.15.

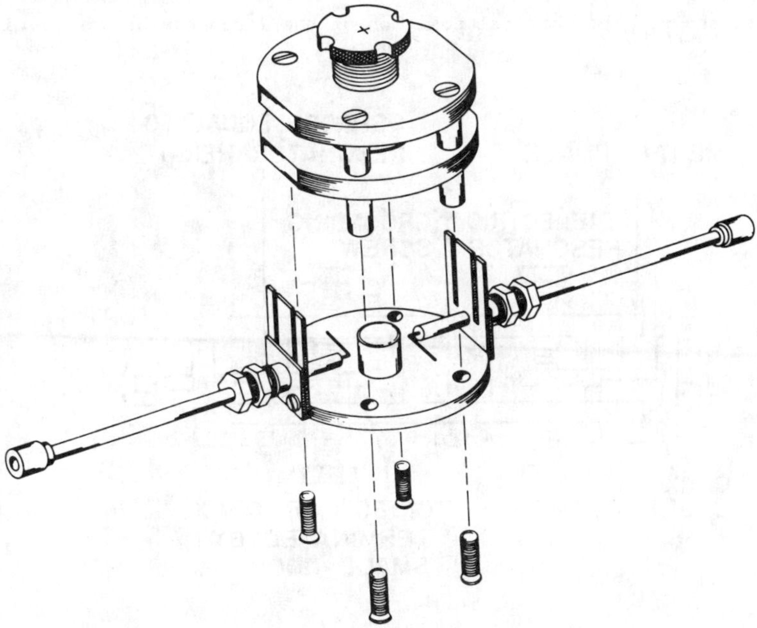

Fig. 3.15 The Courtney holder (reference [9], ©1970 IEEE)

Typical accuracy of the measurement of the dielectric constant ϵ_r is 0.3 %. Courtney estimated this error by assuming that the resonant frequency is measured with an accuracy of 0.1 %, and that the dimensions of the sample are measured with an accuracy of 0.2 % [9].

In the Courtney method, each sample is measured at only one resonant frequency, which is that corresponding to the TE_{011} mode. In addition to this resonant frequency, one can experimentally observe the existence of many other resonant frequencies. If one can positively identify other resonant modes, it is possible to measure the value of ϵ_r from these various measured resonant frequencies, as shown in [13]. By using the resonant modes TE_{011}, TE_{021}, TE_{031}, and TE_{041}, the dielectric constant of a single sample was measured in the range between 4.2 and 10.6 GHz. Furthermore, it was found that the quasi-TE modes are also well suited for this measurement, in particular HEM_{121}, HEM_{221}, and HEM_{141}. The major challenge in using higher resonant modes, however, is the positive identification of each mode.

The parallel-plate resonator used in [13] is shown in Fig. 3.16. The resonant frequency is measured by the transmission method, and the signal input and output are provided by coaxial cables, which have small loops, as shown. The vertical position of the loops can be varied with

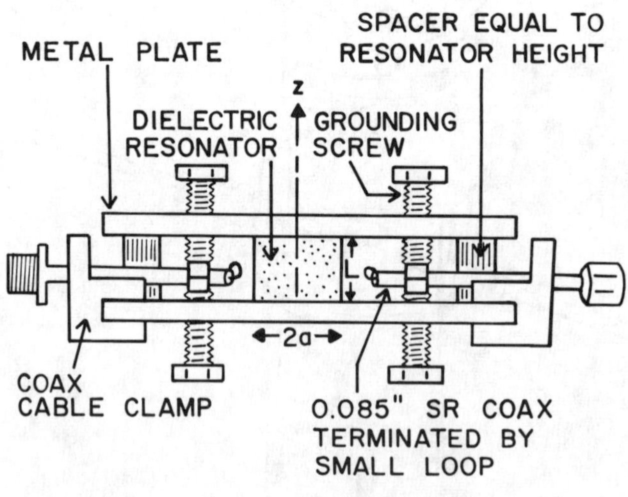

Fig. 3.16(a) Modified Courtney holder, side view (reference [13], ©1985 IEEE)

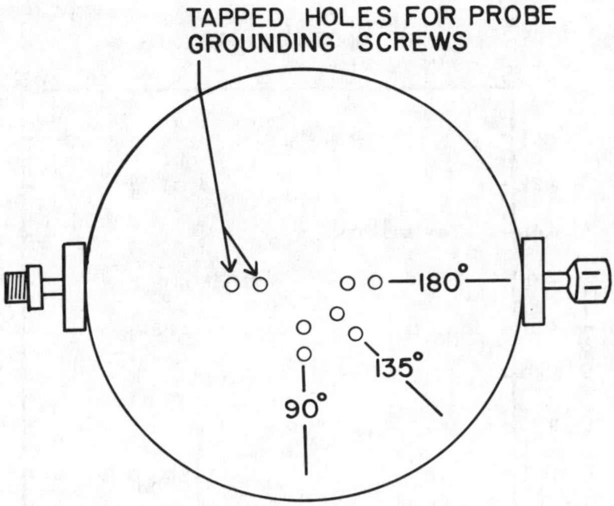

Fig. 3.16(b) Modified Courtney holder, top view (reference [13], ©1985 IEEE)

the use of grounding screws which protrude through each of the two parallel plates. This movement enables the observer to determine the number of standing waves in the z-direction, and, therefore, identify the thrid subscript, p, of the resonant mode.

Another simple indicator of the nature of the resonant mode is the orientation of the coupling loop. By rotating the loop around the axis of the coaxial cable, the magnetic field crossing the loop will induce either a large signal or a small signal in proportion to the intercepted flux over the area of the loop. By studying the field patterns of various modes, it is possible to predict whether the signal of a particular mode should be stronger for a horizontal orientation or a vertical orientation. The third indicator for mode identification is the azimuthal variation of the field. When the first subscript of the resonant mode is $m = 0$, there should be no variation of the field with ϕ. The azimuthal dependence of the field may be observed by moving one of the probes at any of the three different locations, denoted 180°, 135°, and 90°, in Fig. 3.16.

The results of one mode identification study are shown in Fig. 3.17. The observed modes are spaced horizontally at equal distances in order of growing frequency. The vertical axis gives the observed signal level

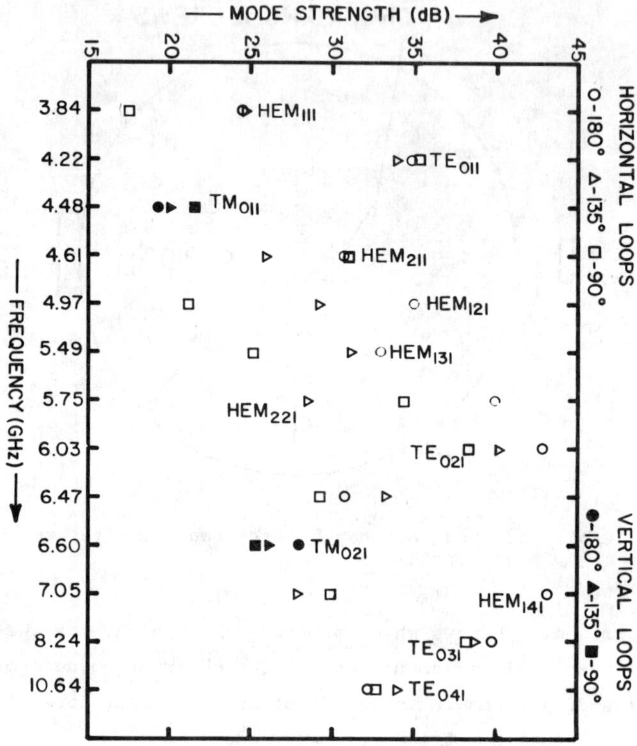

Fig. 3.17 Mode identification chart (reference [13], ©1985 IEEE)

in decibels. Three different symbols are used to denote the relative level of each resonant mode at different azimuthal positions: circles 180°, triangles for 135°, and squares for 90°. Two of the observed resonances, which could not be positively identified, are left without names: one occurring at 6.47 GHz and the other at 7.05 GHz.

It is concluded in [13] that neither the quasi-TM modes, nor the TM modes, are suitable for the measurement of ϵ_r. The main reason is the fact that the presence of a minute air gap between the dielectric sample and the metal plate may considerably alter the resonant frequency, thereby ruining the accuracy of measurement. This fact was already pointed out by Cohn and Kelly [14].

Another reason why the TM and quasi-TM resonant modes are not suitable in the measurement of ϵ_r has been pointed out by Kobayashi and

Katoh [15]. They have shown these resonant modes to be leaky, in the sense that a part of electromagnetic energy is propagating radially outward. The leaky TM_{0np} modes have no radial cut-off frequency, and, consequently, they have a low Q factor. Kobashi and Katoh also provide diagrams for estimating minimum radial dimensions of the metal plates which constitute the parallel-plate resonator operating in the TE_{01p} mode.

Hakki and Coleman [12] derived formulas which also enable one to determine the loss tangent of the dielectric material in the same parallel-plate arrangement. At the resonant frequency of the TE_{011} mode, one needs to measure the unloaded Q factor, Q_0, of the resonator. Since the losses are partially caused by the loss tangent of the dielectric sample, and partially by the resistivity of the two parallel metal plates, one has to subtract the two effects from each other. Quoting [12], the loss tangent is given by

$$\tan\delta = \frac{A}{Q_0} - B \tag{3.83}$$

$$A = 1 + \frac{1}{\epsilon_r} F(x_{01})\, G(y_{01}) \tag{3.84}$$

$$B = \frac{R_s}{2\pi f_{011}^3 \mu_0^2 \epsilon_0 h^3 \epsilon_r}\left[1 + F(x_{01})\, G(y_{01})\right] \tag{3.85}$$

$$F(x_{01}) = \frac{J_1^2(x_{01})}{J_1^2(x_{01}) - J_0(x_{01})\, J_2(x_{01})} \tag{3.86}$$

$$G(y_{01}) = \frac{1}{K_1^2(y_{01})}\left[K_0(y_{01})\, K_2(y_{01}) - K_1^2(y_{01})\right] \tag{3.87}$$

where

$$R_s = \sqrt{\frac{\pi f_{011}\mu_0}{\sigma}} \tag{3.88}$$

is the surface resistivity of the metallic plates [16, Sec. 5.14].

References

[1] C.T. Johnk, Engineering Electromagnetic Fields and Waves. New York: John Wiley & Sons, 1975.

[2] D.K. Cheng, Field and Wave Electromagnetics. Reading, MA: Addison-Wesley, 1983.

[3] W.K. Saunders, "On solutions of Maxwell's equations in an exterior region," Proc. National Academy of Science, vol. 38, pp. 342-348, 1952.

[4] A.W. Snyder, "Asymptotic expressions for eignfunctions and eigenvalues of a dielectric or optical waveguide," IEEE Trans. Microwave Theory Tech., vol. MTT-17, pp. 1130-1138, Dec. 1969.

[5] R.F. Harrington, Time-Harmonic Electromagnetic Fields. New York: McGraw-Hill, 1961.

[6] D. Kajfez, "Basic principles give understanding of dielectric waveguides and resonators," Microwave Systems News, vol. 13, pp. 152-161, May 1983.

[7] D. Kajfez, "Modal field patterns in dielectric rod waveguide," Microwave J., vol. 26, pp. 181-192, May 1983.

[8] E.W. Nichols, Analytic Geometry. New York: Leach, Shewell & Sanborn, 1982.

[9] W.E. Courtney, "Analysis and evaluation of a method of measuring the complex permittivity and permeability of microwave insulators," IEEE Trans. Microwave Theory Tech., vol. MTT-18, pp. 476-485, Aug. 1970.

[10] E. Snitzer, "Cylindrical dielectric waveguide modes," Jour. Optical Soc., vol. 51, pp. 491-498, May 1961.

[11] "IRE standards on antennas and waveguides: definitions of terms, 1953," Proc. IRE, vol. 41, pp. 1721-1728, Dec. 1953.

[12] B.W. Hakki and P.D. Coleman, "A dielectric resonator method of measuring inductive capacities in the millimeter range," IEEE Trans. Microwave Theory Tech., vol. MTT-8, pp. 402-410, July 1960.

[13] P. Wheless and D. Kajfez, "The use of higher resonant modes in measuring the dielectric constant of dielectric resonators," IEEE MTT-S Symposium Dig., pp. 473-476, St. Louis, June 1985.

[14] S.B. Cohn and K.C. Kelly, "Microwave measurement of high-dielectric-constant materials," IEEE Trans. Microwave Theory Tech., vol. MTT-14, pp. 406-410, Sept. 1966.

[15] Y. Kobayashi and M. Katoh, "Microwave measurement of dielectric properties of low-loss materials by the dielectric rod resonator method," IEEE Trans. Microwave Theory Tech., vol. MTT-33, pp. 586-592, July 1985.

[16] S. Ramo, J.R. Whinnery, and T. Van Duzer, Fields and Waves in Communication Electronics. New York: John Wiley & Sons, 1965.

Chapter 4
SIMPLE MODELS
Darko Kajfez

4.1 Introduction

An accurate mathematical description of the electromagnetic field in a dielectric resonator (DR) is considerably more complicated than the field description in a hollow waveguide resonator. If the numerical result for a DR is to be obtained for a given set of dimensions, a large computer is traditionally required because of the large size of the matrices involved and the large number of operations which has to be performed.

It is of great practical interest to approach the solution of the electromagnetic field in the DR in some simplified way that is still capable of giving results which are not too far from the exact values. Two such simple mathematical models of DR to be described here are the Cohn model and the Itoh and Rudokas model. Before the Cohn model is described, a brief review of the notion of the perfect magnetic conductor will be presented in the next two sections.

4.2 Notion of the Magnetic Conductor

Consider a textbook case of the plane wave incident on the interface of two different dielectric media as shown in Fig. 4.1. To make the situation as simple as possible, assume that the direction of the plane wave is normal to the interface, i.e., the incident wave propagation is oriented in the of positive z-direction. Then, a part the energy of the incident wave will be reflected back from the interface as a plane wave traveling in the negative z direction in medium 1. The rest of the energy will cross the interface and propagate as a transmitted wave in medium 2.

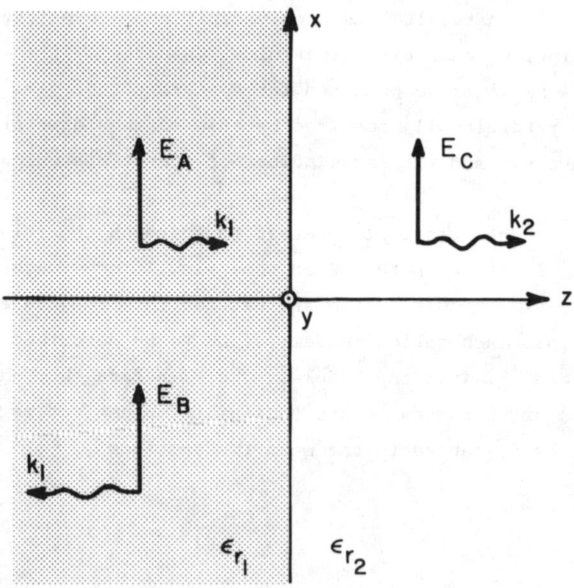

Fig. 4.1 Plane wave incident to the interface of two dielectric media

The electric and the magnetic fields in medium 1 are

$$E_{x1}(z) = E_A\, e^{-jk_1 z} + E_B\, e^{jk_1 z} \qquad (4.1)$$

$$H_{y1}(z) = \frac{1}{\eta_1}\left(E_A\, e^{-jk_1 z} - E_B\, e^{jk_1 z}\right) \qquad (4.2)$$

and in medium 2:

$$E_{x2}(z) = E_C\, e^{-jk_2 z} \tag{4.3}$$

$$H_{y2}(z) = \frac{E_C}{\eta_2}\, e^{-jk_2 z} \tag{4.4}$$

The propagation constants in each medium are

$$k_1 = \omega\sqrt{\mu_0\epsilon_0\epsilon_{r1}} \quad \text{and} \quad k_2 = \omega\sqrt{\mu_0\epsilon_0\epsilon_{r2}} \tag{4.5}$$

The intrinsic impedances in two media are

$$\eta_1 = \sqrt{\frac{\mu_0}{\epsilon_0\epsilon_{r1}}} \quad \text{and} \quad \eta_2 = \sqrt{\frac{\mu_0}{\epsilon_0\epsilon_{r2}}} \tag{4.6}$$

At the interface of two media (coordinate $z = 0$), the tangential electric field must be continuous, such that

$$E_{x1}(0) = E_{x2}(0) \tag{4.7}$$

which leads to

$$E_A + E_B = E_C \tag{4.8}$$

The other requirement at the interface is that the tangential magnetic field must be continuous, such that

$$H_{y1}(0) = H_{y2}(0) \tag{4.9}$$

which results in the following:

$$\frac{1}{\eta_1}(E_A - E_B) = \frac{1}{\eta_2} E_C \tag{4.10}$$

Dividing (4.8) by (4.10) and then solving for the ratio of the reflected to the incident wave amplitude, we obtain

$$\frac{E_B}{E_A} = \frac{\sqrt{\frac{\epsilon_{r1}}{\epsilon_{r2}}} - 1}{\sqrt{\frac{\epsilon_{r1}}{\epsilon_{r2}}} + 1} \tag{4.11}$$

Consider first the situation in which medium 1 has a higher dielectric constant than medium 2. For various ratios $\epsilon_{r1}/\epsilon_{r2}$, the reflection coefficient E_B/E_A will take values such as those listed in Table 4.1.

Table 4.1 PLANE WAVE INCIDENT FROM INSIDE OF THE DIELECTRIC MATERIAL

$\epsilon_{r1}/\epsilon_{r2}$	2.5	10	37	100
E_B/E_A	0.225	0.519	0.717	0.818

When the ratio of dielectric constants is much larger than unity, the reflection coefficient grows toward +1, and most of the energy of the plane wave is reflected back in medium 1. Only a small portion of the energy escapes into medium 2.

The total electric field in medium 1, which is the sum of the incident and the reflected waves, displays a standing wave pattern as indicated in Fig. 4.2a. The figure is drawn for the case where $\epsilon_{r1} = 100$ and $\epsilon_{r2} = 1$. It can be seen that the electric field at the interface is maximum. When the ratio $\epsilon_{r1}/\epsilon_{r2}$ grows to infinity, an open circuit situation is achieved.

The field inside a dielectric resonator is not as simple as the plane wave described above. Nevertheless, any general EM field distribution can be considered as a summation of various plane waves incident under all possible angles. For a plane wave which is incident by an angle other than normal, the reflection coefficient is even larger than the one given by (4.11), as can be verified in textbooks on EM theory (e.g., [1,2]).

For this reason, an approximate computation of the electromagnetic field inside the region made of high dielectric material is possible by assuming that the surface of the dielectric is covered with a perfect

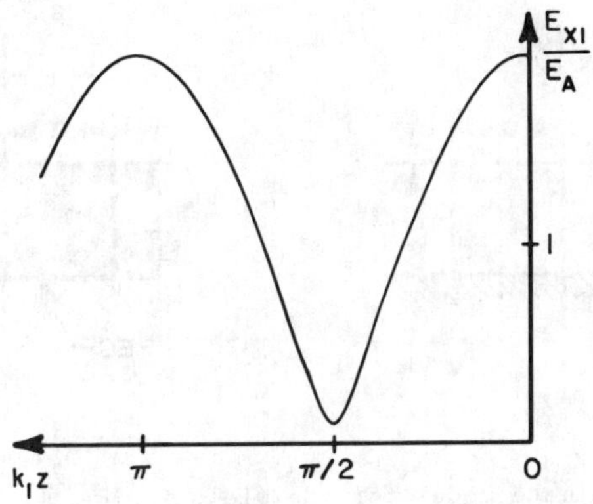

Fig. 4.2(a) Standing wave pattern, $\epsilon_{r1} > \epsilon_{r2}$

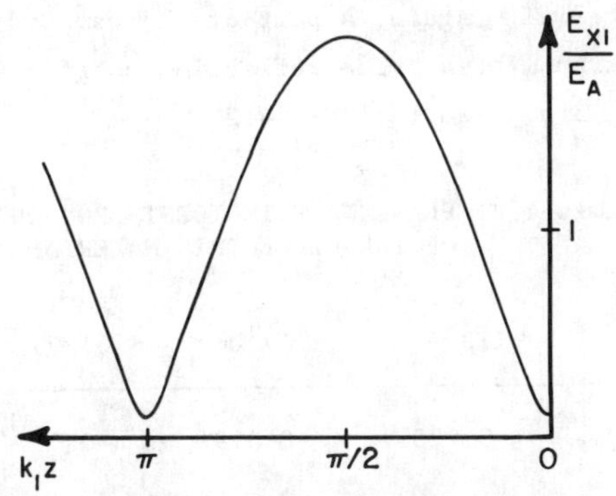

Fig. 4.2(b) Standing wave pattern, $\epsilon_{r1} < \epsilon_{r2}$

magnetic conductor (PMC). This is a nonexistent material which requires the magnetic field tangential to its surface to be zero. The higher the value of the relative dielectric constant ϵ_r, the more accurate are the fields computed by the PMC approximation. An application of this principle is illustrated in Fig. 4.3. Assume that a DR of radius a and length L is resting on a ground plane as shown in Fig. 4.3(a). Then, an approximate model for computing the electromagnetic field inside the dielectric is the idealized cavity shown in Fig. 4.3(b), the bottom wall of which is covered with the PEC, whereas the top and the side walls are covered with the PMC.

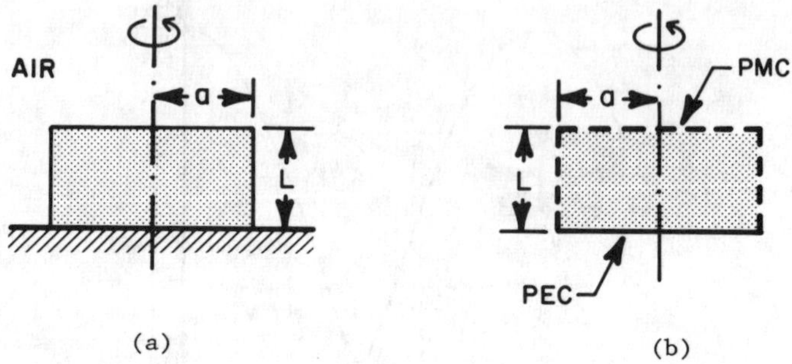

Fig. 4.3 Approximate model for the field inside the dielectric region

In Fig. 4.2(b), the role of dielectrics is reversed, and one observes a plane wave coming from the low-ϵ region upon the interface with a high-ϵ material. Again, a part of the wave is reflected and another part is transmitted. The reflection coefficients for various values of $\epsilon_{r1}/\epsilon_{r2}$ are given in Table 4.2.

Table 4.2 PLANE WAVE INCIDENT FROM OUTSIDE OF THE DIELECTRIC MATERIAL

$\epsilon_{r1}/\epsilon_{r2}$	1/2.5	1/10	1/37	1/100
E_B/E_A	-0.225	-0.519	-0.717	-0.818

It should be noticed that, again, the reflection coefficient has a larger amplitude when the two materials are more dissimilar. When the dielectric constant of the region 2 grows to infinity, the reflection coefficient tends toward negative unity. It is known from transmission line theory that this signifies a short-circuit situation. As seen in Fig. 4.2(b), the total electric field has a minimum at the position of the interface. The figure has been drawn for the case where $\epsilon_{r1} = 1$ and $\epsilon_{r2} = 100$.

Therefore, for an approximate computation of the electromagnetic field outside the region of the high dielectric material, the interface may be covered with a perfect electric conductor (PEC). As an example of this principle, consider a hollow cylindrical cavity which is tuned by means of a dielectric rod, as shown in Fig. 4.4(a). An approximate

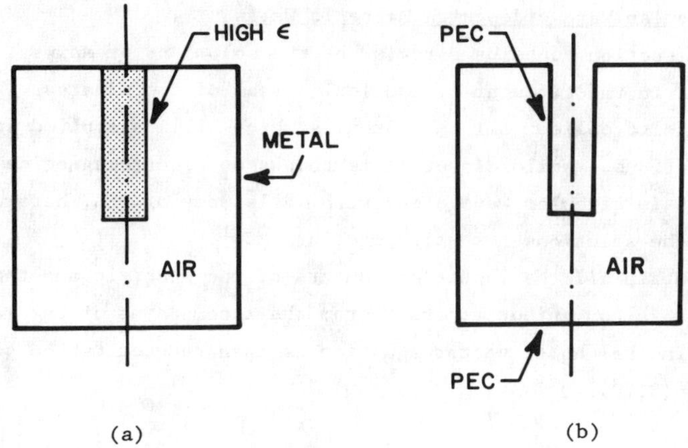

Fig. 4.4 Approximate model for the field outside the dielectric region

computation of the electromagnetic field in the cavity can be made by assuming that the high-ε dielectric rod has been replaced by a PEC rod, as shown in Fig. 4.4(b).

The accuracy of the resonant frequencies for various modes obtained by these two approximate procedures may not be very good, but, nevertheless, one can obtain information on what kinds of modes can be expected, and approximately how the fields look.

4.3 Circular Waveguides with Magnetic Walls

This section contains a review of the solutions of Maxwell's equations in the circular cylindrical system of coordinates. The purpose is to collect all the formulas which will be applied in subsequent sections. While classical textbooks on electromagnetics develop the modes in circular waveguides with walls made of PEC, here we will present the solutions for walls made of PMC.

To obtain all the vector components of the electric and the magnetic field, it is convenient to start from the z component of the magentic field. The Helmholtz vector equation is then reduced to the scalar equation (3.20):

$$\nabla^2 H_z = -k_0^2 \epsilon_r H_z \tag{4.12}$$

The separation-of-variables procedure is based on the assumption that the solution of this differential equation is a product of three functions, as in (3.23):

$$H_z = P(\rho)F(\phi)Z(z) \tag{4.13}$$

The direction of wave propagation is along the z axis so that the Z(z) function takes the following form:

$$Z(z) = A_1 e^{\gamma z} + B_1 e^{-\gamma z} \tag{4.14}$$

When the losses in the waveguide are neglected, the waveguide propagation constant γ may be either pure-real:

$$\gamma = \alpha \qquad \text{(evanescent modes)} \tag{4.15}$$

or pure-imaginary:

$$\gamma = j\beta \qquad \text{(propagating modes)} \tag{4.16}$$

As explained in Ch. 3, the azimuthal variation is described by trigonometric functions,

$$F(\phi) = A_2 \cos(m\phi) + B_2 \sin(m\phi) \tag{4.17}$$

where m is an integer. For the circularly symmetric modes, such as the important TE_{01} mode, the integer m is zero, and the field has no variation in ϕ direction.

The radial variation is given in terms of Bessel functions [1]:

$$P(\rho) = A_3 J_m(k_\rho \rho) + B_3 N_m(k_\rho \rho) \tag{4.18}$$

The radial wave number k_ρ is specified by

$$k_\rho^2 = \gamma^2 + k_0^2 \epsilon_r \tag{4.19}$$

where γ is given by either (4.15) or (4.16). In some situations, the radial dependence is described by modified Bessel functions:

$$P(\rho) = A_4 K_m(k_\rho \rho) + B_4 I_m(k_\rho \rho) \tag{4.20}$$

In that case, k_ρ is defined as

$$k_\rho^2 = -\gamma^2 - k_0^2 \epsilon_r \tag{4.21}$$

Again, γ^2 can be either a positive or a negative real number. For a mode propagating in the z-direction, (4.16) gives

$$\gamma^2 = -\beta^2 \tag{4.22}$$

and, for an evanescent mode, (4.15) gives

$$\gamma^2 = \alpha^2 \tag{4.23}$$

When a solution of the Helmholtz equation for the z component of the H field is determined, the other field components may be obtained by simple differentiation [1]. This procedure will be explained for the example of TE modes ($E_z = 0$). The Maxwell equation:

$$\nabla \times \underline{E} = -j\omega\mu_0 \underline{H} \tag{4.24}$$

splits into the following three scalar equations:

$$-\frac{\partial E_\phi}{\partial z} = -j\omega\mu_0 H_\rho \tag{4.25}$$

$$\frac{\partial E_\rho}{\partial z} = -j\omega\mu_0 H_\phi \tag{4.26}$$

$$\frac{1}{\rho}\frac{\partial}{\partial \rho}(\rho E_\phi) - \frac{1}{\rho}\frac{\partial E_\rho}{\partial \phi} = -j\omega\mu_0 H_z \tag{4.27}$$

The other Maxwell equation:

$$\nabla \times \underline{H} = j\omega\epsilon \underline{E} \tag{4.28}$$

is decomposed similarly into the following set of equations:

$$\frac{1}{\rho}\frac{\partial H_z}{\partial \phi} - \frac{\partial H_\phi}{\partial z} = j\omega\epsilon E_\rho \tag{4.29}$$

$$\frac{\partial H_\rho}{\partial z} - \frac{\partial H_z}{\partial \rho} = j\omega\epsilon E_\phi \tag{4.30}$$

$$\frac{1}{\rho}\frac{\partial}{\partial \rho}(\rho H_\phi) - \frac{1}{\rho}\frac{\partial H_\rho}{\partial \phi} = 0 \tag{4.31}$$

The right-hand side of the last equation is equal to zero because, for the TE modes, E_z is zero by definition.

From (4.25) and (4.30), it is possible to eliminate E_ϕ to obtain

$$H_\rho = \frac{1}{k_\rho^2}\frac{\partial^2 H_z}{\partial z \partial r} \tag{4.32}$$

From (4.26) and (4.29), one can likewise eliminate E_ϕ:

$$H_\phi = \frac{1}{k_\rho^2}\frac{1}{\rho}\frac{\partial^2 H_z}{\partial \phi \partial z} \tag{4.33}$$

When (4.32) is substituted in (4.30) and the separation equation (4.19) is used, E_ϕ is expressed in terms of H_z as follows:

$$E_\phi = \frac{j\omega\mu_0}{k_\rho^2}\frac{\partial H_z}{\partial \rho} \tag{4.34}$$

Also, E_ρ can be expressed in terms of H_z by the use of (4.29) and

(4.33):

$$E_\rho = -\frac{j\omega\mu_0}{k_\rho^2}\frac{1}{\rho}\frac{\partial H_z}{\partial \phi} \tag{4.35}$$

The four equations (4.32) to (4.35) permit one to calculate all the components of the electric and the magnetic TE fields by performing simple differential operations on the H_z component. For the circularly symmetric modes, the derivatives with respect to ϕ are zero, so that all the TE_{on} modes must have

$$E_\rho = 0 \quad \text{and} \quad H_\phi = 0 \tag{4.36}$$

For later reference, the derivatives of Bessel functions appearing in the TE modes are listed below:

$$J_0'(x) = -J_1(x) \tag{4.37}$$

$$J_1'(x) = J_0(x) - \frac{1}{x} J_1(x) \tag{4.38}$$

$$K_0'(x) = -K_1(x) \tag{4.39}$$

$$K_1'(x) = -K_0(x) - \frac{1}{x} K_1(x) \tag{4.40}$$

The integrals which will be needed later are

$$\int_0^a J_0^2(k_\rho\rho)\rho d\rho = \frac{a^2}{2}\left[J_1^2(k_\rho a) + J_0^2(k_\rho a)\right] \tag{4.41}$$

$$\int_0^a J_1^2(k_\rho\rho)\rho d\rho = \frac{a^2}{2}\left[J_0^2(k_\rho a) - \frac{2}{k_\rho a} J_0(k_\rho a)J_1(k_\rho a) + J_1^2(k_\rho a)\right] \tag{4.42}$$

$$\int_a^\infty K_0^2(k_\rho\rho)\rho d\rho = \frac{a^2}{2}\left[K_1^2(k_\rho a) - K_0^2(k_\rho a)\right] \tag{4.43}$$

$$\int_a^{\infty} K_1^2(k_\rho \rho)\rho d\rho = \frac{a^2}{2}\left[K_0^2(k_\rho a) + \frac{2}{k_\rho a} K_1(k_\rho a)K_0(k_\rho a) - K_1^2(k_\rho a)\right] \quad (4.44)$$

It is now possible to construct the fields for various TE_{mn} modes in hollow cylindrical waveguides with PMC walls. The results are very similar to the modes in waveguides with PEC walls. The boundary condition for the PMC waveguide is that H_z must be zero at $\rho = a$. In addition, H_z must be finite at $\rho = 0$, so that the Bessel functions of the second kind must not be used ($B_3 = 0$ in (4.18)). The boundary conditions require H_z to satisfy the following:

$$H_z(\rho{=}a) = AJ_m(k_\rho a)F(\phi)Z(z) = 0 \quad (4.45)$$

The zeros of the Bessel function of the first kind are denoted by x_{mn}, whereby

$$J_m(x_{mn}) = 0, \quad \text{for } n = 1,2,3,\ldots \quad (4.46)$$

For instance, x_{32} is the second zero of the Bessel function of the first kind, of order three. Some of these zeros are listed in Table 4.3, which was compiled from [3].

Table 4.3 ZEROS OF THE BESSEL FUNCTION OF THE FIRST KIND

m	n		
	1	2	3
0	2.4048	5.5201	8.6537
1	3.8317	7.0156	10.1735
2	5.1356	8.4172	11.6198
3	6.3802	9.7610	13.0152

When the TM_{mn} modes in cylindrical waveguides with PMC walls are desired, the boundary conditions require the derivative of the Bessel function to be zero. The zeros of the derivative of the Bessel function of the first kind are denoted by x'_{mn} as follows:

$$J'_m(x'_{mn}) = 0, \quad \text{for } n = 1,2,3,\ldots \quad (4.47)$$

Several lowest zeros, compiled from [3], are listed in Table 4.4.

Table 4.4 ZEROS OF THE DERIVATIVE OF THE BESSEL FUNCTION

m	n		
	1	2	3
0	3.8317	7.0156	10.1735
1	1.8412	5.3314	8.5363
2	3.0542	6.7061	9.9695
3	4.2012	8.0152	11.3459

4.4 The Cohn Model

The electromagnetic field inside a DR with high dielectric constant may be approximately described by assuming that all the surfaces are covered by the perfect magnetic conductor, as explained in Sec. 4.2. Such a "first-order" model of the DR is shown in Fig. 4.5(a). This is nothing other than but a circular cavity resonator, the walls of which are made of the PMC. Using known procedures for the analysis of hollow

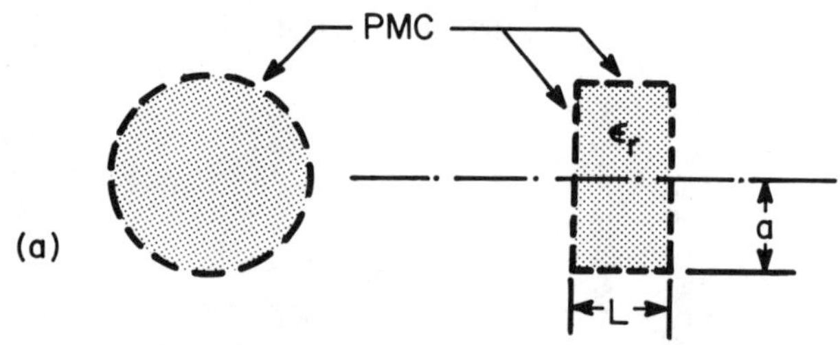

Fig. 4.5(a) First-order model of a DR

resonators, it is possible to compute the resonant frequency of this first-order model. Unfortunately, the computed results are 20 % and more off the measured values so that this model is not very useful for anything other than a homework assignment in a second course on fields and waves.

An improvement of the first-order model is the "second-order" model described by Cohn [4], which is shown in Fig. 4.5(b). The cylindrical PMC shell is retained, but the PMC end caps are removed and replaced by the air-filled hollow waveguides. These two hollow waveguides operate below the cutoff because they are filled with the low dielectric constant. Thus, the modes in these air-filled PMC waveguides are evanescent so that the fields decay exponentially in the z direction away from each end of the resonator.

The model in Fig. 4.5(b) is appropriate for describing the electromagnetic fields of an isolated DR far from any neighboring objects. Such an isolated resonator acts as a miniature antenna, and the energy lost on radiation is manifested in low Q values. The measured Q factor of an isolated $TE_{01\delta}$ resonator is about 50 [5]. In order to keep the

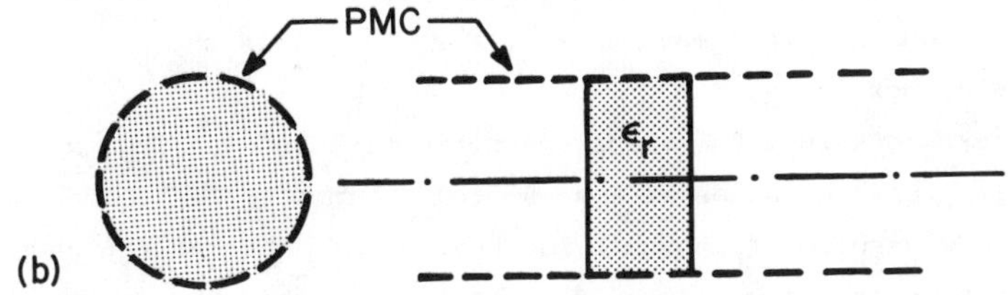

Fig. 4.5(b) Second-order Cohn's model of an isolated DR

unloaded Q factor of a DR reasonably high (a typical requirement is about 5000), it is necessary to prevent radiation by enclosing the DR within a metal shield. An inexpensive solution is to glue the DR to the dielectric substrate, which is used for microstrip interconnections of the microwave system in question. The entire substrate is then placed within a metal box. The box acts as a shield which prevents the external fields from penetrating the system and, by the same token, reduces the loss of energy due to radiation.

The modification of the Cohn model which incorporates the parallel-plate metal enclosure is shown in Fig. 4.5(c). The radius of the dielectric resonator is a, its relative dielectric constant is ϵ_r, and

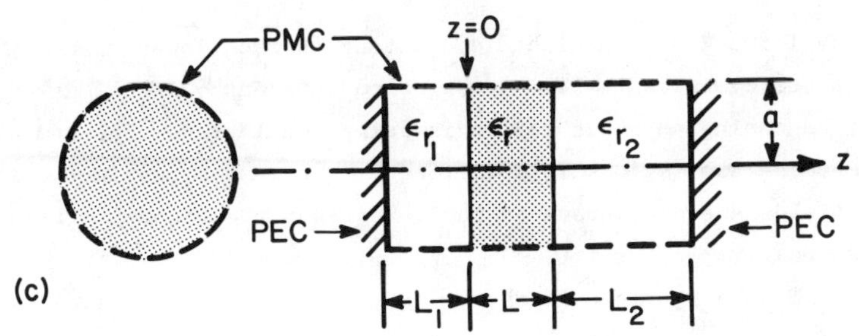

Fig. 4.5(c) Second-order Cohn's model of a shielded DR

its length is L. The region with relative dielectric constant ϵ_{r1} may represent the dielectric substrate on which the resonator is attached. The thickness of the substrate is L_1, and the outside face of the substrate is covered with the perfect electric conductor. The region of the length L_2 then represents the air-filled space above the resonator, and the PEC cap on the right most side of Fig. 4.5(c) is the metal lid of the shielding box.

We will next construct the electromagnetic field of the TE_{01} mode which satisfies the boundaries of Fig. 4.5(c). Since $\epsilon_r >> 1$, the mode within the PMC waveguide of length L is above cutoff so that the propagation constant γ becomes equal to $j\beta$. The field within the region $0 < z < L$ is, then,

$$H_z = (Ae^{j\beta z} + Be^{-j\beta z})J_0(k_\rho \rho) \tag{4.48}$$

The other field components follow from (4.32) and (4.34):

$$H_\rho = -\frac{j\beta}{k_\rho}(Ae^{j\beta z} - Be^{-j\beta z})J_1(k_\rho \rho) \tag{4.49}$$

$$E_\phi = -\frac{j\omega\mu_0}{k_\rho}(Ae^{j\beta z} + Be^{-j\beta z})J_1(k_\rho \rho) \tag{4.50}$$

These are the only non-zero components for the TE_{01} mode.

The radial propagation constant of this mode is fixed by the requirement (4.46), which gives

$$k_\rho a = x_{01} = 2.4048\ldots \tag{4.51}$$

In regions 1 and 2, the dielectric constant is much lower than ϵ_r and the modes are evanescent. The propagation constant γ is now given by α_1 or α_2, depending on which region is being considered. The radial variation in these two regions is specified by the same k_ρ as in (4.51), so that the fields are continuous at the interfaces $z = 0$ and $z = L$ (as will be shown next).

For region 1, the H_z field is thus,

$$H_{z1} = (C_1 e^{\alpha_1 z} + D_1 e^{-\alpha_1 z})J_0(k_\rho \rho) \tag{4.52}$$

E_ϕ is obtained from (4.34):

$$E_{\phi 1} = -\frac{j\omega\mu_0}{k}(C_1 e^{\alpha_1 z} + D_1 e^{-\alpha_1 z}) J_1(k_\rho \rho) \quad (4.53)$$

At $z = -L_1$, $E_{\phi 1}$ must vanish because of the conducting wall. This requirement results in

$$D_1 = -C_1 e^{-2\alpha_1 L_1} \quad (4.54)$$

$E_{\phi 1}$ field may be now expressed in terms of the hyperbolic sine function:

$$E_{\phi 1} = -\frac{j\omega\mu_0}{k_\rho} 2C_1 e^{-\alpha_1 L_1} \sinh \alpha_1(z + L_1) J_1(k_\rho \rho) \quad (4.55)$$

The components of the magnetic field in region 1 are then given by

$$H_{z1} = 2C_1 e^{-\alpha_1 L_1} \sinh \alpha_1(z + L_1) J_0(k_\rho \rho) \quad (4.56)$$

$$H_{\rho 1} = -\frac{\alpha_1}{k_\rho} 2C_1 e^{-\alpha_1 L_1} \cosh \alpha_1(z + L_1) J_1(k_\rho \rho) \quad (4.57)$$

In an analogous manner, the fields in region 2 are formulated as follows:

$$H_{z2} = 2C_2 e^{\alpha_2(L_2 + L)} \sinh \alpha_2(z - L_2 - L) J_0(k_\rho \rho) \quad (5.58)$$

$$E_{\phi 2} = -\frac{j\omega\mu_0}{k_\rho} 2C_2 e^{\alpha_2(L_2 + L)} \sinh \alpha_2(z - L_2 - L) J_1(k_\rho \rho) \quad (5.59)$$

$$H_{\rho 2} = -\frac{\alpha_2}{k_\rho} 2C_2 e^{\alpha_2(L_2 + L)} \cosh \alpha_2(z - L_2 - L) J_1(k_\rho \rho) \quad (5.60)$$

The separation constants are

$$\beta^2 = k_0^2 \epsilon_r - \left(\frac{x_{01}}{a}\right)^2 \quad (4.61)$$

$$\alpha_1^2 = \left(\frac{x_{01}}{a}\right)^2 - k_0^2 \epsilon_{r1} \quad (4.62)$$

$$\alpha_2^2 = \left(\frac{x_{01}}{a}\right)^2 - k_0^2\epsilon_{r2} \tag{4.63}$$

The next step in setting up the field model is to ensure that the tangential components of the electric, as well as the magnetic field, are continuous at the interface. At z = 0, we require $E_{\phi 1} = E_\phi$ and $H_{\rho 1} = H_\rho$. This provides two equations:

$$2C_1 e^{-\alpha_1 L_1} \sinh \alpha_1 L_1 = A + B \tag{4.64}$$

$$\alpha_1 2C_1 e^{-\alpha_1 L_1} \cosh \alpha_1 L_1 = j\beta(A - B) \tag{4.65}$$

Dividing (4.64) by (4.65), C_1 is eliminated, thus,

$$\tanh \alpha_1 L_1 = \frac{\alpha_1}{j\beta} \cdot \frac{A + B}{A - B} \tag{4.66}$$

Similarly, by specifying $E_\phi = E_{\phi 2}$ and $H_\rho = H_{\rho 2}$ at z = L, we obtain

$$\tanh(-\alpha_2 L_2) = \frac{\alpha_2}{j\beta} \cdot \frac{Ae^{j\beta L} + Be^{-j\beta L}}{Ae^{j\beta L} - Be^{-j\beta L}} \tag{4.67}$$

The two constants A and B have been selected to describe the amplitudes of the forward and reverse traveling waves inside the resonator. For a complete standing wave pattern, the amplitudes of A and B must be the same. Their ratio is then a complex number of unity magnitude, and as of now, undetermined phase

$$\frac{B}{A} = e^{j\phi} \tag{4.68}$$

Then the right-hand side fraction in (4.66) becomes

$$\frac{A + B}{A - B} = j \cot \frac{\phi}{2} \tag{4.69}$$

and the right-hand side in (4.67) gives

$$\frac{Ae^{j\beta L} + Be^{-j\beta L}}{Ae^{j\beta L} - Be^{-j\beta L}} = -j \cot (\beta L - \frac{\phi}{2}) \tag{4.70}$$

When (4.69) is substituted in (4.66) and solved for ϕ, the result becomes

$$\frac{\phi}{2} = \frac{\phi_1}{2} = \tan^{-1} \left(\frac{\alpha_1}{\beta} \coth \alpha_1 L_1\right) \tag{4.71}$$

where the value of phase angle ϕ, expressed in terms of the parameters of region 1 only, was denoted by ϕ_1. Similarly, from (4.70) and (4.67) the following is obtained:

$$\beta L - \frac{\phi_1}{2} = \tan^{-1} \left(\frac{\alpha_2}{\beta} \coth \alpha_2 L_2\right) \tag{4.72}$$

It is now convenient to denote the right-hand side of (4.72) by $\phi_2/2$, in analogy with (4.71):

$$\frac{\phi_2}{2} = \tan^{-1} \left(\frac{\alpha_2}{\beta} \coth \alpha_2 L_2\right) \tag{4.73}$$

The resonance condition (4.72) is now written as

$$\beta L = \frac{\phi_1}{2} + \frac{\phi_2}{2} + \ell\pi, \qquad \ell = 0,1,2,3,... \tag{4.74}$$

For generality, the added angle $\ell\pi$ gives all possible resonance conditions. When $\ell = 0$, the mode is called $TE_{01\delta}$, where δ signifies a non-integer number, smaller than unity:

$$\delta = \frac{1}{\pi} \left(\frac{\phi_1}{2} + \frac{\phi_2}{2}\right) \tag{4.75}$$

In general, the modes may be denoted by TE_{01p}, where

$$p = \ell + \delta \qquad \text{for } \ell = 0,1,2,3, \ ... \tag{4.76}$$

In traditional hollow waveguide resonators, the third subscript denotes a number of half-wavelength variations of the field as a function of coordinate z. In that case, the subscript is always an integer. As we have just seen, for the $TE_{01\delta}$ mode, the field variation in a DR undergoes less than one half-wavelength variation within the resonator length L. The difference is caused by the end effects at $z = 0$ and $z = L$.

Consider a special case of $L_1 = 0$ and $L_2 = 0$. This represents a DR sandwiched between two parallel plates which are touching each resonator end. From (4.71) and (4.73), one obtains $\phi_1/2 = \pi/2$ and $\phi_2/2 = \pi/2$. The lowest resonance ($\ell = 0$) is then given by $\beta L = \pi$.

As another extreme, assume the two metal plates are moved to infinity ($L_1 = \infty$ and $L_2 = \infty$). Equations (4.71) and (4.73) then reduce to

$$\frac{\phi_1}{2} = \tan^{-1}\left(\frac{\alpha_1}{\beta}\right) \quad \text{and} \quad \frac{\phi_2}{2} = \tan^{-1}\left(\frac{\alpha_2}{\beta}\right) \tag{4.77}$$

This represents an isolated resonator in the free space. In this case, $\epsilon_{r1} = \epsilon_{r2} = 1$, and the resonance condition (4.74) becomes

$$\beta L = 2\tan^{-1}\left(\frac{\alpha}{\beta}\right) \tag{4.78}$$

with α and β given by (4.61) and (4.63).

As an example, take an isolated DR in free space ($\epsilon_{r1} = \epsilon_{r2} = 1$). The resonator material is $\epsilon_r = 38$, its radius is $a = 5.25$ mm, and its length is $L = 4.6$ mm. From the exact numerical solution, it is known that the resonant frequency is 4.82 GHz [5]. From the Cohn "second-order" model, a numerical solution of the transcendental equation (4.78) gives 4.60 GHz (an error of −4.8 %). This is certainly an improvement over the Cohn "first-order" model which, for the same DR, estimates the resonant frequency to be 6.37 GHz (an error of +32 %). However, for practical applications, it would be desirable to have an even more accurate model than the Cohn second-order one.

4.5 Perturbational Correction to the Cohn Model

The electromagnetic field in the Cohn model of the DR, shown in Fig. 4.5(b) and 4.5(c), is zero everywhere outside the PMC wall (i.e., for $\rho > a$). In reality, the tangential field outside the cylindrical surface of the resonator is of the same strength as the tangential field on the inner side of that surface, and then gradually decreases when an observer moves radially away from that surface. Thus, a part of the total stored electric, as well as magnetic, field energy also exists in the region $\rho > a$, and this part of the energy is entirely neglected in the Cohn second-order model.

To improve the model, we retain the same electric and magnetic fields inside $\rho < a$ as for the PMC model from Fig. 4.5(c). However, we then remove the PMC wall and postulate that the outside tangential electric field at $\rho = a$ must be continuous with the inside tangential field. The expanded DR model consists of six regions, as shown in Fig. 4.6. Note that the region inside the high dielectric is now denoted as region 6. The electric field in this region is the same as in the Cohn model:

$$E_{\phi 6} = E_0 J_1(k_\rho \rho) \cos\left(\beta z - \frac{\phi_1}{2}\right) \qquad (4.79)$$

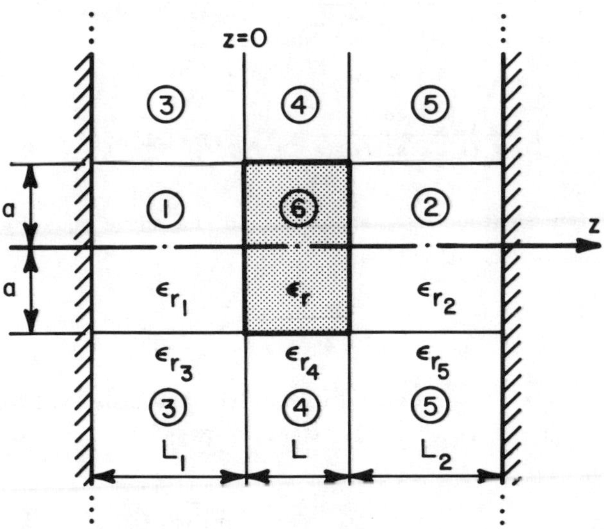

Fig. 4.6 Shielded DR

The electric fields in regions 1 and 2 are

$$E_{\phi 1} = E_0 \frac{\cos \frac{\phi_1}{2}}{\sinh \alpha_1 L_1} J_1(k_\rho \rho) \sinh \alpha_1 (z + L_1) \tag{4.80}$$

$$E_{\phi 2} = E_0 \frac{\cos \frac{\phi_2}{2}}{\sinh \alpha_2 L_2} J_1(k_\rho \rho) \sinh \alpha_2 (z - L_2 - L) \tag{4.81}$$

The multiplicative factors have been selected in such a way that E_ϕ is continuous over the interfaces between regions 1 and 6, and between regions 2 and 6.

The field in the outer regions 3, 4, and 5 will now be selected in such a way that the radial dependence will be determined by the modified Bessel functions, which are monotonically decaying with increasing radius ρ. The axial dependence should be the same as in the corresponding inner regions 1, 6, and 2. Thus, the electric fields are

$$E_{\phi 4} = E_0 \frac{J_1(k_\rho a)}{K_1(k_{\rho 2} a)} K_1(k_{\rho 2}\rho) \cos \left(\beta z - \frac{\phi_1}{2}\right) \tag{4.82}$$

$$E_{\phi 3} = E_0 \frac{J_1(k_\rho a)\cos \frac{\phi_1}{2}}{K_1(k_{\rho 2} a)\sinh \alpha_1 L_1} K_1(k_{\rho 2} r) \sinh \alpha_1 (z + L_1) \tag{4.83}$$

$$E_{\phi 5} = E_0 \frac{J_1(k_\rho a)\cos \frac{\phi_2}{2}}{K_1(k_{\rho 2} a)\sinh \alpha_2 L_2} K_1(k_{\rho 2}\rho) \sinh \alpha_1 (z - L_2 - L) \tag{4.84}$$

The separation constant $k_{\rho 2}$, appearing in the above equations, is obtained from (4.21) as follows:

$$k_{\rho 2}^2 = \beta^2 - k_0^2 \tag{4.85}$$

A convenient formula for computing $k_0 a$ from the known values of frequency f and radius a is the following [6]:

$$k_0 a = \frac{\pi}{150} a_{mm} f_{GHz} \tag{4.86}$$

The multiplicative constants $J_1(k_\rho a)$ etc. ensure the continuity of the electric field at the interface $\rho = a$. For instance, when (4.79) and (4.82) are compared at $\rho = a$, it is seen that $E_{\phi 4}$ is an identical function of z as $E_{\phi 6}$. The same is true at interfaces of the other regions.

Figure 4.7(a) shows the computed distribution of the electric field vs. radial distance for a DR with $\epsilon_r = 38$, a = 5.25 mm, L = 4.6 mm. The variation of the field in region 6 (r < a) is described by the function $J_1(k_c r)$. It is seen that it starts from zero at the origin, reaches a maximum at $\rho/a \simeq 0.75$ and then, at $\rho = a$, decreases to a value about 89 % of the maximum. In region 4, the field is given by the function $K_1(k_{\rho 2}\rho)$ which behaves somewhat like a decaying exponential function.

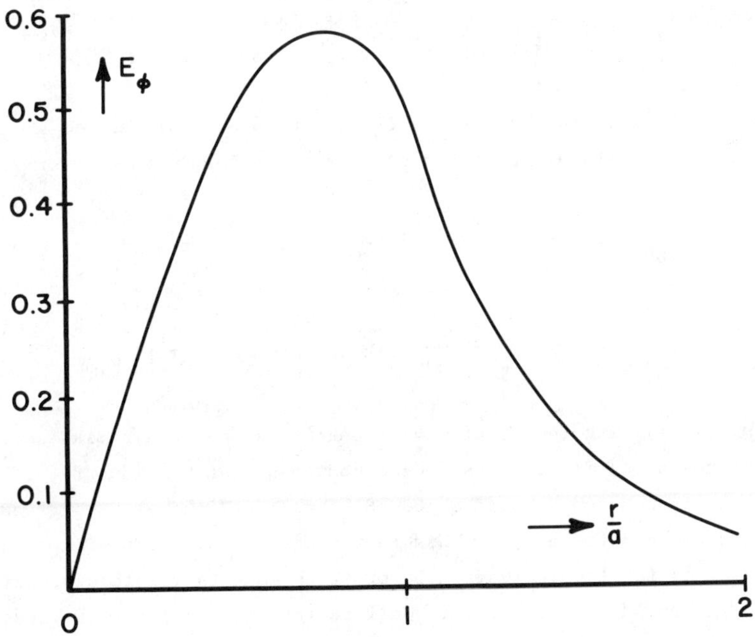

Fig. 4.7(a) Electric field vs. radial distance

The magnetic field in each region may be computed from the electric field by using the Maxwell equation

$$\underline{H} = \frac{j}{\omega\mu_0} \nabla \times \underline{E} \tag{4.87}$$

For example, the magnetic field in region 6 is given by

$$H_{z6} = j \frac{k_\rho}{\omega\mu_0} E_0 J_0(k_\rho \rho) \cos\left(\beta z - \frac{\phi_1}{2}\right) \tag{4.88}$$

$$H_{\rho 6} = j \frac{\beta}{\omega\mu_0} E_0 J_1(k_\rho \rho) \sin\left(\beta z - \frac{\phi_1}{2}\right) \tag{4.89}$$

In the resonator, the magnetic field is in time quadrature with the electric field (note the factor j). As a function of ρ, the radial component of H is specified by $J_1(k_\rho \rho)$, which means that $H_{\rho 6}$ behaves like $E_{\phi 6}$ in Fig. 4.7(a) (except for the multiplicative constant). On the other hand, the z component of the magnetic field is described by the function $J_0(k_\rho \rho)$, which vanishes at $\rho = a$ (see Fig. 4.7(b)) in accordance with the location of the PMC wall.

When the H field is computed by (4.87) in an outside region, such as region 4, it consists again of two components as follows:

$$H_{z4} = -\frac{jk_{\rho 2}}{\omega\mu_0} E_0 \frac{J_1(k_\rho a)}{K_1(k_{\rho 2} a)} K_0(k_{\rho 2} \rho) \cos\left(\beta z - \frac{\phi_1}{2}\right) \tag{4.90}$$

$$H_{\rho 4} = \frac{j\beta}{\omega\mu_0} E_0 \frac{J_1(k_\rho a)}{K_1(k_{\rho 2} a)} K_1(k_{\rho 2} \rho) \sin\left(\beta z - \frac{\phi_1}{2}\right) \tag{4.91}$$

Comparing (4.91) with (4.89) at $\rho = a$, one finds that the radial component of the magnetic field is indeed continuous at the interface. However, H_z is not continuous because $H_{z6}(\rho{=}a) = 0$, whereas H_{z4} given by (4.90) has no zero for any z. This fact clearly demonstrates that the field model is far from perfect. In order to make the field expressions complete, it would be necessary to utilize infinite series of various functions in each region. This can certainly be accomplished, as has been done for an open resonator by Tsuji et al. [7]. For an enclosure similar to the one in Fig. 4.6, complete field expansions were developed by Hong and Jansen [8]. However, the field expansions of this kind require considerably more computational effort than we wish to invest here.

There is one simple way to make the magnetic field continuous, however. It can be postulated that the z component of the magnetic field is zero for $\rho > a$:

$$H_{z3} \equiv 0 , \quad H_{z4} \equiv 0 , \quad \text{and} \quad H_{z5} \equiv 0 \tag{4.92}$$

Such an assumption makes H_z continuous as represented in Fig. 4.7(b). It is amazing how this crude approximation improves the numerical results.

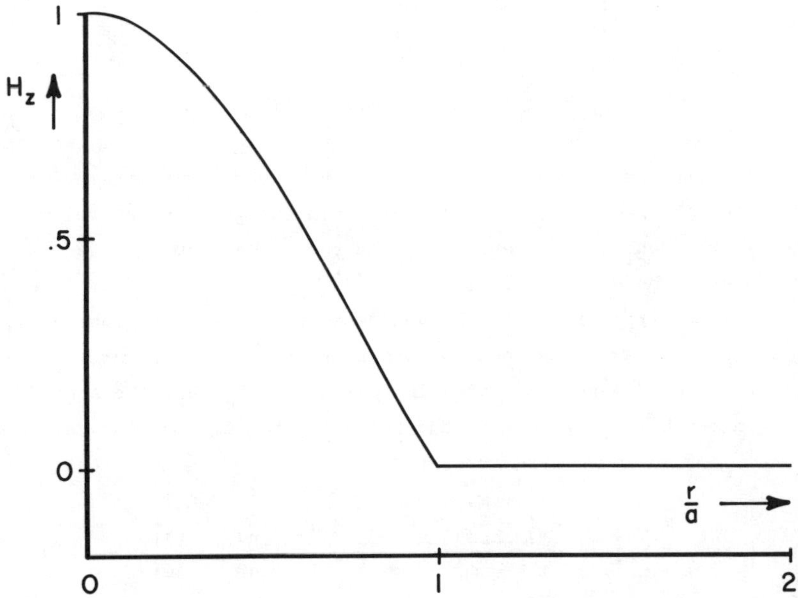

Fig. 4.7(b) Axial magnetic field vs. radial distance

The electric and the magnetic fields in regions 1 through 6 will next be used for determination of the resonant frequency by the use of the perturbational principle. Consider a general resonant cavity with PEC walls such as shown in Fig. 4.8(a). At the resonant frequency ω_0, the fields inside the cavity volume are denoted by $\underline{E}_0$ and $\underline{H}_0$. If now the metal wall of the cavity is pushed inward for a volume ΔV, as in Fig. 4.8(b), the resonant frequency changes by the following amount [9, sec. 7.2]:

$$\frac{\omega - \omega_0}{\omega_0} \simeq \frac{\iiint_{\Delta V} (\mu|\underline{H}_0|^2 - \epsilon|\underline{E}_0|^2)dV}{\iiint_{V} (\mu|\underline{H}_0|^2 + \epsilon|\underline{E}_0|^2)dV} \tag{4.93}$$

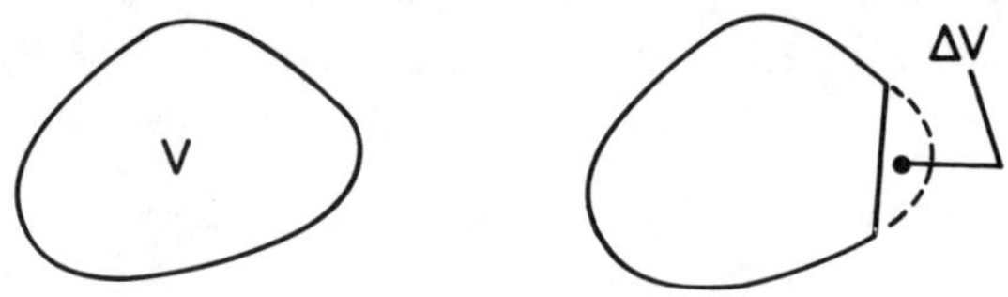

Fig. 4.8 Cavity perturbation

In the above expression, we recognize that the numerator represents the difference of the stored magnetic and electric energies in the volume ΔV, whereas the denominator represents the sum of both energies integrated over the entire volume V.

In our model from Fig. 4.6, the volume ΔV consists of regions 3, 4, and 5, and the PMC wall has been moved outward instead of inward. Denoting the stored magnetic energy in region i by W_{mi} and the stored electric energy by W_{ei}, we obtain the following formula for the resonant frequency ω_r of the perturbed resonator:

$$\omega_r = \omega_0 \left(1 + \frac{W_{m3} + W_{m4} + W_{m5} - W_{e3} - W_{e4} - W_{e5}}{W_{m1} + W_{m2} + W_{m6} + W_{e1} + W_{e2} + W_{e6}} \right) \tag{4.94}$$

In the above, ω_0 is the resonant frequency of the Cohn model before perturbation.

The evaluation of integrals for the stored energies in individual regions can be accomplished in analytical form. The resulting formulas are not very illuminating, and they will not be listed here. When programmed on a personal computer, (4.94) gave an improved accuracy in comparison with the Cohn model. For instance, for the example of the isolated DR described at the end of Sec. 4.4, the perturbational correction gave the resonant frequency as 4.85 GHz, which differs only 0.5 % from the value computed by the surface integral equation method.

Unfortunately, the perturbational formula is not as accurate for shielded resonators. When compared with measured results from the literature [10,11], the largest discrepancy was found to be 1.5 % (see Table 4.5).

The instructions for use of the computer program of the Cohn model with perturbational correction are given in Appendix 4.A. The program called DRESP is written in BASIC language and is implemented on the IBM PC. Besides computing the resonant frequency, it also provides a table of the energy distribution in various regions, and it plots the field distribution as a function of the z-coordinate.

Several diagrams of the field distribution computed by DRESP will be presented here in order to acquire a better familiarity with the behavior of dielectric resonators. The horizontal coordinate represents the distance in the z-direction, and the vertical coordinate represents the relative field amplitude. One can observe in Figs. 4.9 to 4.13 that H_z component has the same sign everywhere in space, while the H_ρ

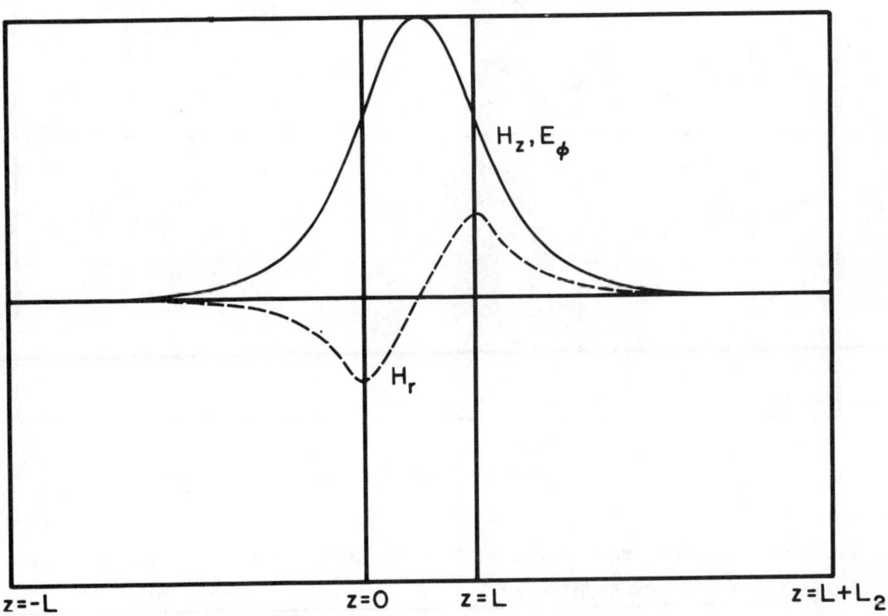

Fig. 4.9 Field distribution in an isolated DR

component changes sign at about the center of the DR.

The field distribution in an isolated DR made of material with $\epsilon_r = 38$ can be seen in Fig. 4.9. It is sufficient to select the distance to the shielding PEC plate to be about three times as large as the resonator length ($L_1 = 3L$ and $L_2 = 3L$) in order to make the effect of the plates negligible as far as the resonant frequency is concerned. Observe that the H_z component is an even function of z, while the H_ρ component is an odd function of z. The greatest intensity of either field is concentrated within the resonator, and only a rapidly decaying field exists outside of the dielectric region. In Fig. 4.10, the left-hand PEC wall is brought close to the resonator by selecting $L_1 = L/2$. The maximum of the H_z field is no longer at the center of the DR, but

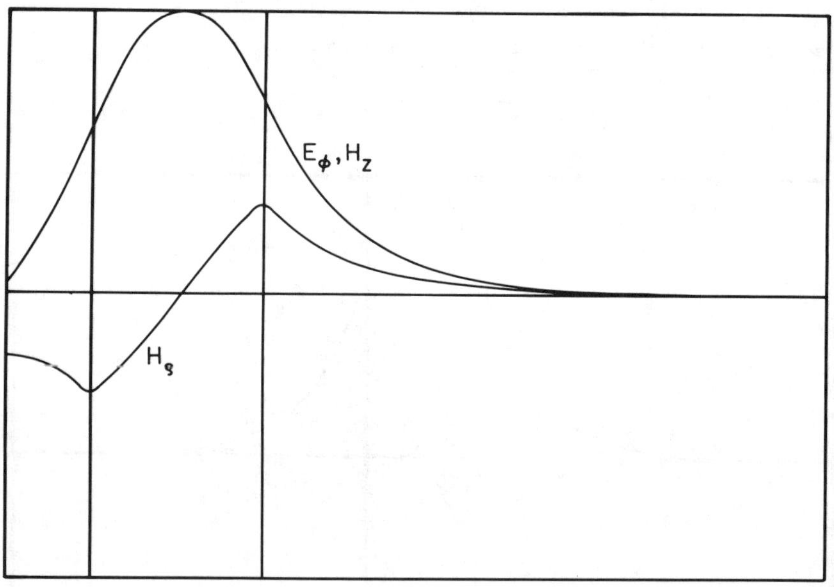

Fig. 4.10 Field distribution, $L_1 = \frac{L}{2}$, $\epsilon_{r1} = 1$

has moved slightly to the right. In Fig. 4.11, the distances are the same, but a dielectric slab of $\epsilon_r = 10$ has been placed between the resonator and the PEC. This situation should imitate a DR resting on an alumina substrate. There is very little difference between Fig. 4.10

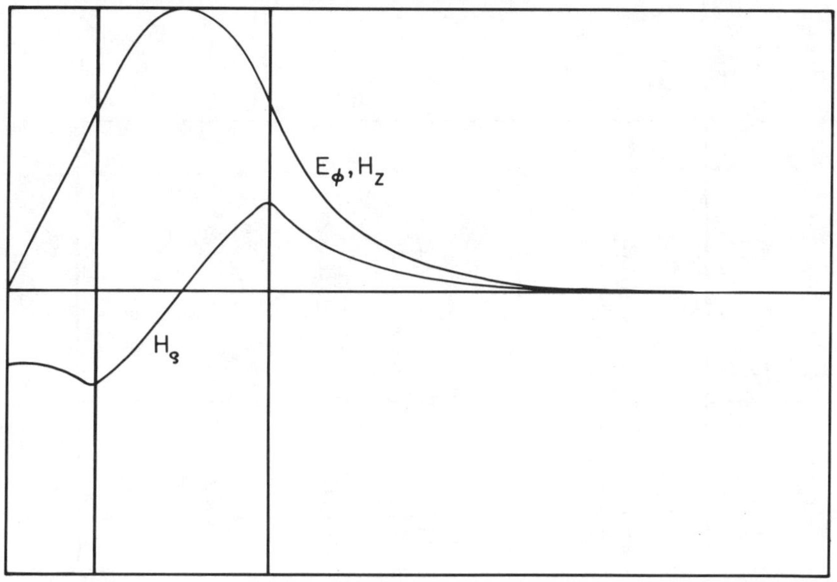

Fig. 4.11 Field distribution, $L_1 = \frac{L}{2}$, $\epsilon_{r1} = 10$

and Fig. 4.11, except that Fig. 4.10 has slightly stronger fields in region 1. Another slight difference is a change in frequency from 5.00 GHz (Fig. 4.10) to 4.95 GHz (Fig. 4.11).

Figure 4.12 depicts the DR placed directly on the metal wall. Note that the solid line, which is proportional to either E_ϕ or H_z, is zero on the left face of the resonator. The reason for this is the boundary condition requiring that the tangential electric field on the surface of the PEC must vanish. Figure 4.13 shows the situation in which both metal walls are touching the resonator. This is the situation for the TE_{011} resonator in the Courtney holder used for the measurement of the dielectric constant (see Sec. 3.8). In this mode, the field variation in the z-direction consists of exactly one-half wavelength.

In all the previously shown cases, the variation within the resonator consisted of less than one-half wavelength. Finally, Fig. 4.14 shows a resonant mode $TE_{01,1+\delta}$, which exhibits more than one-half wavelength variation in the z-direction. The change in mode of operation has a great influence on the operating frequency. The resonant frequency of the $TE_{01\delta}$ mode in Fig. 4.9 is 4.85 GHz, whereas the mode $TE_{01,1+\delta}$ in the same resonator has a resonant frequency of 8.19 GHz.

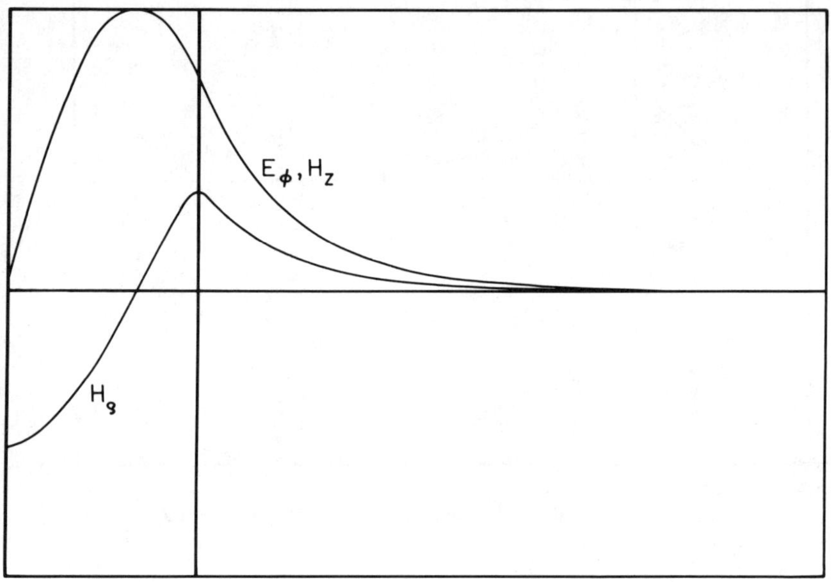

Fig. 4.12 Field distribution, $L_1 = 0$

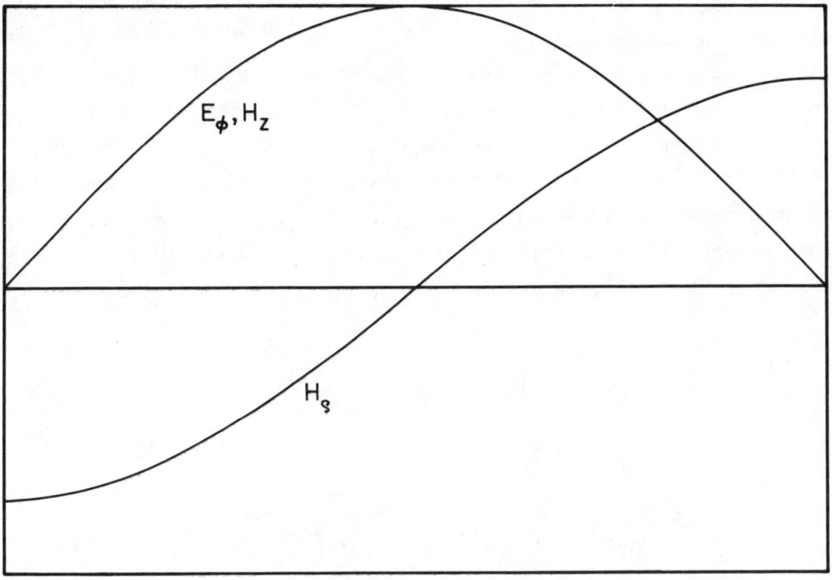

Fig. 4.13 Field distribution, $L_1 = 0$ and $L_2 = 0$

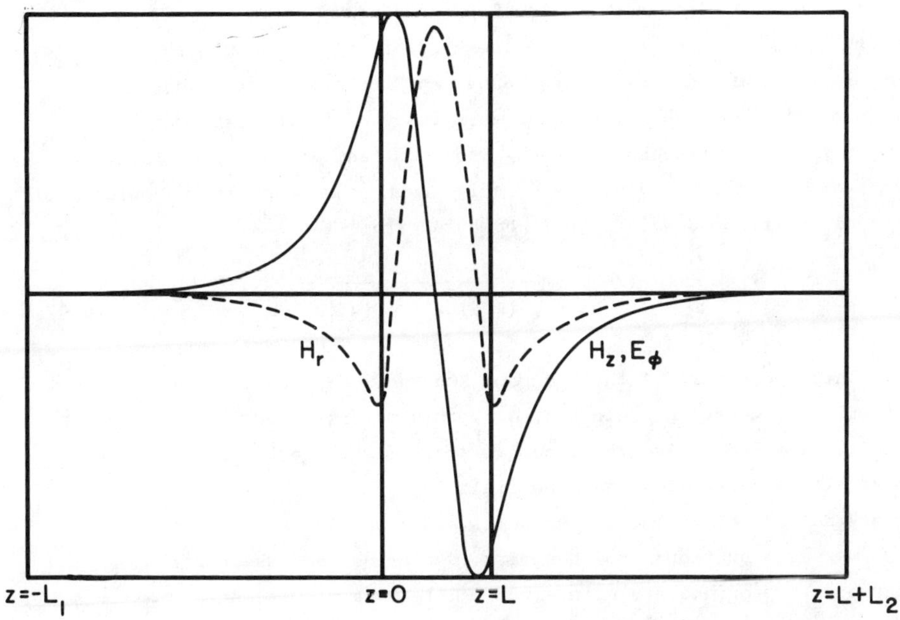

Fig. 4.14 Field distribution, isolated DR, mode $TE_{01,1+\delta}$

4.6 The Itoh and Rudokas Model

Instead of using the idealized waveguide with PMC walls like in the Cohn model, Itoh and Rudokas [13] start with the more realistic dielectric rod waveguide, such as the one described in Ch. 3. Therefore, the continuity of both the electric and the magnetic fields tangential to the interface between regions 6 and 4 is ensured. As shown in Ch. 3, the requirement for the continuity of fields in the dielectric rod waveguide leads to the following eigenvalue equation for the TE_{on} modes:

$$\frac{J_0(k_{\rho 1}a)}{J_1(k_{\rho 1}a)} = -\frac{k_{\rho 2}a}{k_{\rho 1}a} \cdot \frac{K_0(k_{\rho 2}a)}{K_1(k_{\rho 2}a)} \tag{4.95}$$

where the argument $k_{\rho 2}a$ is given by

$$k_{\rho 2}a = \sqrt{(k_0a)^2(\epsilon_{r6} - \epsilon_{r4}) - (k_{\rho 1}a)^2} \tag{4.96}$$

and where region 6 denotes the inside of the rod, while region 4 denotes the outside of the rod as in Fig. 4.6. The number $k_{\rho 1}a$, which satisfies (4.95), is called the eigenvalue of the TE_{on} mode. For the PMC hollow waveguide, the eigenvalue of the TE_{01} mode was denoted x_{01}, its value being 2.4048. Here, the eigenvalue $k_{\rho 1}a$ of the dielectric rod waveguide depends on the rod radius, frequency, and the dielectric constant ϵ_r. Figure 4.15 shows the eigenvalue as a function of the parameter k_0a, where k_0 is the free-space propagation constant and a is the radius of the rod. When the eigenvalue $k_{\rho 1}a$ is known, the propagation constant of the dielectric rod waveguide is computed from:

$$\beta a = \sqrt{(k_0a)^2\epsilon_{r6} - (k_{\rho 1}a)^2} \tag{4.97}$$

The middle part of the Itoh and Rudokas model consists of a dielectric rod waveguide. In Fig. 4.6, this part is comprised of regions 6 and 4. The PEC walls are placed at z = -L, and z = L + L_2. Furthermore, it is postulated that the field in corner regions 3 and 5 is zero everywhere. The fields in regions 1 and 2 should be selected so that the Maxwell equations, the boundary conditions, and the continuity between the regions are maintained. All these requirements are satisfied by choosing the followig electric fields in various regions:

$$E_{\phi 1} = E_0 \frac{\cos \frac{\phi_1}{2}}{\sinh \alpha_1 L_1} J_1(k_{\rho 1}\rho) \sinh \alpha_1(z + L_1) \tag{4.98}$$

$$E_{\phi 2} = E_0 \frac{\cos \frac{\phi_1}{2}}{\sinh \alpha_2 L_2} J_1(k_{\rho 1}\rho) \sinh \alpha_2(z - L - L_2) \tag{4.99}$$

$$E_{\phi 4} = E_0 \frac{J_1(k_{\rho 1}a)}{K_1(k_{\rho 2}a)} K_1(k_{\rho 2}\rho) \cos (\beta z - \frac{\phi_1}{2}) \tag{4.100}$$

$$E_{\phi 6} = E_0 J_1(k_{\rho 1}\rho) \cos (\beta z - \frac{\phi_1}{2}) \tag{4.101}$$

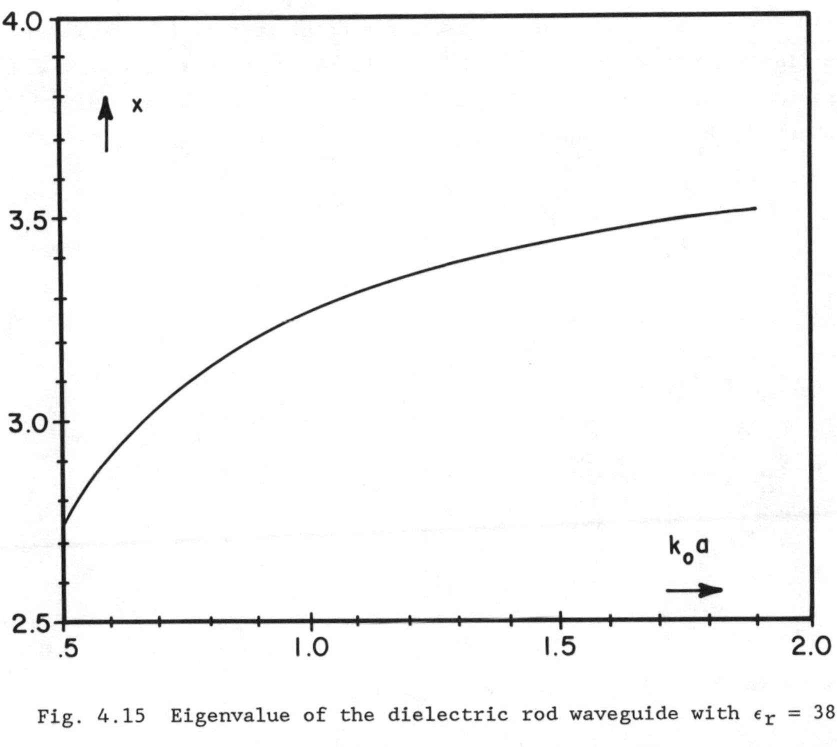

Fig. 4.15 Eigenvalue of the dielectric rod waveguide with $\epsilon_r = 38$

The radial and the axial components of the corresponding magnetic field in various regions may be obtained from the Maxwell equation (4.87) as was done in the Cohn model, but it turns out that they are not needed in the approach to be used here. The symbols α_1, α_2, ϕ_1, ϕ_2 in the above equations have the same meaning as defined earlier in the Cohn model. The continuity of the electric and magnetic fields leads to the same equation (4.74) as in the Cohn model. The difference is that β is now determined by (4.97) in terms of the eigenvalue $k_{\rho 1}a$, which must satisfy the transcendental equation (4.95), whereas in the Cohn model, β was given by (4.61) and the eigenvalue was equal to a constant $x_{01} = 2.4048$.

The two models differ in their field behavior as a function of radial distance. Consider, for example, an open DR with $\epsilon_r = 35$, a = 5 mm, and L = 5 mm. The eigenvalue obtained by the Itoh and Rudokas model is $k_{\rho 1}a = 2.8476$, while in the Cohn model $k_\rho a = x_{01} = 2.4048$. Therefore, in the Cohn model, the H_z field, which is given by the Bessel function $J_0(k_\rho r)$, vanishes at $\rho = a$ (see Fig. 4.7(b)). On the other hand, $k_{\rho 1}a$ for the Itoh and Rudokas model is larger, so that the H_z field turns out to be negative at the resonator surface (as the Bessel function has passed the first zero). Outside of the dielectric, the radial dependence of H_z is specified by the monotonically decaying function $K_0(k_{\rho 2}\rho)$, as shown in Fig. 4.16.

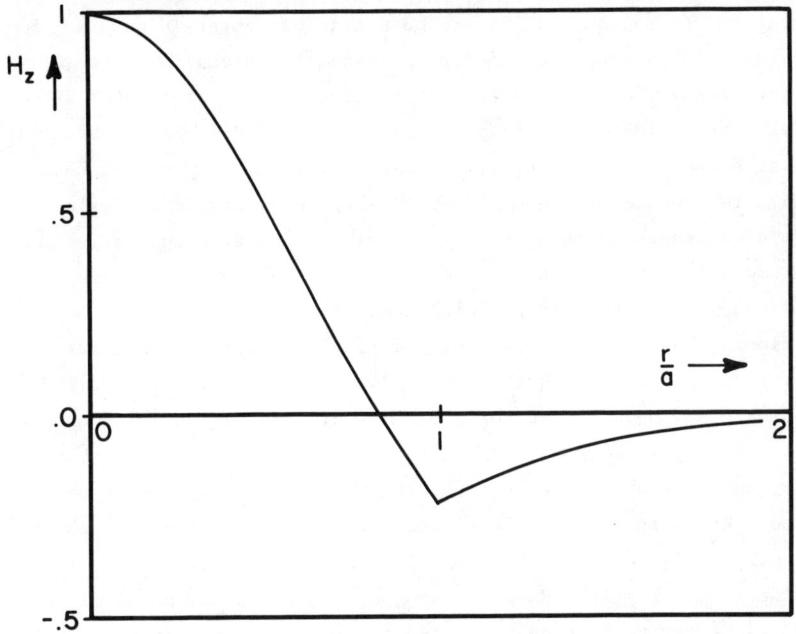

Fig. 4.16 H_z field variation vs. radial distance for Itoh and Rudokas model

4.7 Numerical Solution of the Pair of Transcendental Equations

Equations (4.95) and (4.74) are both transcendental in nature, the former involving ordinary and modified Bessel functions, and the latter involving ordinary and hyperbolic tangent functions. At first glance, it appears that a mainframe computer may be required in order to obtain numerical answers. Fortunately, the use of a large computer can be avoided. Two such approaches will be described here: an approximate procedure, which requires only a programmable pocket calculator, and another more accurate procedure, which is suitable for programming in BASIC language of a typical personal computer.

A simple strategy for obtaining numerical answers to a similar system of transcendental equations has been outlined by Iveland [14]. His approach may be utilized for the DR of circular shape, although he originally analyzed a DR of rectangular, rather than circular, shape. Iveland started his analysis by assuming a value for the operating frequency, and then he was able to substitute the numerical values in the required formulas one after another, ending with a value for the resonator length L. If L does not come out equal to the physical length of the resonator in question, the assumed value of the frequency must be corrected (increased or decreased), and a new value of L must be computed. Since the small changes in frequency produce almost directly proportional changes in the computed value of L, one can find the proper frequency after only a few iterations.

The procedure may be implemented on a pocket calculator if a reasonable approximation is found for the Bessel functions appearing in (4.95). Such an approximation is described in [15]. First, for the assumed value of frequency, an auxiliary constant y_0 is computed:

$$y_0 = \sqrt{(k_0 a)^2(\epsilon_{r6} - \epsilon_{r4}) - x_{01}^2} \tag{4.102}$$

Then, an approximate eigenvalue is obtained as follows:

$$k_{\rho 1} a = 2.405 + \frac{y_0}{2.405\left(1 + \dfrac{2.43}{y_0} + 0.291 y_0\right)} \tag{4.103}$$

The attenuation constants in regions 1 and 2 are

$$\alpha_1 = \sqrt{k_{\rho 1}^2 - k_0^2 \epsilon_{r1}} \tag{4.104}$$

$$\alpha_2 = \sqrt{k_{\rho 1}^2 - k_0^2 \epsilon_{r2}} \tag{4.105}$$

The propagation constant common to both region 6 and region 4 is

$$\beta = \sqrt{k_0^2 \epsilon_{r6} - k_{\rho 1}^2} \tag{4.106}$$

Finally, the resonator length is

$$L = \frac{1}{\beta} \left(\tan^{-1}\left(\frac{\alpha_1}{\beta} \coth \alpha_1 L_1 \right) + \tan^{-1}\left(\frac{\alpha_2}{\beta} \coth \alpha_2 L_2 \right) \right) \tag{4.107}$$

In order to get a better feel for the error involved in the above computational procedure, we will apply it to a practical situation. Maj and Modelski [11] report on a measurement of a DR in an environment such as shown in Fig. 4.6. The measured data are as follows:

$$\epsilon_{r6} = 34.19 , \quad \epsilon_{r1} = 9{,}6 , \quad \epsilon_{r2} = 1 , \quad f = 4.348 \text{ GHz} ,$$

$$a = 7.49 \text{ mm} , \quad L = 7.48 \text{ mm} , \quad L_1 = .7 \text{ mm} , \quad L_2 = .72 \text{ mm}$$

When these data (except L) are substituted in the above equations with $f = 4.348$ GHz taken as the input frequency, the computed result comes out to be $L = 7.18$ mm (instead of the desired result, 7.48 mm). In order to obtain a larger value of L, the input frequency must be decreased. After only two more iterations, the following is obtained:

$$L = 7.481 \text{ mm at } f = 4.267 \text{ GHz}$$

The error in frequency is -1.8 %. Comparison with other measured results of the shielded DR shows typical errors being smaller than 2 % [15]. However, for an isolated DR the error of this procedure is much larger. For example, the isolated DR taken from reference [5] is

$$\epsilon_{r6} = 38 , \quad a = 5.25 , \quad L = 4.6 \text{ mm}$$

The measured resonant frequency was 4.85 GHz. Substituted into the same set of equations, this value gives too large result for L (5.6 mm instead of 4.6 mm). After two more iterations, one obtains L = 4.612 mm for f = 5.14 GHz. Thus, the error in frequency is +5.8 %.

Another approach for simultaneously solving (4.74) and (4.95), which is suitable for solution on a personal computer, is the following. For the $TE_{01\delta}$ resonance, the ratio of Bessel functions appearing in (4.95) has a zero at $x_0 = 2.4048$ and a pole at $x_1 = 3.8317$. The suitable approximation is, then,

$$\frac{J_0(x)}{J_1(x)} = \frac{1}{x - x_1} \sum_{n=1}^{5} a_n (x - x_0)^n \tag{4.108}$$

The coefficients a_n can be obtained by fitting the above equation to the tabulated values of Bessel functions [3]. The values used here are

$$a_1 = 1.4282, \quad a_2 = -0.7160, \quad a_3 = 0.2571, \quad a_4 = -0.1177, \quad a_5 = 0.0282$$

The error of the above formula is smaller than 0.04 % in the range $x_0 < x < 3.3$.

The ratio of modified Bessel functions which appear in the right-hand side of (4.95) may be approximated by the following power series:

$$\frac{K_0(x)}{K_1(x)} = \frac{1}{1 + \sum_{n=1}^{5} \frac{a_n}{x^n}} \tag{4.109}$$

The coefficients a_n are given below

$$a_1 = 0.49907, \; a_2 = -0.11226, \; a_3 = 0.06539, \; a_4 = -0.02679, \; a_5 = 0.00445$$

The error of this formula is smaller than 0.04 % for $x > 0.7$.

Using (4.108) and (4.109) it is possible to circumvent the drawback of the personal computer, namely, the lack of computer library in BASIC language, which would contain the higher mathematical functions, such as Bessel functions.

The two transcendental equations are then simultaneously solved as follows. For simplicity of notation, denote the two implicit functions by f(x,y) and g(x,y), where x and y are two independent variables. We are looking for a point (x,y) at which both

$$f(x,y) = 0 \quad \text{and} \quad g(x,y) = 0 \tag{4.110}$$

In the vicinity of the solution, f and g will be approximated by the linear functions:

$$f(x,y) \simeq ax + by + c \tag{4.111}$$

$$g(x,y) \simeq Ax + By + C \tag{4.112}$$

When f = 0, (4.111) gives a straight line:

$$y = -\frac{ax}{b} - \frac{c}{b} \tag{4.113}$$

Similarly, g = 0 results in another straight line:

$$y = -\frac{Ax}{B} - \frac{C}{B} \tag{4.114}$$

The intersection of these two straight lines is at

$$x = \frac{Bc - bC}{Ab - aB} \tag{4.115}$$

and

$$y = \frac{aC - Ac}{Ab - aB} \tag{4.116}$$

Therefore, if the coefficients a, b, c, A, B, C are known, the zero of the system of equations (4.110) can be computed by (4.115) and (4.116). What remains is to evaluate these coefficients.

The implicit function f(x,y) can be interpreted as a three-dimensional surface over the x,y plane. For a sufficiently small range of variables x and y, the surface is approximated by a plane. By evaluating three points on this plane, the location of the plane is entirely specified. This will be done here as follows.

Denote the starting point by $x = x_2$, $y = y_2$. The corresponding value of f is denoted by f_2. The next point is chosen as $x_1 = x_2 + \Delta x$, $y_1 = y_2$. The corresponding value of f is denoted by f_1. The third point is selected as $x_3 = x_2$ and $y_3 = y_2 + \Delta y$, and the corresponding function is denoted by f_3. Hence, the linear coefficients from (4.111) are

$$a = \frac{f_1 - f_2}{\Delta x} \tag{4.117}$$

$$b = \frac{f_3 - f_2}{\Delta y} \tag{4.118}$$

$$c = f_2 - ax_2 - by_2 \tag{4.119}$$

Similar three-point evaluation at the same three points in the x,y plane gives the linear coefficients of the function g:

$$A = \frac{g_1 - g_2}{\Delta x} \tag{4.120}$$

$$B = \frac{g_3 - g_2}{\Delta y} \tag{4.121}$$

$$C = g_2 - Ax_2 - By_2 \tag{4.122}$$

The search consists of evaluating functions f and g at three close points, computing the linear coefficients from (4.117) to (4.122), and then moving to the new point given by (4.115) and (4.116). There, the whole procedure is repeated again.

In the case of interest here, the two variables are selected to be

$$x = k_{\rho 1}a \quad \text{and} \quad y = k_0 a \tag{4.123}$$

The argument of the modified Bessel functions is denoted by z:

$$z = k_{\rho 2}a \tag{4.124}$$

The two implicit functions are then defined in accordance with (4.95) and (4.74) as

$$f = \frac{J_0(x)}{J_1(x)} + \frac{z}{x}\frac{K_0(z)}{K_1(z)} \tag{4.125}$$

and

$$g = \beta L - \frac{\phi_1}{2} - \frac{\phi_2}{2} - \ell\pi \tag{4.126}$$

The search is very rapid, and only three or four iterations are typically required for an accuracy of 10^{-5}.

When the starting point is far from the correct solution, functions f and g may depart from the linear model; the first step computed by (4.115) and (4.116) may be too large, and the point may fall outside of the feasible range. This situation may be recognized by an attempt to evaluate the square root of a negative number in (4.96). If this happens, the search algorithm reduces the step by half and the procedure is repeated.

For isolated resonators, the Itoh and Rudokas system of transcendental equations gives a frequency which is too high by about 5 %. For closer and closer spacing of the two metal walls, the solution becomes more and more accurate, and when L_1 and L_2 both tend toward zero, the system of equations becomes exact.

4.8 Variational Improvement of the Itoh and Rudokas Model

When the field in some resonant cavity is known only approximately, the best way to compute the resonant frequency of this cavity is through the use of some variational formula. All the variational expressions contain some ratio of energies stored inside the resonator volume. The advantage of the variational formulas is that the small inaccuracies in the field distribution have a negligible effect on the resulting value of frequency.

Variational formulas for the resonant frequency can be classified to be of the E-type, H-type, or mixed type. Since we have simple expressions (4.98) to (4.101) for the electric field of the Itoh and Rudokas model, we will use the E-field variational formula.

The electric field in regions 1, 2, 4, and 6 satisfies the Helmholtz wave equation, so that the tangential components of the field are continuous over the interfaces between individual regions. The magnetic field in these regions can be obtained by using (4.87), and it also has continuous tangential component over each interface, as long as $k_{\rho 1}$ and $k_{\rho 2}$ satisfy the eigenvalue equation (4.95).

In corner regions 3 and 5, the original Itoh and Rudokas model assumes both the electric and the magnetic field to be equal to zero. This assumption is convenient in simplifying the mathematical procedure for computing the resonant frequency. Unfortunately, such an assumption creates a discontinuity of the electric as well as magnetic fields at the interfaces between the corner regions and inner regions.

An improvement of the Itoh and Rudokas model can be achieved by choosing the electric field in the corner regions according to (4.83) and (4.84), as it was done in Sec. 4.5. Then, the electric field in regions 3 and 5 becomes continuous over the interfaces with inner regions. Unfortunately, (4.83) and (4.84) do not satisfy the Helmholtz wave equation, so that a self-consistent magnetic field cannot be constructed from these two equations. For this reason, the magnetic field in regions 3 and 5 is simply left to be zero. However, to justify the sudden jump of the tangential magnetic field from zero (e.g. in region 3) to a finite value in the neighboring regions (e.g. regions 1 and 4), surface electric currents J_s are added on the interfaces, as indicated in Fig. 4.17.

The variational formula for computing the resonant frequency of the model in Fig. 4.17 will be derived by using Rumsey's reaction concept in

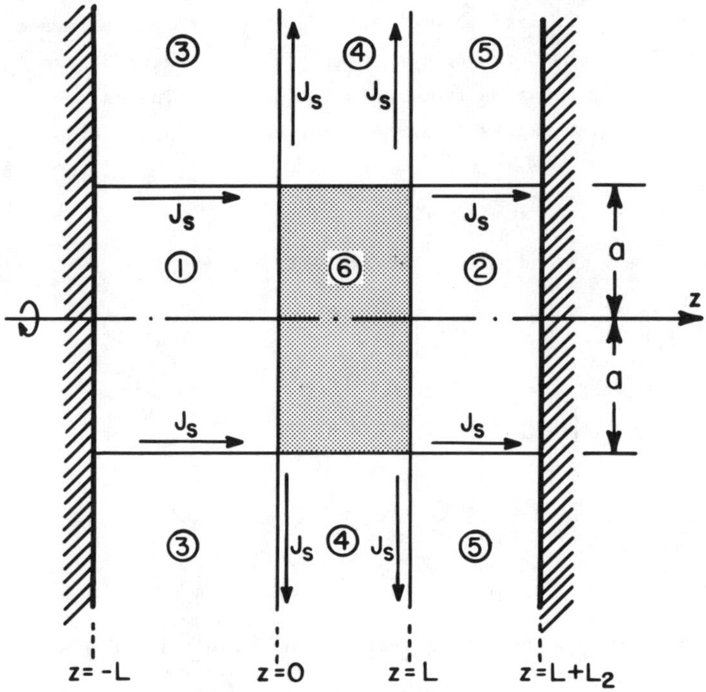

Fig. 4.17 Improved model of the shielded DR

a manner described by Harrington [9, Sec. 7.7]. The E-field variational formula is obtained from the self-reaction

$$\iiint_V \underline{E} \cdot \underline{J} \, dV = 0 \tag{4.127}$$

The source $\underline{J}$ inside volume V of the resonant cavity is obtained from [9, eq. 7.68]:

$$\underline{J} = -j\omega\epsilon\underline{E} - \frac{1}{j\omega} \nabla \times \mu^{-1}\nabla \times \underline{E} \tag{4.128}$$

On the interfaces where a jump in the tangential magnetic field occurs, the electric surface currents have the value

$$\underline{J}_s = \underline{H} \times \hat{n} \tag{4.129}$$

where $\hat{n}$ is the unit vector, normal to the interface, pointing out of the inner region under consideration. For example, when the magnetic field in region 1 is considered on the interface with region 3, the unit vector points radially outwards from region 1 ($\hat{n} = \hat{\rho}$). The magnetic field in (4.129) is next expressed in terms of the electric field:

$$\underline{J}_s = -\frac{1}{j\omega\mu} \nabla \times \underline{E} \times \hat{n} \tag{4.130}$$

Substituting (4.128) and (4.130) in (4.127), one obtains

$$-j\omega \iiint_V \epsilon\underline{E} \cdot \underline{E}\, dV - \frac{1}{j\omega} \iiint_V \underline{E} \cdot (\nabla \times \mu^{-1}\nabla \times \underline{E})\, dV - \frac{1}{j\omega} \iint_S \underline{E} \cdot \mu^{-1}(\nabla \times \underline{E} \times \hat{n})\, dS = 0 \tag{4.131}$$

Using the vector identity

$$\underline{A} \cdot \underline{B} \times \underline{C} = \underline{B} \cdot \underline{C} \times \underline{A} \tag{4.132}$$

for the last integral, the following variational formula is obtained:

$$\omega^2 = \frac{\iiint_V \underline{E} \cdot (\nabla \times \mu^{-1}\nabla \times \underline{E})\, dV + \iiint_S (\hat{n} \times \underline{E}) \cdot \mu^{-1}(\nabla \times \underline{E})\, dS}{\iiint_V \epsilon\, \underline{E} \cdot \underline{E}\, dV} \tag{4.133}$$

The surface integral in the numerator of (4.133) has been originated by the surface electric currents J_s. Therefore, it is necessary to evaluate this integral only on those interfaces in Fig. 4.17, on which J_s exists.

The evaluation of integrals in (4.133) is performed separately for each of the regions. All of the integrals can be evaluated analytically using formulas (4.41) to (4.44). Considerable savings can be achieved by realizing that for regions 1, 2, 4, and 6, the volume integral in the numerator is identical with the volume integral in the denominator, except for the multiplicative constant. Namely, for isotropic dielectric materials, the Helmholtz wave equation is [9, p. 37]

$$\nabla \times \nabla \times \underline{E} = k^2\underline{E} \tag{4.134}$$

where the propagation constant in the medium is

$$k^2 = \omega_0^2 \mu_0 \epsilon_0 \epsilon_r \tag{4.135}$$

Note that the angular frequency ω_0 signifies the frequency of the original Itoh and Rudokas model, and not the corrected resonant frequency ω_r obtained by the variational formula. The individual volume integrals in the denominator are next denoted by D_i (subscript i stands for the region i = 1 to 6):

$$D_i = \iiint_{V_i} \epsilon_i |\underline{E}_i|^2 dV \tag{4.136}$$

Similarly, the individual volume integrals in the numerator of (4.133) are denoted N_i:

$$N_i = \iiint_{V_i} \underline{E}_i \cdot \nabla \times \mu_i^{-1} \nabla \times \underline{E}_i \, dV \tag{4.137}$$

From (4.134) it follows that N_i is related to D_i as follows:

$$N_i = \omega_0^2 D_i \quad \text{for } i = 1,2,4,6 \tag{4.138}$$

This enables considerable savings in the analytical as well as the computational effort.

After cancelling a common factor,

$$\epsilon_0 2\pi J_1^2(k_{\rho 1} a) a^2 E_0^2 / 4 \tag{4.139}$$

the results of integration take the following form:

$$D_1 = \epsilon_{r1} \left(\frac{\cos \frac{\phi_1}{2}}{\sinh \alpha_1 L_1} \right)^2 (\sigma_1 - 1) L_1 T_r(k_{\rho 1} a) \tag{4.140}$$

$$D_2 = \epsilon_{r2} \left(\frac{\cos \frac{\phi_1}{2}}{\sinh \alpha_2 L_2} \right)^2 (\sigma_2 - 1) L_2 T_r(k_{\rho 1} a) \tag{4.141}$$

$$D_4 = \epsilon_{r4} L(1 + \theta) P_r(k_{\rho 2} a) \tag{4.142}$$

$$D_6 = \epsilon_{r6} L(1 + \theta) T_r(k_{\rho 1} a) \tag{4.143}$$

Symbols σ_1 and σ_2 denote the result of integration of hyperbolic functions:

$$\sigma_i = \frac{\sinh(2\alpha_i L_i)}{2\alpha_i L_i} \qquad \text{for } i = 1,2 \tag{4.144}$$

and θ is the result of integration of the trigonometric functions:

$$\theta = \frac{\sin \phi_1 + \sin \phi_2}{\phi_1 + \phi_2} \tag{4.145}$$

The integration of Bessel functions results in the following:

$$T_r(x) = \frac{J_0^2(x)}{J_1^2(x)} - \frac{2}{x} \cdot \frac{J_0(x)}{J_1(x)} + 1 \tag{4.146}$$

$$P_r(x) = \frac{K_0^2(x)}{K_1^2(x)} + \frac{2}{x} \cdot \frac{K_0(x)}{K_1(x)} - 1 \tag{4.147}$$

Corner regions 3 and 5 do not satisfy (4.134); for these regions, the volume integration in the numerator must be performed by evaluating the double curl operation as it stands. Nevertheless, the analytical evaluation of these integrals is straightforward, and the results are:

$$D_3 = \epsilon_{r3} \frac{\cos^2 \frac{\phi_1}{2}}{\sinh^2 \alpha_1 L_1} L_1(\sigma_1 - 1) \; P_r(k_{\rho 2} a) \tag{4.148}$$

$$N_3 = - \frac{\alpha_1^2 + k_{\rho 2}^2}{k_0^2} \frac{\cos^2 \frac{\phi_1}{2}}{\sinh^2 \alpha_1 L_1} L_1(\sigma_1 - 1) \; P_r(k_{\rho 2} a) \tag{4.149}$$

$$D_5 = \epsilon_{r5} \frac{\cos^2 \frac{\phi_2}{2}}{\sinh^2 \alpha_2 L_2} L_2(\sigma_2 - 1) P_r(k_{\rho 2} a) \tag{4.150}$$

$$N_5 = - \frac{\alpha_2^2 + k_{\rho 2}^2}{k_0^2} \frac{\cos^2 \frac{\phi_2}{2}}{\sinh^2 \alpha_2 L_2} L_2(\sigma_2 - 1) P_r(k_{\rho 2} a) \tag{4.151}$$

The surface integration terms are denoted by N_H and N_V (subscripts H and V signifying the horizontal and the vertical interfaces, respectively, in Fig. 4.17):

$$N_H = 2 \frac{k_{\rho 1} a J_0(k_{\rho 1} a)}{(k_0 a)^2 J_1(k_{\rho 1} a)} \left[\left(\frac{\cos \frac{\phi_1}{2}}{\sinh \alpha_1 L_1} \right)^2 L_1(\sigma_1 - 1) + \left(\frac{\cos \frac{\phi_2}{2}}{\sinh \alpha_2 L_2} \right)^2 L_2(\sigma_2 - 1) \right] \tag{4.152}$$

$$N_V = - \frac{\beta}{k_0^2} (\sin \phi_1 + \sin \phi_2) P_r(k_{\rho 2} a) \tag{4.153}$$

Both N_H and N_V, as well as N_3 and N_5, are real negative numbers, thus reducing the resonant frequency from the value ω_0 evaluated by the original Itoh and Rudokas model to a lower, corrected value ω_r, computed as follows:

$$\omega_r^2 = \omega_0^2 \frac{D_1 + D_2 + N_3 + D_4 + N_5 + D_6 + N_H + N_V}{D_1 + D_2 + D_3 + D_4 + D_5 + D_6} \tag{4.154}$$

The corrected values of the resonant frequency are much closer to the published measured resonant frequencies of shielded resonators reported by various authors [10,11,12]. Typical errors are listed in Table 4.5 where the results obtained by (4.154) are compared with the measured data.

Table 4.5 COMPARISON OF PERTURBATIONAL AND VARIATIONAL METHODS WITH PUBLISHED DATA

Case	Measured	Perturbational Result		Variational Result	
	f(GHz)	f(GHz)	Error	f(GHz)	Error
1	7.94	7.821	-1.49%	7.836	+1.31%
2	4.348	4.286	-1.42%	4.297	-1.17%
3	2.131	2.126	-0.23%	2.130	-0.03%
4	4.85	4.855	-0.10%	4.893	+0.88%
5	9.11	9.033	-0.84%	9.110	+0.00%

Table 4.6 provides the detailed dimensions and dielectric constants for each of the five cases listed in Table 4.5.

Table 4.6 DATA ON PUBLISHED DR CASES

Case	1	2	3	4	5
ϵ_{r1}	1.0	9.6	1.0	1.0	1.0
ϵ_{r2}	1.0	1.0	1.0	1.0	1.0
ϵ_{r6}	36.2	34.19	63.7	38.0	38.0
a(mm)	3.015	7.49	14.29	5.25	2.86
L_1(mm)	3.41	0.70	0.0	∞	∞
L_2(mm)	3.41	0.72	0.0	∞	∞
L(mm)	4.16	7.48	11.25	4.6	2.38
Computed f(GHz)	7.94	-	-	4.829	9.13
Measured f(GHz)	7.94	4.348	2.131	4.85	9.11
Reference	[10]	[11]	[12]	[5]	[7]

The computer program, named DRESV2, evaluates the approximate resonant frequency by variational formula (4.154). The instructions for use and the listing of the program, written in BASIC, are given in Appendix 4.B. Besides the resonant frequency, the program also evaluates the Q factor due to resistive losses in the two shielding plates.

4.9 Mechanical Tuning of a DR Mounted on Microstrip

In many applications, dielectric resonator is mounted on a microstrip substrate, and the tuning of the resonant frequency is provided by a tuning screw coming through the top cover. This situation is shown in Fig. 4.18.

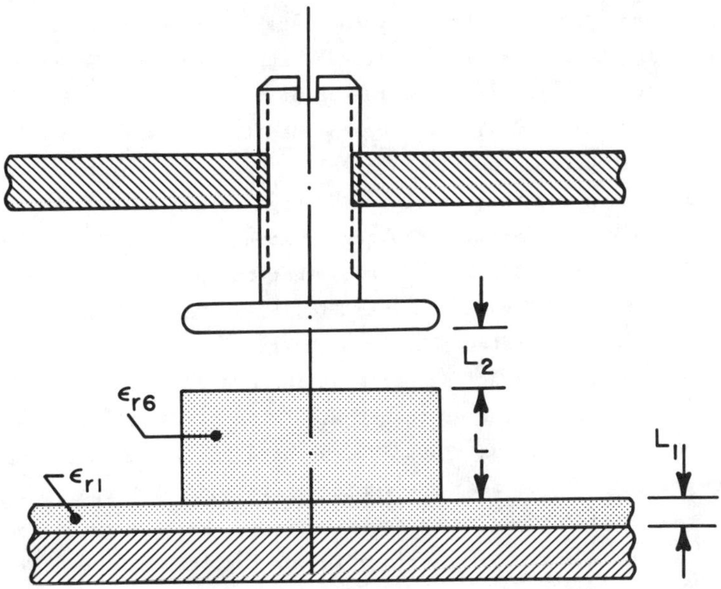

Fig. 4.18 Mechanical tuning of a dielectric resonator

The operation of the tuning mechanism can be explained by the perturbational principle (4.93). When a metal wall of any resonant cavity is moved inward, the change in resonant frequency is proportional to the difference in stored magnetic and electric energies within the displaced volume ΔV. In particular, if the stored magnetic energy in the volume ΔV is larger than the stored electric energy in the volume ΔV, then the resonant frequency will increase after the wall has moved inward.

The resonant system in Fig. 4.18 possesses a rotational symmetry. Thus, the electric field of the mode $TE_{01\delta}$ has only the ϕ-component. This E_ϕ field is oriented tangentially to the surface of the metal plate at the end of the tuning screw. In the immediate vicinity of the plate, the boundary condition requires the tangential electric field to be zero. Therefore, for the $TE_{01\delta}$ mode, the stored electric energy in the vicinity of the metal plate is zero. When the screw is lowered, the only displaced stored energy is the magnetic energy, and the resonant frequency must increase in accordance with (4.93).

In designing the tuning mechanism, it is of interest to determine the change of frequency as a function of the distance L_2. Obviously, the smaller L_2, the larger is the increase in resonant frequency that can be obtained. However, bringing the metal surface of the tuning screw close to the resonator will produce appreciable surface currents, which will, in turn, reduce the Q factor of the resonator. By knowing how much the overall Q will deteriorate when frequency is tuned by a given amount, it becomes possible to select materials and their dimensions in such a way that an optimum design is achieved.

Suppose we want to analyze the tuning mechanism from Fig. 4.18 with the aid of the program DRESV2. We know that the program is based on a slightly different configuration (see Fig. 4.17) in which the metal plate, located at the distance L_2 from the resonator, is of infinite extent. On the other hand, the plate at the end of the tuning screw has a diameter which is not larger than that of the dielectric resonator. Thus, we realize that by using DRESV2, we have in fact assumed the electromagnetic field between the tuning plate and the actual metal cover of the shielding box to be equal to zero. We hope this assumption does not introduce significant errors in our results.

The dielectric resonator to be analyzed is Murata-Erie model DRD077UC034B, which is made of dielectric $\epsilon_r = 37.7$; its radius is $a = 3.85$ mm and its length is $L = 3.41$ mm. The resonator is placed on a dielectric substrate $\epsilon_{r1} = 2.54$ of thickness $L_1 = 0.254$ mm (10 mil). The metal plate below the substrate is made of aluminum. When these data are entered in the program, and when the distance L_2 is varied from 0.001 mm (to avoid entering a zero) to about 6 mm, the frequency and the Q factor computed by the program are such as shown in Fig. 4.19.

Observing the frequency as a function of distance L_2, we conclude that increasing the distance beyond 6 mm does not affect the frequency in any appreciable way. The lowest frequency is about 7.56 GHz. When the distance L_2 is reduced to less than about 2 mm, frequency starts increasing rapidly. When the tuning plate is touching the resonator, the resonant frequency reaches 9.08 GHz.

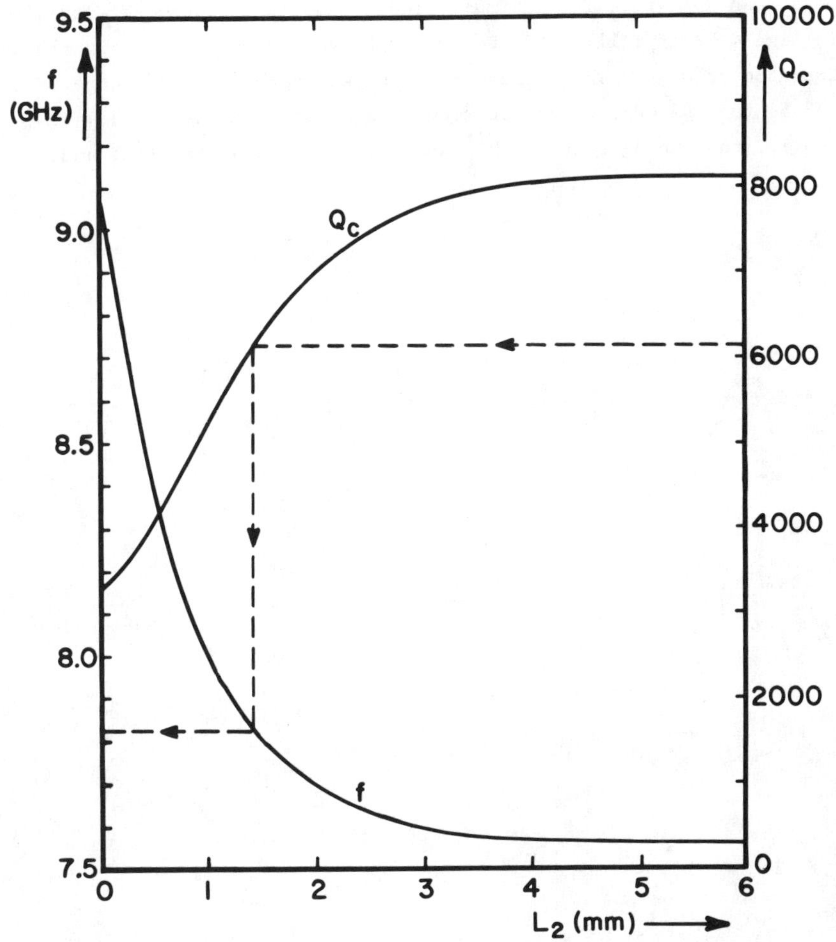

Fig. 4.19 Resonant frequency and Q factor versus tuning distance L_2

At the same time as frequency is increased, the Q factor is lowered. The program DRESV2 computes the Q factor by using the incremental frequency rule, which was explained in Sec. 2.7. The Q factor computed by this method takes into account the conductor losses in the two metal plates. Hence, it is denoted by Q_c. When the tuning plate is at

L_2 = 6 mm, a high value Q_c = 8149 is obtained. When the tuning plate is touching the resonator, Q_c drops to 3265.

It is now possible to select the tuning range in which the resonator operates in a satisfactory manner. Suppose we do not allow Q_c to drop by more than 25 % from its maximum value. Thus, the lowest permissible Q_c in Fig. 4.19 is 6111. This occurs at L_2 = 1.42 mm, which corresponds to frequency f = 7.82 GHz. Therefore, the acceptable tuning range is from 7.56 to 7.82 GHz, which constitutes a relative change of 3.4 %. Typically, the tuning ranges achieved in practice are between 1 and 5 %.

Appendix 4.A PROGRAM DRESP

Dimensions of the Resonator

The dielectric resonator is situated between two parallel plates of infinite extent as shown in Fig. 4.6. The radius of the resonator is a and its length is L. The material of the resonator is ϵ_r. L_1 and L_2 are distances to the shielding plates. When both of these distances are about three times larger than the radius a, the resonant frequency of the resonator is very nearly the same as if the resonator were located in free space.

Example: $\epsilon_r = 38$, a = 5.25 mm, L = 4.6 mm

$\epsilon_{r1} = \epsilon_{r2} = \epsilon_{r3} = \epsilon_{r4} = \epsilon_{r5} = 1$

$L_1 = 15$ mm, $L_2 = 15$ mm

The resonator dimensions and dielectric constants are entered through the data lines 90 and 100. If one lists lines 90 and 100, one finds the following:

```
90 DATA   38, 5.25, 4.6
100 DATA   1, 15, 1, 15
```

For a different resonator, one has to edit these two lines appropriately.

The dielectric constants of the outer regions 3, 4, and 5 are defined by the program lines 1530 to 1550:

```
1560  ER(3) = ER(1)
1570  ER(4) = 1
1580  ER(5) = ER(2)
```

These statements make the dielectric constants of regions 3 and 1 equal to each other, and likewise for regions 5 and 2. The dielectric constant in region 4 is set to unity (for air dielectric). For some special applications, the user may want to change these three statements according to his needs.

Computation of the Resonant Frequency

The perturbational numerical procedure selected here is performed in two steps. First, an estimate of resonant frequency is made on the basis of the cylindrical waveguide model with perfect magnetic wall. Then the wall is removed, and the field outside the resonator is made continuous. The resonant frequency is afterwards computed from the total stored electric and magnetic energies. The computed results have been checked against the method of moments solution for isolated dielectric resonators, and they were found to agree within one percent. For shielded dielectric resonators, the agreement is not so good.

To compute the resonant frequency of the resonator from the previous example, the procedure is as follows:

- Put the DOS disk in drive A and the dielectric resonator program in drive B.
- Switch the computer on, enter the date and the time, and wait for the prompt A>. Then, type

  GRAPHICS (return)

  BASICA (return)
- When the prompt OK is received, type

  LOAD "B:DRESP" (return)
- The computer answers with OK, then type

  RUN (key F2)

The screen will display the resonator dimensions, and then it will solve the transcendental equation for the Cohn model [4] in the first step of the computation. Each time the search increment is reduced 10 times, another digit of accuracy is achieved, and the screen display looks as follows:

```
counting ...  1
counting ...  2
counting ...  3
counting ...  4
counting ...  5
Cohn's model
freq(Cohn) = 4.600455 GHz
perturbational result
freq(pert) = 4.855179 GHz
want to plot the field? (y or n)
```

Plotting the field distribution

If the plot is not wanted, one types letter n and this part of the program is skipped. Assume that we want the field plotted, then we type letter y and hit the return key. The prompt comes:

how many points in each region? (n1, n, n2)

About twenty points or less is appropriate; here, we type

10, 10, 10

and hit the return key. The graphical display of the field will appear on the screen as shown in Fig. 4.20. The two vertical lines in the center of the figure indicate the left and the right faces of the dielectric resonator. The vertical lines on the left and right edges of the figure indicate the position of the metal plates. The horizontal line in the center is the abscissa, indicating zero field amplitude.

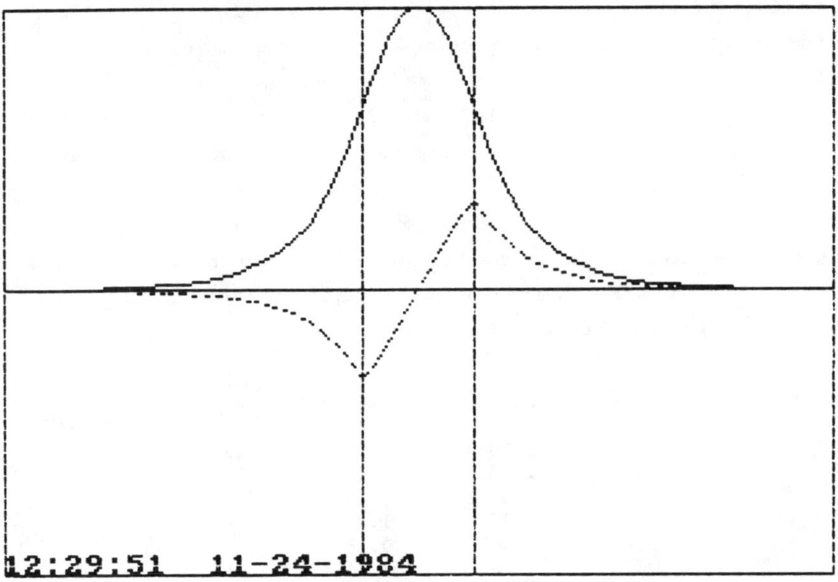

Fig. 4.20 Field distribution in an isolated DR, solid line: E_ϕ and H_z, broken line: H_r

The solid line is proportional to the H_z and E_ϕ components of the field, and the broken line is proportional to the H_r component. It can be seen that the H_z field is maximum inside the resonator, while the H_r is zero at the center of the resonator. Both fields decay exponentially outside the resonator.

The time and the date appear at the bottom of the diagram. The program is now in the WAIT position, for the case where one wants to have the printout of the diagram. This can be accomplished by pressing the PRTSC key. When the print is finished, press

CTRL BREAK

To get in the normal letter mode, type

SCREEN 2 (return)

F10 (function key) (return)

KEY ON (return)

Table of the Energy Distribution

In the perturbational procedure, the stored electric and magnetic energies are computed for each of the six regions shown in Fig. 4.6. The relative distributions of the energy can be useful in deciding how to couple to the resonator. In order to print the table, proceed as follows:

RUN (key F2)

After the previously decribed results appear on the screen, the last line is

want to plot the field? (y or n)

This time, we type n and hit the return key. The next question appears:

want the energy distribution table? (y or n)

Type y, return, and the following appears

j	eps(j)	we(j) %	wm(j) %
1.00	1.00	0.33	13.22
2.00	1.00	0.33	13.22
3.00	1.00	0.16	3.19
4.00	1.00	1.27	4.88
5.00	1.00	0.16	3.19
6.00	38.00	97.74	62.29

It can be seen that 97.74 % of the stored electric energy is located within the resonator (region 6), and only 62.29 % of the magnetic energy is stored within the resonator. Therefore, the best way to couple strongly to the resonator would be to place some magnetic coupling mechanism in region 1 or 2 because in each of these two regions the magnetic field is relatively strong (13 % of the total magnetic energy).

The example computed above was for the $TE_{01\delta}$ mode. The program can also compute the $TE_{01,\ell+\delta}$ modes. The integer ℓ is normally set to zero. This information is stored in line 110 of the program:

110 LMODE = 0

For instance, if we wish to compute the resonant frequency of the resonator with the same dimensions as before, but operating in the $TE_{01,1+\delta}$ mode, we have to make LMODE = 1. Running the program for this mode, we get:

$$freq(Cohn) = 7.684845 \text{ GHz}$$

$$freq(pert) = 8.19373 \text{ GHz}$$

```
10 REM ************** file dresp ***********************************
20 REM          dielectric resonator analysis by perturbational method
30 REM          based on Cohn's model
40 REM          copyright kajfez, november 1984
50 DIM ER(6),NUM(6),DEN(6)
60 DIM WW(3),DWE(3),DWM(3)
70 P01=2.40483
80 P012=P01*P01
90 DATA 38,5.25,4.6
100 DATA 1,15,1,15
110 LMODE=0
120 ACC=5
130 PI=3.141593
140 READ ER(6),A,L
150 READ ER(1),L1,ER(2),L2
160 PRINT "input data"
170 PRINT "er=";ER(6);"a=";A;"L=";L;"LMODE=";LMODE
180 PRINT "er1=";ER(1);"L1=";L1
190 PRINT "er2=";ER(2);"L2=";L2
200 PRINT "accuracy of solving transc. eq. : ";ACC+1;" digits"
210 FOLD=150*P01/(PI*A*SQR(ER(6)))
220 FFNEW=FOLD*1.001
230 GOSUB 560
240 DF=FOLD*.1
250 REM**************** solving the transcendental equation *****
260 FFNEW=FOLD+DF
270 FUNCOLD=FUNC
280 GOSUB 560
290 SIGN=FUNC*FUNCOLD
300 FOLD=FFNEW
310 NTRIAL=NTRIAL+1
320 IF SIGN<0 GOTO 350
330 IF NTRIAL > 50 THEN GOTO 390
340 GOTO 260
350 DF=-DF*.1
360 COUNT=COUNT+1
370 PRINT "counting... ";COUNT
380 IF COUNT<ACC GOTO 260
390 SH1=(X1-XI1)/2
400 SH2=(X2-XI2)/2
410 ETA=120*PI
420 REM                 component amplitudes
430 HZ10=COS(TH1)/SH1
440 HR10=-AL1A*HZ10/P01
450 HRR=BA/P01
460 HZ20=-COS(TH2)/SH2
470 HR20=-AL2A*HZ20/P01
480 GOSUB 1550
490 PRINT "want to plot the field? (y or n)"
500 INPUT A$
510 IF A$="y" THEN GOSUB 810
520 PRINT "want the energy distribution table? (y or n)
530 INPUT A$
540 IF A$="y" THEN GOSUB 2010
550 END
560 REM *************** transcendental function ****************
570 KOA=PI*FFNEW*A/150
```

```
580 KOA2=KOA*KOA
590 RADIC=P012-KOA2*ER(1)
600 AL1A=SQR(RADIC)
610 RADI=P012-KOA2*ER(2)
620 AL2A=SQR(RADI)
630 RADA=KOA2*ER(6)-P012
640 BA=SQR(RADA)
650 AL1L1=AL1A*L1/A
660 AL2L2=AL2A*L2/A
670 X1=EXP(AL1L1)
680 XI1=1/X1
690 X2=EXP(AL2L2)
700 XI2=1/X2
710 CT1=(X1+XI1)/(X1-XI1)
720 CT2=(X2+XI2)/(X2-XI2)
730 ARG1=AL1A*CT1/BA
740 ARG2=AL2A*CT2/BA
750 TH1=ATN(ARG1)
760 IF TH1 < 0 THEN PRINT "negative length th1"
770 TH2=ATN(ARG2)
780 IF TH2 < 0 THEN PRINT "negative length th2"
790 FUNC=(TH1+TH2+LMODE*PI)/BA-L/A
800 RETURN
810 REM*************** computation of field distribution *******
820 PRINT "how many points in each region? (n1,n,n2)
830 INPUT N1,N,N2
840 NTOT=N1+N+N2+1
850 DIM ZZ(NTOT),HZ(NTOT),HR(NTOT)
860 REM                    region 1
870 DZ=L1/N1
880 Z=-L1
890 FOR I1=1 TO N1
900 AEX1=AL1A*(Z+L1)/A
910 XP=EXP(AEX1)
920 XIP=1/XP
930 SHA1=(XP-XIP)/2
940 CHA1=(XP+XIP)/2
950 ZZ(I1)=Z
960 HZ(I1)=HZ10*SHA1
970 HR(I1)=HR10*CHA1
980 Z=Z+DZ
990 NEXT I1
1000 REM                   region inside
1010 DZ=L/N
1020 FOR I=N1+1 TO N1+N
1030 ANG=BA*Z/A-TH1
1040 ZZ(I)=Z
1050 HZ(I)=COS(ANG)
1060 HR(I)=HRR*SIN(ANG)
1070 Z=Z+DZ
1080 NEXT I
1090 REM                   region 2
1100 DZ=L2/N2
1110 IF HZ(N1+N) < 0 THEN HZ20=-HZ20
1120 IF HR(N1+N) < 0 THEN HR20=-HR20
1130 FOR I2=N1+N+1 TO NTOT
1140 AEX2=AL2A*(Z-L2-L)/A
```

```
1150 SP=EXP(AEX2)
1160 SPI=1/SP
1170 SHA2=(SP-SPI)/2
1180 CHA2=(SP+SPI)/2
1190 ZZ(I2)=Z
1200 HZ(I2)=HZ20*SHA2
1210 HR(I2)=HR20*CHA2
1220 Z=Z+DZ
1230 NEXT I2
1240 REM ****************  plotting the field distribution *************
1250 CLS
1260 KEY OFF
1270 SCREEN 1
1280 LOCATE 25,1
1290 PRINT TIME$+"  "+DATE$
1300 LINE (0,0)-(319,199),,B
1310 VV=99
1320 VVV=199
1330 HH=319
1340 LENGTH=L1+L+L2
1350 SCALE=319/LENGTH
1360 L1H=L1*SCALE
1370 LLH=(L1+L)*SCALE
1380 LINE (L1H,0)-(LLH,VVV),,B
1390 LINE (HH,VV)-(0,VV)
1400 FOR I=2 TO NTOT
1410 ZL=(ZZ(I)+L1)*SCALE
1420 HHL=VV*(1-HZ(I))
1430 LINE -(ZL,HHL)
1440 NEXT I
1450 VAM=.5
1460 HHL=VV*(1-VAM*HR(1))
1470 PSET (0,HHL)
1480 FOR I=2 TO NTOT
1490 ZL=(ZZ(I)+L1)*SCALE
1500 HHL=VV*(1-VAM*HR(I))
1510 LINE -(ZL,HHL),,,&HCCCC
1520 NEXT I
1530 WAIT 1,0 '(for printer), then press ctrl break to get out
1540 RETURN
1550 REM ************** perturbational formula *************************
1560 ER(3)=ER(1)
1570 ER(4)=1
1580 ER(5)=ER(2)
1590 KC4A2=KOA2*(ER(6)-ER(4))-P012
1600 KC4A=SQR(KC4A2)
1610 HZ102=HZ10*HZ10
1620 HZ202=HZ20*HZ20
1630 KOB=1/(1+(.4832-.0511/KC4A)/KC4A)
1640 IF KC4A < .8 THEN PRINT "Bessel approximation not accurate !"
1650 KOB2=KOB*KOB
1660 PRX=KOB2+2*KOB/KC4A-1
1670 QRX=1-KOB2
1680 SIF1=SIN(TH1*2)
1690 SIF2=SIN(TH2*2)
1700 THET=(SIF1+SIF2)*.5/(TH1+TH2+LMODE*PI)
1710 ARG1=2*AL1L1
```

```
1720 XH1=EXP(ARG1)
1730 XHI1=1/XH1
1740 SIG1=(XH1-XHI1)*.25/AL1L1
1750 ARG2=2*AL2L2
1760 XH2=EXP(ARG2)
1770 XHI2=1/XH2
1780 SIG2=(XH2-XHI2)*.25/AL2L2
1790 A2=A*A
1800 WW(1)=ER(1)*HZ102*A2*L1*(SIG1-1)
1810 WW(2)=ER(2)*HZ202*A2*L2*(SIG2-1)
1820 WW(3)=ER(6)*A2*L*(1+THET)
1830 DWE(1)=ER(3)*HZ102*A2*L1*PRX*(SIG1-1)
1840 DWE(2)=ER(4)*A2*L*PRX*(1+THET)
1850 DWE(3)=ER(5)*HZ202*A2*L2*PRX*(SIG2-1)
1860 DWM(1)=(RADIC/KOA2)*HZ102*A2*L1*PRX*(SIG1+1)
1870 DWM(2)=(RADA/KOA2)*A2*L*PRX*(1-THET)
1880 DWM(3)=(RADI/KOA2)*HZ202*A2*L2*PRX*(SIG2+1)
1890 FOR I=1 TO 3
1900 SWW=SWW+WW(I)
1910 SDWE=SDWE+DWE(I)
1920 SDWM=SDWM+DWM(I)
1930 NEXT I
1940 PERT=(SDWE-SDWM)*.5/SWW
1950 PERFRE=FOLD*(1-PERT)
1960 PRINT "Cohn's model"
1970 PRINT "freq(Cohn)=",FOLD," GHz"
1980 PRINT "perturbational result"
1990 PRINT "freq(pert)=",PERFRE,"GHz"
2000 RETURN
2010 REM ************** ENERGY DISTRIBUTION TABLE **********************
2020 DIM WE(6),WM(6)
2030 PRINT "    j     eps(j)  we(j)%  wm(j)%"
2040 PRINT
2050 SUME=SWW+SDWE
2060 SUMM=SWW+SDWM
2070 WE(1)=WW(1)/SUME
2080 WE(2)=WW(2)/SUME
2090 WE(3)=DWE(1)/SUME
2100 WE(4)=DWE(2)/SUME
2110 WE(5)=DWE(3)/SUME
2120 WE(6)=WW(3)/SUME
2130 TEM=(RADIC/KOA2)*(SIG1+1)+(P012/KOA2)*(SIG1-1)
2140 WM(1)=TEM*HZ102*A2*L1/SUMM
2150 TEM=(RADA/KOA2)*(1-THET)+(P012/KOA2)*(1+THET)
2160 WM(6)=TEM*L*A2/SUMM
2170 TEM=(RADI/KOA2)*(SIG2+1)+(P012/KOA2)*(SIG2-1)
2180 WM(2)=TEM*HZ202*L2*A2/SUMM
2190 WM(3)=DWM(1)/SUMM
2200 WM(4)=DWM(2)/SUMM
2210 WM(5)=DWM(3)/SUMM
2220 FOR J=1 TO 6
2230 WM(J)=100*WM(J)
2240 WE(J)=100*WE(J)
2250 PRINT USING "#####.##"; J,ER(J),WE(J),WM(J)
2260 NEXT J
2270 RETURN
```

Appendix 4.B PROGRAM DRESV2

Computation of Resonant Frequency

Program DRESV2 computes the resonant frequency of the same resonator shown in Fig. 4.6, but the program provides neither a graphical display nor a table of energy distribution. It has been observed that, for shielded resonators, the accuracy of the results is better than that of DRESP.

The input data are entered through data lines 90 and 100. Line 110 contains the value of LMODE, as in DRESP. The dielectric constants in regions 3, 4, and 5 are entered in lines 140 to 160:

```
140 ER(3) = ER(1)
150 ER(4) = 1
160 ER(5) = ER(2)
```

These lines may be changed if other values of dielectric constants are desired. For example, the input data for a dielectric resonator with $\epsilon_{r6} = 37.6$, $a = 4.25$ mm, $h = 3.7$ mm, $\epsilon_{r1} = \epsilon_{r2} = 1$, $L_1 = L_2 = 2.91$ mm will be entered in the following way

```
90 DATA 37.6, 4.25, 3.7
100 DATA 1, 2.91, 1, 2.91
```

After the command RUN, the following output appears:

```
*************************************************************
             dresv2  01-27-1986          15:52:24
*************************************************************
input data
er= 37.6 a= 4.25 L= 3.7 LMODE= 0
er1= 1 L1= 2.91
er2= 1 L2= 2.91
searching for the eigenvalue...
 1           koa=              .7214197     eigx=          2.9
 2           koa=              .565719      eigx=          2.880314
 3           koa=              .5877552     eigx=          2.893725
 4           koa=              .5887659     eigx=          2.893954
 5           koa=              .5887722     eigx=          2.893961
Itoh & Rudokas model
freq(I&R)=    6.614539      GHz
variational result
freq(var)=    6.1765        GHz
want the percent error in frequency? (y or n)
```

In the steps 1 to 5 (fewer steps may appear in other examples) the program searches for the solution of the transcendental system of equations of the Itoh and Rudokas model [13]. Afterwards, the variational procedure is used to compute the frequency. When the exact value of frequency is known for a particular resonator, it is possible to check the percent error of the result computed by DRESV2. Suppose we have measured the resonant frequency of this resonator to be 6.39 GHz. Then, we type

y

The prompt comes:

enter the exact frequency in GHz

After entering the value 6.39, the answer appears:

-3.34 %

When L_1 and L_2 tend to zero, the computed frequency approaches the exact value. Such a result may be useful for predicting the resonant frequency of the resonator between two metal planes (Courtney holder). For example, we enter the following data:

```
90 DATA 63.7,  14.28,   11.252
100 DATA 1.,   0.0001,  1.,    0.0001
```

The result is 2.130866 GHz. The value given by Courtney [12] is 2.131 GHz. Therefore, the computed value agrees within -0.01 % with the measured value.

Computation of Q Factor

The program computes the Q factor due to losses in conducting plates (located at the extreme left and the extreme right sides in Fig. 4.6). The computation is performed by the "incremental frequency rule," described in Sec. 2.7. The application of this rule consists of moving the conductor surfaces for the length of one skin depth and computing the Q from the change in resonant frequency due to this move.

The program is ready for computation of Q after the following prompt:

want to compute the Q factor? (y or n)

After answering y, the next prompt is

shield: copper, aluminum, brass, or other? (c,a,b, or o)

Suppose the walls are made of brass. After answering b, the following appears:

```
skin depth=   1.616215      microns
searching for the eigenvalue...
 1          koa=            .5887721      eigx=           2.89396
 2          koa=            .5887803      eigx=           2.893973
Itoh & Rudokas model
freq(I&R)=    6.614631      GHz
variational result
freq(var)=    6.176669      GHz
Q(due to shield losses)=    36384.99
```

Therefore, the Q factor due to the losses in two metal walls is about 36,000. This means that these losses are almost negligible. If the Q factor of the dielectric material is 5000, we may compute the overall Q of the shielded resonator as follows:

```
PRINT 1/(1/5000 + 1/36385)
```

The overall Q comes out to be 4395.917.

```
10 REM ************** file dresv2 *********************************
20 REM          dielectric resonator analysis by variational method
30 REM          based on Itoh and Rudokas model
40 REM          q factor computed by the incremental frequency rule
50 REM          copyright Kajfez, january 1986
60 DIM ER(6),NUM(6),DEN(6)
70 DIM XX(3),KK(3),ALF(2),FIH(2),LL(2),FCT(3),GCT(3)
80 REM          lengths should be entered in mm
90 DATA 37.6,4.25,3.7
100 DATA 1,2.91,1,2.91
110 LMODE=0
120 READ ER(6),A,L
130 READ ER(1),LL(1),ER(2),LL(2)
140 ER(3)=ER(1)
150 ER(4)=1
160 ER(5)=ER(2)
170 PRINT "************************************************************"
180 PRINT "             dresv2  ";DATE$,TIME$
190 PRINT "************************************************************"
200 PRINT "input data"
210 PRINT "er=";ER(6);"a=";A;"L=";L;"LMODE=";LMODE
220 PRINT "er1=";ER(1);"L1=";LL(1)
230 PRINT "er2=";ER(2);"L2=";LL(2)
240 PI=3.141593
250 NQ=0
260 REM **************  2-dimensional search for the solution *********
270 REM **************  of the transcendental equation       *********
280 XX(2)=2.9
290 IF ER(2)-ER(1)>0 THEN EMAX=ER(2) ELSE EMAX=ER(1)
300 KMIN=XX(2)/SQR(ER(6)-ER(4))
310 KMAX=XX(2)/SQR(EMAX)
320 KK(2)=(9*KMIN+KMAX)/10
330 DXX=.00001
340 DKK=.00001
350 ITER=0
360 PRINT "searching for the eigenvalue..."
370 XX(1)=XX(2)+DXX
380 KK(1)=KK(2)
390 XX(3)=XX(2)
400 KK(3)=KK(2)+DKK
410 FOR ITI=1 TO 3
420 X=XX(ITI)
430 KO=KK(ITI)
440 K02=KO*KO
450 XIT2=X*X
460 GOSUB 1940
470 RA=K02*(ER(6)-ER(4))-XIT2
480 IF RA > 0 GOTO 550
490 STEPX=STEPX/2
500 STEPK=STEPK/2
510 XX(2)=XX(2)-STEPX
520 KK(2)=KK(2)-STEPK
530 PRINT "step too large, start again with 1/2 smaller step"
540 GOTO 350
550 YY=SQR(RA)
560 KC4A=YY
570 GOSUB 2000
```

```
580 FCT(ITI)=JOB+YY*KOB/X
590 BA=SQR(KO2*ER(6)-XIT2)
600 FOR JIT=1 TO 2
610 ALF(JIT)=SQR(XIT2-KO2*ER(JIT))
620 POW=ALF(JIT)*LL(JIT)/A
630 IF POW>8 GOTO 680
640 EP=EXP(POW)
650 EI=1/EP
660 AGU=(EP+EI)/(EP-EI)
670 GOTO 690
680 AGU=1
690 AGU=AGU*ALF(JIT)/BA
700 FIH(JIT)=ATN(AGU)
710 NEXT JIT
720 GCT(ITI)=FIH(1)+FIH(2)-BA*L/A+LMODE*PI
730 NEXT ITI
740 AL=(FCT(1)-FCT(2))/DXX
750 AU=(GCT(1)-GCT(2))/DKK
760 BL=(FCT(3)-FCT(2))/DXX
770 BU=(GCT(3)-GCT(2))/DKK
780 CL=FCT(2)-AL*XX(2)-BL*KK(2)
790 CU=GCT(2)-AU*XX(2)-BU*KK(2)
800 DENO=AU*BL-AL*BU
810 XNEW=(CL*BU-CU*BL)/DENO
820 KNEW=(CU*AL-CL*AU)/DENO
830 STEPX=XNEW-XX(2)
840 STEPK=KNEW-KK(2)
850 STEP2=STEPX^2+STEPK^2
860 PRINT ITER+1,"koa=",KK(2),"eigx=",XX(2)
870 XX(2)=XNEW
880 KK(2)=KNEW
890 IF STEP2<1E-12 THEN 960
900 ITER=ITER+1
910 IF ITER>10 THEN 930
920 GOTO 370
930 PRINT "solution not found after 10 iterations"
940 GOTO 1870
950 REM          if the search is successful, re-evaluate the constants
960 KOA=KK(2)
970 FIR=KOA*150/(PI*A)
980 PRINT "Itoh & Rudokas model"
990 PRINT "freq(I&R)=",FIR," GHz"
1000 EIGX=XX(2)
1010 KOA2=KOA*KOA
1020 EIG2=EIGX*EIGX
1030 RADIC=EIG2-KOA2*ER(1)
1040 AL1A=SQR(RADIC)
1050 RADI=EIG2-KOA2*ER(2)
1060 AL2A=SQR(RADI)
1070 RADA=KOA2*ER(6)-EIG2
1080 BA=SQR(RADA)
1090 AL1L1=AL1A*LL(1)/A
1100 AL2L2=AL2A*LL(2)/A
1110 IF AL1L1>8 THEN GOTO 1170
1120 Z1=EXP(AL1L1)
1130 ZI1=1/Z1
1140 CT1=(Z1+ZI1)/(Z1-ZI1)
```

```
1150 SIH1=(Z1-ZI1)*.5
1160 GOTO 1180
1170 CT1=1
1180 IF AL2L2>8 THEN GOTO 1240
1190 Z2=EXP(AL2L2)
1200 ZI2=1/Z2
1210 CT2=(Z2+ZI2)/(Z2-ZI2)
1220 SIH2=(Z2-ZI2)*.5
1230 GOTO 1250
1240 CT2=1
1250 ARG1=AL1A*CT1/BA
1260 ARG2=AL2A*CT2/BA
1270 TH1=ATN(ARG1)
1280 TH2=ATN(ARG2)
1290 REM ************** variational formula ***************************
1300 KC4A2=RADA-KOA2*ER(4)
1310 KC4A=SQR(KC4A2)
1320 GOSUB 2000
1330 X=EIGX
1340 GOSUB 1940
1350 J0B2=J0B*J0B
1360 TRX=J0B2-2*J0B/EIGX+1
1370 K0B2=K0B*K0B
1380 PRX=K0B2+2*K0B/KC4A-1
1390 SIF1=SIN(TH1*2)
1400 SIF2=SIN(TH2*2)
1410 THET=(SIF1+SIF2)*.5/(TH1+TH2+LMODE*PI)
1420 CO12=A*(COS(TH1)^2)/AL1A
1430 IF AL1L1>8 THEN 1470
1440 SECN=AL1L1/(SIH1*SIH1)
1450 PARM1=CT1-SECN
1460 GOTO 1480
1470 PARM1=1
1480 COPAM1=CO12*PARM1
1490 CO22=A*(COS(TH2)^2)/AL2A
1500 IF AL2L2 > 8 THEN 1540
1510 SECN=AL2L2/(SIH2*SIH2)
1520 PARM2=CT2-SECN
1530 GOTO 1550
1540 PARM2=1
1550 COPAM2=CO22*PARM2
1560 NUM(1)=ER(1)*COPAM1*TRX
1570 NUM(2)=ER(2)*COPAM2*TRX
1580 NUM(3)=-(RADIC+KC4A2)*PRX*COPAM1/KOA2
1590 NUM(4)=ER(4)*L*(1+THET)*PRX
1600 NUM(5)=-(RADI+KC4A2)*PRX*COPAM2/KOA2
1610 NUM(6)=ER(6)*L*(1+THET)*TRX
1620 DEN(1)=NUM(1)
1630 DEN(2)=NUM(2)
1640 DEN(3)=ER(3)*PRX*COPAM1
1650 DEN(4)=NUM(4)
1660 DEN(5)=ER(5)*PRX*COPAM2
1670 DEN(6)=NUM(6)
1680 DENSUM-0
1690 SURVER=-BA*A*(SIF1+SIF2)*PRX/KOA2
1700 SURHOR=X*J0B*2*(COPAM1+COPAM2)/KOA2
1710 NUMSUM=SURHOR+SURVER
```

```
1720 FOR J=1 TO 6
1730 DENSUM=DENSUM+DEN(J)
1740 NUMSUM=NUMSUM+NUM(J)
1750 NEXT J
1760 VARKOA=KOA*SQR(NUMSUM/DENSUM)
1770 VARFRE=VARKOA*150/(PI*A)
1780 PRINT "variational result"
1790 PRINT "freq(var)=",VARFRE," GHz"
1800 IF NQ=1 THEN RETURN
1810 PRINT "want the percent error in frequency? (y or n)"
1820 INPUT A$
1830 IF A$="y" THEN GOSUB 1880
1840 PRINT "want to compute the Q factor? (y or n)"
1850 INPUT B$
1860 IF B$="y" THEN GOSUB 2060
1870 END
1880 REM ************** percent error ***********************************
1890 PRINT "enter the exact frequency in GHz"
1900 INPUT EXCFRE
1910 PERC=100*(VARFRE/EXCFRE-1)
1920 PRINT USING "+##.## _%";PERC
1930 RETURN
1940 REM *************  function J0B=J0(X)/J1(X) ***********************
1950 XMX0=X-2.4048
1960 TEM=(.0282*XMX0-.1177)*XMX0+.2571
1970 TEM=(TEM*XMX0-.716)*XMX0+1.4282
1980 J0B=TEM*XMX0/(X-3.8317)
1990 RETURN
2000 REM *************  function K0B=K0(KC4A)/K1(KC4A) *****************
2010 KI=1/KC4A
2020 TEM=(.00445*KI-.02679)*KI+.06539
2030 TEM=(TEM*KI-.11226)*KI+.49907
2040 K0B=1/(1+TEM*KI)
2050 RETURN
2060 REM ************** q factor ***********************************
2070 NQ=1
2080 F0=VARFRE
2090 PRINT "shield : copper, aluminum, brass, or other? (c,a,b,or o)"
2100 INPUT A$
2110 IF A$="c" THEN SIGMA=5.8E+07
2120 IF A$="a" THEN SIGMA=3.72E+07
2130 IF A$="b" THEN SIGMA=1.57E+07
2140 IF A$>< "o" GOTO 2180
2150 PRINT "enter conductivity in Siemens/meter"
2160 INPUT SIGMA
2170 IF SIGMA<.1 GOTO 2300
2180 SKIN=50/(SQR(F0*SIGMA)*PI)
2190 PRINT "skin depth=",1000*SKIN," microns"
2200 LL(1)=LL(1)-SKIN
2210 LL(2)=LL(2)-SKIN
2220 GOSUB 350
2230 DF=VARFRE-F0
2240 IF DF/F0<.000001 GOTO 2280
2250 Q=F0/DF
2260 PRINT "Q(due to shield losses)=",Q
2270 RETURN
2280 PRINT "insignificant losses in the shield, quit"
```

```
2290 RETURN
2300 PRINT "conductivity too small, quit"
2310 RETURN
```

References

[1] S. Ramo, J.R. Whinnery, and T. Van Duzer, Fields and Waves in Communication Electronics, New York: John Wiley & Sons, 1965.

[2] E.C. Jordan and K.G. Balmain, Electromagnetic Waves and Radiating Systems, Englewood Cliffs, NJ: Prentice-Hall, 1968.

[3] M. Abramowitz and I.A. Stegun (ed.), Handbook of Mathematical Functions, Washington, D.C.: National Bureau of Standards, Third printing with corrections, March 1965.

[4] S.B. Cohn, "Microwave bandpass filters containing high-Q dielectric resonators," IEEE Trans. Microwave Theory Tech., vol. MTT-16, pp. 218-227, April 1968.

[5] A.W. Glisson, D. Kajfez, and J. James, "Evaluation of modes in dielectric resonators using a surface integral equation formulation," IEEE Trans. Microwave Theory Tech., vol. MTT-31, pp. 1023-1029, Dec. 1983.

[6] D. Kajfez, "Basic principles give understanding of dielectric waveguides and resonators," Microwave Syst. News, vol. 13, pp. 152-161, May 1983.

[7] M. Tsuji, H. Shigesawa, and K. Takiyama, "Analytical and experimental investigations on several resonant modes in open dielectric resonators," IEEE Trans. Microwave Theory Tech., pp. 628-633, June 1984.

[8] U.S. Hong and R.H. Jansen, "Numerical analysis of shielded dielectric resonators including substrate, support disc and tuning post," Electron. Lett., vol. 18, no. 23, pp. 1000-1002, 11 Nov. 1982.

[9] R.F. Harrington, Time-Harmonic Electromagnetic Fields, New York: McGraw-Hill, 1961.

[10] M. Jaworski and M.W. Pospieszalski, "An accurate solution of the cylindrical dielectric resonator problem," IEEE Trans. Microwave Theory Tech., vol. MTT-27, pp. 639-643, July 1979.

[11] S. Maj and J.W. Modelski, "Application of a dielectric resonator on microstrip line for a measurement of complex permittivity," IEEE MTT-S Int. Microwave Symp. Dig., pp. 525-527, San Francisco, May 30 - June 1, 1984.

[12] W.E. Courtney, "Analysis and evaluation of a method of measuring the complex permittivity and permeability of microwave insulators," IEEE Trans. Microwave Theory Tech., vol. MTT-18, pp. 476-485, Aug. 1970.

[13] T. Itoh and R.S. Rudokas, "New method for computing the resonant frequencies of dielectric resonators," IEEE Trans. Microwave Theory Tech., vol. MTT-25, pp. 52-54, Jan. 1977.

[14] T.D. Iveland, "Dielectric resonator filters for application in microwave integrated circuits," IEEE Trans. Microwave Theory Tech., vol. MTT-19, pp. 643-652, July 1971.

[15] D. Kajfez, "Elementary functions procedure simplifes dielectric resonators' design", Microwave Syst. News, vol. 12, pp. 133-140, June 1982.

Chapter 5
RIGOROUS ANALYSIS METHODS

Krzysztof A. Michalski

5.1 Introduction

The simple cylindrical dielectric resonator (DR) models described in Ch. 4 are approximate, since they postulate non-Maxwellian fields as the point of departure. Although these simple models are useful as aids in understanding the principles of operation of DRs, and they may be helpful at initial stages of the analysis or design process, they fail to provide the high accuracy usually required in present-day microwave circuit design. As a result, more sophisticated (and accurate), rigorous techniques have been recently developed which allow one to take into account the influence of the environment on the characteristics of a DR. These techniques are rigorous in the sense that the solution is obtained in the form of successive approximations converging toward the exact solution. Therefore, at least in principle, they allow one to compute both the resonant frequency and the field distribution to any desired accuracy.

The purpose of this chapter is twofold: first, to survey the rigorous techniques available for the analysis of DRs and, second, to present selected numerical results obtained by these techniques. We only consider circular-cylindrical (pillbox or disc) and tubular (ring) resonators, since these are the shapes most often encountered in practice. Also, we direct most of the attention to the analysis of shielded DRs, i.e., resonators placed in a parallel-plate waveguide or in a cylindrical cavity (possibly on a substrate), as illustrated in Fig. 5.1. (Since this structure exhibits rotational symmetry with respect to the z axis, only one half of the meridian (ϕ = constant) cross section is shown in the figure.) More information on open (isolated) DRs, i.e., resonators in infinite, homogeneous space, can be found in Ch. 6.

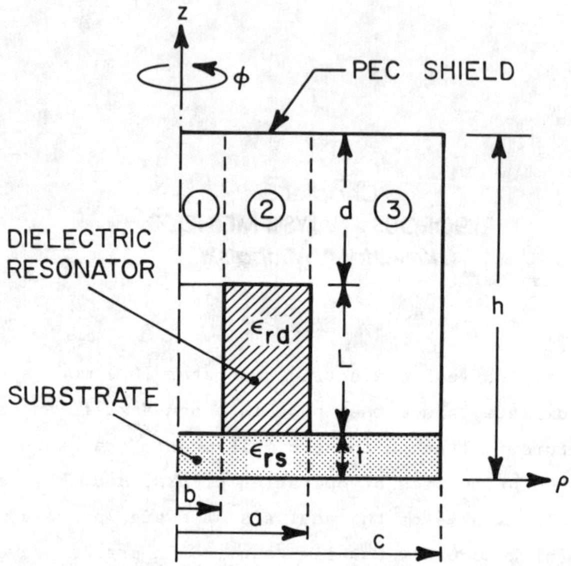

Fig. 5.1 Dielectric ring resonator on a substrate in a cylindrical cavity

The point of departure in any rigorous analysis of a DR consists of Maxwell's equations, which, in the absence of sources, can be stated as

$$\underline{\nabla} \times \underline{H} = j\omega\epsilon_0\epsilon_r\underline{E} \tag{5.1a}$$

$$\underline{\nabla} \times \underline{E} = -j\omega\mu_0\underline{H} \tag{5.1b}$$

$$\underline{\nabla} \cdot \underline{H} = 0 \tag{5.1c}$$

$$\underline{\nabla} \cdot (\epsilon_r\underline{E}) = 0 \tag{5.1d}$$

We assume here that the region of interest is characterized by permeability μ_0 and permittivity $\epsilon = \epsilon_0\epsilon_r$, where ϵ_r, the relative permittivity (or the relative dielectric constant), can vary with position. Referring to the resonator shown in Fig. 5.1, we have $\epsilon_r = \epsilon_{rs}$ in the substrate, $\epsilon_r = \epsilon_{rd}$ inside the DR, and $\epsilon_r = 1$ in the remaining region bounded by the metallic walls, which are assumed to be perfectly electrically conducting (PEC).

We should emphasize here, that in a rigorous analysis the DR proper, i.e., the high-ϵ dielectric sample of pillbox or ring shape, must be considered as an integral part of its environment. Hence, from this point of view, the geometry shown in Fig. 5.1 is nothing else but an inhomogeneously filled cavity. In this chapter, where there is no danger of confusion, we will somewhat loosely apply the term "resonator" when referring to both the complete resonant system and the DR proper.

It is often necessary to determine the resonant frequencies and associated field distributions (modes) of the resonator (Fig. 5.1) in relation to the permittivity of the DR, its position within the PEC enclosure, and other parameters. If we define the operators L and M and a vector $\underline{\Phi}$, as [1, Ch.9]

$$L = \begin{bmatrix} 0 & \underline{\nabla}\times \\ \underline{\nabla}\times & 0 \end{bmatrix}, \qquad M = \begin{bmatrix} \epsilon_0 \epsilon_r \underset{\sim}{1} & 0 \\ 0 & -\mu_0 \underset{\sim}{1} \end{bmatrix}, \qquad \underline{\Phi} = \begin{bmatrix} \underline{E} \\ \underline{H} \end{bmatrix} \tag{5.2}$$

where $\underset{\sim}{1}$ is the identity tensor of rank two, this problem can be reduced to the determination of the eigenvalues and eigenfunctions of the following boundary value problem [1]:

$$L\underline{\Phi} = j\omega M\underline{\Phi} \tag{5.3a}$$

$$\hat{n} \times \underline{E} = \underline{0} \quad \text{on } S \tag{5.3b}$$

where S is the PEC cavity boundary with unit normal $\hat{n}$. We note that the first two Maxwell equations are incorporated in (5.3a). It can be shown that the operators L and M are self-adjoint with the inner product [1]:

$$\langle \underline{\Phi}_1, \underline{\Phi}_2 \rangle = \iiint\limits_V (\underline{E}_1 \cdot \underline{E}_2 + \underline{H}_1 \cdot \underline{H}_2) dV \tag{5.4}$$

where V is the region bounded by S. Following the method of moments [1], we expand the unknown solution $\underline{\Phi}$ in a series:

$$\underline{\Phi} = \sum_j \alpha_j \underline{f}_j \tag{5.5}$$

where α_j are constant coefficients to be determined and $\underline{f}_j$ are suitable basis functions. We then substitute the series (5.5), truncated to N terms, into (5.3a) and take the inner product of this equation with $\underline{f}_i$,

$i = 1,\ldots,N$. As a result of this step, which is often referred to as "testing," we obtain the matrix eigenvalue equation:

$$\underset{\sim}{\ell}|\alpha\rangle = j\omega\, \underset{\sim}{m}|\alpha\rangle \tag{5.6}$$

with the elements of the N×N matrices $\underset{\sim}{\ell}$ and $\underset{\sim}{m}$ defined, respectively, by

$$\ell_{ij} = \langle \underline{f}_i, L\underline{f}_j \rangle \tag{5.7}$$

and

$$m_{ij} = \langle \underline{f}_i, M\underline{f}_j \rangle \tag{5.8}$$

If a complete set of the basis functions $\underline{f}_j$ is used, exact solutions to the original problem can be obtained, at least in principle, by letting $N \to \infty$, and the resonant frequencies of the resonator can be determined from the eigenvalues $j\omega$. This procedure, which is referred to as the Galerkin method, since the same functions are employed as expansion and testing functions, is intimately related to the Rayleigh-Ritz method [1], which minimizes the functional:

$$j\omega = \frac{\langle \underline{\Phi}, L\underline{\Phi} \rangle}{\langle \underline{\Phi}, M\underline{\Phi} \rangle} \tag{5.9}$$

obtained by taking the inner product of (5.3a) with $\underline{\Phi}$. As is well known [2], this functional is stationary about the eigenfunctions of (5.3) and the stationary values are then the corresponding eigenvalues. If we substitute the expansion (5.5) into (5.9) and apply the constraints:

$$\frac{\partial\omega}{\partial\alpha_i} = 0\ , \quad i = 1,\ldots,N \tag{5.10}$$

then (5.6) is obtained.

The important problem, which must be addressed when using the Galerkin-Rayleigh-Ritz method summarized above is the proper choice of the set of basis functions in (5.5). The usual choice of the empty-cavity modes for the expansion functions [1] results in a slow convergence of the method when applied to the DR problem [3]. The rate of convergence can be significantly improved by using the modes of the dielectric post resonator [4]. The price one pays for this improvement, however, is the necessity to solve the auxiliary eigenvalue problem of a dielectric-post resonator as an intermediate step. Perhaps because of these difficulties, this method has not been used in the analysis of DRs, with the exception of Krupka's work [4], which is limited to the TE axisymmetric modes of a DR on a substrate in a parallel-plate waveguide.

The solution of electromagnetic boundary value problems is often facilitated by the introduction of vector potentials (or Hertz vectors). The magnetic vector potential, $\underline{A}$, is defined by

$$\underline{H} = \frac{1}{\mu_0}\underline{\nabla} \times \underline{A} \tag{5.11}$$

which is certainly consistent with (5.1c). Upon substituting (5.11) into the first two Maxwell equations, one arrives at

$$\underline{E} = -\frac{j\omega}{k_0^2}[k_0^2\,\underline{A} + (\underline{\nabla}\epsilon_r^{-1})\underline{\nabla}\cdot\underline{A} + \epsilon_r^{-1}\underline{\nabla\nabla}\cdot\underline{A}] \tag{5.12}$$

with $\underline{A}$ given by

$$\underline{\nabla}^2\underline{A} - \epsilon_r^{-1}(\underline{\nabla}\epsilon_r)\underline{\nabla}\cdot\underline{A} + k_0^2\epsilon_r\underline{A} = 0 \tag{5.13}$$

where $k_0^2 = \omega^2\mu_0\epsilon_0$. Obviously, the introduction of the vector potential did not result in much analytical progress, since the solution of (5.13) is not any easier than the solution of the original problem. However, we will see that the use of potentials is productive in cases where the resonator cross section can be divided into partial regions in which the field can be decomposed into constituents TM or TE to z. Approaches based on this procedure are sometimes referred to as partial region methods (PRM). We remark that although (5.13) is strictly valid only for continuous ϵ_r, we will be able to specialize it to the case of a piecewise-constant permittivity.

Exact solutions of (5.3) are only available for the cases where the DR completely fills two of the cavity dimensions (the dielectric parallel-plate resonator discussed in Ch. 3 is one such case). In other cases, one must resort to numerical methods to find approximate solutions. Simple as it may appear, the problem (5.3) poses a formidable challenge, because it is vector in nature and because the relative permittivity ϵ_r, in spite of being piecewise-constant, is a function of both ρ and z in the region where the solution is sought. With the PRM approach, it is often possible to reduce the vector eigenvalue problem

(5.3) to two simpler, scalar problems.

In the case of pillbox or ring resonators placed concentrically in a cylindrical cavity, in a parallel-plate waveguide, or in free space (and we limit attention to these cases), it is always possible to break the resonant field into constituents (modes) with azimuthal variation given by cos $m\phi$ or sin $m\phi$ (as usual in problems with circular symmetry, there is a twofold eigenvalue degeneracy here), $m = 0,1,2,\ldots$, and to analyze them independently, thus rendering the problem two-dimensional. As discussed in Ch. 3, in the axisymmetric case ($m = 0$), the set of modes can be further divided into transverse electric to z, designated TE, or H modes, and transverse magnetic to z, designated TM, or E modes. If $m > 0$, only hybrid electromagnetic modes (having non-zero E_z and H_z) exist. These modes are denoted variously as HEM, HE, or EH modes. The use of both HE and EH to denote hybrid modes usually implies that the modes designated HE are considered "H like" and the modes designated EH are considered "E like." This method of designation, which is analogous to that used in cylindrical dielectric waveguides [5], is quite arbitrary, for it depends on what one chooses as the criterion of "likeness." For example, in one proposed scheme, the mode is designated EH if in a homogeneously filled cavity it ultimately becomes an E mode, etc. [6].

In the case of open DRs, or DRs placed symmetrically in a parallel-plate waveguide or a cavity, the modes can be further classified as either odd or even with respect to the (equatorial) symmetry plane. The odd (even) modes are those which are unaffected by inserting a PEC (PMC) wall in the symmetry plane. (Here, PMC stands for perfect magnetic conductor.) Some authors introduce additional symbols to distinguish between these modes [7,8].

Different modes from the same class are distinguished by two or (more often) three subscripts. The first subscript, m, almost always refers to the azimuthal dependence of the mode as either cos $m\phi$ or sin $m\phi$. In the case of a DR in free space, a second subscript, n, and a third subscript, p, are usually introduced, which refer, respectively, to the number of field extrema within the DR in the radial and axial directions. Often, the index p is replaced by $\ell + \delta$, where $\ell = 0,1,\ldots$, and $0 < \delta < 1$ [9]. This notation means that there are ℓ and a fraction half-period field variations in the DR along the z-axis.

In the case of shielded DRs, a consensus has not yet emerged as to the meaning of the second and third indices. This is due to the fact that, depending on the value of ϵ_{rd} and the relative dimensions of the dielectric sample and the PEC shield, the modes of the resonator can be considered as the modes of the cavity perturbed by the presence of the dielectric inhomogeneity, or the modes of the DR proper perturbed by the presence of the cavity. The former, referred to as the cavity-type, or exterior modes, are those whose energy is predominantly concentrated outside the dielectric sample, and whose resonant frequencies depend strongly on the size of the cavity, but are only weakly affected by the change in the size of the DR proper. The latter, referred to as the DR-type, or interior modes, are those whose energy is predominantly concentrated in the dielectric sample, and whose resonant frequencies are not drastically affected by changing the dimensions of the shield. Hence, in the case of the exterior modes it would make sense to follow the nomenclature used in the classification of the modes of the cavity forming the shield of the DR. Similarly, one could designate the interior modes according to the scheme adopted for the modes of the DR in free space. Unfortunately, this relatively clear picture is blurred by the occurrence of "mixed-type modes," whose energy is almost equally divided between the dielectric sample and its exterior in the cavity. Possibly for this reason, Zaki and Chen [8] have recently proposed to use only two subscripts in the mode designation. In their system, the first subscript still refers to the angular variation of the mode and the second simply classifies the resonant frequencies in increasing order without any reference to the modal field distribution in the DR. Since most practical applications of DRs exploit the interior modes, the three-subscript notation does seem to have some merit, however.

Until recently, DRs have been almost exclusively operated in the axisymmetric modes, the $TE_{01\delta}$ being the most popular. Consequently, most of the literature on the subject of dielectric resonators limits treatment to the axisymmetric case. As the DRs gain popularity, however, new applications are likely to appear, which will exploit the hybrid modes. For example, the $HEM_{11\delta}$ mode has been recently employed in a dual-mode filter (see Sec. 9.8). Hence, efficient analysis methods capable of solving the problem (5.3) in the general, $m \neq 0$ case are becoming increasingly important.

In the next sections, we review various rigorous methods of solving the problem posed in (5.3). Our approach here is to describe in considerable detail one representative method and to review other rigorous methods in a more cursory way. Hence, we begin in Sec. 5.2 with a detailed treatment of a PRM approach based on a partition of the resonator in complementary annular regions, such as regions 1, 2, and 3 in Fig. 5.1. This method belongs to a group of mode matching methods, which seem to be especially well-suited for the analysis of shielded resonators. A typical procedure in this category can be summarized as follows. The resonant system is divided into a number of complementary regions sufficiently regular to ensure the separability of the wave equation. The fields in each partial region are represented in terms of a series of the appropriate waveguide modes with yet unknown coefficients. When the fields of the adjacent regions are "matched" to satisfy the appropriate continuity conditions, an infinite homogeneous system of simultaneous linear equations is obtained for the expansion coefficients. This system has nontrivial solutions only when its determinant vanishes. Hence, the resonant frequencies can be found by searching for the zeros of the determinant. In practice, the system is truncated to a finite size $N \times N$, for example, and computations are repeated with increasing N until convergence is achieved. Since in the procedure described in Sec. 5.2 we match the fields of radial waveguide sections, we refer to it as the radial mode-matching method. We illustrate this method on the example of a ring resonator (Fig. 5.1), because it comprises a pillbox resonator as a special case when $a = 0$. For the sake of clarity and completeness, some of the tedious details are included in the appendices.

Other mode-matching methods are briefly reviewed in Sec. 5.3. A prominent member in this group is a method based on the division of the resonator into a number of horizontal layers, which can be considered sections of dielectric-loaded cylindrical waveguides. We refer to this procedure as the axial mode matching method.

In Sec. 5.4, we review finite-element and finite-difference methods. These methods, which have only been applied to axisymmmetric cases, seek to solve (5.3) without first dividing the resonator cross section into more regular partial regions. The governing differential equation is either approximated in terms of finite differences over the whole resonator cross section or the unknown field is expanded in terms of finite elements, leading to a matrix eigenvalue problem, which can be solved by standard procedures. Hence, the search for zeros of a matrix determinant is usually not required.

In Sec. 5.5, we discuss perturbational methods based on the expansion of the fields in asymptotic series in inverse powers of $\sqrt{\epsilon_r}$, where ϵ_r is the relative dielectric constant of the DR. Although these methods are exact only in the limit as $\epsilon_r \to \infty$, we choose to include them in this review because they are semi-analytic in nature and explicitly give the dependence of various parameters on ϵ_r. Hence, the numerical solution does not have to be repeated for every new value of ϵ_r. The perturbational methods have only been applied to axisymmetric cases.

In Sec. 5.6, we discuss a group of methods based on the solution of surface or volume integral equations. These methods are applicable in cases where the suitable Green's function, which comprises the kernel of the integral equation, can be found. In the case of a DR in infinite space, which is discussed in detail in Ch. 6, the free-space Green's function is employed. As in the mode-matching methods, the resonant frequencies are found as zeros of a matrix determinant.

In Sec. 5.7, we present selected numerical results which illustrate how changing various parameters affects the resonant frequencies of various modes of DRs. This section should also help the reader interpret the results presented in research papers and to alert him or her to the diverse notation used by various authors.

We conclude this chapter in Sec. 5.8 with a brief discussion and assessment of the rigorous analysis methods.

We should point out before leaving this section that the classification of methods introduced in this review is by no means the only one possible. One could easily devise other meaningful classifications based on different distinguishing features shared by various methods.

5.2 Radial Mode Matching Method

In the radial mode matching method, the resonator cross section is divided into complementary regions, as indicated in Fig. 5.1. We observe that in each region the permittivity is independent of ρ, i.e., $\epsilon_r = \epsilon_r(z)$. In fact, ϵ_r is a piecewise-constant function of z. A typical region with three dielectric layers is illustrated in Fig. 5.2. We

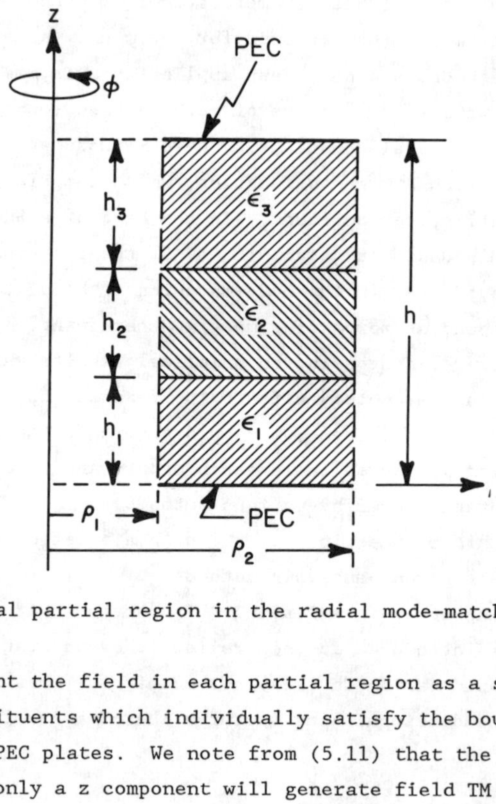

Fig. 5.2 Typical partial region in the radial mode-matching method

seek to represent the field in each partial region as a superposition of TM and TE constituents which individually satisfy the boundary conditions at the PEC plates. We note from (5.11) that the magnetic vector potential with only a z component will generate field TM to $\hat{z}$. Hence, we postulate

$$\underline{A} = \hat{z}\ \psi^e(\rho,z) \begin{Bmatrix} \cos m\phi \\ \sin m\phi \end{Bmatrix} \tag{5.14}$$

which reduces (5.13) to the scalar equation:

$$\frac{1}{\rho}\frac{\partial}{\partial\rho}\left(\rho\frac{\partial\psi^e}{\partial\rho}\right) - \frac{m^2}{\rho^2}\psi^e + \epsilon_r\frac{\partial}{\partial z}\left(\epsilon_r^{-1}\frac{\partial\psi^e}{\partial z}\right) + k_0^2\epsilon_r\psi^e = 0 \tag{5.15}$$

Upon substituting (5.14) into (5.11) and (5.12), one can express the magnetic and electric field components in terms of ψ^e as

$$H_\rho^e = \frac{m}{\mu_0\rho}\psi^e \begin{Bmatrix} -\sin m\phi \\ \cos m\phi \end{Bmatrix} \tag{5.16a}$$

$$H_\phi^e = -\frac{1}{\mu_0}\frac{\partial\psi^e}{\partial\rho}\begin{Bmatrix} \cos m\phi \\ \sin m\phi \end{Bmatrix} \tag{5.16b}$$

$$H_z^e = 0 \tag{5.16c}$$

$$E_\rho^e = -\frac{j\omega}{k_0^2\epsilon_r}\frac{\partial^2\psi^e}{\partial\rho\partial z}\begin{Bmatrix} \cos m\phi \\ \sin m\phi \end{Bmatrix} \tag{5.17a}$$

$$E_\phi^e = -\frac{j\omega m}{k_0^2\epsilon_r\rho}\frac{\partial\psi^e}{\partial z}\begin{Bmatrix} -\sin m\phi \\ \cos m\phi \end{Bmatrix} \tag{5.17b}$$

$$E_z^e = -\frac{j\omega}{k_0^2}\left[\frac{\partial}{\partial z}\epsilon_r^{-1}\frac{\partial\psi^e}{\partial z} + k_0^2\psi^e\right]\begin{Bmatrix} \cos m\phi \\ \sin m\phi \end{Bmatrix} \tag{5.17c}$$

As anticipated, the potential (5.14) generates a TM field, which is sometimes termed an E field (hence, the superscript e on ψ and the field components). Since E_ρ and E_ϕ must vanish on the PEC plates, we have from (5.17):

$$\frac{\partial\psi^e}{\partial z} = 0 \quad \text{at} \quad z = 0 \quad \text{and} \quad z = h \tag{5.18}$$

Also, if a region is adjacent to a cavity wall, we require that

$$\psi^e = 0 \quad \text{at} \quad \rho = c \tag{5.19}$$

where c is the radius of the cylindrical shield. If c = ∞ (parallel-plate waveguide), we replace (5.19) by the radiation condition. If the region borders on the z axis, we must also require that ψ^e be bounded at

$\rho = 0$.

We note that (5.15) can be written as

$$D_\rho \psi^e + D_z^e \psi^e = 0 \tag{5.20}$$

where D_ρ and D_z^e are differential operators acting on functions of ρ and z, respectively. We may expand ψ^e in each partial region in terms of the eigenfunctions Z_p^e of the operator D_z^e as

$$\psi^e(\rho, z) = \sum_{p=1}^{\infty} R_p^e(\rho) Z_p^e(z) \tag{5.21}$$

where R_p^e are radial functions chosen to satisfy (5.15). The eigenfunctions Z_p^e and the corresponding eigenvalues λ_p^e are determined by solving the Sturm-Liouville problem:

$$\frac{d}{dz}\left(\epsilon_r^{-1} \frac{d}{dz} Z_p^e\right) + \left(k_0^2 + \lambda_p^e \epsilon_r^{-1}\right) Z_p^e = 0 \tag{5.22a}$$

$$\frac{d}{dz} Z_p^e = 0 \quad \text{at} \quad z = 0 \quad \text{and} \quad z = h \tag{5.22b}$$

For real and piecewise-continuous ϵ_r, the system (5.22) is self-adjoint and has an infinite number of real, discrete eigenvalues [10]. The eigenfunctions Z_p^e can also be made real-valued by multiplying them by appropriate non-zero constants. When properly normalized, they are orthonormal with a weight ϵ_r^{-1}, i.e.,

$$\int_0^h \epsilon_r^{-1}(z)\, Z_p^e(z)\, Z_q^e(z)\, dz \equiv \langle \epsilon_r^{-1} Z_p^e, Z_q^e \rangle = \delta_{pq} \tag{5.23}$$

where δ_{pq} is the Kronecker delta. (We will find the inner product notation introduced in (5.23) useful in later developments.) With reference to Fig. 5.2, we have

$$\epsilon_r(z) = \begin{cases} \epsilon_1, & 0 < z < h_1 \\ \epsilon_2, & h_1 < z < h_1 + h_2 \\ \epsilon_3, & h_1 + h_2 < z < h \end{cases} \tag{5.24}$$

Since the determination of Z_p^e and λ_p^e is quite tedious in this case, we defer these tasks to Appendices 5.A and 5.B, respectively.

Upon substituting (5.20) into (5.15) and using (5.22a), we find that the radial functions R_p^e must satisfy

$$\frac{1}{\rho}\frac{d}{d\rho}\left(\rho\frac{d}{d\rho}R_p^e\right) - \left(\lambda_p^e + \frac{m^2}{\rho^2}\right)R_p^e = 0 \tag{5.25}$$

which is the modified Bessel equation of order n, with linearly independent solutions $I_m(\sqrt{\lambda_p^e}\,\rho)$ and $K_m(\sqrt{\lambda_p^e}\,\rho)$ [11]. It is shown in Appendix 5.B that a finite number of the eigenvalues λ_p^e are negative. Hence, if we introduce the notation:

$$\sqrt{\lambda_p^e} = \begin{cases} j\zeta_p^e, & \lambda_p^e < 0 \\ \xi_p^e, & \lambda_p^e > 0 \end{cases} \tag{5.26}$$

where ξ_p^e and ζ_p^e are positive, we may choose $J_m(\zeta_p^e\rho)$ and $Y_m(\zeta_p^e\rho)$ as the linearly independent pair of solutions of (5.25) when $\lambda_p^e < 0$, and $I_m(\xi_p^e\rho)$ and $K_m(\xi_p^e\rho)$ when $\lambda_p^e > 0$. If the partial region is radially infinite, we may also have to use the special combination of $J_m(\zeta_p^e\rho)$ and $Y_m(\zeta_p^e\rho)$, which represents an outgoing cylindrical wave, i.e., $H_m^{(2)}(\zeta_p^e\rho)$, the Hankel function of the second kind. Thus, we define, for notational efficiency,

$$P_p^e(\rho) = \begin{cases} J_m(\zeta_p^e\rho), & \lambda_p^e < 0 \\ I_m(\xi_p^e\rho), & \lambda_p^e > 0 \end{cases} \tag{5.27}$$

$$Q_p^e(\rho) = \begin{cases} Y_m(\zeta_p^e\rho), & \lambda_p^e < 0 \\ K_m(\xi_p^e\rho), & \lambda_p^e > 0 \end{cases} \tag{5.28}$$

$$W_p^e(\rho) = \begin{cases} H_m^{(2)}(\zeta_p^e\rho), & \lambda_p^e < 0 \\ K_m(\xi_p^e\rho), & \lambda_p^e > 0 \end{cases} \tag{5.29}$$

We may now express the radial function in a typical partial region, such as region 2 in Fig. 5.1, as

$$\text{(Region 2)} \qquad R_p^e(\rho) = A_p^e\, P_p^e(\rho) + B_p^e\, Q_p^e(\rho) \tag{5.30}$$

In region 1, which borders on the z axis, we reject the Neumann and Macdonald functions, since they are unbounded at $\rho = 0$. Hence, we simply have

$$\text{(Region 1)} \qquad R_p^e(\rho) = A_p^e\, P_p^e(\rho) \tag{5.31}$$

If region 3 is finite, we select for $R_p^e(\rho)$ a combination of Bessel functions which is consistent with (5.19). If $c = \infty$, however, we choose a solution of (5.25) which either represents a radially outgoing wave, or an evanescent wave. Hence, if we introduce the notation:

$$S_p^e(\rho) = \begin{cases} P_p^e(\rho)Q_p^e(c) - P_p^e(c)Q_p^e(\rho)\ , & c \text{ finite} \\ W_p^e(\rho)\ , & c = \infty \end{cases} \tag{5.32}$$

we can write

$$\text{(Region 3)} \qquad R_p^e(\rho) = A_p^e\, S_p^e(\rho) \tag{5.33}$$

We note that if $c = \infty$ and at a resonant frequency $\lambda_p^e > 0$, there is no radiation of energy in the radial direction and the resonant mode is trapped. On the other hand, if $\lambda_p^e < 0$, then radiation occurs and the resonant mode becomes leaky. The resonant frequencies of leaky modes are complex-valued.

Upon substituting the expansion (5.21) into (5.16) and (5.17), one finally obtains, for each partial region,

$$H_\rho^e(\rho,\phi,z) = \frac{m}{\mu_0 \rho} \sum_{p=1}^{\infty} R_p^e(\rho) Z_p^e(z) \begin{Bmatrix} -\sin m\phi \\ \cos m\phi \end{Bmatrix} \tag{5.34a}$$

$$H_\phi^e(\rho,\phi,z) = -\frac{1}{\mu_0} \sum_{p=1}^{\infty} R_p^{e\,\prime}(\rho) Z_p^e(z) \begin{Bmatrix} \cos m\phi \\ \sin m\phi \end{Bmatrix} \tag{5.34b}$$

$$H_z^e(\rho,\phi,z) = 0 \tag{5.34c}$$

$$E_\rho^e(\rho,\phi,z) = -\frac{j\omega}{k_0^2\epsilon_r}\sum_{p=1}^{\infty} R_p^{e\prime}(\rho)Z_p^{e\prime}(z)\begin{Bmatrix}\cos m\phi\\ \sin m\phi\end{Bmatrix} \tag{5.35a}$$

$$E_\phi^e(\rho,\phi,z) = -\frac{j\omega m}{k_0^2\epsilon_r\rho}\sum_{p=1}^{\infty} R_p^{e}(\rho)Z_p^{e\prime}(z)\begin{Bmatrix}-\sin m\phi\\ \cos m\phi\end{Bmatrix} \tag{5.35b}$$

$$E_z^e(\rho,\phi,z) = -\frac{j\omega}{k_0^2\epsilon_r}\sum_{p=1}^{\infty} R_p^{e}(\rho)\lambda_p^e\, Z_p^{e\prime}(z)\begin{Bmatrix}\cos m\phi\\ \sin m\phi\end{Bmatrix} \tag{5.35c}$$

In these equations, primes over quantities denote derivatives with respect to the argument.

To be able to represent an arbitrary field in each partial region, we also need the TE part. Hence, by analogy with (5.11) and (5.14), which generate a TM field, we introduce the electric vector potential, $\underline{F}$, as

$$\underline{E} = -\frac{1}{\epsilon_0}\,\underline{\nabla}\times\underline{F} \tag{5.36}$$

with

$$\underline{F} = \hat{z}\,\psi^h(\rho,z)\begin{Bmatrix}-\sin m\phi\\ \cos m\phi\end{Bmatrix} \tag{5.37}$$

where the form of azimuthal dependence is chosen for consistency with (5.14). It may seem at first that (5.36) is inconsistent with (5.1d), which in expanded form is

$$\epsilon_r\underline{\nabla}\cdot\underline{E} + \underline{E}\cdot\underline{\nabla}\epsilon_r = 0 \tag{5.38}$$

However, if we note that $\underline{E}$ is TE to z and, in each partial region, $\epsilon_r = \epsilon_r(z)$, we see that (5.38) gives $\underline{\nabla}\cdot\underline{E} = 0$, as required.

Upon substituting (5.36) and (5.37) into Maxwell's equations (5.1a) and (5.1b), one determines that ψ^h must satisfy

$$\frac{1}{\rho}\frac{\partial}{\partial\rho}\left(\rho\frac{\partial\psi^h}{\partial\rho}\right) - \frac{m^2}{\rho^2}\psi^h + \frac{\partial^2\psi^h}{\partial z^2} + k_0^2\epsilon_r\psi^h = 0 \tag{5.39}$$

or, in operator notation,

$$D_\rho \psi^h + D_z^h \psi^h = 0 \tag{5.40}$$

The magnetic and electric field components can be expressed in terms of ψ^h as

$$H_\rho^h = -\frac{j\omega}{k_0^2} \frac{\partial^2 \psi^h}{\partial \rho \partial z} \begin{Bmatrix} -\sin m\phi \\ \cos m\phi \end{Bmatrix} \tag{5.41a}$$

$$H_\phi^h = \frac{j\omega m}{k_0^2 \rho} \frac{\partial \psi^h}{\partial z} \begin{Bmatrix} \cos m\phi \\ \sin m\phi \end{Bmatrix} \tag{5.41b}$$

$$H_z^h = -\frac{j\omega}{k_0^2} \left(\frac{\partial^2 \psi^h}{\partial z^2} + k_0^2 \epsilon_r \psi^h \right) \begin{Bmatrix} -\sin m\phi \\ \cos m\phi \end{Bmatrix} \tag{5.41c}$$

$$E_\rho^h = \frac{m}{\epsilon_0 \rho} \psi^h \begin{Bmatrix} \cos m\phi \\ \sin m\phi \end{Bmatrix} \tag{5.42a}$$

$$E_\phi^h = \frac{1}{\epsilon_0} \frac{\partial \psi^h}{\partial \rho} \begin{Bmatrix} -\sin m\phi \\ \cos m\phi \end{Bmatrix} \tag{5.42b}$$

$$E_z^h = 0 \tag{5.42c}$$

It is easy to see that in the present case the counterparts of (5.18) and (5.19) are, respectively,

$$\psi^h = 0 \quad \text{at} \quad z = 0 \quad \text{and} \quad z = h \tag{5.43}$$

and

$$\frac{\partial \psi^h}{\partial \rho} = 0 \quad \text{at} \quad \rho = c \tag{5.44}$$

We expand ψ^h in terms of eigenfunctions Z_p^h of the operator D_z^h as

$$\psi^h(\rho, z) = \sum_{p=1}^{\infty} R_p^h(\rho) Z_p^h(\rho) \tag{5.45}$$

The eigenfunctions Z_p^h and the corresponding eigenvalues λ_p^h are determined from

$$\frac{d^2}{dz^2} Z_p^h + (k_0^2\epsilon_r + \lambda_p^h) Z_p^h = 0 \tag{5.46a}$$

$$Z_p^h = 0 \quad \text{at} \quad z = 0 \quad \text{and} \quad z = h \tag{5.46b}$$

When properly normalized, the eigenfunctions Z_p^h possess the orthonormality property:

$$< Z_p^h \,,\, Z_q^h > = \delta_{pq} \tag{5.47}$$

The determination of Z_p^h and λ_p^h for a typical partial region shown in Fig. 5.2 is discussed in Appendix 5.A. The choice of the radial functions R_p^h is dictated by similar considerations as in the TM case. Hence, the expressions for $R_p^h(\rho)$ in regions 1 and 2 are as those given in (5.31) and (5.30), respectively, with the superscripts e replaced by h. There is a difference in region 3, however, since now the radial functions must be consistent with (5.44). If we introduce the notation:

$$T_p^h(\rho) = \begin{cases} P_p^h(\rho)Q_p^{h'}(c) - P_p^{h'}(c)Q_p^h(\rho) \,, & c \text{ finite} \\ W_p^h(\rho) \,, & c = \infty \end{cases} \tag{5.48}$$

we may express $R_p^h(\rho)$ in region 3, as

$$R_p^h(\rho) = A_p^h \, T_p^h(\rho) \tag{5.49}$$

Finally, upon substituting (5.45) into (5.41) and (5.42), we obtain

$$H_\rho^h(\rho,\phi,z) = -\frac{j\omega}{k_0^2} \sum_{p=1}^{\infty} R_p^{h'}(\rho) Z_p^{h'}(z) \begin{Bmatrix} -\sin m\phi \\ \cos m\phi \end{Bmatrix} \tag{5.50a}$$

$$H_\phi^h(\rho,\phi,z) = \frac{j\omega m}{k_0^2 \rho} \sum_{p=1}^{\infty} R_p^h(\rho) Z_p^{h'}(z) \begin{Bmatrix} \cos m\phi \\ \sin m\phi \end{Bmatrix} \tag{5.50b}$$

$$H_z^h(\rho,\phi,z) = \frac{j\omega}{k_0^2} \sum_{p=1}^{\infty} R_p^h(\rho) \lambda_p^h \, Z_p^h(z) \begin{Bmatrix} -\sin m\phi \\ \cos m\phi \end{Bmatrix} \tag{5.50c}$$

$$E_\rho^h(\rho,\phi,z) = \frac{m}{\epsilon_0 \rho} \sum_{p=1}^{\infty} R_p^h(\rho) Z_p^h(z) \begin{Bmatrix} \cos m\phi \\ \sin m\phi \end{Bmatrix} \tag{5.51a}$$

$$E_\phi^h(\rho,\phi,z) = \frac{1}{\epsilon_0} \sum_{p=1}^{\infty} R_p^{h\prime}(\rho) Z_p^h(z) \begin{Bmatrix} -\sin m\phi \\ \cos m\phi \end{Bmatrix} \tag{5.51b}$$

$$E_z^h(\rho,\phi,z) = 0 \tag{5.51c}$$

The next step in the radial mode matching procedure is to enforce the continuity of tangential electric and magnetic fields across the cylindrical boundaries between regions 1 and 2 ($\rho = b$) and between regions 2 and 3 ($\rho = a$). Hence, if we distinguish the fields of the three partial regions by superscripts (1), (2), and (3) (we will omit the parentheses if there is no danger of confusing the superscript with a power), we can state the continuity conditions as

$$E_z^{e1} - E_z^{e2} = 0 \,, \quad \rho = b \tag{5.52a}$$

$$H_\phi^{e1} - H_\phi^{e2} + H_\phi^{h1} - H_\phi^{h2} = 0 \,, \quad \rho = b \tag{5.52b}$$

$$E_z^{e2} - E_z^{e3} = 0 \,, \quad \rho = a \tag{5.52c}$$

$$H_\phi^{e2} - H_\phi^{e3} + H_\phi^{h2} - H_\phi^{h3} = 0 \,, \quad \rho = a \tag{5.52d}$$

$$H_z^{h1} - H_z^{h2} = 0 \,, \quad \rho = b \tag{5.52e}$$

$$E_\phi^{e1} - E_\phi^{e2} + E_\phi^{h1} - E_\phi^{h2} = 0 \,, \quad \rho = b \tag{5.52f}$$

$$H_z^{h2} - H_z^{h3} = 0 \,, \quad \rho = a \tag{5.52g}$$

$$E_\phi^{e2} - E_\phi^{e3} + E_\phi^{h2} - E_\phi^{h3} = 0 \,, \quad \rho = a \tag{5.52h}$$

It is noted that both the TM and TE field constituents in each partial region are incorporated in these equations. As the next step, the field expansions (5.34) to (5.35) and (5.50) to (5.51) are substituted in

(5.52), and the resulting equations are "tested." Here, testing means multiplying an equation by a member of a suitably chosen set of functions and integrating it in z between z = 0 and z = h. The natural choice for the testing functions are in our case the functions $Z_p^{ei}(z)$ and $Z_p^{hi}(z)$, i = 1,2,3, because in each partial region they have the orthogonality properties (5.23) and (5.47). Hence, we test (5.52a) with $Z_q^{e1}(z)$, (5.52b) and (5.52d) with $Z_q^{e2}(z)/\epsilon_r^{(2)}(z)$, (5.52c) with $Z_q^{e3}(z)$, (5.52e) with $Z_q^{h1}(z)$, (5.52f) and (5.52h) with $Z_q^{h2}(z)$, and (5.52g) with $Z_q^{h3}(z)$, where q = 1,2,... As a result, we obtain an infinite homogeneous set of algebraic equations for the coefficients A_p^{e1}, A_p^{e2}, B_p^{e2}, A_p^{e3}, A_p^{h1}, A_p^{h2}, B_p^{h2}, and A_p^{h3}, where p = 1,2,... If, in a partial region i, we truncate the infinite series (5.34) to (5.35) after N^{ei} terms, and the series (5.50) to (5.51) after N^{hi} terms, this set reduces to a finite system of $N^{e1} + 2N^{e2} + N^{e3} + N^{h1} + 2N^{h2} + N^{h3}$ equations with the same number of unknowns. This set has nontrivial solutions only at frequencies where the determinant of the system matrix vanishes. Hence, by searching for the zeros of the determinant, one can compute the resonant frequencies of the DR. Since the system matrix has many zero and unit submatrices (a result of the orthogonality of the expansion and testing functions), one can easily reduce its size by eliminating some unknowns in terms of the others. Hence, after straightforward but tedious algebra, we obtain the reduced N × N system given below, where $N = 2N^{e2} + 2N^{h2}$.

$$
\left[\begin{array}{cc||cc}
\underset{\sim}{e} - \underset{\sim}{d}\underset{\sim}{a}^{-1}\underset{\sim}{b} & \underset{\sim}{f} - \underset{\sim}{d}\underset{\sim}{a}^{-1}\underset{\sim}{c} & \underset{\sim}{h} - \underset{\sim}{g}\underset{\sim}{A}^{-1}\underset{\sim}{B} & \underset{\sim}{i} - \underset{\sim}{g}\underset{\sim}{A}^{-1}\underset{\sim}{C} \\
\hline
\underset{\sim}{m} - \underset{\sim}{p}\underset{\sim}{\ell}^{-1}\underset{\sim}{j} & \underset{\sim}{n} - \underset{\sim}{p}\underset{\sim}{\ell}^{-1}\underset{\sim}{k} & \underset{\sim}{r} - \underset{\sim}{t}\underset{\sim}{L}^{-1}\underset{\sim}{J} & \underset{\sim}{s} - \underset{\sim}{t}\underset{\sim}{L}^{-1}\underset{\sim}{K} \\
\hline\hline
\underset{\sim}{E} - \underset{\sim}{D}\underset{\sim}{a}^{-1}\underset{\sim}{b} & \underset{\sim}{F} - \underset{\sim}{D}\underset{\sim}{a}^{-1}\underset{\sim}{c} & \underset{\sim}{H} - \underset{\sim}{G}\underset{\sim}{A}^{-1}\underset{\sim}{B} & \underset{\sim}{I} - \underset{\sim}{G}\underset{\sim}{A}^{-1}\underset{\sim}{C} \\
\hline
\underset{\sim}{M} - \underset{\sim}{P}\underset{\sim}{\ell}^{-1}\underset{\sim}{j} & \underset{\sim}{N} - \underset{\sim}{P}\underset{\sim}{\ell}^{-1}\underset{\sim}{k} & \underset{\sim}{R} - \underset{\sim}{T}\underset{\sim}{L}^{-1}\underset{\sim}{J} & \underset{\sim}{S} - \underset{\sim}{T}\underset{\sim}{L}^{-1}\underset{\sim}{K}
\end{array}\right]
\left[\begin{array}{c}
|A^{e2}\rangle \\ \hline |B^{e2}\rangle \\ \hline\hline |A^{h2}\rangle \\ \hline |B^{h2}\rangle
\end{array}\right]
=
\left[\begin{array}{c}
|0\rangle \\ \hline |0\rangle \\ \hline\hline |0\rangle \\ \hline |0\rangle
\end{array}\right]
\qquad (5.53)
$$

The elements of the submatrices appearing in (5.53) are defined as

$$a_{pq} = P_q^{e1}(b)\lambda_p^{e1}\delta_{pq} \qquad (5.54)$$

$$b_{pq} = -\,P_p^{e2}(b)\lambda_p^{e2}\langle Z_q^{e1}, Z_p^{e2}/\epsilon_r^{(2)}\rangle \tag{5.55}$$

$$c_{pq} = -\,Q_p^{e2}(b)\lambda_p^{e2}\langle Z_q^{e1}, Z_p^{e2}/\epsilon_r^{(2)}\rangle \tag{5.56}$$

$$d_{pq} = -\,P_p^{e1\prime}(b)\langle Z_q^{e2}, Z_p^{e1}/\epsilon_r^{(2)}\rangle \tag{5.57}$$

$$e_{pq} = P_q^{e2\prime}(b)\delta_{pq} \tag{5.58}$$

$$f_{pq} = Q_p^{e2\prime}(b)\delta_{pq} \tag{5.59}$$

$$g_{pq} = \frac{jm}{\omega\epsilon_0 b} P_p^{h1}(b)\langle Z_q^{e2}, Z_p^{h1\prime}/\epsilon_r^{(2)}\rangle \tag{5.60}$$

$$h_{pq} = -\,\frac{jm}{\omega\epsilon_0 b} P_p^{h2}(b)\langle Z_q^{e2}, Z_p^{h2\prime}/\epsilon_r^{(2)}\rangle \tag{5.61}$$

$$i_{pq} = -\,\frac{jm}{\omega\epsilon_0 b} Q_p^{h2}(b)\langle Z_q^{e2}, Z_p^{h2\prime}/\epsilon_r^{(2)}\rangle \tag{5.62}$$

$$j_{pq} = P_p^{e2}(a)\lambda_p^{e2}\langle Z_q^{e3}, Z_p^{e2}/\epsilon_r^{(2)}\rangle \tag{5.63}$$

$$k_{pq} = Q_p^{e2}(a)\lambda_p^{e2}\langle Z_q^{e3}, Z_p^{e2}/\epsilon_r^{(2)}\rangle \tag{5.64}$$

$$\ell_{pq} = -\,S_q^{e3}(a)\lambda_q^{e3}\delta_{pq} \tag{5.65}$$

$$m_{pq} = -\,P_q^{e2\prime}(a)\delta_{pq} \tag{5.66}$$

$$n_{pq} = -\,Q_q^{e2\prime}(a)\delta_{pq} \tag{5.67}$$

$$p_{pq} = S_p^{e3\prime}(a)\langle Z_q^{e2}, Z_p^{e3}/\epsilon_r^{(2)}\rangle \tag{5.68}$$

$$r_{pq} = \frac{jm}{\omega\epsilon_0 a} P_p^{h2}(a)\langle Z_q^{e2}, Z_p^{h2\prime}/\epsilon_r^{(2)}\rangle \tag{5.69}$$

$$s_{pq} = \frac{jm}{\omega\epsilon_0 a} Q_p^{h2}(a)<Z_q^{e2}, Z_p^{h2'}/\epsilon_r^{(2)}> \tag{5.70}$$

$$t_{pq} = -\frac{jm}{\omega\epsilon_0 a} T_p^{h3}(a)<Z_q^{e2}, Z_p^{h3'}/\epsilon_r^{(2)}> \tag{5.71}$$

$$A_{pq} = P_q^{h1}(b)\lambda_q^{h1}\delta_{pq} \tag{5.72}$$

$$B_{pq} = -P_p^{h2}(b)\lambda_p^{h2}<Z_q^{h1}, Z_p^{h2}> \tag{5.73}$$

$$C_{pq} = -Q_p^{h2}(b)\lambda_p^{h2}<Z_q^{h1}, Z_p^{h2}> \tag{5.74}$$

$$D_{pq} = -\frac{jm}{\omega\mu_0 b} P_p^{e1}(b)<Z_q^{h2}, Z_p^{e1'}/\epsilon_r^{(2)}> \tag{5.75}$$

$$E_{pq} = \frac{jm}{\omega\mu_0 b} P_p^{e2}(b)<Z_q^{h2}, Z_p^{e2'}/\epsilon_r^{(2)}> \tag{5.76}$$

$$F_{pq} = \frac{jm}{\omega\mu_0 b} Q_p^{e2}(b)<Z_q^{h2}, Z_p^{e2'}/\epsilon_r^{(2)}> \tag{5.77}$$

$$G_{pq} = P_p^{h1'}(b)<Z_q^{h2}, Z_p^{h1}> \tag{5.78}$$

$$H_{pq} = -P_q^{h2'}(b)\delta_{pq} \tag{5.79}$$

$$I_{pq} = -Q_q^{h2'}(b)\delta_{pq} \tag{5.80}$$

$$J_{pq} = P_p^{h2}(a)\lambda_p^{h2}<Z_q^{h3}, Z_p^{h2}> \tag{5.81}$$

$$K_{pq} = Q_p^{h2}(a)\lambda_p^{h2}<Z_q^{h3}, Z_p^{h2}> \tag{5.82}$$

$$L_{pq} = -T_q^{h3}(a)\lambda_q^{h3}\delta_{pq} \tag{5.83}$$

$$M_{pq} = -\frac{jm}{\omega\mu_0 a} P_p^{e2}(a)<Z_q^{h2}, Z_p^{e2'}/\epsilon_r^{(2)}> \tag{5.84}$$

$$N_{pq} = - \frac{jm}{\omega\mu_0 a} Q_p^{e2}(a)\langle Z_q^{h2}, Z_p^{e2'}/\epsilon_r^{(2)}\rangle \tag{5.85}$$

$$P_{pq} = \frac{jm}{\omega\mu_0 a} S_p^{e3}(a)\langle Z_q^{h2}, Z_p^{e3'}/\epsilon_r^{(2)}\rangle \tag{5.86}$$

$$R_{pq} = P_q^{h2'}(a)\delta_{pq} \tag{5.87}$$

$$S_{pq} = Q_q^{h2'}(a)\delta_{pq} \tag{5.88}$$

$$T_{pq} = - T_p^{h3'}(a)\langle Z_q^{h2}, Z_p^{h3}\rangle \tag{5.89}$$

The coefficients not appearing in (5.53) can be determined from

$$\left(\begin{array}{c} |A^{e1}\rangle \\ \hline |A^{e3}\rangle \end{array}\right) = - \left(\begin{array}{c|c} \underset{\sim}{a}^{-1}\underset{\sim}{b} & \underset{\sim}{a}^{-1}\underset{\sim}{c} \\ \hline \underset{\sim}{\ell}^{-1}\underset{\sim}{j} & \underset{\sim}{\ell}^{-1}\underset{\sim}{k} \end{array}\right) \left(\begin{array}{c} |A^{e2}\rangle \\ \hline |B^{e2}\rangle \end{array}\right) \tag{5.90}$$

$$\left(\begin{array}{c} |A^{h1}\rangle \\ \hline |A^{h3}\rangle \end{array}\right) = - \left(\begin{array}{c|c} \underset{\sim}{A}^{-1}\underset{\sim}{B} & \underset{\sim}{A}^{-1}\underset{\sim}{C} \\ \hline \underset{\sim}{L}^{-1}\underset{\sim}{J} & \underset{\sim}{L}^{-1}\underset{\sim}{K} \end{array}\right) \left(\begin{array}{c} |A^{h2}\rangle \\ \hline |B^{h2}\rangle \end{array}\right) \tag{5.91}$$

The matrix inversions in (5.53) and (5.90) to (5.91) are trivial, since they only involve diagonal matrices. We also note that the inner products appearing in the matrix elements involve only trigonometric functions (Appendix 5.A) and can be integrated analytically. The expressions are tedious, however, and are not included here.

In the case of axisymmetric modes ($m = 0$) the lower left and upper right quarters of the matrix of the system (5.53) become zero, hence it can be broken into two independent systems of lower dimensions, corresponding to TM and TE modes of the DR. In that case, the zeros of the determinant of the upper left quarter of the matrix yield the resonant frequencies of the TM modes, and the zeros of the determinant of the lower right quarter of the matrix give the resonant frequencies of the TE modes.

In the absence of losses, resonant frequencies of a shielded DR are real (Q factors are infinite). In that case, the zeros of the matrix determinant of the system (5.53) can be determined by the secant method [12], or a similar iterative procedure which does not require the knowledge of the derivative of the function whose zeros are being sought. The iteration is terminated when the change between two consecutive computed frequencies is less than some prescribed small value. As a rule, these procedures must be supplied with two values of frequency (starting points) reasonably close to the resonant frequency of the mode of interest. These starting points can be obtained by a preliminary analysis of the DR using one of the simple, approximate models (Ch. 4), or simply by plotting the determinant as a function of frequency in a specified frequency range and detecting its sign changes.

As was mentioned earlier, resonant frequencies of leaky modes of a parallel-plate waveguide resonator are complex-valued and, consequently, they can be found by searching for the zeros of the determinant of (5.53) in the complex plane by means, for example, of Muller's method [12]. However, since the shield is usually designed to prevent the leakage of energy by radiation, the leaky modes are of little practical importance.

The determinant of the matrix in (5.53) can be computed by the Gaussian elimination with partial pivoting (row interchanges) [12]. After the matrix is triangularized at a computed resonant frequency by the Gaussian procedure, the last diagonal element is typically several orders of magnitude less than the other diagonal elements (theoretically, this element should be zero). Hence, one can discard the last equation, assign an arbitrary value to the last unknown coefficient and compute the remaining unknowns by back substitution. If desired, the computed coefficients can then be renormalized, for example, so that the maximum coefficient magnitude is one. We are at liberty to do so because at a resonant frequency the matrix in (5.53) is singular and the solution is only determined to within a multiplicative constant. The coefficients not included in (5.53) are obtained from (5.90) and (5.91). Of course, these coefficients are only needed if the modal field distribution is to be computed. The modal fields can be determined by adding, in each partial region, the corresponding field components given by (5.34) to (5.35) and (5.50) to (5.51). The plots of the modal fields are often indispensable when one tries to classify the mode, since only the first mode subscript, m, is known a priori.

The radial mode matching approach was first applied to the analysis of cylindrical dielectric resonators by Kobayashi, et al. [7] and, independently, by Crombach and Michelfeit [6]. The first group of authors consider a pillbox resonator located at the center of a cylindrical PEC cavity or placed symmetrically in a parallel-plate waveguide. The symmetry of the structure is exploited in the formulation and modes which are odd and even with respect to the symmetry plane are computed separately. The TE, TM, and hybrid modes are considered, but the results presented are for $\epsilon_r > 90$. More recently, Kobayashi and Miura [13] used the radial mode matching technique [7] to determine the optimum dimensions for obtaining the best separation of the resonant frequencies of the neighboring modes from the resonant frequency of the mode of interest for ring and pillbox DRs. They did this for the $TE_{01\delta}$ and $HE_{11\delta}$ modes, which are often used in microwave filters.

The Crombach and Michelfeit [6] analysis is not limited to a symmetric resonator and allows for the presence of the substrate. These authors consider both the pillbox and ring resonators enclosed by a cylindrical cavity in the general, $m \neq 0$ case. Field plots are presented in [6] for the quasi-TE_{021} mode of the cavity containing a pillbox resonator on a substrate for several different radii of the DR. Mode charts are also given showing the dependence of the resonant frequencies of several lowest-order modes of a shielded DR on the radius of the dielectric sample for constant cavity dimensions. Similar plots are given for the $TE_{01\delta}$ mode of a pillbox DR and a ring DR with the height of the cavity and the radius of the inner hole as parameters, respectively. Some of the results are quoted in Sec. 5.7.

Convergence studies presented by Kobayashi et al. [7,13] indicate a rapid convergence of the resonant frequency for the $TE_{01\delta}$ mode, and much slower convergence rates for the $TM_{01\delta}$ and $EH_{11\delta}$ modes. This phenomenon can be attributed to the predominance in the latter group of the E_z field component, which is singular near the edges of the dielectric sample [14,15]. Since the functions used in the expansion of the field at the side surface of the DR are continuous in z, they are not well-suited for representing this singular behavior and, consequently, an excessive number of terms is required to approximate the field distribution. This has been recognized in a recent paper by Kuznetsov et al. [15], who expand E_z on the boundary between the partial regions in terms of Gegenbauer polynomials with a weighting function, which takes into account the special features of the field behavior. Data presented in the paper indicate that with this approach resonant frequencies of all modes converge at a similar, fast rate. Several plots are presented in [15] showing the frequency variation of the lowest TE, TM, EH, and HE modes as functions of the resonator dimensions and the size of the cavity. The effect of the substrate is taken into account.

Finally, we mention the papers by Komatsu and Murakami [16], Maystre et al. [17], and by Maj and Pospieszalski [18], which only consider the axisymmetric TE modes. The first two authors [16] use the radial mode-matching method as an intermediate step in the analysis of the coupling between a microstrip line and a pillbox DR on a substrate in a parallel-plate waveguide.

Maystre et al. [17] limit their attention to a parallel-plate resonator without substrate. Only a few numerical results are presented, showing the effect of the air gap on the resonant frequency of the $TE_{01\delta}$ mode. These results compare favorably with experimental data, which are also included.

Maj and Pospieszalski [18] allow in their analysis for up to three partial regions (cf. Fig. 5.1), which makes it possible to treat ring as well as pillbox resonators. In each partial region, they allow for an arbitrary number of layers, hence the substrate and double resonators can be accommodated. Numerical and experimental results are presented in the paper for the lowest TE mode for a double resonator in a parallel-plate waveguide.

5.3 Axial Mode Matching and Other Methods

The radial mode matching method described in detail in the previous section is based on the partition of the resonator cross section into partial regions in which the permittivity ϵ_r is independent of the radial coordinate ρ (Fig. 5.1). Each partial region is then considered as a section of a dielectrically loaded radial waveguide. In a complementary procedure, which will be referred to as the axial mode-matching method, the resonator is divided into partial regions in which ϵ_r is independent of the axial coordinate (i.e., z), as illustrated in Fig. 5.3. It is noted that all partial regions so defined may be regarded

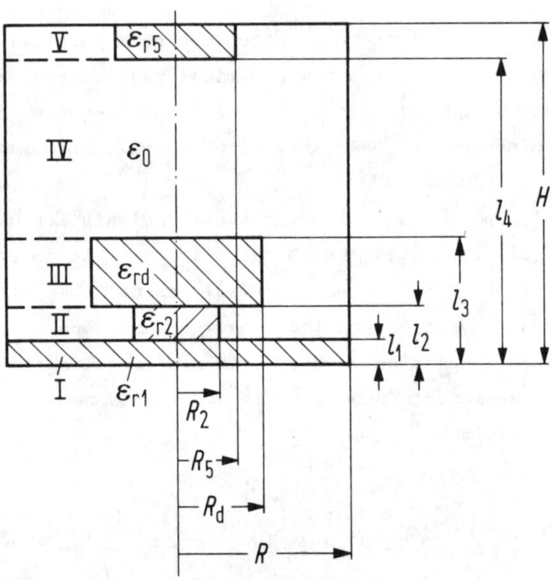

Fig. 5.3 Shielded dielectric resonator (III) including substrate (I), support (II), and tuning post (V) (reference [23], ©1984 AEU)

as sections of dielectrically-loaded (regions II, III, and V) or homogeneous (regions I and IV) cylindrical waveguides. Therefore, the transverse fields in each partial region i, i = I,...,V, can be expanded in terms of cylindrical waveguide modes as [19]

$$\underline{E}^i_t = \sum_{n=1}^{N} [A^i_n \cos\beta^i_n z + B^i_n \sin\beta^i_n z]\ \underline{E}^i_{tn} \qquad (5.92a)$$

$$\underline{H}_t^i = -j \sum_{n=1}^{N} [A_n^i \cos\beta_n^i z - B_n^i \sin\beta_n^i z] \, \underline{H}_{tn}^i \qquad (5.92b)$$

where β_n^i are the modal propagation constants, $\underline{E}_{tn}^i$ and $\underline{H}_{tn}^i$ are the transverse field expansion functions, and A_n^i and B_n^i are the field coefficients. At a given frequency, the propagation constants β_n^i must be found from a transcendental equation involving Bessel functions of order m and their derivatives, obtained by enforcing the continuity of the tangential field components inside each partial region and the condition of vanishing tangential electric field on the metallic enclosure. This equation, which also arises in the analysis of an infinite, partially filled cylindrical waveguide [20,21], is quite complicated due to the fact that, except for rotationally symmetric fields, the modes are neither TE or TM to any cylindrical coordinate. Hence, unlike in the radial mode matching method (Sec. 5.2), one has to deal with hybrid modes in all partial regions except for those which are homogeneous.

When the field expansions (5.92) for adjacent partial regions are matched at common planar interfaces, a system of simultaneous algebraic equations is obtained for the expansion coefficients. The resonant frequencies of the DR can then be found as zeros of the system determinant, as in the radial mode matching method.

The axial mode-matching method was first described by Hong and Jansen [19], but they only implemented it for the axisymmetric, m = 0 case. According to the authors, this restriction was imposed to avoid complications introduced by the possible occurrence in some frequency ranges of complex-valued (i.e., having both the real and imaginary parts) propagation coefficients in some partial regions, as predicted by Clarricoats and Taylor [22]. Results are presented in [19] and in a later paper [23], showing the influence of the substrate, the shielding, and a dielectric disc support on the frequencies of various TE and TM modes of the resonator shown in Fig. 5.3. We quote some of these results in Sec. 5.7.

More recently, Zaki and Atia [21] and Zaki and Chen [8] have applied the axial mode matching method to determine the TE, TM, and hybrid modes of a DR placed symmetrically in a cylindrical cavity. The same geometry was previously analyzed by Kobayashi et al. [7] by the radial mode-matching method. Zaki and Chen present plots of field distributions of several low-order modes in the equatorial symmetry plane and in the end plane of the cavity. A mode chart is also given in [8] showing the dependence of the resonant frequencies of various modes as a function of the cavity height. The computed results are shown to agree very well with measured data. The authors claim that their method converges faster than the radial mode matching method used by Kobayashi et al. [7].

Another rigorous approach, which can be classified as a mode-matching method (although the author himself refers to it as a differential method), is due to Vincent [24]. This method is apparently applicable to both the TE and TM axisymmetric modes of a shielded resonator, but has only been implemented for the former. The only restriction on the shape of the DR is that it must be a body of revolution with respect to the z-axis. Furthermore, the permittivity ϵ_{rd} of the dielectric sample can be a function of both ρ and z. In this procedure, the cavity is divided into two complementary partial regions by an artificial cylindrical surface of radius R, which encloses the dielectric sample and extends from the bottom to the top plate of the shield. The E_ϕ and H_z components of the fields are then expanded in both regions in Fourier sine series in z. In the exterior region, which is homogeneous, the expansion coefficients can be expressed in terms of Bessel functions. In the inhomogeneous interior region, these coefficients are shown to satisfy a set of coupled first order ordinary differential equations, which is derived by expanding $\epsilon_r(\rho,z)$ in a Fourier cosine series in z and by exploiting the orthogonality of Fourier harmonics. Subsequently, the coefficients of the interior region obtained by N simultaneous integrations of this system by a predictor-corrector formula starting from N arbitrary linearly independent values on the z axis are matched at $\rho = R$ with the corresponding coefficients of the exterior region, resulting in an $N \times N$ set of homogeneous algebraic equations. The zeros of the system determinant yield the resonant frequencies of the resonator. A plot is presented in the paper [24] showing the variation of the resonant frequency of the $TE_{01\delta}$ mode of a disc DR in a parallel-plate waveguide as a function of the width of the air gap. Also, plots are given showing the change in the resonant frequency of a dielectric sphere versus the distance between the plates and versus the radius of the sphere. The author claims that this method is highly accurate for resonators with dielectric constants not exceeding 35. The presented results were obtained with N not exceeding 11 and with 50 or less integration steps.

We conclude this section with a brief description of the approach of Tsuji et al. [25,26] to the analysis of isolated DRs. These authors expand the fields inside and outside a pillbox resonator in terms of spherical wave functions [20] and match the tangential field components $\underline{E}_t$ and $\underline{H}_t$ of the two regions at the DR's surface. Since this surface is not spherical, the orthogonality of the modes cannot be exploited in the computation of the modal coefficients. Therefore, rather than trying to match the fields exactly, these authors fit the fields to the boundary conditions in the least-squares sense. For this purpose, the mean square error E is defined as

$$E = \int_{\Gamma} \left(|\underline{E}_{t1} - \underline{E}_{t2}|^2 + \eta_1^2 |\underline{H}_{t1} - \underline{H}_{t2}|^2 \right) d\ell \qquad (5.93)$$

where the subscripts 1 and 2 refer, respectively, to the interior and exterior regions, and η_1 is the intrinsic impedance of region 1. It is noted that, because of the axial symmetry of the structure, the integral over the whole surface of the DR could be reduced in (5.93) to just an integral along the boundary contour Γ of the resonator at an arbitrary ϕ-coordinate. The field expansions, truncated to N terms, are substituted into (5.93) and the derivatives of E with respect to the modal coefficients are set to zero, as in the Rayleigh-Ritz procedure [1]. Nontrivial solutions of the resulting system of homogeneous algebraic equations yield the coefficients which minimize the error E. The resonant frequencies can be found by searching for the zeros of the system determinant in the complex angular-frequency plane. Having found a complex zero $s = \sigma + j\omega$, one computes the resonant frequency f_0 and the intrinsic quality factor Q_0 due to radiation loss as

$$f_0 = \frac{|s|}{2\pi}, \qquad Q_0 = \frac{|s|}{2\sigma} \qquad (5.94)$$

Using this technique, Tsuji et al. [25,26] computed resonant frequencies and Q factors of several lower-order TE, TM, HE, an EH modes for pillbox resonators with different aspect ratios. Their results compare very well with measured data, which are also included. Convergence studies reported in [26] indicate that as few as ten terms in the series are required for convergence in the case of the HE modes, and about twice as many in the case of the EH modes. The slower convergence rate for the latter mode group is attributed to the predominance in the EH modes of the E_z field component, which is singular at the edges of the DR [14,15]. This method is also applicable to shapes other than pillbox, provided they are axisymmetric.

5.4 Finite-Element and Finite-Difference Methods

Finite-element and finite-difference methods have only been applied in the case of the ϕ-independent TM and TE modes. The field equations for those modes can be easily obtained by putting m = 0 in (5.16) and (5.41), respectively. It is customary, however, to express the field components of the TM (TE) modes in terms of H_ϕ (E_ϕ), rather than ψ^e (ψ^h). Hence, in the TM case we have, from (5.15) and (5.16),

$$E_\rho = -\frac{1}{j\omega\epsilon_0\epsilon_r}\frac{\partial}{\partial z}H_\phi \tag{5.95a}$$

$$E_z = \frac{1}{j\omega\epsilon_0\epsilon_r\rho}\frac{\partial}{\partial\rho}(\rho H_\phi) \tag{5.95b}$$

$$(D + k_0^2\epsilon_r)H_\phi = 0 \tag{5.95c}$$

where we defined for future reference operator D as

$$D \equiv \frac{1}{\rho}\frac{\partial}{\partial\rho}\left(\rho\frac{\partial}{\partial\rho}\right) - \frac{1}{\rho^2} + \frac{\partial^2}{\partial z^2} \tag{5.96}$$

The equations (5.95) are valid in each region over which ϵ_r is constant, but do not hold at interfaces across which ϵ_r changes discontinuously [27]. However, we may solve for the fields in each homogeneous region and match the tangential components at the interfaces to obtain a solution that is valid everywhere.

In the TE case, we have, from (5.40) and (5.41),

$$H_\rho = \frac{1}{j\omega\mu_0}\frac{\partial}{\partial z}E_\phi \tag{5.97a}$$

$$H_z = -\frac{1}{j\omega\mu_0\rho}\frac{\partial}{\partial\rho}(\rho E_\phi) \tag{5.97b}$$

and

$$(D + k_0^2\epsilon_r)E_\phi = 0 \tag{5.97c}$$

In the remainder of this section we will limit attention to the TE case, since the development for the TM case is similar.

In the finite-element method [28,29], the resonator cross section (Fig. 5.1) is subdivided into a finite number of patches or "finite elements," usually of triangular shape, and in each patch the unknown is represented as a linear combination of interpolatory polynomials N_i. Hence, if we put $\alpha = E_\phi$ for notational simplicity, the value of α anywhere within a triangular finite element may be written as

$$\alpha(\rho,z) \simeq \sum_{i=1}^{M} c_i N_i(\rho,z) \tag{5.98}$$

where c_i are the unknown coefficients, which are the values of α at the M specified points (nodes) of the triangle. Owing to the rotational symmetry of the resonator (Fig. 5.1), it is sufficient to triangularize only one-half of the meridian cross section. The permittivity inside each element must be constant. The polynomials N_i are constructed such that α is continuous across the element boundary, so that the field continuity conditions are automatically satisfied. If first-order polynomials are employed (M = 3 in that case), (5.98) is called a first-order element. Usually, it is more efficient to use second- or higher-order finite elements [29].

There exist several different finite-element formulations, the variational method and the weighted-residual method being the most popular. In the former, use is made of the functional:

$$F(\alpha) = - \iiint_V \alpha \, (D + k_0^2 \epsilon_r)\alpha \, dV \tag{5.99}$$

which is stationary about the solution of (5.97c) [2]. Here, the integral is over the resonator volume V. Since α is ϕ-independent, the integration in ϕ is trivial and the remaining double integral can be transformed by integration by parts as

$$F(\alpha) = \iint_C \left\{ \left(\frac{\partial \alpha}{\partial \rho}\right)^2 + \left(\frac{\partial \alpha}{\partial z}\right)^2 + \left(\frac{1}{\rho^2} - k_0^2 \epsilon_r\right)\alpha^2 \right\} \rho d\rho dz$$

$$+ \int_c \alpha \frac{\partial \alpha}{\partial n} \rho \, dc \tag{5.100}$$

where we omitted a multiplicative factor of 2π, which is of no consequence. The first integral in (5.100) is over the resonator cross section C, and the second is along its boundary contour c, which has a unit normal $\hat{n}$. In the case of a DR in a PEC cavity (Fig. 5.1), the second integral is identically zero, because $E_\phi = 0$ on c. For an open resonator, this term vanishes likewise on account of the radiation condition, but the first integral is over the infinite space.

Substituting (5.98) into the functional (5.100) and imposing the Rayleigh-Ritz conditions [1]:

$$\partial F(\alpha)/\partial c_i = 0 \tag{5.101}$$

leads to the generalized matrix eigenvalue problem:

$$\underset{\sim}{A}|c\rangle = k_0^2\, \underset{\sim}{B}|c\rangle \tag{5.102}$$

where $\underset{\sim}{A}$ and $\underset{\sim}{B}$ are square matrices of known coefficients. The derivatives in (5.101) are only taken with respect to the coefficients of interior nodes, since the coefficients of the boundary nodes and the nodes on the z-axis are known to be zero. The system (5.102) can be solved for the eigenvalues k_0^2 and the corresponding eigenvectors $|c\rangle$ by using a standard computer subroutine. The resonant frequencies of the resonator are then easily obtained from the eigenvalues, and the eigenvectors yield the coefficients, which can be used in (5.98) to recover the corresponding modal field distributions.

In the weighted-residual formulation, the approximation (5.98) is substituted into the differential equation (5.97c) and the residual is minimized with respect to a set of suitably chosen weighting functions by the Rayleigh-Ritz procedure. When the same functions N_i are selected for the expansion and weight functions, this method is called the Galerkin weighted-residual approach, and can be shown to be equivalent to the variational method described above.

To achieve sufficient accuracy, hundreds of finite elements are typically required. However, the matrices in (5.102) are sparse (i.e., most of the entries are zeros) due to the fact that the element of the ith row and jth column is non-zero only if i and j correspond to nodes of the same triangle.

The finite-element method was applied in the present context by Gil and Gismero [30], Gil and Perez [31], and by Kooi et al. [32]. Gil et al. [30,31] have used first- and higher-order rectangular elements to solve the problem of TE and TM modes of a shielded DR with the effects of the substrate, the dielectric support (spacer), and the tuning screw taken into account. They present results for a double resonator, ring resonator, and pillbox resonator, which are in excellent agreement with the results of previous analyses [18,13,17] by mode-matching methods. They also present new results showing the change in the resonant frequency of the $TE_{01\delta}$ mode versus the diameter and depth of penetration of a tuning screw.

The Kooi et al. [32] analysis is limited to the fundamental TE mode of a shielded pillbox DR on a substrate. They employ only first-order triangular elements, hence the computed results are not very accurate. The resonant frequency of the $TE_{01\delta}$ mode is plotted as a function of many parameters, such as the substrate thickness, the resonator height, and the size of the cavity. Some measured data are also included for comparison.

In the finite-difference method, the operator itself, rather than its domain, as in the finite-element method, is discretized. These two techniques bear many similarities, however. In applying the finite-difference method to the resonator of Fig. 5.1, we introduce a rectangular grid of nodes (i,j) in one half of the meridian cross section and replace the differential operator in (5.97c) by the usual finite difference approximation [2]. If we distinguish quantities associated with a node (i,j) by the subscripts ij in a region with dielectric constant ϵ_r, we obtain

$$\frac{\alpha_{i+1,j} - 2\alpha_{ij} + \alpha_{i-1,j}}{(\Delta\rho)^2} + \frac{\alpha_{i+1,j} - \alpha_{i-1,j}}{2\rho_i\Delta\rho} + \frac{\alpha_{i,j+1} - 2\alpha_{ij} + \alpha_{i,j-1}}{(\Delta z)^2} + \left(k_0^2\epsilon_r - \frac{1}{\rho_i^2}\right)\alpha_{ij} = 0 \qquad (5.103)$$

where the distance between adjacent points of the grid is taken to be $\Delta\rho$ along the radial direction, and Δz along the z-axis. If we introduce

$$\rho_i = i\Delta\rho \ , \quad R = \frac{\Delta\rho}{\Delta z} \ , \quad \text{and} \quad \lambda = k_0^2(\Delta\rho)^2 \qquad (5.104)$$

we can rewrite (5.103) as

$$\alpha_{ij} \frac{1 + 2i^2(R^2 + 1)}{i^2 \epsilon_r} - \alpha_{i+1,j} \frac{1 + 2i}{2i\epsilon_r} - \alpha_{i-1,j} \frac{2i - 1}{2i\epsilon_r}$$

$$- \alpha_{i,j+1} \frac{R^2}{\epsilon_r} - \alpha_{i,j-1} \frac{R^2}{\epsilon_r} = \lambda \alpha_{ij} \qquad (5.105)$$

These equations must be properly modified for grid points adjacent to the metallic cavity walls, to account for the short-circuit boundary conditions there [33,34]. When the equations for all grid points (i,j) are collected, we obtain the matrix eigenvalue equation:

$$\underset{\sim}{A}|\alpha\rangle = \lambda|\alpha\rangle \qquad (5.106)$$

where $\underset{\sim}{A}$ is a non-symmetric band matrix. This equation can be solved for the eigenvalues λ and the corresponding eigenvectors $|\alpha\rangle$ by standard methods.

The finite-difference procedure summarized here was used by Guillon and his coworkers [33,34,35]. Computed results are presented in [33,34] for the TM and TE modes of pillbox, ring, and double DRs in a cylindrical cavity. In several cases, experimental data are included and are shown to compare very well with the computed results.

5.5 Perturbational Methods

In discussing the perturbational-asymptotic series technique, we will refer to the ring dielectric resonator illustrated in Fig. 5.4 (of which the pillbox is a special case). Only one half of the meridian cross section of the rotationally symmetric structure is shown in the figure. The DR resides in free space and is characterized by the

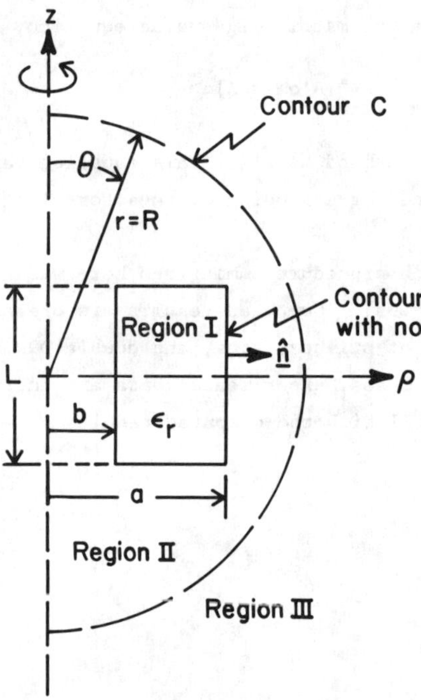

Fig. 5.4 Ring resonator in free space

relative dielectric constant ϵ_r. It will also be convenient in the subsequent discussion to also introduce the index of refraction N as

$$N = \sqrt{\epsilon_r}\ , \quad k_0 = \frac{k}{N} \tag{5.107}$$

The meridian plane is subdivided into regions I, II, and III, as illustrated in the Fig. 5.4. It should be noted that regions II and III have the same dielectric constant of free space, hence the circular boundary C separating them is merely an artifice introduced in the analysis.

We will limit attention to the axisymmetric TE modes. Hence, putting $\alpha = E_\phi$ and referring to (5.97c), we see that α must satisfy

$$\begin{cases} D\alpha + k^2\alpha = 0 \,, & \text{in region I} \qquad (5.108a) \\ D\alpha + \dfrac{k^2}{N^2}\,\alpha = 0 \,, & \text{in regions II and III} \qquad (5.108b) \end{cases}$$

In addition to satisfying (5.108), α and $\partial\alpha/\partial n$ must be continuous on the generating contour Γ, which has an outward normal $\hat{n}$. Also, α must be zero on the z-axis and at infinity. Equations (5.108), subject to these boundary conditions, constitute an eigenvalue problem, which must be solved for the resonant wave number k and the corresponding modal function α (the number of such solutions is infinite, but we will be only interested in one or two lowest, dominant modes). Having found k, one can determine the resonant frequency from

$$f_0 = \frac{c}{2\pi}\,\frac{k}{N} \qquad (5.109)$$

where c is the velocity of light in vacuum. The corresponding modal function α is substituted in (5.97) to obtain the fields, from which one can determine the Q factor due to radiation losses from (see Ch. 2)

$$Q = \frac{kc}{N}\,\frac{W}{P_r} \qquad (5.110)$$

where W is the total stored field energy and P_r is the radiated power.

In the perturbational approach, which is sometimes referred to as the Van Bladel's method after its originator [36], the eigenvalue problem (5.108) is solved iteratively by expanding α and k in the asymptotic power series in inverse powers of N as

$$\alpha = \alpha_0 + \frac{1}{N^2}\,\alpha_2 + \frac{1}{N^4}\,\alpha_4 + \cdots \qquad (5.111a)$$

$$k^2 = k_0^2 + \frac{1}{N^2}\,(k^2)_2 + \frac{1}{N^4}\,(k^2)_4 + \cdots \qquad (5.111b)$$

where the functions $\alpha_0, \alpha_2, \ldots,$ and the coefficients $k_0^2, (k^2)_2, \ldots,$ are independent of N. It can be shown [37] that only even inverse powers of N are present in (5.111). The coefficient k_0 appearing in (5.111b) is the zero order approximation to k, and must not be confused with the free-space wave number, which is not explicitly used in the remainder of this section. The series (5.111) is inserted into (5.108) and the coefficients of equal powers of 1/N on both sides are equated, yielding the system of equations which the expansion coefficients must satisfy. Thus, equating the coefficients of zero powers of 1/N leads to the following equation for α_0 [37]:

$$\begin{cases} D\alpha_0 + k_0^2\alpha_0 = 0\,, & \text{in region I} \qquad (5.112a) \\ D\alpha_0 = 0\,, & \text{in regions II and III} \qquad (5.112b) \end{cases}$$

Similarly, equating the coefficients of $1/N^2$ yields the equation for α_2 [37]:

$$\begin{cases} D\alpha_2 + k_0^2\alpha_2 = -\,(k^2)_2\alpha_0\,, & \text{in region I} \qquad (5.113a) \\ D\alpha_2 = -\,k_0^2\alpha_0\,, & \text{in regions II and III} \qquad (5.113b) \end{cases}$$

Using the expansions (5.111), one can represent the Q factor (5.110) as [37]

$$Q = Q_0\left(1 + \frac{1}{N^2}\frac{Q_2}{Q_0} + \cdots\right) \tag{5.114}$$

with the leading term given by [37,38]

$$Q_0 = \frac{12N^3}{k_0^5}\,\frac{\displaystyle\iint_I \alpha_0^2\rho\,d\rho\,dz}{\left\{\displaystyle\iint_I \alpha_0\rho^2\,d\rho\,dz\right\}^2} \tag{5.115}$$

where the integration is over the DR cross section (region I in Fig. 5.4).

Equations (5.112) can be solved by the finite-element method discussed in the previous section. In the present case, with $N \to \infty$, the functional (5.100) takes the form:

$$F(\alpha) = \iint_{I} \left\{ \left(\frac{\partial\alpha}{\partial\rho}\right)^2 + \left(\frac{\partial\alpha}{\partial z}\right)^2 + \left(\frac{1}{\rho^2} - k^2\right)\alpha^2 \right\} \rho d\rho dz$$

$$+ \iint_{II+III} \left\{ \left(\frac{\partial\alpha}{\partial\rho}\right)^2 + \left(\frac{\partial\alpha}{\partial z}\right)^2 + \frac{\alpha^2}{\rho^2} \right\} \rho d\rho dz \qquad (5.116)$$

The application of the finite-element method to the configuration of Fig. 5.4 is complicated by the fact that the region of interest is of infinite extent. To overcome this difficulty, Verplanken and Van Bladel [38] postulate zero field outside a finite spherical region of radius R, where R is chosen large compared with the dimensions of the DR (R = 10a, for example [39]). Consequently, they triangularize only regions I and II, and set α_0 to zero in region III (Fig. 5.4). This rather crude approximation may result in poor accuracy of the solution. In a more refined approach, DeSmedt [37] approximates the field in region III by a finite series of static spherical harmonics of the form:

$$\alpha_0 = \sum_{i=1}^{m} \frac{A_i}{r^{i+1}} \frac{P_i^1(\cos\theta)}{\sin\theta}, \quad \text{in region III} \qquad (5.117)$$

The unknown coefficients A_i are obtained by enforcing the continuity of this series and the finite-element basis functions on the contour C (Fig. 5.4) (the boundary integral appearing in (5.100) must be retained in the functional (5.116) in this case).

Having solved the system (5.112) for k_0 and α_0, one can compute the leading terms in the asymptotic expansions for the resonant frequency and the Q factor from (5.109) and (5.115), respectively. Using this method, Verplanken and Van Bladel obtained results for the TM [39] and TE [38] fundamental modes of ring and pillbox DRs. In the TE case, this zero-order theory does not yield sufficient accuracy for the values of ϵ_r commonly used in DRs. Comparison with the known analytical solution of the spherical resonator shows the accuracy to be, for the lowest mode and ϵ_r = 100, on the order of 1 % for the resonant frequency and 10 % for the quality factor. For ϵ_r = 35, which is a value often used in practice, the accuracies worsened by a factor on the order of two [38].

Recently, DeSmedt [37] extended the limits of applicability of the perturbational approach by evaluating higher-order (correction) terms in the asymptotic series (5.111) and (5.114). For example, the coefficient $(k^2)_2$ in (5.112b) is shown to be given by [37]

$$(k^2)_2 = -k_0^2 \frac{\iint_{II+III} \alpha_0^2 \rho d\rho dz}{\iint_{I} \alpha_0^2 \rho d\rho dz} \tag{5.118}$$

With α_0, k_0^2, and $(k^2)_2$ determined, the finite-element method can be employed to solve the system (5.113) for α_2, which gives the second term in the asymptotic series for the modal function α. The solution of (5.113) is unique if one imposes the proper orthogonality condition on α_0 and α_2 [37]. With α_2 determined, the higher order coefficients $(k^2)_4$ and Q_2 can be computed from rather lengthy formulas given in [37], and can be used in (5.109) and (5.114) to compute the corrected values of f_0 and Q, respectively. DeSmedt [37] reports that for the spherical DR, using one term, two terms, or three terms in the expansion (5.111b) for k, respectively, gives a relative error of about 2.1, 0.54, or 0.17 percent at $\epsilon_r = 39$. Also, including one or two terms in the expansion (5.114) for Q yields accuracies of 19 % or 1.3 %, respectively, for the same DR. To achieve these accuracies, a total of 70 third-order finite elements in regions I and II, and nine terms in the expansion (5.117) in region III were required [37].

The perturbational method has the remarkable feature that the calculations do not have to be repeated for every new value of ϵ_r. Therefore, once the systems (5.112) and (5.113) are solved for the specified dimensions a, b, and L (Fig. 5.4), one can compute the resonant frequency from (5.109) and the Q factor from (5.114) for any (but sufficiently large) value of ϵ_r. Also, one can quickly investigate, without additional computations, the effect of variations of ϵ_r on the properties of the resonator. For example, from (5.109) and (5.111b), one can easily derive the relative shift of the resonant frequency due to a variation in ϵ_r to be [37]

$$\frac{\Delta f_0}{f_0} \simeq -\frac{1}{2}\left(1 + \frac{(k^2)_2}{\epsilon_r k_0^2} + \cdots\right)\frac{\Delta\epsilon_r}{\epsilon_r} \tag{5.119}$$

The field structure of the lowest, ϕ-independent TE mode of a DR closely resembles that generated by a magnetic dipole [38]. For this mode, E_ϕ, and thus α, is an even function with respect to the equatorial plane (Fig. 5.4), i.e., $\partial\alpha/\partial z = 0$ at $z = 0$, which is the magnetic wall condition. The Q factor due to radiation of the magnetic dipole mode is proportional to N^3, as can be seen from (5.115). The lowest TE mode for which $\alpha = 0$ at $z = 0$, i.e., the mode which satisfies the electric wall condition on the equatorial plane, has a field structure similar to that of a magnetic quadrupole. It can be shown [37] that the Q factor of the magnetic quadrupole mode is proportional to N^5. DeSmedt has found both the magnetic dipole and quadrupole modes for a DR in free space [37], above an electric or magnetic wall [40], and inside a cylindrical waveguide [41]. We quote some his results for the free space case in Sec. 5.7.

The TE modes are sometimes termed the "unconfined modes" [36], because in the limit, as $\epsilon_r \to \infty$, their magnetic field does not vanish on the surface of the resonator. In contrast to the TE modes, the ϕ-independent TM modes are of the "confined" type, because as $\epsilon_r \to \infty$, their magnetic field is zero on the DR's surface. The lowest modes of this variety have a field structure similar to that of an electric dipole or electric quadrupole [39]. The electric dipole mode is of interest for applications because its Q factor is proportional to N^5.

Before leaving this section, we mention another perturbational technique based on an asymptotic series expansion in terms of inverse powers of N, recently developed by Gol'berg and Penzyakov [42]. In contrast to Van Bladel's approach described above, which is based on the solution of a partial differential equation, the Gol'berg and Penzyakov method is based on a volume integral equation for the electric field in the DR. Therefore, their technique has the advantage that the fields outside the dielectric sample do not have to be considered. On the other hand, the integral equation technique leads to a full matrix, whereas the matrix in the differential equation method is sparse. Gol'berg and Penzyakov [42] considered the ϕ-independent TE_{0np} modes and computed resonant frequencies and Q factors of several modes for DRs of various shapes (spherical, pillbox, ring, double), and for composite and inhomogeneous DRs.

5.6 Green's Function - Integral Equation Methods

As the name indicates, the integral equation methods are based on the solution of an integral equation, rather than a differential equation. The integral equation is derived by means of a suitable Green's function, which constitutes its kernel. There are two integral equation formulations possible [43]. In one, which employs a volume integral equation, the dielectric inhomogeneity (i.e., the DR) is replaced by equivalent polarization currents. In the other, the DR is replaced by equivalent electric and magnetic surface currents, leading to a surface integral equation. The success of either of these methods depends on one's ability to find the suitable Green's function, which is usually a dyadic [44]. The situation is simpler in the case of an open resonator, since only the free-space Green's function is required [49] (see also Ch. 6).

In the case of shielded resonators, the determination of the Green's function tends to be difficult, particularly for more complex resonators. This task can be somewhat simplified by subdividing the resonator into two regular partial regions and by erecting a PMC or PEC wall at the boundary between the regions [45,46]. Magnetic or electric current sheets are then postulated on this wall, maintaining the correct field in both regions. By enforcing the field continuity condition between the partial regions, an integral equation is obtained for the unknown current. This equation is then discretized by the method of moments [1], employing suitable expansion and testing functions. The resonant frequencies are obtained as zeros of the matrix determinant, as in the mode matching methods. Jaworski and Pospieszalski [45] have employed this method in the case of axisymmetric TE modes of a pillbox resonator placed symmetrically in a parallel-plate waveguide. The ϕ-independent TM modes of a DR in a cylindrical cavity were analyzed by this technique by Kapustin [46].

The volume integral equation method was used by Omar and Schunemann [47,48] to analyze DRs in a cylindrical cavity and in a rectangular waveguide.

5.7 Sample Numerical Results

We present in this section selected numerical results obtained by various rigorous methods which have been discussed in this review. These results illustrate how changing various parameters affects the resonant frequencies of various modes of open and shielded DRs. This section should also help the reader interpret the data presented in research papers and to alert him or her to the diverse notation used by various authors in classifying the modes. For more results the reader is referred to the papers listed in the bibliography at the end of this chapter.

The data presented in Figs. 5.5 through 5.9 are for a ring resonator in free space (see Fig. 5.4 for the geometry) and were obtained by DeSmedt [37] by the perturbational approach discussed in Sec. 5.5. The pillbox resonator case is included in these results as a special case when b/a = 0. These data show the dependence of the resonant wave number and the Q factor on the aspect ratio L/(2a) and the (normalized) size of the inner hole b/a. The solid lines pertain to the $TE_{01\delta}$ (magnetic dipole) mode and the dashed lines to the $TE_{011+\delta}$ (magnetic quadrupole) mode. Figure 5.5 gives the zero-order normalized resonant

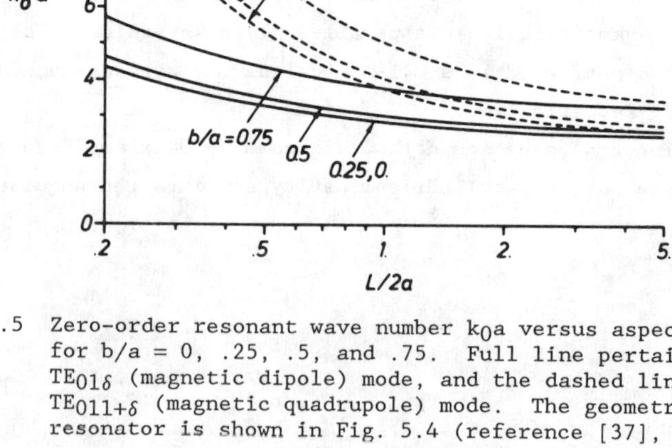

Fig. 5.5 Zero-order resonant wave number k_0a versus aspect ratio L/(2a) for b/a = 0, .25, .5, and .75. Full line pertains to the $TE_{01\delta}$ (magnetic dipole) mode, and the dashed line to the $TE_{011+\delta}$ (magnetic quadrupole) mode. The geometry of the resonator is shown in Fig. 5.4 (reference [37], ©1984 IEEE)

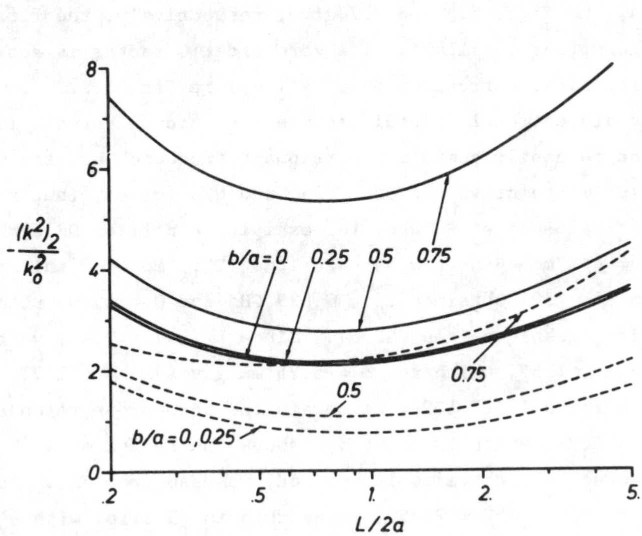

Fig. 5.6 First correction of the resonant wave number $-(k^2)_2/k_0^2$ versus aspect ratio L/(2a) for b/a = 0, .25, .5, and .75. Full line pertains to the $TE_{01\delta}$ (magnetic dipole) mode, and the dashed line to the $TE_{011+\delta}$ (magnetic quadrupole) mode. The geometry of the resonator is shown in Fig. 5.4 (reference [37], ©1984 IEEE)

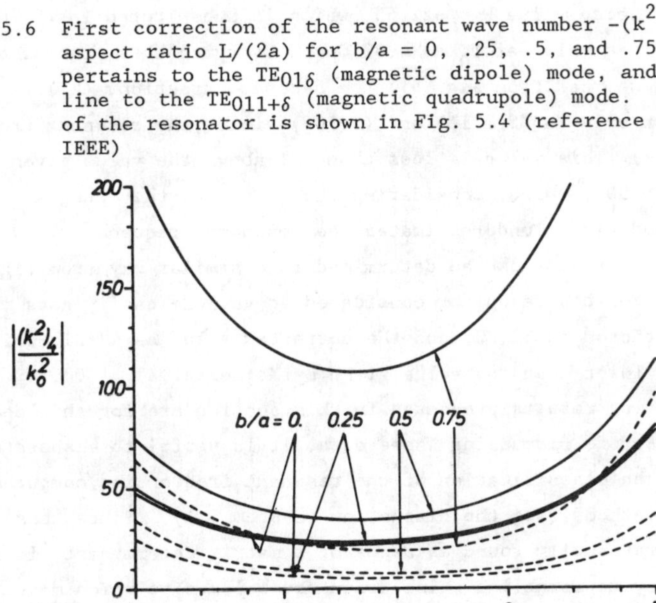

Fig. 5.7 Second correction of the resonant wave number $|(k^2)_4/k_0^2|$ versus aspect ratio L/(2a) for b/a = 0, .25, .5, and .75. Full line pertains to the $TE_{01\delta}$ (magnetic dipole) mode, and the dashed line to the $TE_{011+\delta}$ (magnetic quadrupole) mode. The correction is positive for the dipole mode and negative for the quadrupole mode. The geometry of the resonator is shown in Fig. 5.4 (reference [37], ©1984 IEEE)

wave number k_0a, and Figs. 5.6 and 5.7 give, respectively, the first and second corrections (see (5.111b)). The zero-order Q factor is shown in Fig. 5.8, and the first correction (see (5.114)) in Fig. 5.9.

These plots are extremely useful because they are universal, i.e., they can be used to easily compute the resonant frequency and the Q factor for a wide range of values of L/(2a) and b/a for any (but sufficiently large) value of ϵ_r. Take, for example, a pillbox DR with a = 5.25 mm, L = 4.6 mm, and $\epsilon_r = 38$. For the $TE_{01\delta}$ mode of this resonator Kajfez et al. [50] obtained $f_0 = 4.829$ GHz and Q = 45.8, using the integral equation technique (see Ch. 6). Since L/(2a) = 0.44, we read from Fig. 5.5 $k_0a = 3.37$, which for a = 5.25 mm gives $k_0 = 641.91$. Using this value for k in (5.109), we obtain the zero-order resonant frequency as 4.97 GHz, which is about 3 % above the Kajfez et al. [50] result. If a more accurate value is desired, we read from Fig. 5.6 the first correction, $-(k^2)_2/k_0^2 = 2.25$. Using this in (5.111b) with $\epsilon_r = 38$ gives the corrected value k = 622.65, which is substituted in (5.109) to give the improved value of f_0 as 4.82 GHz. Carrying this process one step farther, we read from Fig. 5.7 the second correction, $(k^2)_4/k_0^2 = 25$, which when used in (5.111b) and (5.109) yields the resonant frequency as 4.87 GHz. This value is less than 1 % above the value given by Kajfez et al. [50], which, considering the fact that the integral equation method always underestimates the resonant frequency, is very accurate. The Q factor can be determined in a similar way from Figs. 5.8 and 5.9. For the resonator considered above, one easily gets the zero-order Q factor as 32.80, and the corrected value as 42.31, which is less than 8 % lower than the value given by Kajfez et al. [50].

The remaining results presented in this section are for shielded resonators. Before discussing these data, it is useful to elaborate some more on the classification of the resonant frequencies computed by the rigorous methods (see the discussion in Sec. 5.1). Since these frequencies are usually found as zeros of a matrix determinant, it is not always easy to establish which frequency belongs to which mode. Perhaps the most reliable procedure, but at the same time the most tedious one, is to make detailed plots of the field structure at the given resonant frequency. From such plots, one can determine the number of field extrema within the dielectric sample in the radial and axial directions, and assign the proper set of indices to the mode in

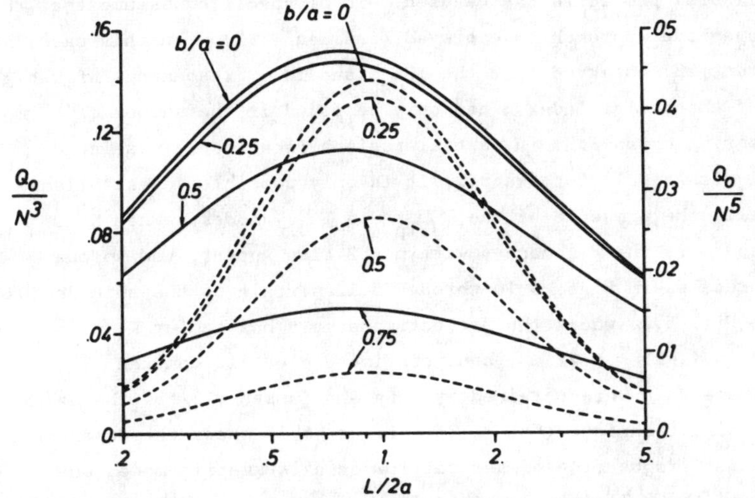

Fig. 5.8 Zero-order Q factor versus aspect ratio L/(2a) for b/a = 0, .25, .5, and .75. Full line shows Q_0/N^3 for the $TE_{01\delta}$ (magnetic dipole) mode, and the dashed line Q_0/N^5 for the $TE_{011+\delta}$ (magnetic quadrupole) mode. The geometry of the resonator is shown in Fig. 5.4 (reference [37], ©1984 IEEE)

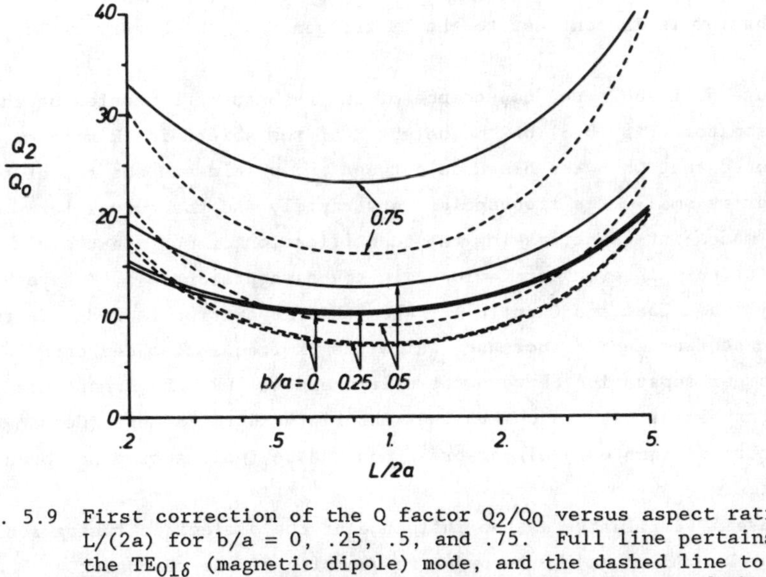

Fig. 5.9 First correction of the Q factor Q_2/Q_0 versus aspect ratio L/(2a) for b/a = 0, .25, .5, and .75. Full line pertains to the $TE_{01\delta}$ (magnetic dipole) mode, and the dashed line to the $TE_{011+\delta}$ (magnetic quadrupole) mode. The geometry of the resonator is shown in Fig. 5.4 (reference [37], ©1984 IEEE)

question. If mode matching techniques are employed, a much easier (but less reliable) procedure can be used. To be specific, assume that the axial waveguide approach is employed (see Sec. 5.3). In this case, Hong and Jansen [23] observed that the TE_{0np} and TM_{0np} resonances with higher values of the radial index n can only be found if the number of terms N in the series expansions in partial regions (see (5.92)) is large enough. With N = 1, for example, it is only possible to establish numerically the presence of the TE_{01p} and TM_{01p} modes. When N is increased to 2, the resonances with n = 2 also appear, and so on.

The results in Figs. 5.10 through 5.12 pertain to the shielded DR shown in Fig. 5.3, where the dielectric sample has radius R_d = 3.03 mm, height $H_d = \ell_3 - \ell_2$ = 4.22 mm, and relative dielectric constant ϵ_{rd} = 36.2. These data were obtained by Hong and Jansen [23] by the axial mode-matching technique (Sec. 5.3). Figure 5.10 shows the dependence of the resonant frequencies of several lowest axisymmetric modes on the radius R of the cylindrical shield in the case where the pillbox resonator lies directly on the substrate. The subscripts D and H denote, respectively, the DR-type and cavity-type resonances (H stands for "Hohlraum," which means cavity in German). It can be observed in Fig. 5.10 that if the distance of the side wall of the shield is not too close to the DR, the TE resonant frequencies only weakly depend on R. This behavior is in contrast to the TM frequencies, which vary rapidly with R.

Figure 5.11 shows the dependence of the resonant frequencies of the same resonator (Fig. 5.3) on the height H of the shield for R = 15 mm. It is noted that when the distance between the shield and the top of the resonator is small, the frequencies vary rapidly and the curves for various modes intersect, making the identification of modes extremely difficult (this is even more evident in the data presented by Kobayashi et al. [7] and Zaki and Chen [8]). When this separation is reduced, the TE resonant frequencies increase, while the TM frequencies decrease. For a larger separation between the shield and the DR, the dependence of the TE resonances on H is weak. The subscripted word "höher" (German for higher) designates a higher-order mode whose indices have not been identified.

Figure 5.12 illustrates the influence of the dielectric tuning rod of radius R_5 = 3 mm on the resonant frequencies of the resonator of Fig. 5.3 for H = 8.9 mm and R = 15 mm. Data are presented for two values of

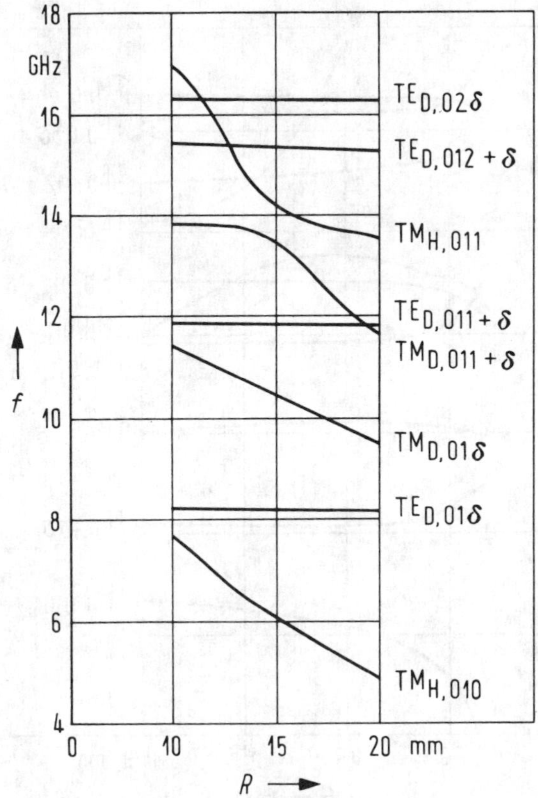

Fig. 5.10 Resonant frequencies of the shielded DR of Fig. 5.3 versus the radius R of the cylindrical shield for R_d = 3.03 mm, $H_d = \ell_3 - \ell_2$ = 4.22 mm, ϵ_{rd} = 36.2, $\ell_1 = \ell_2$ = 0.7 mm, $H = \ell_4$ = 8.9 mm, and ϵ_{r1} = 9.5. The subscripts D and H denote, respectively, interior and exterior modes (reference [23], ©1983 AEU)

the dielectric constant ϵ_{r5} of the tuning rod: the solid lines are for ϵ_{r5} = 10, and the dashed lines for ϵ_{r5} = 36.2 (which is the same as the dielectric constant of the resonator). It can be seen that all resonant frequencies decrease with increased depth of penetration of the rod. For ϵ_{r5} = 36.2, the resonant frequency of the $TE_{01\delta}$ mode can be changed by about 2.5 % (from 8.199 GHz to 7.988 GHz) by increasing the length of the rod $H - \ell_4$ from 0 to 3 mm (when $H - \ell_4$ = 3 mm, the separation between the top of the DR and the tuning rod is about 1 mm).

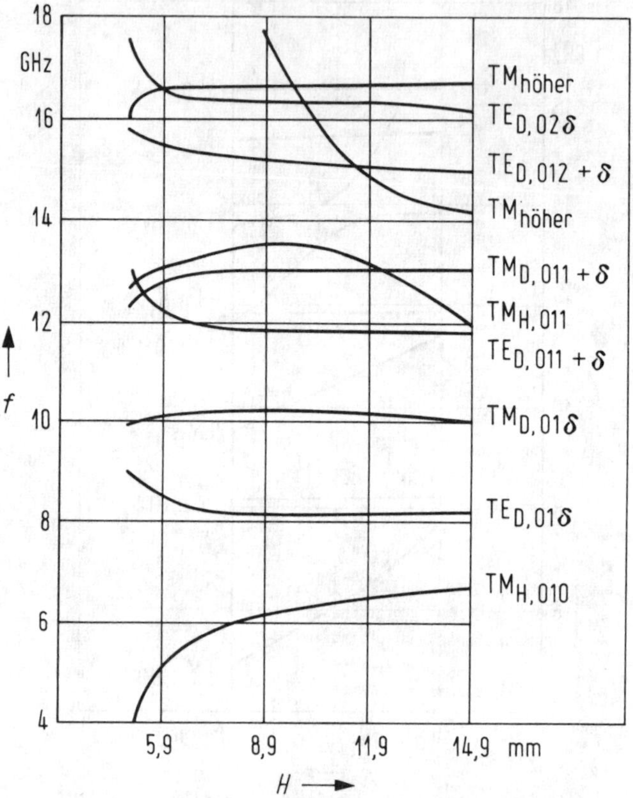

Fig. 5.11 Resonant frequencies of the shielded DR of Fig. 5.3 versus the height H of the cylindrical shield for $R_d = 3.03$ mm, $H_d = \ell_3 - \ell_2 = 4.22$ mm, $\epsilon_{rd} = 36.2$, $\ell_1 = \ell_2 = 0.7$ mm, $H = \ell_4$, $R = 15$ mm, and $\epsilon_{r1} = 9.5$. The subscripts D and H denote, respectively, interior and exterior modes (reference [23], ©1983 AEU)

The last two figures (Figs. 5.13 and 5.14) show data obtained by Crombach and Michelfeit [6] by the radial mode matching technique (see Sec. 5.2). These authors classify the axisymmetric TE (H) modes as interior (DR-type), exterior (cavity-type), or mixed, using a different and perhaps less ambiguous criterion than the energy distribution criterion discussed in Sec. 5.1. According to Crombach and Michelfeit, a mode is of the interior type if all vortices of the magnetic field in a ϕ = constant plane are inside the DR, and of the exterior type if all magnetic field vortices are outside. If some of the vortices are

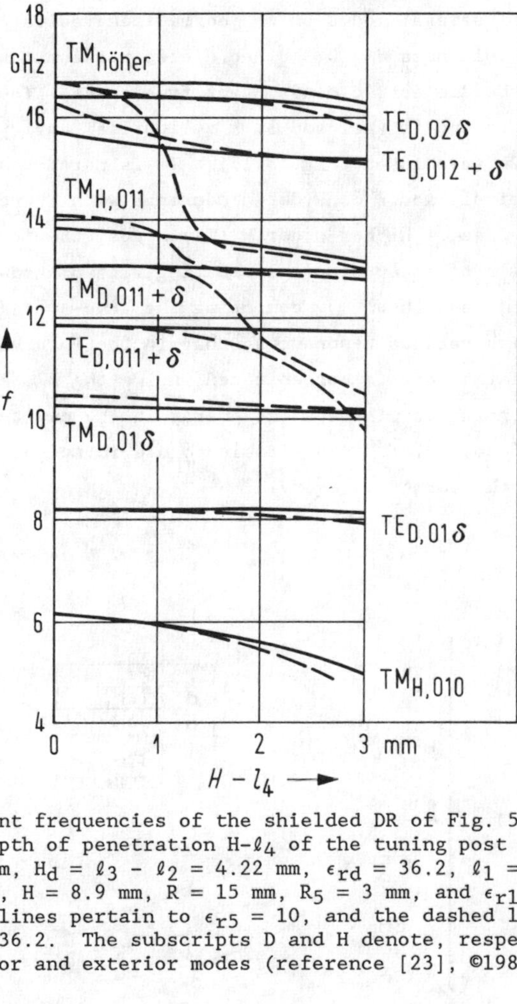

Fig. 5.12 Resonant frequencies of the shielded DR of Fig. 5.3 versus the depth of penetration $H-\ell_4$ of the tuning post for R_d = 3.03 mm, $H_d = \ell_3 - \ell_2$ = 4.22 mm, ϵ_{rd} = 36.2, $\ell_1 = \ell_2$ = 0.7 mm, H = 8.9 mm, R = 15 mm, R_5 = 3 mm, and ϵ_{r1} = 9.5. The solid lines pertain to ϵ_{r5} = 10, and the dashed lines to ϵ_{r5} = 36.2. The subscripts D and H denote, respectively, interior and exterior modes (reference [23], ©1983 AEU)

inside and some are outside, the mode is of the mixed type. In this nomenclature, the second index denotes the number of vortices in the radial direction and the third index denotes the number of vortices in the axial direction. The interior and exterior modes are distinguished, respectively, by subscripted and superscripted indices. The mixed-type modes carry both the subscripts and superscripts. For example, the mixed H_{011}^{021} mode has in the radial direction one vortex inside the DR and two vortices outside.

Figure 5.13 shows the dependence of the normalized resonant wave numbers bk_0 of several modes on the normalized radius ρ_A/b for a pillbox resonator of thickness d = 0.35b and dielectric constant $\epsilon_2 = 30\epsilon_0$, where ϵ_0 is the dielectric constant of free space. The DR resides on a substrate layer and is enclosed in a cylindrical cavity of height b and radius ρ_S = 2b (see inset, Fig. 5.13). It is noted that resonant frequencies of all modes considered decrease with increasing ρ_A and that, in the case of higher-order H (TE) modes, the frequency curves exhibit intervals of steep slope alternating with plateaus. Crombach and Michelfeit [6] have shown, by computing the time-average energy density associated with various resonances, that in the intervals of steep slope the energy of the mode is concentrated inside the DR, and in the intervals corresponding to plateaus the energy in the cavity outside the DR prevails. Of course, it is desirable to use in technical applications the modes in the former category.

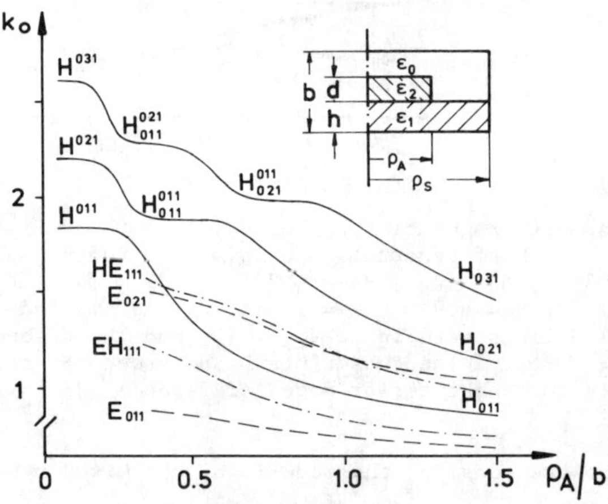

Fig. 5.13 Lowest resonant wave numbers bk_0 of the shielded DR (see the inset) versus the normalized radius ρ_A/b of the dielectric sample for ρ_S = 2b, h = 0.3b, d = 0.35b, $\epsilon_1 = 9.5\epsilon_0$, and $\epsilon_2 = 30\epsilon_0$, where ϵ_0 and k_0 are the free-space dielectric constant and wave number, respectively (reference [6], ©1981 Frequenz)

Figure 5.14 shows the influence of the radius ρ_I of the inner hole of a shielded ring resonator (see inset, Fig. 5.14) on the resonant frequency f_0 of the $TE_{01\delta}$ mode for two values of the dielectric constant, (a) $\epsilon_2 = 30\epsilon_0$, and (b) $\epsilon_2 = 40\epsilon_0$. It is seen that initially f_0 varies slowly with increasing ρ_I. When the ratio of the inner and outer radii of the resonator reaches 1/3, f_0 increases by only 1.2 %

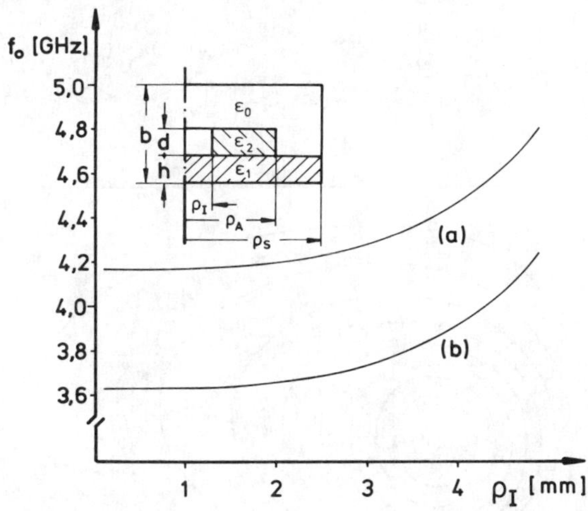

Fig. 5.14 Resonant frequency of the $TE_{01\delta}$ mode versus the radius ρ_I of the inner hole for a shielded ring resonator (see the inset) with $\rho_S = 17.5$ mm, $\rho_A = 7.5$ mm, h = 3 mm, d = 6 mm, b = 29 mm, $\epsilon_1 = 2.5\epsilon_0$, for (a) $\epsilon_2 = 30\epsilon_0$, and (b) $\epsilon_2 = 40\epsilon_0$, where ϵ_0 is the free-space dielectric constant (reference [6], ©1981 Frequenz)

with respect to the value for $\rho_I = 0$ (pillbox) in case (a), and by 2 % in case (b). A further increase of the inner radius causes the resonant frequency to increase more rapidly. In this range of values of ρ_I, the ring resonator can be effectively tuned by inserting a dielectric screw in the inner hole. In a recent paper, Crombach et al. [53] show that the resonant frequency can be varied up to 20 % by a tuning ring placed over the resonator.

Finally, in Figs. 5.15 through 5.18 are shown magnetic field plots of the non-leaky TE_{0np} modes for a typical DR in a parallel-plate waveguide. These plots were obtained from a computer program, which implements the radial mode matching method described in Sec. 5.2. The geometry of the resonator is as illustrated in Fig. 5.15. The high-ϵ dielectric sample of radius R = 2.79 mm and height b = 2.24 mm resides on a substrate layer of thickness a = 0.635 mm. The distance between the DR top and the upper metallic plate is c = 2.2 mm. The relative dielectric constants of the substrate (ϵ_1), the DR proper (ϵ_2), and the surrounding medium (ϵ_3) are, respectively, 10, 29.8, and 1. A resonator with these parameters was used by Komatsu and Murakami [16] in the theoretical and experimental analysis of the coupling between a DR and a microstrip line.

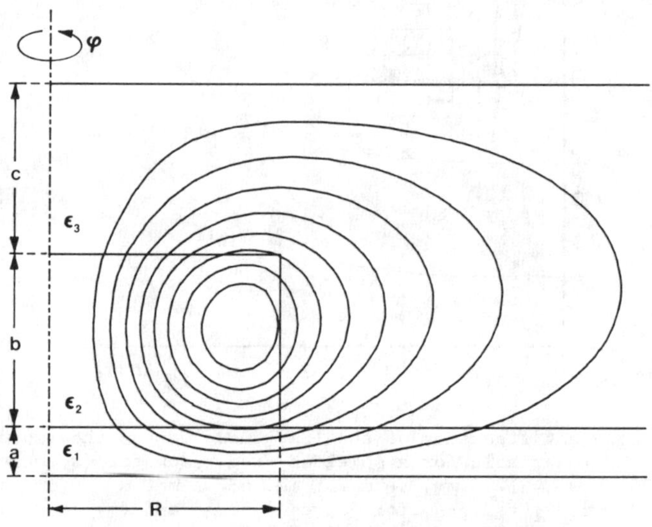

Fig. 5.15 Magnetic field lines in a ϕ=const. plane of the $TE_{01\delta}$ mode of a resonator with a = 0.635 mm, b = 2.24 mm, c = 2.2 mm, ϵ_1 = 10, ϵ_2 = 29.8, and ϵ_3 = 1. The resonant frequency f_0 = 11.68 GHz.

The first resonant mode (Fig. 5.15), which has the resonant frequency f_0 = 11.68 GHz, can be classified as the $TE_{01\delta}$ mode (see the discussion on the designation of modes in Sec. 5.1). The next (Fig. 5.16), with f_0 = 19.42 GHz, is the $TE_{011+\delta}$ resonator mode (which is obviously distorted by the presence of the shield and the substrate). The third (Fig. 5.17), with f_0 = 20.01 GHz, can be classified as the $TE_{02\delta}$ mode

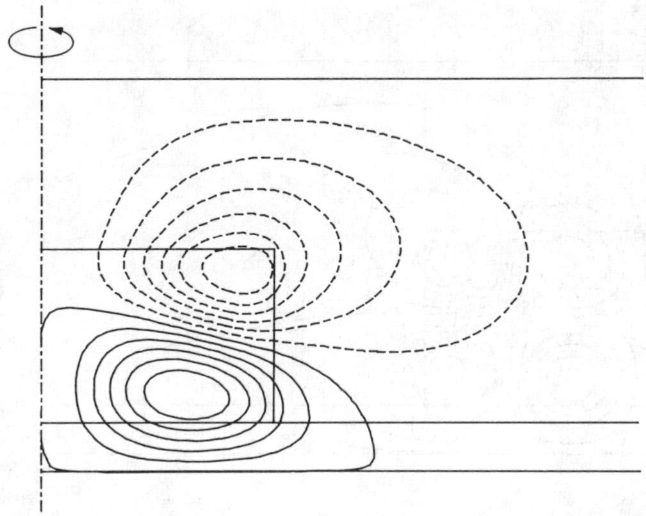

Fig. 5.16 Magnetic field lines in a ϕ = const. plane of the $TE_{011+\delta}$ mode of the resonator of Fig. 5.15. The resonant frequency f_0 = 19.42 GHz.

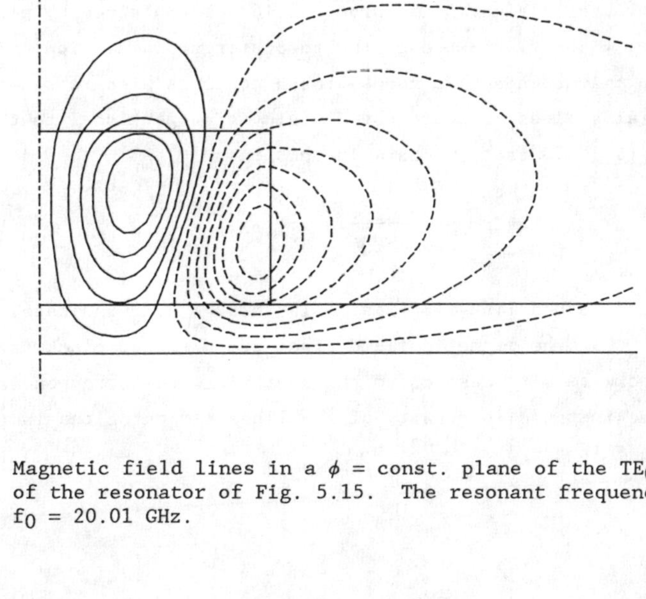

Fig. 5.17 Magnetic field lines in a ϕ = const. plane of the $TE_{02\delta}$ mode of the resonator of Fig. 5.15. The resonant frequency f_0 = 20.01 GHz.

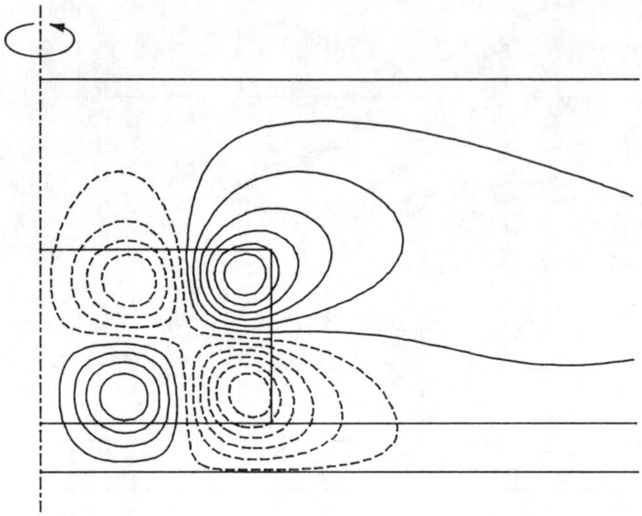

Fig. 5.18 Magnetic field lines in a ϕ = const. plane of the $TE_{021+\delta}$ mode of the resonator of Fig. 5.15. The resonant frequency f_0 = 26.25 GHz.

(observe the small difference in the resonant frequencies of this mode and the previous one). Finally, the mode of Fig. 5.18, which has the resonant frequency f_0 = 26.25 GHz, can be classified as the $TE_{021+\delta}$ mode. All higher modes of this resonator are leaky (propagating).

The contours in Figs. 5.15 through 5.18 were obtained by plotting the $\rho E_\phi(\rho,z)$ = const. lines over the resonator cross section. In the case of the ϕ-independent TE modes, these contours also represent the magnetic fields lines, because the H-field is proportional to the gradient of ρE_ϕ [6]. In fact, we have in this case

$$\underline{H} = \frac{1}{j\omega\mu_0\rho} \hat{\phi} \times \underline{\nabla}(\rho E_\phi) \tag{5.120}$$

The solid and dashed lines in Figs. 5.15 through 5.18 indicate, respectively, that the magnetic field lines are directed clockwise or counter clockwise with respect to the center of the corresponding vortex. The increase in density of the lines indicates the increase in the field magnitude.

5.8 Assessment of Methods and Summary

Having reviewed the major rigorous analysis methods for both shielded and open DRs, we are now in a position to discuss their merits and shortcomings. In the case of shielded resonators, two competing techniques have emerged, which are applicable to both axisymmetric ($m = 0$) and hybrid ($m \geq 1$) modes: the radial mode-matching method (Sec. 5.2) and the axial mode matching method (Sec. 5.3). For simplicity, we will refer to these methods in the following discussion as the radial and axial methods, respectively. When applied to a typical DR in a cylindrical cavity (Fig. 5.1), each of these two methods will give sufficiently accurate results, but one of them may require less effort than the other. Systematic convergence studies comparing various techniques have not appeared in the literature, so we have little guidance in this respect. As was discussed in Sec. 5.2, the radial method suffers from a relatively slow convergence for modes with a strong axial component of the electric field, which is singular at the edges of the DR [13,15]. We may speculate that similar difficulty will be encountered in the axial method for modes with strong radial component of the electric field. The source of the difficulty seems to be the fact that in both methods we seek to represent potentially singular functions in terms of continuous waveguide modes.

The relative efficiency of these methods is also likely to vary from case to case, i.e., what is the "best" method for a particular DR may not be as good when the dimensions of the resonator are changed. For example, the axial approach is not applicable to the case of a DR in a parallel-plate waveguide, and we expect it to be inferior for large values of c/h (Fig. 5.1). In this case, many cylindrical waveguide modes (i.e., many terms in the series (5.92)) in each partial region will be required to represent the field in the resonator accurately. By a similar argument, we conclude that the axial method would be preferred if $h/c \to \infty$ (however, this situation does not appear to be very common).

The convergence of the field representations in the partial regions is not the only factor determining the efficiency of the given method. The computational effort will also strongly depend on the number of partial regions required by each method. In the case of the particular DR shown in Fig. 5.1, there are three regions for both methods. We note, however, that adding another layer of dielectric (a spacer, for example) or another DR (to form a double resonator) will not increase the number of partial regions in the radial method, but it will add one or more regions in the axial approach. Therefore, taking the number of partial regions as the sole criterion, we would choose the radial method for a double resonator, and the axial method for a ring resonator.

Another important factor in comparing the two mode matching techniques is the kind of the transcendental eigenvalue equation which one has to solve. In the radial method, this equation involves trigonometric functions, while in the axial approach it contains Bessel functions. Since the trigonometric equation is well understood (see Appendix 5.B), the former method appears to have an edge over the latter in this respect. This is particularly true for ring DRs, where the eigenvalue equation is extremely complex in the axial method. There is also the issue of the possible occurrence of complex-valued propagation constants in some partial regions, which discouraged some researchers [19,23] from using the axial method in the case of hybrid modes. However, Zaki and Chen [8] obtained excellent results by this technique, in spite of the fact that they only searched for real roots.

In the case of ϕ-independent modes ($m = 0$) of shielded resonators, there are several other rigorous methods at our disposal, in addition to the two mode matching techniques already discussed. Hence, there is the Galerkin-Rayleigh-Ritz method (Sec. 5.1), the differential mode matching method (Sec. 5.3), the finite-element and finite-difference methods (Sec. 5.4), and the Green's function - integral equation methods (Sec. 5.6). Depending on the problem considered, some of these methods may have advantages over the axial or radial mode matching techniques. For example, the finite-element method usually offers more flexibility in modifying the structure, which is important in the design process. The principles involved in this method are quite simple (Sec. 5.4), and most of the difficulties are relegated to the programming stage. Usually, hundreds of finite elements must be employed, hence the efficiency of this technique depends to a large degree on the availability of sophisticated subroutine packages, which exploit the sparseness of the matrices.

It is perhaps worth mentioning that the finite-element method and the Galerkin-Rayleigh-Ritz method of Sec. 5.1 are closely related through the use of the same variational principle (the Rayleigh-Ritz procedure). As a result, both methods always overestimate the resonant frequencies, i.e., they provide upper bounds for the true resonant frequencies. In contrast, the mode matching methods and the Green's function method always underestimate the true resonant frequencies, i.e., they provide the lower bounds. In the case of the Green's function method [45,46], this is the consequence of it being related to the Weinstein variational method of intermediate problems, which is known to yield underestimated eigenvalues [51].

The Green's function technique [45,46] does enjoy one important advantage over the mode matching methods: owing to the fact that the partial regions in this method are always bounded by perfectly conducting (PEC or PMC) walls, there always exist two different, but equivalent, representations of the Green's function in each region. Hence, one can pick a representation with better convergence properties for a particular geometry, or the representation, which leads to a simple eigenvalue equation. However, the determination of the Green's functions, particularly for more complex resonators, tends to be difficult. This is probably the reason why these methods have only been applied to axisymmetric cases.

For isolated DRs (i.e., DRs in free space), the available rigorous techniques include the integral equation method (Ch. 6), the mode-matching/Rayleigh-Ritz method (Sec. 5.3), and the perturbational methods (Sec. 5.4). The first two methods have been employed to determine the axisymmetric, as well as hybrid, modes of pillbox resonators, and both can be extended to treat rotationally symmetric DRs of arbitrary cross section. No systematic studies of the relative efficiency of the two methods are available. Both are computationally expensive, as they lead to full matrices with elements determined by numerical integration. Since the integral equation method employs subsectional basis functions, it will typically result in a matrix of larger size than the mode-matching method. On the other hand, this method is not plagued by the convergence problems encountered in the mode-matching technique.

The perturbational methods (Sec. 5.5) are based on the asymptotic expansion of the fields in terms of the inverse powers of ϵ_r, and are only accurate in the limit as $\epsilon_r \to \infty$. The importance of these methods lies in the fact that they are semi-analytic in nature and explicitly give the dependence of various parameters on ϵ_r. The numerical part involves either an iterative solution of a system of partial differential equations in an infinite domain, or a solution of an integral equation over the DR cross section. The numerical solution, however, does not have to be repeated for every new value of ϵ_r. The perturbational methods have only been applied to the axisymmetric modes. For the commonly encountered values of ϵ_r, they offer accuracies comparable to that achieved by other methods described here only if higher-order correction terms are included.

We remark, in conclusion, that the survey of rigorous analysis methods of DRs presented here is not meant to be exhaustive, and that, in view of the vast literature of the subject, some omissions are unavoidable. Nevertheless, we hope that the content of this chapter is representative of the available techniques, and that it will serve as a useful introduction to the rapidly developing discipline of dielectric resonators.

Appendix 5.A DERIVATION OF THE EIGENFUNCTIONS AND EIGENVALUES OF THE OPERATORS D_z^e AND D_z^h

The mode matching procedure described in Sec. 5.2 is valid for partial regions with any number of dielectric layers provided that the eigenvalues λ_n and the corresponding eigenfunctions Z_n for both the TM and TE cases are available. In this appendix, we solve the eigenvalue problems (5.22) and (5.46) in the case of a partial region with three dielectric layers (Fig. 5.2), which will allow us to consider a dielectric resonator on a substrate (Fig. 5.1). The results can be readily specialized to the simpler case of a region with two layers, such as partial regions 1 and 3 in Fig. 5.1.

We first turn attention to the eigenvalue problem (5.22). Since within each dielectric layer the eigenfunctions will have different functional forms, it is convenient to introduce the notation:

$$Z_n^e(z) = \frac{1}{\sqrt{C_n^e}} \begin{cases} Z_{n1}^e(z) , & z \in [0, h_1] \\ Z_{n2}^e(z) , & z \in [h_1, h_1 + h_2] \\ Z_{n3}^e(z) , & z \in [h_1 + h_2, h] \end{cases} \tag{5.A1}$$

where C_n^e are normalization coefficients chosen to render the eigenfunctions orthonormal with the inner product (5.23). It is now easy to see from (5.22a) that in the kth layer, where k = 1,2, or 3, the function Z_{nk}^e must satisfy the equation:

$$\frac{d^2}{dz^2} Z_{nk}^e + \left(k_0^2\epsilon_k + \lambda_n^e\right) Z_{nk}^e = 0 \tag{5.A2}$$

where ϵ_k is the relative dielectric constant of kth layer (cf. eq. (5.24)). In view of (5.22b), Z_{n1}^e and Z_{n2}^e must also obey the boundary conditions:

$$\frac{d}{dz} Z_{n1}^e = 0 \quad \text{for} \quad z = 0 \tag{5.A3}$$

$$\frac{d}{dz} Z_{n3}^e = 0 \quad \text{for} \quad z = h \tag{5.A4}$$

From (5.16a) and (5.16b), we see that the continuity of tangential magnetic field across the interfaces between the dielectric layers demands that

$$Z^e_{n1} = Z^e_{n2} \quad \text{at} \quad z = h_1 \tag{5.A5}$$

$$Z^e_{n2} = Z^e_{n3} \quad \text{at} \quad z = h_1 + h_2 \tag{5.A6}$$

Similarly, in view of (5.17a) and (5.17b), the continuity of the tangential electric field requires that

$$\frac{1}{\epsilon_1}\frac{d}{dz}Z^e_{n1} = \frac{1}{\epsilon_2}\frac{d}{dz}Z^e_{n2} \quad \text{at} \quad z = h_1 \tag{5.A7}$$

$$\frac{1}{\epsilon_2}\frac{d}{dz}Z^e_{n2} = \frac{1}{\epsilon_3}\frac{d}{dz}Z^e_{n3} \quad \text{at} \quad z = h_1 + h_2 \tag{5.A8}$$

If we define, for notational convenience,

$$\gamma^e_{nk} = \sqrt{k_0^2\epsilon_k + \lambda^e_k} \tag{5.A9}$$

we can readily construct the functions Z^e_{nk} as

$$Z^e_{n1}(z) = \epsilon_1 \cos\gamma^e_{n1}z \tag{5.A10a}$$

$$Z^e_{n2}(z) = \epsilon_2 \frac{\gamma^e_{n1}\sin\gamma^e_{n1}h_1}{\gamma^e_{n2}}\left[\cot\Phi^e_n \cos\gamma^e_{n2}(z - h_1) - \sin\gamma^e_{n2}(z - h_1)\right] \tag{5.A10b}$$

$$Z^e_{n3}(z) = \epsilon_2 \frac{\gamma^e_{n1}\sin\gamma^e_{n1}h_1\sin\gamma^e_{n2}h_2}{\gamma^e_{n2}\cos\gamma^e_{n3}h_3}\left[\cot\Phi^e_n \cot\gamma^e_{n2}h_2 - 1\right]\cos\gamma^e_{n3}(h - z) \tag{5.A10c}$$

where we also defined

$$\cot\Phi^e_n = \frac{\epsilon_1\gamma^e_{n2}}{\epsilon_2\gamma^e_{n1}}\cot\gamma^e_{n1}h_1 \tag{5.A11}$$

Written in this form, the functions Z^e_{nk} explicitly satisfy all the required conditions except for (5.A8). If we subject (5.A10b) and (5.A10c) to (5.A8) and introduce the notation:

$$\cot\Psi^e_n = \frac{\epsilon_3\gamma^e_{n2}}{\epsilon_2\gamma^e_{n3}}\cot\gamma^e_{n3}h_3 \tag{5.A12}$$

we obtain

$$\tan\Phi^e_n + \tan\Psi^e_n + \left(1 - \tan\Phi^e_n\tan\Psi^e_n\right)\tan\gamma^e_{n2}h_2 = 0 \tag{5.A13}$$

which is equivalent to

$$\tan\left(\Phi^e_n + \Psi^e_n\right) + \tan\gamma^e_{n2}h_2 = 0 \tag{5.A14}$$

This transcendental equation must be numerically solved for the eigenvalues λ^e_n. The corresponding eigenfunctions are thus given by (5.A1), where Z^e_{nk} are defined by (5.A10). From (5.23) and (5.24), the normalization coefficients are found as

$$C^e_n = \frac{1}{2}(\gamma^e_{n1})^2\sin^2\gamma^e_{n1}h_1\left\{\frac{\epsilon_1 h_1}{(\gamma^e_{n1})^2}\left(1 + \cot^2\gamma^e_{n1}h_1\right) + \frac{\epsilon_2 h_2}{(\gamma^e_{n2})^2}\left(1 + \cot^2\Phi^e_n\right)\right.$$

$$+ \frac{\epsilon_2\cot\Phi^e_n}{\gamma^e_{n2}}\frac{(\gamma^e_{n2})^2 - (\gamma^e_{n1})^2}{(\gamma^e_{n1})^2(\gamma^e_{n2})^2} + \sin^2\gamma^e_{n2}h_2\left(\cot\Phi^e_n + \cot\gamma^e_{n2}h_2\right)^2$$

$$\left.\cdot\left(\frac{\epsilon_3 h_3}{(\gamma^e_{n3})^2}\left(1 + \cot^2\gamma^e_{n3}h_3\right) + \frac{\epsilon_2\cot\Psi^e_n}{\gamma^e_{n2}}\frac{(\gamma^e_{n2})^2 - (\gamma^e_{n3})^2}{(\gamma^e_{n3})^2(\gamma^e_{n2})^2}\right)\right\} \tag{5.A15}$$

The eigenvalue problem (5.46) can be solved by a procedure similar to that given above. Hence, if we introduce functions Z^h_{nk} by analogy to (5.A1), we will find from (5.46a) that they must satisfy (5.A2). The boundary conditions (5.46b) dictate

$$Z^h_{n1} = 0 \quad \text{at} \quad z = 0 \tag{5.A16}$$

$$Z^h_{n3} = 0 \quad \text{at} \quad z = h \tag{5.A17}$$

From (5.41a) and (5.41b), we see that the continuity of the tangential magnetic field across the interfaces between the dielectric layers demands that

$$\frac{d}{dz} Z^h_{n1} = \frac{d}{dz} Z^h_{n2} \quad \text{at} \quad z = h_1 \tag{5.A18}$$

$$\frac{d}{dz} Z^h_{n2} = \frac{d}{dz} Z^h_{n3} \quad \text{at} \quad z = h_1 + h_2 \tag{5.A19}$$

while the continuity of tangential electric field components, given by (5.42a) and (5.42b), requires that

$$Z^h_{n1} = Z^h_{n2} \quad \text{at} \quad z = h_1 \tag{5.A20}$$

$$Z^h_{n2} = Z^h_{n3} \quad \text{at} \quad z = h_1 + h_2 \tag{5.A21}$$

We can easily construct the functions Z^h_{nk} as

$$Z^h_{n1}(z) = \frac{\sin\gamma^h_{n1} z}{\gamma^h_{n1}} \tag{5.A22a}$$

$$Z^h_{n2}(z) = \frac{\cos\gamma^h_{n1} h_1}{\gamma^h_{n2}} \left[\tan\Phi^h_n \cos\gamma^h_{n2}(z - h_1) + \sin\gamma^h_{n2}(z - h_1)\right] \tag{5.A22b}$$

$$Z^h_{n3}(z) = \frac{\cos\gamma^h_{n1} h_1 \cos\gamma^h_{n2} h_2}{\gamma^h_{n3} \cos\gamma^h_{n3} h_3} \left[\tan\Phi^h_n \tan\gamma^h_{n2} h_2 - 1\right] \sin\gamma^h_{n3}(h - z) \tag{5.A22c}$$

where we introduced

$$\tan\Phi^h_n = \frac{\gamma^h_{n2}}{\gamma^h_{n1}} \tan\gamma^h_{n1} h_1 \tag{5.A23}$$

These functions satisfy all the required conditions except for (5.A21).

If we introduce

$$\tan\Psi_n^h = \frac{\gamma_{n2}^h}{\gamma_{n3}^h} \tan\gamma_{n3}^h h_3 \tag{5.A24}$$

and subject Z_{n2}^h and Z_{n3}^h to (5.A21), we obtain

$$\tan\Phi_n^h + \tan\Psi_n^h + \left(1 - \tan\Phi_n^h \tan\Psi_n^h\right) \tan\gamma_{n2}^h h_2 = 0 \tag{5.A25}$$

which can be written as

$$\tan\left(\Phi_n^h + \Psi_n^h\right) + \tan\gamma_{n2}^h h_2 = 0 \tag{5.A26}$$

Solving this transcendental equation yields the eigenvalues λ_n^h. The normalization constants C_n^h which make $Z_n^h(z)$ an orthonormal set with the inner product (5.47) are found as

$$C_n^h = \frac{1}{2}\cos^2\gamma_{n1}^h h_1 \left\{ \frac{h_1}{(\gamma_{n1}^h)^2}\left(1 + \tan^2\gamma_{n1}^h h_1\right) + \frac{h_2}{(\gamma_{n2}^h)^2}\left(1 + \tan^2\Phi_n^h\right) \right.$$

$$- \frac{\tan\Phi_n^h}{\gamma_{n2}^h}\,\frac{(\gamma_{n2}^h)^2 - (\gamma_{n1}^h)^2}{(\gamma_{n1}^h)^2\,(\gamma_{n2}^h)^2} + \cos^2\gamma_{n2}^h h_2\left(1 - \tan\Phi_n^h \tan\gamma_{n2}^h h_2\right)^2$$

$$\left. \cdot \left[\frac{h_3}{(\gamma_{n3}^h)^2}\left(1 + \tan^2\gamma_{n3}^h h_3\right) - \frac{\tan\Psi_n^h}{\gamma_{n2}^h}\,\frac{(\gamma_{n2}^h)^2 - (\gamma_{n3}^h)^2}{(\gamma_{n3}^h)^2\,(\gamma_{n2}^h)^2} \right] \right\} \tag{5.A27}$$

The solution of the transcendental equations (5.A14) and (5.A26) for the eigenvalues λ_n^e and λ_n^h, respectively, is discussed in Appendix 5.B.

It is of interest to point out that by following a procedure similar to that given above we could, with little extra effort, find the eigenvalues and eigenfunctions for a partial region with the lower PEC wall (Fig. 5.2) replaced by a perfect magnetic conductor (PMC). This would allow us to analyze symmetric resonators more efficiently by separately considering modes that are even and odd in z. This would also give us the capability of analyzing symmetric double resonators.

Appendix 5.B SOLUTION OF THE EIGENVALUE EQUATIONS

We discuss in this appendix the solution of the transcendental equation (5.A14) and (5.A26) for the eigenvalues λ_n^e and λ_n^h, respectively. Equations of this type, which are commonly encountered in the analysis of inhomogeneously filled waveguides, must be solved numerically. In view of the fact that the transcendental functions involved have poles interspersed between the zeros, it is rather difficult to write a reliable computer program for the computation of the eigenvalues. As Maystre et al. [17] point out, missing just one eigenvalue can lead to completely erroneous final results in the mode-matching method. These authors, who only consider the TE case, describe a systematic procedure for the solution of (5.A26), which is, however, more complicated and less efficient than the procedure given below.

Referring to Fig. 5.2, we assume that the middle layer has the highest dielectric constant, i.e., $\epsilon_2 > \epsilon_1$ and $\epsilon_2 > \epsilon_3$. Since in the mode matching procedure the middle region will correspond to the high-ϵ dielectric sample, this condition will be satisfied in most practical cases.

Turning attention to the solution of (5A.14), we consider the integral:

$$I_n^e = \int_0^h Z_n^e \frac{d}{dz}\left(\epsilon_r^{-1} \frac{d}{dz} Z_n^e\right) dz = -\int_0^h \epsilon_r^{-1} \left[Z_n^{e\prime}\right]^2 dz \tag{5.B1}$$

where, in the integration by parts, the integrated terms vanish in view of (5.22b). (Here, as elsewhere in this chapter, the prime denotes a derivative with respect to the argument.) Evidently, since the eigenfunctions Z_n^e are real, $I_n^e < 0$. On the other hand, by using (5.22a) we can express I_n^e as

$$I_n^e = -\int_0^h (k_0^2 + \lambda_n^e \epsilon_r^{-1})(Z_n^e)^2 dz \tag{5.B2}$$

Hence, in view of (5.24), we obtain the result:

$$(k_0^2\epsilon_1 + \lambda_n^e)\int_0^{h_1} (Z_n^e)^2 dz + (k_0^2\epsilon_2 + \lambda_n^e)\int_{h_1}^{h_1+h_2} (Z_n^e)^2 dz$$

$$+ (k_0^2\epsilon_3 + \lambda_n^e)\int_{h_1+h_2}^{h} (Z_n^e)^2 dz > 0 \qquad (5.B3)$$

which proves the important property of the eigenvalues [17]:

$$\lambda_n^e > - k_0^2\epsilon_2 \qquad (5.B4)$$

Since (5.22) has an infinite number of real distinct eigenvalues, we conclude that, at most, a finite number of them will be negative, as illustrated in Fig. 5.B1.

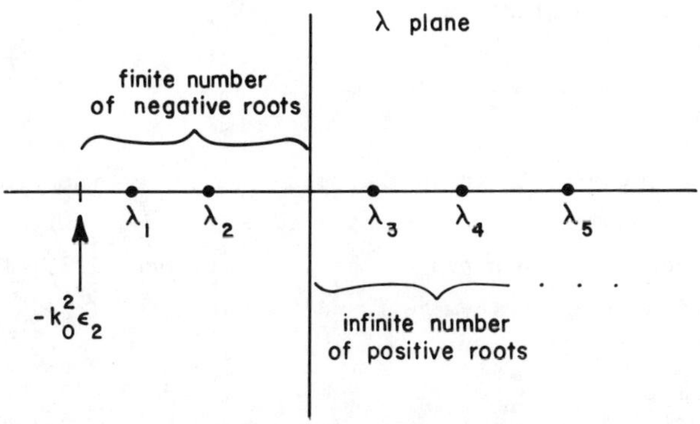

Fig. 5.B1 Distribution of eigenvalues in the radial mode-matching method

The important consequence of (5.B4) is that γ_{n2}^e is always positive real (cf. (5A.9)). Hence, it will be convenient to express γ_{n1}^e and γ_{n3}^e in terms of γ_{n2}^e as

$$\gamma_{ni}^e = \sqrt{(\gamma_{n2}^e)^2 - k_0^2\,(\epsilon_2 - \epsilon_i)}\ , \quad i = 1,3 \qquad (5.B5)$$

Referring to (5.A11) and (5.A12), we observe that although γ_{n1}^e and γ_{n3}^e may be imaginary, Φ_n^e and Ψ_n^e are always real. Thus, (5.A14) is equivalent to

$$\Phi_n^e + \Psi_n^e + \gamma_{n2}^e h_2 = (n - 1)\pi , \quad n = 1,2,... \tag{5.B6}$$

or

$$\arctan\left(\frac{\epsilon_2 \gamma_{n1}^e}{\epsilon_1 \gamma_{n2}^e} \tan\gamma_{n1}^e h_1\right) + \arctan\left(\frac{\epsilon_2 \gamma_{n3}^e}{\epsilon_3 \gamma_{n2}^e} \tan\gamma_{n3}^e h_3\right)$$

$$+ \gamma_{n2}^e h_2 - (n - 1)\pi = 0 , \quad n = 1,2,... \tag{5.B7}$$

With γ_{n1}^e and γ_{n3}^e given by (5.B5), (5.B7) is a transcendental equation for γ_{n2}^e. For values of γ_{n2}^e at which γ_{ni}^e, i = 1,3, are imaginary, we replace in (5.B7) $\gamma_{ni}^e \tan\gamma_{ni}^e h_i$ by $-|\gamma_{ni}^e|\tanh|\gamma_{ni}^e h_i|$. Since arctan is a multivalued function with the principal branch having values in the range $(-\pi/2, \pi/2)$, we also need a criterion of branch selection in (5.B7). A simple analysis indicates that $m\pi$ should be added to the principal value of the arctan, if

$$(2m - 1)\frac{\pi}{2} < \gamma_{ni}^e h_i < (2m + 1)\frac{\pi}{2} \tag{5.B8}$$

With this choice, the expression on the left side of the equation (5.B7) is, for each n, a monotonically increasing function of γ_{n2}^e with only one zero, which can easily be computed by the secant method [12]. The corresponding eigenvalue can then be found from

$$\lambda_n^e = (\gamma_{n2}^e)^2 - k_0^2 \epsilon_2 \tag{5.B9}$$

It appears that at least one of the eigenvalues λ_n^e will be negative at all frequencies (see Fig. 5.B1), which means that the corresponding radial mode is always propagating. Hence, the dominant mode in the TM case does not possess a low-frequency cutoff. This has been proved by Felsen and Marcuvitz [52] in the case of two dielectric layers.

The TE eigenvalues, λ_n^h, can be found from (5.A26) by a procedure analogous to that given above. By following the steps leading to (5.B6) and (5.B7), we obtain in this case

$$\Phi_n^h + \Psi_n^h + \gamma_{n2}^h h_2 = n\pi \ , \qquad n = 1,2,\ldots \tag{5.B10}$$

and

$$\arctan\left(\frac{\gamma_{n2}^h}{\gamma_{n1}^h}\tan\gamma_{n1}^h h_1\right) + \arctan\left(\frac{\gamma_{n2}^h}{\gamma_{n3}^h}\tan\gamma_{n3}^h h_3\right)$$
$$+ \gamma_{n2}^h h_2 - n\pi = 0 \ , \qquad n = 1,2,\ldots \tag{5.B11}$$

Given n, equation (5.B11) can be solved for γ_{n2}^h by the secant method and the corresponding eigenvalue λ_n^h can be determined from an equation analogous to (5.B9). It may happen in the present case that there are no negative eigenvalues, hence the TE radial modes possess a low-frequency cutoff. It may also be recognized that (5B.11) is closely related to the resonance condition (4.74), utilized in the computation of resonant frequencies of the two approximate DR models described in Ch. 4.

References

[1] R.F. Harrington, Field Computation by Moment Methods. New York: Macmillan, 1968.

[2] J. Van Bladel, Electromagnetic Fields. New York: Hemisphere, 1985.

[3] J. Krupka, "Optimization of an electrodynamic basis for determination of the resonant frequencies of microwave cavities partially filled with a dielectric," IEEE Trans. Microwave Theory Tech., vol. MTT-31, pp. 302-305, March 1983.

[4] J. Krupka, "Computations of frequencies and intrinsic Q factors of TE_{0nm} modes of dielectric resonators," IEEE Trans. Microwave Theory Tech., vol. MTT-33, pp. 274-277, March 1985.

[5] E. Snitzer, "Cylindrical dielectric waveguide modes," J. Opt. Soc. Am., vol. 51, pp. 491-498, May 1961.

[6] U. Crombach and R. Michelfeit, "Resonanzfrequenzen und Feldstarken in geschirmten dielektrischen Scheiben- und Ringresonatoren," Frequenz, vol. 35, no. 12, pp. 324-328, 1981.

[7] Y. Kobayashi, N. Fukuoka, and S. Yoshida, "Resonant modes for a shielded dielectric rod resonator," Electronics and Communications in Japan, vol. 64-B, no. 11, pp. 46-51, 1981.

[8] K.A. Zaki and C. Chen, "New results in dielectric loaded resonators," IEEE Trans. Microwave Theory Tech., to be published.

[9] S.B. Cohn, "Microwave bandpass filters containing high Q dielectric resonators," IEEE Trans. Microwave Theory Tech., vol. MTT-16, pp. 210-217, April 1968.

[10] B. Friedman, Principles and Techniques of Applied Mathematics. New York: John Wiley & Sons, 1956.

[11] M. Abramowitz and I.A. Stegun (Eds.), Handbook of Mathematical Functions. New York: Dover, 1965.

[12] S.D. Conte and C. de Boor, Elementary Numerical Analysis. New York: McGraw-Hill, 1980.

[13] Y. Kobayashi and M. Miura, "Optimum design of shielded dielectric rod and ring resonators for obtaining the best mode separation," IEEE MTT-S Int. Microwave Symp. Dig., pp. 184-186, San Francisco, CA, May 30-June 1, 1984.

[14] J.B. Andersen and V.V. Solodukhov, "Field behavior near a dielectric wedge," IEEE Trans. Antennas Propag., vol. AP-26, pp. pp. 598-602, July 1978.

[15] V.A. Kuznetsov, A.M. Lerer, and V.S. Mikhalevskiy, "Resonant frequencies of disk dielectric resonators," Soviet J. Comm. Tech. Electr., vol. 30, Jan. 1985.

[16] Y. Komatsu and Y. Murakami, "Coupling coefficient between microstrip line and dielectric resonator," IEEE Trans. Microwave Theory Tech., vol. MTT-31, pp. 34-40, Jan. 1983.

[17] D. Maystre, P. Vincent, and J.C. Mage, "Theoretical and experimental study of the resonant frequency of a cylindrical dielectric resonator," IEEE Trans. Microwave Theory Tech., vol. MTT-31, pp. 846-848, October 1983.

[18] Sz. Maj and M. Pospieszalski, "A composite, multilayered cylindrical dielectric resonator," IEEE MTT-S Int. Microwave Symp. Dig., pp. 190-192, San Francisco, CA, May 30 - June 1, 1984.

[19] U.S. Hong and R.H. Jansen, "Numerical analysis of shielded dielectric resonators including substrate, support disc and tunning post," Electron. Lett., vol. 18, pp. 1000-1002, Nov. 1982.

[20] R.F. Harrington, Time-Harmonic Electromagnetic Fields. New York: McGraw-Hill, 1961.

[21] K.A. Zaki and A.E. Atia, "Modes in dielectric-loaded waveguides and resonators," IEEE Trans. Microwave Theory Tech., vol. MTT-31, pp. 1039-1045, December 1983.

[22] P.J.B. Clarricoats and B.C. Taylor, "Evanescent and propagating modes of dielectric-loaded circular waveguide," Proc. IEE, vol. 111, pp. 1951-1956, Dec. 1964.

[23] U.S. Hong and R.H. Jansen, "Veranderung der Resonanzfrequenzen dielektrischer Resonatoren in Mikrostripschaltungen durch Umgebungsparameter," Arch. Elek. Ubertragung (AEU), vol. 38, pp. 106-112, 1984.

[24] P. Vincent, "Computation of the resonant frequency of a dielectric resonator by a differential method," Appl. Phys., vol. A-31, pp. 51-54, 1983.

[25] M. Tsuji, H. Shigesawa, and K. Takiyama, "On the complex resonant frequency of open dielectric resonators," IEEE Trans. Microwave Theory Tech., vol. MTT-31, pp. 392-396, May 1983.

[26] M. Tsuji, H. Shigesawa, and K. Takiyama, "Analytical and experimental investigations on several resonant modes in open dielectric resonators," IEEE Trans. Microwave Theory Tech., vol. MTT-32, pp. 628-633, June 1984.

[27] R.E. Collin, Field Theory of Guided Waves. New York: McGraw-Hill, 1960.

[28] M.V.K. Chari and P.P. Silvester (Eds.), Finite Elements in Electrical and Magnetic Field Problems. New York: John Wiley & Sons, 1980.

[29] P.P. Silvester and R.L. Ferrari, Finite Elements For Electrical Engineers. New York: Cambridge University Press, 1983.

[30] F.H. Gil and J. Gismero, "Finite element analysis of dielectric resonators on microstrip structures," XXIst General Assembly of URSI, Florence, Italy, Aug. 28-Sept. 5, 1984.

[31] F.H. Gil and J. Perez, "Analysis of dielectric resonators with tuning screw and supporting structure," IEEE MTT-S Int. Microwave Symp. Dig., pp. 485-488, St. Louis, MO, June 4-6, 1985.

[32] P.S. Kooi, M.S. Leong, and A.L.S. Prakash, "Finite-element analysis of the shielded cylindrical dielectric resonator," Proc. IEE, Pt. H, vol. 132, pp. 7-16, Feb. 1985.

[33] J. Delaballe, P. Guillon, and Y, Garault, "Local complex permittivity measurement od MIC substrates," Arch. Elektron. Ubertragung (AEU), vol. 35, no. 2, pp. 80-83, 1981.

[34] P. Guillon, J.P. Balabaud, and Y. Garault, "TM_{01p} tubular and cylindrical dielectric resonator mode," IEEE MTT-S Int. Microwave Symp. Dig., pp. 163-166, Los Angeles, CA, June 15-19, 1981.

[35] P. Guillon, D. Rousset, and Y. Garault, "A new determination for the resonant frequencies and field patterns of dielectric resonators, " Proc. Europ. Microwave Conf., Brighton, England, Sept. 1979.

[36] J. Van Bladel, "On the resonances of a dielectric resonator of very high permittivity," IEEE Trans. Microwave Theory Tech., vol. MTT-23, pp. 199-208, Feb. 1975.

[37] R. DeSmedt, "Correction due to a finite permittivity for a ring resonator in free space," IEEE Trans. Microwave Theory Tech., vol. MTT-32, pp. 1288-1293, Oct. 1984.

[38] M. Verplanken and J. Van Bladel, "The magnetic-dipole resonances of ring resonators of very high permittivity," IEEE Trans. Microwave Theory Tech., vol. MTT-27, pp. 328-333, April 1979.

[39] M. Verplanken and J. Van Bladel, "The electric-dipole resonances of ring resonators of very high permittivity," IEEE Trans. Microwave Theory Tech., vol. MTT-24, pp. 108-112, Feb. 1976.

[40] R. DeSmedt, "Dielectric resonator above an electric or magnetic wall," Arch. Elek. Ubertragung (AEU), vol. 37, pp. 6-14, Jan. 1983.

[41] R. DeSmedt, "Dielectric resonator inside a circular waveguide," Arch. Elek. Ubertragung (AEU), vol. 38, pp. 113-120, March 1984.

[42] L.B. Gol'berg and V.V. Penzyakov, "Calculation of axially-symmetric H modes in dielectric resonators using the integral equation method," Radio Eng. Electronics Physics, vol. 27, pp. 60-66, Sept. 1982.

[43] R. Mittra (Ed.), Computer Techniques for Electromagnetics. New York: Pergamon, 1973.

[44] C.-T. Tai, Dyadic Green's Functions in Electromagnetic Theory. Scranton: Intext, 1971.

[45] M. Jaworski and M.W. Pospieszalski, "An accurate solution of the cylindrical dielectric resonator problem," IEEE Trans. Microwave Theory Tech., vol. MTT-27, pp. 639-643, July 1979; correction: ibid., vol. MTT-28, p. 673, June 1980.

[46] Yu.G. Kapustin, "Design of circular waveguide resonator with dielectric disk," Izvestiya Vuzov, Radiofizika, vol. 25, pp. 1337-1344, Nov. 1982.

[47] A.S. Omar and K. Schunemann, "Scattering by dielectric obstacles inside guiding structures," IEEE MTT-S Int. Microwave Symp. Dig., pp. 321-323, San Francisco, CA, May 30 - June 1, 1984.

[48] A.S. Omar and K. Schunemann, "Scattering by material and conducting bodies inside waveguides, Part I: Theoretical formulations," IEEE Trans. Microwave Theory Tech., vol. MTT-34, pp. 266-272, Feb. 1986.

[49] A.W. Glisson, D. Kajfez, and J. James, "Evaluation of modes in dielectric resonators using a surface integral equation formulation," IEEE Trans. Microwave Theory Tech., vol. MTT-31, pp. 1023-1029, Dec. 1983.

[50] D. Kajfez, A.W. Glisson, and J. James, "Computed modal field distributions for isolated dielectric resonators," IEEE Trans. Microwave Theory Tech., vol. MTT-32, pp. 1609-1616, Dec. 1984.

[51] S.H. Gould, Variational Methods for Eigenvalue Problems. London: Oxford University Press, 1957.

[52] L.B. Felsen and N. Marcuvitz, Radiation and Scattering of Waves. Englewood Cliffs, NJ: Prentice Hall, 1973.

[53] U. Crombach, R. Gesche, and N. Lochel, "Abstimmbare dielektrische Ringresonatoren," Frequenz, vol. 39, no. 1-2, pp. 45-49, 1985.

Chapter 6
INTEGRAL EQUATION TECHNIQUES

Allen W. Glisson

6.1 Introduction

In practical applications a dielectric resonator is expected to operate within a given frequency range and the electromagnetic field in the vicinity of the resonator is expected to exhibit behavior particular to the desired mode of operation (often $TE_{01\delta}$). One of the disadvantages associated with the use of dielectric resonators, however, is that one may find resonant frequencies of undesired modes in proximity to the resonant frequency of the desired mode. It is, therefore, of great practical importance to be able to determine the resonant frequencies and field patterns of the unwanted modes of a resonator, as well as those of the desired mode, so that proper operation of, and coupling to, the resonator can be obtained. The proximity of the resonant frequencies and the field patterns of the various modes are, of course, influenced by the surroundings of the resonator, such as metal cavity walls, metal tuning screws, or dielectric tuning rods. If the resonant frequencies and field patterns of the resonator structure can be reliably and accurately computed, it should be possible to determine the most appropriate way to modify the environment of the resonator so as to alter the resonant frequencies of the interfering modes or to suppress the excitation of these modes.

One approach to the analysis of dielectric resonators that has the potential to provide such information under a wide variety of circumstances is a surface integral equation (SIE) approach. In this chapter we present a description of a very accurate numerical procedure

---

Much of the material in this chapter is adapted, with permission, from papers appearing in IEEE Transactions on Microwave Theory and Techniques (references [19] and [22]).

developed at the University of Mississippi for the analysis of dielectric resonators that is based upon a SIE formulation. The method described here has been applied to compute resonant frequencies and Q factors due to radiation for rotationally symmetric, isolated dielectric resonators with a high degree of accuracy. It has also been used to compute the electric and magnetic field distributions associated with each resonant mode. The procedure can be generalized in several ways to treat more complex geometries as well.

Many different approaches to the analysis of dielectric resonators have been described in the literature [1-21] and some of these methods are described briefly in preceding chapters of this book. Some of the methods are based on simplifications of the desired geometry, such as the introduction of a magnetic wall (e.g., [1,2,5]). In the method of Itoh and Rudokas approximations are used for the fields for a particular geometry [6], while Van Bladel has employed asymptotic expansions valid for high permittivity materials [3]. Still other methods, such as those of Jaworski and Pospieszalski [9] and Tsuji et al. [16], represent rigorous formulations for particular geometries which are solved numerically. Until recently, methods utilizing rigorous formulations had been presented only for resonator modes having no azimuthal variation (modes with first subscript m = 0, e.g., TE_{0np}) and were applicable only to resonators which conform to constant-coordinate surfaces in cylindrical coordinates. Results for some of the hybrid electromagnetic modes have been presented in [13], [19], and [20].

In this chapter we utilize a surface integral equation formulation and the method of moments for the analysis of dielectric resonators. This approach may be used not only to determine resonant frequencies and Q factors, but also to compute electromagnetic field distributions in the vicinity of the resonator [19,22]. A general discussion of surface integral equation formulations can be found in [23], while method of moments procedures are described in [24]. The numerical method is applied here to dielectric bodies that are situated in open space and are rotationally symmetric, but which may have arbitrary cross sections. All resonant modes, including the hybrid modes with $m \neq 0$, can be included in the analysis. The SIE approach offers several computational advantages over finite difference equation or volume integral equation approaches, particularly when the resonator is not enclosed in a metal shield. The volume integral equation (VIE) approach, however, can be implemented in a similar fashion (compare, for example, [25] and [26]). A VIE approach would retain most of the advantages of the SIE approach for open resonators and would permit study of resonators consisting of an inhomogeneous material. We employ the SIE formulation for resonators made of a homogeneous material because fewer unknowns are generally required for the numerical solution than for a VIE solution. Both the VIE and the SIE methods also admit to further extensions which might include complete or partial metal shields, additional resonators, tuning screws, etc., which are rotationally symmetric and have the same axis of symmetry, as well as extensions to include some non-rotationally symmetric bodies [27]. For example, one should be able to accurately model coupling between a microstrip transmission line and an adjacent dielectric resonator using these methods, although the solution would be more complex than that presented here.

In the next two sections we describe the use of the equivalence principle in the formulation of the SIE for the dielectric resonator. In Sec. 6.4 the method of moments is applied to the surface integral equation to obtain a homogeneous system of simultaneous equations in which the unknowns are the coefficients of expansion functions used to represent the equivalent surface currents, which are actually just components of the electric and magnetic fields tangential to the body surface. The procedure used to determine resonant frequencies and Q factors by locating the zeros in the complex frequency plane of the determinant of the homogeneous equation set is described in Sec. 6.5. Numerical and experimental results for resonant frequencies and Q factors are presented in Sec. 6.6, along with a universal mode chart for a particular cylindrical resonator.

Once a resonant frequency for a particular resonator is found, one can compute the actual field structure for the mode and assign an appropriate name and set of mode indices, such as $TE_{01\delta}$. In the remaining sections of this chapter, we briefly describe the procedure used to compute the equivalent surface currents and the field structure for the resonator. The procedure for determining the equivalent surface currents is presented in Sec. 6.7. Computation of the equivalent surface currents is a prerequisite for calculation of the fields, but it also provides a relatively rapid means of preliminary mode identification, as is indicated in Sec. 6.8. The use of the equivalent currents in the computation of the fields inside and outside the resonator is described in a Sec. 6.9. Knowledge of the actual field distribution permits accurate mode identification and is desirable for designing better mechanisms for coupling to the resonator or for the design of structures to suppress unwanted modes. A catalog of field plots is presented in Sec. 6.10, along with an indication as to how the plots may be used in the design process. Finally, in Sec. 6.11 we briefly discuss the application of the SIE approach to complex resonator shapes and to resonators in environments other than isolated space.

6.2 Mathematical Model of an Isolated Dielectric Resonator

In this section we apply the equivalence principle as described by Harrington [28] to develop equivalent mathematical models for the dielectric resonator which lead to the formulation of a surface integral equation characterizing the resonator. The models for a resonator in open space are developed by considering the situation in which a homogeneous dielectric body with surface S and unit surface normal $\underline{\hat{n}}$ is immersed in a homogeneous medium as shown in Fig. 6.1. Regions 1 and 2 shown in the figure designate the regions exterior and interior to S, respectively, and are characterized by media parameters $(\mu_1, \epsilon_1, \sigma_1)$ and $(\mu_2, \epsilon_2, \sigma_2)$. The fields in regions 1 and 2 are given by $(\underline{E}^s, \underline{H}^s)$ and $(\underline{E}, \underline{H},)$, respectively.

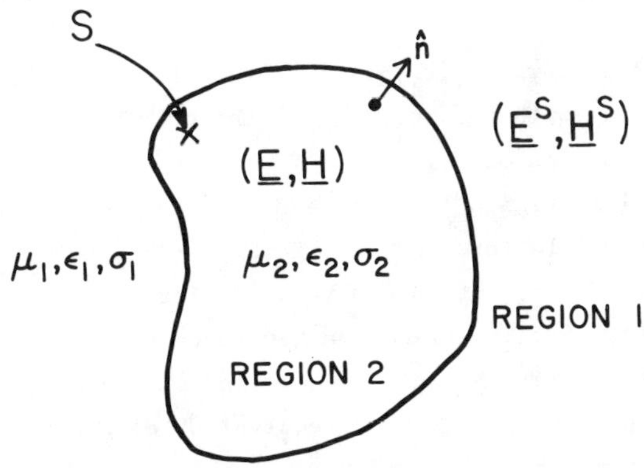

Fig. 6.1 Homogeneous dielectric body located in an infinite homogeneous medium

Many different situations can be constructed via the equivalence principle in which the field in some region of space is equivalent to the field in the same region of the original situation of Fig. 6.1. Particularly useful equivalent situations for the development of an SIE model are those in which the field in one of the regions, either region 1 or region 2, is equivalent to the field which would exist in that region in the original situation, but in which both regions consist of the same uniform material. An equivalent model for region 1 is shown in Fig. 6.2(a). We begin construction of the model by delineating a phantom, or imaginary, surface S in a homogeneous medium having the

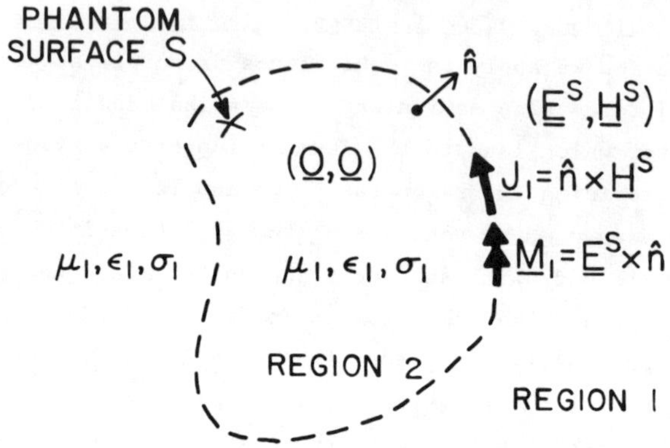

Fig. 6.2(a) A situation which is equivalent to the original situation of Fig. 6.1 for an observation point in region 1

constitutive parameters $(\mu_1, \epsilon_1, \sigma_1)$ of region 1. The phantom surface is chosen to coincide with the position of the original body surface S in the original situation of Fig. 6.1. To create a situation equivalent to that of Fig. 6.1 in region 1, the field exterior to S must be chosen to be $(\underline{E}^s, \underline{H}^s)$. The field interior to S in Fig. 6.2(a) should be a source-free, Maxwellian field, but is otherwise arbitrary. The zero field is a convenient choice. If the fields we have chosen for this equivalent situation are to satisfy Maxwell's equations at the phantom surface S, however, it is necessary to include equivalent electric and magnetic surface currents, $J_1 = \hat{\underline{n}} \times \underline{H}^s$ and $\underline{M}_1 = \underline{E}^s \times \hat{\underline{n}}$, to support the jump in the field across S. With reference to uniqueness principles, we are then assured that the currents $\underline{J}_1$ and $\underline{M}_1$ radiate in homogeneous space to produce the fields indicated in Fig. 6.2(a). The field $(\underline{E}^1, \underline{H}^1)$ anywhere in the homogeneous space of Fig. 6.2(a) may be computed from knowledge of $\underline{J}_1$ and $\underline{M}_1$ by using (6.1) and (6.2) below (with i = 1) and the homogeneous-region electric and magnetic vector and scalar potentials:

$$\underline{E}^i(\underline{r}) = -\, j\omega \underline{A}^i(\underline{r}) - \nabla\Phi^i(\underline{r}) - \frac{1}{\epsilon_i} \nabla \times \underline{F}^i(\underline{r}) \tag{6.1}$$

$$\underline{H}^i(\underline{r}) = -\, j\omega \underline{F}^i(\underline{r}) - \nabla\Psi^i(\underline{r}) + \frac{1}{\mu_i} \nabla \times \underline{A}^i(\underline{r}) \tag{6.2}$$

where the potentials are defined by the surface integrals:

$$\underline{A}^i(\underline{r}) = \frac{\mu_i}{4\pi} \int_S \underline{J}_i(\underline{r}')G^i(\underline{r},\underline{r}')\, dS' \tag{6.3}$$

$$\underline{F}^i(\underline{r}) = \frac{\epsilon_i}{4\pi} \int_S \underline{M}_i(\underline{r}')G^i(\underline{r},\underline{r}')\, dS' \tag{6.4}$$

$$\Phi^i(\underline{r}) = \frac{1}{4\pi\epsilon_i} \int_S \rho_i^e(\underline{r}')G^i(\underline{r},\underline{r}')\, dS' \tag{6.5}$$

$$\Psi^i(\underline{r}) = \frac{1}{4\pi\mu_i} \int_S \rho_i^m(\underline{r}')G^i(\underline{r},\underline{r}')\, dS' \tag{6.6}$$

with

$$G^i(\underline{r},\underline{r}') = \frac{e^{-jk_iR}}{R}, \quad R = |\underline{r}-\underline{r}'| \tag{6.7}$$

In (6.1) to (6.7), a time dependence of $\exp(j\omega t)$ is assumed and suppressed. Vectors locating the observation and source coordinates in a global coordinate system are $\underline{r}$ and $\underline{r}'$, respectively, and the wave numbers of the homogeneous media indicated in Figs. 6.2(a) and 6.2(b) are defined by $k_i = \omega\sqrt{\mu_i\epsilon_i}$, $i = 1,2$. These equations are valid for lossy media ($\sigma_i \neq 0$) as well, if one simply interprets ϵ_i in the equations as the complex permittivity $(\epsilon - j\sigma/\omega)$. The quantities ρ_i^e and ρ_1^m appearing in (6.5) and (6.6) are the equivalent electric and magnetic surface charge densities, and are related to the surface currents through the continuity equations:

$$\rho_i^e(\underline{r}') = \frac{j}{\omega}\left[\underline{\nabla}_s' \cdot \underline{J}_i(\underline{r}')\right] \tag{6.8}$$

$$\rho_i^m(\underline{r}') = \frac{j}{\omega}\left[\underline{\nabla}_s' \cdot \underline{M}_i(\underline{r}')\right] \tag{6.9}$$

When the observation point $\underline{r}$ is in region 1, the field $(\underline{E}^1,\underline{H}^1)$ computed via (6.1) and (6.2) is $(\underline{E}^s,\underline{H}^s)$, which is the correct field for region 1 in the original situation. For $\underline{r}$ in region 2, (6.1) and (6.2) will yield zero for the field $(\underline{E}^1,\underline{H}^1)$. The situation shown in Fig. 6.2(a) is, therefore, equivalent to the original situation of Fig. 6.1 when the observation point $\underline{r}$ is located in region 1.

Of course, (6.1) and (6.2) are useful to determine ($\underline{E}^s,\underline{H}^s$) in region 1 only if we know $\underline{J}_1$ and $\underline{M}_1$, as these quantities are the sources of the potential functions used to compute the field. We do not yet have sufficient information to determine $\underline{J}_1$ and $\underline{M}_1$, however, since these currents are defined in terms of the tangential components to S of the unknown field ($\underline{E}^s,\underline{H}^s$). To obtain the required additional information, we complete our model development by constructing a situation which is equivalent to the original situation in region 2 (the interior of the dielectric body). Such an equivalent situation is shown in Fig. 6.2(b).

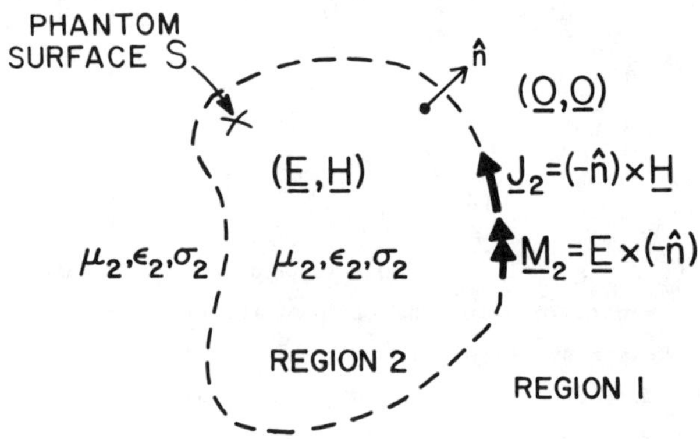

Fig. 6.2(b) A situation which is equivalent to the original situation of Fig. 6.1 for an observation point in region 2

The equivalent model for region 2 is constructed in a manner analogous to that for region 1. The phantom surface S is placed in a homogeneous medium having constitutive parameters of region 2 and the field interior to S is chosen to be the correct field ($\underline{E},\underline{H}$), while a convenient choice for the exterior region is the zero field. Equivalent surface currents $\underline{J}_2 = (-\hat{\underline{n}}) \times \underline{H}$ and $\underline{M}_2 = \underline{E} \times (-\hat{\underline{n}})$ are included to support the jump in the field at S. The field ($\underline{E}^2,\underline{H}^2$) may then be computed anywhere in the homogeneous space of Fig. 6.2(b) using (6.1) and (6.2) as before, but with $i = 2$. For an observation point $\underline{r}$ in region 2, the field ($\underline{E}^2,\underline{H}^2$) computed in this manner is ($\underline{E},\underline{H}$), which is the correct field for region 2 in the original situation. The zero field is obtained when the observation point $\underline{r}$ is in region 1. The situation shown in Fig. 6.2(b) is, therefore, equivalent to the original situation of

Fig. 6.1 when the observation point $\underline{r}$ is in region 2. $\underline{J}_2$ and $\underline{M}_2$, however, are not yet useful because they are defined in terms of the unknown field $(\underline{E},\underline{H})$. In the next section, boundary conditions are applied at S to develop equations which can be solved for all the equivalent currents.

6.3 Formulation of the Surface Integral Equations

The two equivalent situations depicted in Fig. 6.2 may be used to compute the correct fields in the appropriate regions once the equivalent currents $\underline{J}_i$, $\underline{M}_i$, i = 1,2, are known. The equivalent situations, rather than the original material body configuration of Fig. 6.1, are employed for the computation of the fields because they enable one to use the homogeneous-region potential functions defined in (6.3) to (6.6). The current sources for these potentials are defined in terms of the unknown fields $(\underline{E},\underline{H})$ and $(\underline{E}^s,\underline{H}^s)$. These unknown fields, however, must satisfy the boundary conditions

$$\hat{\underline{n}} \times \underline{E} = \hat{\underline{n}} \times \underline{E}^s \tag{6.10}$$

$$\hat{\underline{n}} \times \underline{H} = \hat{\underline{n}} \times \underline{H}^s \tag{6.11}$$

at the surface S in the original problem. Application of these conditions to the definitions of the equivalent currents eliminates two of the equivalent currents as unknowns:

$$\underline{J}_2 = (-\hat{\underline{n}}) \times \underline{H} = -\hat{\underline{n}} \times \underline{H}^s = -\underline{J}_1 \equiv -\underline{J} \tag{6.12}$$

$$\underline{M}_2 = \underline{E} \times (-\hat{\underline{n}}) = -\underline{E} \times \hat{\underline{n}} = -\underline{M}_1 \equiv -\underline{M} \tag{6.13}$$

where for simplicity we have defined the unsubscripted currents $\underline{J}$ and $\underline{M}$.

The fields in (6.10) and (6.11) can be expressed as functions of the equivalent currents through (6.1) and (6.2), so the boundary conditions also lead to the equations:

$$\hat{\underline{n}} \times [\underline{E}^s(\underline{J}_1,\underline{M}_1;\underline{r}) - \underline{E}(\underline{J}_2,\underline{M}_2;\underline{r})] = \underline{0}$$

$$\hat{\underline{n}} \times [\underline{H}^s(\underline{J}_1,\underline{M}_1;\underline{r}) - \underline{H}(\underline{J}_2,\underline{M}_2;\underline{r})] = \underline{0}\ , \quad \underline{r} \in S$$

or, using (6.12), (6.13), and the linearity of the fields,

$$\hat{\underline{n}} \times [\underline{E}^s(\underline{J},\underline{M};\underline{r}) + \underline{E}(\underline{J},\underline{M};\underline{r})] = \underline{0} \tag{6.14}$$

$$\hat{\underline{n}} \times [\underline{H}^s(\underline{J},\underline{M};\underline{r}) + \underline{H}(\underline{J},\underline{M};\underline{r})] = \underline{0}\ , \quad \underline{r} \in S \tag{6.15}$$

where the dependence of the fields on the currents and on the observation point is explicitly indicated. In (6.14) and (6.15) the fields $(\underline{E},\underline{H})$ and $(\underline{E}^s,\underline{H}^s)$ must be evaluated in the limit as $\underline{r}$ approaches S from the inside and from the outside, respectively. The exterior-region field $(\underline{E}^s,\underline{H}^s)$ can be computed via the equivalent situation of Fig. 6.2(a) using (6.1) through (6.9) with i = 1 and with $\underline{J}_1,\underline{M}_1$ replaced by $\underline{J},\underline{M}$. The field $(\underline{E},\underline{H})$ appearing in (6.14) and (6.15) represents the negative of the true interior-region field and can be computed similarly via the equivalent situation of Fig. 6.2(b) using (6.1) through (6.9) with i = 2 and $\underline{J}_2,\underline{M}_2$ replaced by $\underline{J},\underline{M}$. Equations (6.14) and (6.15) can then be represented as two coupled integral equations which are valid on the surface S and have the form

$$\left\{ j\omega\left[\underline{A}^1(\underline{r}) + \underline{A}^2(\underline{r})\right] + \nabla\left[\Phi^1(\underline{r}) + \Phi^2(\underline{r})\right] + \nabla \times \left[\frac{1}{\epsilon_1}\underline{F}^1(\underline{r}) + \frac{1}{\epsilon_2}\underline{F}^2(\underline{r})\right]\right\}_{\tan} = \underline{0} \qquad (6.16)$$

$$\left\{ j\omega\left[\underline{F}^1(\underline{r}) + \underline{F}^2(\underline{r})\right] + \nabla\left[\Psi^1(\underline{r}) + \Psi^2(\underline{r})\right] - \nabla \times \left[\frac{1}{\mu_1}\underline{A}^1(\underline{r}) + \frac{1}{\mu_2}\underline{A}^2(\underline{r})\right]\right\}_{\tan} = \underline{0}\,, \quad \underline{r} \in S \qquad (6.17)$$

where the subscript "tan" denotes the components of the vectors tangential to the surface S and where the sources of the potentials are the unsubscripted currents $\underline{J}$ and $\underline{M}$ and their associated surface charge densities.

The coupled equations (6.16) and (6.17) can be solved numerically to obtain the equivalent currents $\underline{J}$ and $\underline{M}$. Because the equations represent source-free field solutions for the geometry of Fig. 6.1, however, solutions will exist only for a set of discrete (complex) frequencies. Once these frequencies are determined, the equations may be solved numerically for the modal current distributions $\underline{J}$ and $\underline{M}$, and these currents can in turn be used to compute the field distribution in either region 1 or region 2 using the equivalent situations shown in Fig. 6.2.

6.4 Numerical Solution of the Surface Integral Equations

For the analysis of dielectric resonators, it is convenient to restrict the class of geometries to be considered to those involving rotationally symmetric dielectric bodies (bodies of revolution). Subsequent comments in this chapter are, therefore, applicable to a body of revolution which is formed by rotating a planar curve C, the so-called generating arc, about an axis which is chosen to be the z-axis of a Cartesian coordinate system (Fig. 6.3). As in Fig. 6.1, region 1, exterior to the body, and region 2, interior to the body, are characterized by media parameters $(\mu_1, \epsilon_1, \sigma_1)$ and $(\mu_2, \epsilon_2, \sigma_2)$, respectively. Surface coordinates (t, ϕ) are introduced on S, where t is the arc length along the generating curve and ϕ is the azimuthal angle measured from the x-z plane. The components of the orthogonal right-handed triad of unit vectors, $(\hat{\underline{n}}, \hat{\underline{\phi}}, \hat{\underline{t}})$, are normal to S and tangent to the ϕ and t

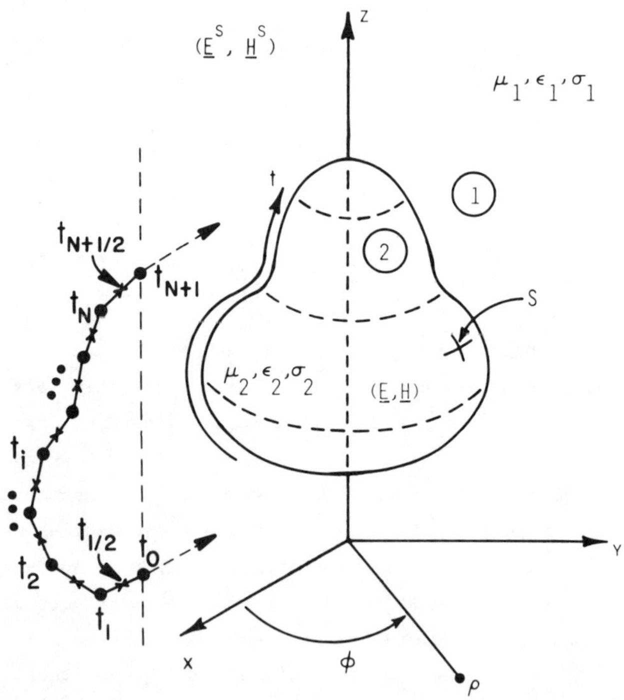

Fig. 6.3 Geometry and discretization of the generating arc for a homogeneous dielectric body of revolution (reference [19], ©1983 IEEE)

coordinate lines, respectively. For numerical purposes, the generating arc is approximated by a sequence of linear segments as shown in the figure.

The surface integral equation approach has been used by various authors for treating problems involving electromagnetic scattering by dielectric bodies of revolution [29-32]. In this chapter we employ for the analysis of rotationally symmetric dielectric resonators the approach presented in [32]. The basic numerical procedures used are the same as those employed for scattering problems and are described elsewhere [32-34]. However, for completeness, as well as to aid in description of the modal solution, the numerical procedures are briefly outlined in this section.

To take advantage of the rotational symmetry of the body, we expand all currents and scalar Green's functions in Fourier series in ϕ. For example:

$$\underline{J}(t',\phi') = \sum_{m=-\infty}^{\infty} \underline{J}_m(t')e^{jm\phi'} \tag{6.18}$$

$$G^i(t,t',\phi-\phi') = \frac{e^{-jk_iR}}{R} = \frac{1}{2\pi}\sum_{m=-\infty}^{\infty} G^i_m(t,t')e^{jm(\phi-\phi')} \tag{6.19}$$

where

$$R = |\underline{r}-\underline{r}'| = [\rho^2 + \rho'^2 - 2\rho\rho'\cos(\phi-\phi') + (z-z')^2]^{1/2} \tag{6.20}$$

$$G^i_m(t,t') = \int_{-\pi}^{\pi} G^i(t,t',\alpha)\cos(m\alpha)\, d\alpha \tag{6.21}$$

The Fourier series expansion of the kernel $G^i(t,t',\phi-\phi')$ is possible because the distance R is periodic in the variable $(\phi-\phi')$. The magnetic current $\underline{M}$ is expanded in a similar manner. Expansion of the source and field quantities in Fourier series leads to equations which can be decoupled with respect to the angular variation, so that each Fourier component pair $(\underline{J}_m(t),\underline{M}_m(t))$ can be computed independently. The Fourier component current pair $(\underline{J}_m(t),\underline{M}_m(t))$ can then be used to compute the associated electric and magnetic fields, which will also behave as $\exp(jm\phi)$. Furthermore, current components with Fourier index -m can be related to those with index m, and then combined to describe the "real" field behavior (e.g., $\cos(m\phi)$). Because the currents for each index m are independent, we may solve the equations for a particular value of the Fourier index m, which amounts to choosing a particular azimuthal variation of the field in advance.

While the variation of the currents (and fields) is specified in the ϕ direction in advance, their variation along the surface S in the t direction remains to be determined. We model the variation in the t direction by applying the method of moments to equations (6.16) and (6.17) to obtain (for the Fourier component m of interest) a set of simultaneous equations which may be represented in matrix form as

$$\underset{\sim}{Z}_m |I_m\rangle = |0\rangle \tag{6.22}$$

where $\underset{\sim}{Z}_m$ is the moment matrix and $|I_m\rangle$ is a column vector containing the coefficients of the surface current expansion for the mth Fourier component. The generating arc is approximated as a sequence of linear segments with the t-coordinate discretized as shown in Fig. 6.3 for application of the method of moments. The t variations of the orthogonal vector components J_t and J_ϕ of electric current for the mth Fourier component are expanded in basis functions $\Pi_t^i(t)$ and $\Pi_\phi^i(t)$, respectively, where the superscript i now refers to the coordinates t_i, rather than to the interior or exterior region, as follows:

$$\underline{J}_m(t) \simeq \hat{\underline{t}} \sum_{i=1}^{N} J_t^{mi} \Pi_t^i(t) + \hat{\underline{\phi}} \sum_{i=1}^{N+1} J_\phi^{mi} \Pi_\phi^i(t) \tag{6.23}$$

where

$$\Pi_t^i(t) = \begin{cases} \dfrac{\rho_i}{\rho}, & t_{i-1/2}<t<t_{i+1/2} \\ 0, & \text{otherwise} \end{cases} \tag{6.24}$$

$$\Pi_\phi^i(t) = \begin{cases} 1, & t_{i-1}<t<t_i \\ 0, & \text{otherwise} \end{cases} \tag{6.25}$$

The basis functions $\Pi_t^i(t)$ and $\Pi_\phi^i(t)$ are illustrated in Fig. 6.4 for a portion of a generating arc along which ρ is a constant (e.g., along the side of a cylinder). The expansion (6.23) is substituted into (6.18) to

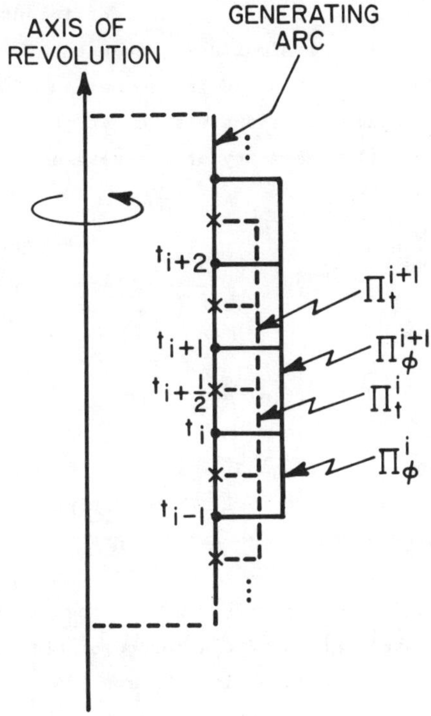

Fig. 6.4 Basis functions $\Pi_t^i(t)$ and $\Pi_\phi^i(t)$ for a portion of generating arc along which ρ is constant (reference [19], ©1983 IEEE)

obtain the full expansion in both the t and ϕ variables for the surface current $\underline{J}$ on S. The electric charge density $\rho_m^e(t)$ for the mth Fourier component is approximated from the continuity equation (6.8) as

$$\rho_m^e(t) \simeq \frac{j}{\omega} \sum_{i=1}^{N+1} \left\{ \frac{\rho_i J_t^{mi} - \rho_{i-1} J_t^{m,i-1}}{\rho_{i-1/2}\, \Delta t_i} + \frac{jm}{\rho_{i-1/2}} J_\phi^{mi} \right\} \Pi^i(t) \qquad (6.26)$$

where, on the right side of (6.26), $\rho_\nu = \rho(t_\nu)$, $\Delta t_i = t_i - t_{i-1}$, and $\Pi^i(t) = \Pi_\phi^i(t)$, and where $\rho_0 J_t^{m,0} = \rho_{N+1} J_t^{m,N+1} = 0$. Representations for $\underline{M}_m$ and ρ_m^m follow from (6.23) and (6.26) by replacing electric source quantities by the corresponding magnetic source quantities.

Expansions for all the equivalent source quantities are substituted into (6.16) and (6.17). The result is a set of coupled integral equations which depend on the unknown coefficients of the current expansions. A system of simultaneous equations, which may be solved for these coefficients, is obtained in the method of moments procedure by "testing" the coupled equations with a set of testing functions. For the body of revolution geometry of interest here, we choose the testing functions to be

$$T_t^{pq}(t,\phi) = \frac{\Pi_t^q(t)}{\rho} e^{-jp\phi} \tag{6.27}$$

and

$$T_\phi^{pq}(t,\phi) = \frac{\Delta t_q}{\rho} \delta(t-t_{q-1/2})e^{-jp\phi} \tag{6.28}$$

The t-components of (6.16) and (6.17) are tested with (6.27) (i.e., the t components of (6.16) and (6.17) are multiplied by the function in (6.27) and the resulting terms are integrated over the surface S), while the ϕ-components are tested with (6.28). The result is a set of simultaneous equations of the form (6.22) for the mth Fourier component. Details of the application of the method of moments to obtain the moment matrix in (6.22) may be found in [33] and [34].

Under the source-free conditions assumed for the analysis of the dielectric resonator, of course, the matrix equation (6.22) has a solution only when the determinant of the moment matrix $\underset{\sim}{Z}_m$ is zero:

$$\det(\underset{\sim}{Z}_m) = 0 \tag{6.29}$$

The next step in the solution procedure is, therefore, to determine the particular values of complex frequency for which (6.29) holds. At such frequencies, the matrix equation (6.22) can be solved for the current coefficient vector $|I_m\rangle$, which represents the t variation of the surface current distribution having Fourier index m, or equivalently, of the distribution at the surface S of the tangential components of the electric and magnetic fields with Fourier index m.

6.5 Computation of Resonant Frequency and Q Factor

In this section we consider the problem of determining the frequencies at which solutions of (6.16) and (6.17) exist in the absence of impressed sources, i.e., the frequencies at which the source-free modal fields exist. This problem is analogous to the problem of determining the natural frequencies of a passive, lumped-element circuit, where a circuit "response" can exist even though the external voltage and current sources have been set to zero [35]. In the numerical analysis of the dielectric resonator, it corresponds to the determination of frequencies for which (6.29) holds. For the isolated resonator geometry indicated by Fig. 6.1, however, there will be no real frequencies at which source-free fields can exist, such as there are within perfectly conducting cavities, because energy must be lost in the form of radiation. Hence, (6.16) and (6.17) have no solution for real frequencies. The exponential decay in time of the fields due to radiation, however, can be included by defining the complex frequency:

$$s = \sigma + j\omega \tag{6.30}$$

where ω is the actual angular frequency of oscillation and σ is related to the time decay rate of the field [35]. The symbol σ used to denote the real part of the complex frequency in (6.30) is not to be confused with the conductivity σ used in Fig. 6.1. The meaning of the symbol should be clear from the context in which it is used.

When (6.30) is used, the time dependence of the field is assumed to be of the form exp(st) and is suppressed. The equations preceding (6.30) remain valid with the introduction of the complex frequency if one replaces the term $j\omega$ by s everywhere it appears, including in the definition of the square of the wave number k ($k^2 = -(j\omega)^2\mu\epsilon$) and in the complex permittivity if the medium is lossy. With the frequency generalized to be complex in this manner, source-free field solutions to (6.16) and (6.17) exist at discrete complex frequencies. These frequencies can be determined numerically by searching for the frequencies at which the determinant of the moment matrix $\underset{\sim}{Z}_m$ is zero as indicated in (6.29). The roots of (6.29) in the complex frequency plane are designated by

$$s_{m,\nu} = \sigma_{m,\nu} + j\omega_{m,\nu} \tag{6.31}$$

where $\omega_{m,\nu}$ is the resonant frequency of the mode (m,ν) and $\sigma_{m,\nu}$ is the decay time constant of the mode. The subscript ν in this notation is an integer used to denote unique complex frequency values for which (6.29) is satisfied. For each value of m in (6.29), there will be solutions for $\nu = 1,2,3,\ldots$ These solutions each correspond to a different resonant mode. Once the electromagnetic field distribution is determined for a particular solution, the index ν may be replaced by two subscripts (n and p) which denote the radial and axial behaviors of the field, if this is appropriate for the resonator shape under consideration (such as for a cylindrical resonator). The quantity $\sigma_{m,\nu}$ in (6.31) is inversely proportional to the radiation Q factor for the mode (see also Ch. 2):

$$Q_{m,\nu} = -\frac{\omega_{m,\nu}}{2\sigma_{m,\nu}} \tag{6.32}$$

The roots (6.31) of (6.29) may be found by searching in the complex frequency plane using one of several available search techniques. For a dielectric resonator, however, the search in the complex frequency plane for the roots of (6.29) can be made fairly efficient because the Q factor for the modes of interest is usually relatively large. Thus, it is generally practical to search along the imaginary axis (where $s = j\omega$) for crude values of the resonant frequency $\omega_{m,\nu}$ (i.e., $2\pi f_{m,\nu}$). A plot of the moment matrix determinant along the imaginary axis of the complex frequency plane is shown in Fig. 6.5 for a case in which the fields are azimuthally symmetric (m = 0). The dielectric resonator in this example is a cylindrical "pillbox" which has relative permittivity $\epsilon_r = 35$, radius a = 5 mm, and length h = 5 mm. The absolute value, the real part, and the imaginary part of the determinant are plotted for the Fourier component m = 0. For the numerical model in this example, the generating arc of the body of revolution is described by only 7 points (N = 5 in (6.23)), and the resulting matrix is of size 22 × 22. In the range between 2 GHz and 8 GHz the absolute value of the determinant in Fig. 6.5 shows two distinct minima, one at 5.1 GHz and the other at 7.6 GHz. With the help of diagrams from Gelin et al. [11], the two resonant modes can be tentatively identified as $TE_{01\delta}$ and $TM_{01\delta}$. Identification of modes on the basis of their field distributions is discussed in Sec. 6.10.

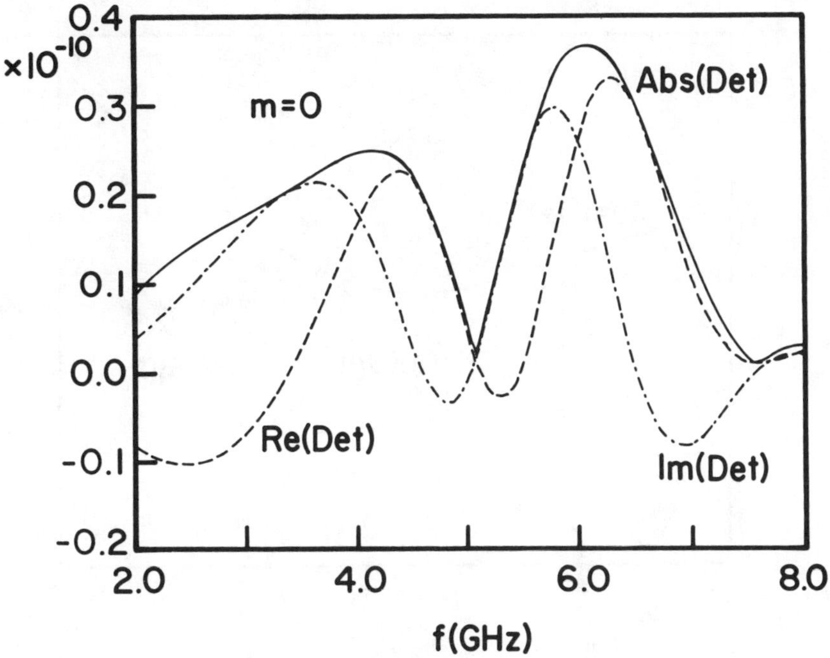

Fig. 6.5 Moment matrix determinant for azimuthally symmetric Fourier component (m = 0) along the imaginary axis ($s = j\omega$) of the complex frequency plane for a cylindrical dielectric resonator with $\epsilon_r = 35$, radius a = 5 mm, and h = 5 mm (reference [19], ©1983 IEEE)

The use of relatively few unknowns, and hence a small matrix, as in the preceding example, provides a fairly rapid means of locating crude resonant frequency values. To determine more accurate values of the resonant frequencies, as well as the values of the corresponding Q factors, we must extend the search for roots of (6.29) off the imaginary axis of the complex frequency plane. In Fig. 6.6 is shown an example of the behavior of the moment matrix determinant along a straight line path that is perpendicular to the imaginary axis and begins at an imaginary-axis minimum of the determinant. It is observed that, while the absolute value shows a broad minimum, the real and imaginary parts almost appear to be two straight lines, each going through zero at a different point. This curve suggests that the real and imaginary parts of the determinant may each be approximated by a linear function of the complex frequency in the vicinity of the complex root. With this

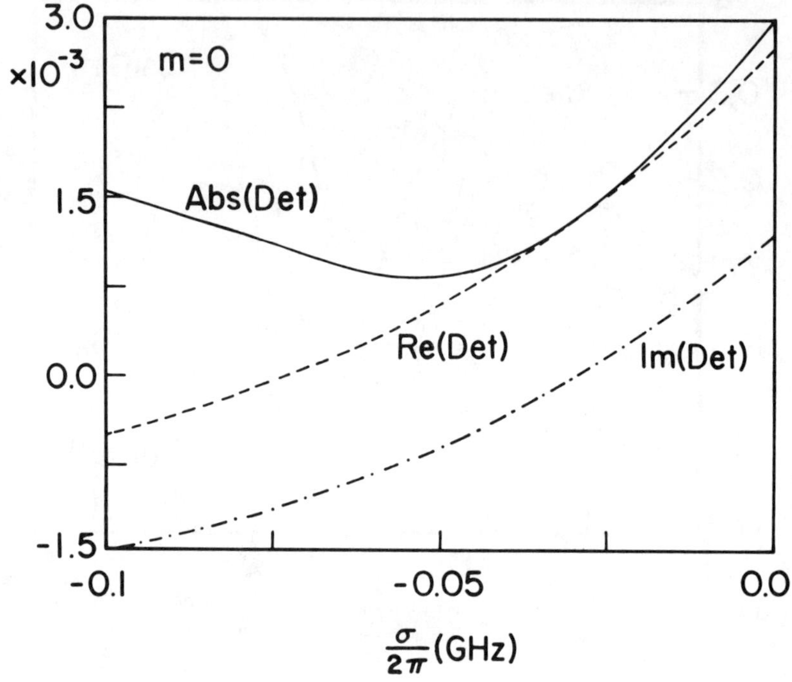

Fig. 6.6 Example of moment matrix determinant along a path beginning at an imaginary-axis minimum of the determinant and proceeding along a constant-ω cut in the complex frequency plane (reference [19], ©1983 IEEE)

approximation, it is possible to employ a simple iterative search procedure in which each iteration requires the evaluation of the moment matrix at only three points in the complex plane [36]. A simple two-point per step iterative procedure can also be implemented via a complex-plane Taylor series representation of the determinant. For efficiency, the iterative search may begin using only a few unknowns to model the resonator. As the search proceeds, the model accuracy may be improved by increasing the number of expansion functions for the equivalent surface currents, so that a more accurate resonant frequency value is obtained. Since each point in the search is obtained by computing the determinant of, for example, a 22×22 (or larger) matrix, the need for economy of computer time is evident. It has been found that three to five iterative steps are usually necessary for an accurate determination of the complex root.

The dependence of the resonant frequency and the Q factor on the accuracy of the numerical model is indicated in Fig. 6.7. This figure shows the convergence of the computed resonant frequency and Q factor as the number of points modeling the generating arc (N + 2) is increased from 7 (22 unknowns) to 37 (142 unknowns). The results are plotted versus 1/(N + 2), for the $TE_{01\delta}$ resonance of the same resonator as in the previous example (ϵ_r = 35, a = 5 mm, h = 5 mm). One observes that

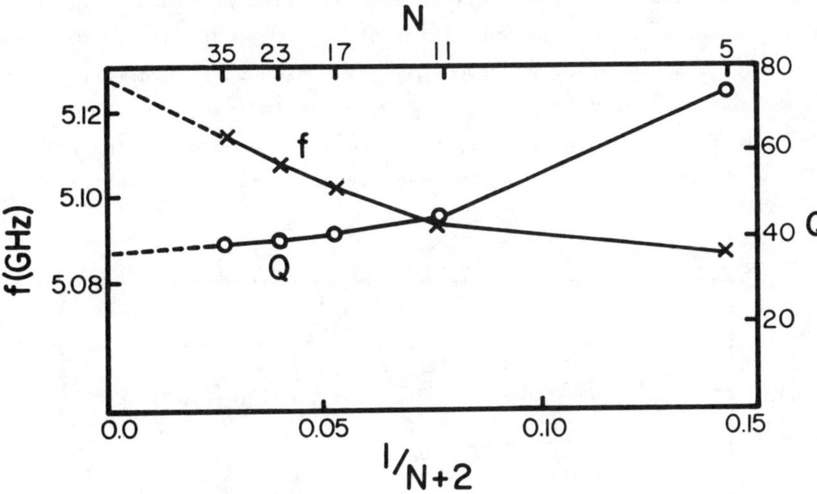

Fig. 6.7 Resonant-frequency and Q-factor values as functions of a number of unknowns used in the numerical model for a cylindrical dielectric resonator with ϵ_r = 35, a = 5 mm, and h = 5 mm (reference [19], ©1983 IEEE)

the resonant frequency and the Q factor show good convergence as N is increased. Note that the resonant frequency computed using N = 5 differs by less than 1 % from the value of the resonant frequency which would be obtained by extrapolating the curves (dashed lines) to represent an infinite number of points. The Q factor generally converges somewhat more slowly than does the resonant frequency, as is evident in Fig. 6.7. The extrapolated value of resonant frequency for $N \to \infty$ agrees with Gelin et al. [11] within 1 %, while the extrapolated value of Q is lower, coming closer to the values given by Verplanken and Van Bladel [4]. The resonant-frequency and Q-factor values shown in Fig. 6.7 also agree well with the predicted values of Q = 40 and f = 5.1 GHz given by Tsuji et al. [16].

As previously indicated, we may choose in advance the azimuthal variation of our solutions by picking the Fourier index m. If the index is changed to m = 1, we obtain the so-called hybrid electromagnetic (HEM) field solutions with variation exp(jϕ). A plot of the determinant of the moment matrix along the imaginary axis ($\sigma = j\omega$) of the complex frequency plane is shown in Fig. 6.8 for the case in which m = 1 and for the same resonator as in the previous example. The absolute value of the determinant has two apparent minima below 8 GHz, one at 6.3 GHz and another at 7.1 GHz. Based on mode charts for dielectric rod waveguides, such as those presented in Ch. 3, the two modes indicated by the determinant minima in Fig. 6.8 may be tentatively denoted as the $HEM_{11\delta}$ and $HEM_{12\delta}$ modes, where the hybrid modes of a dielectric resonator are

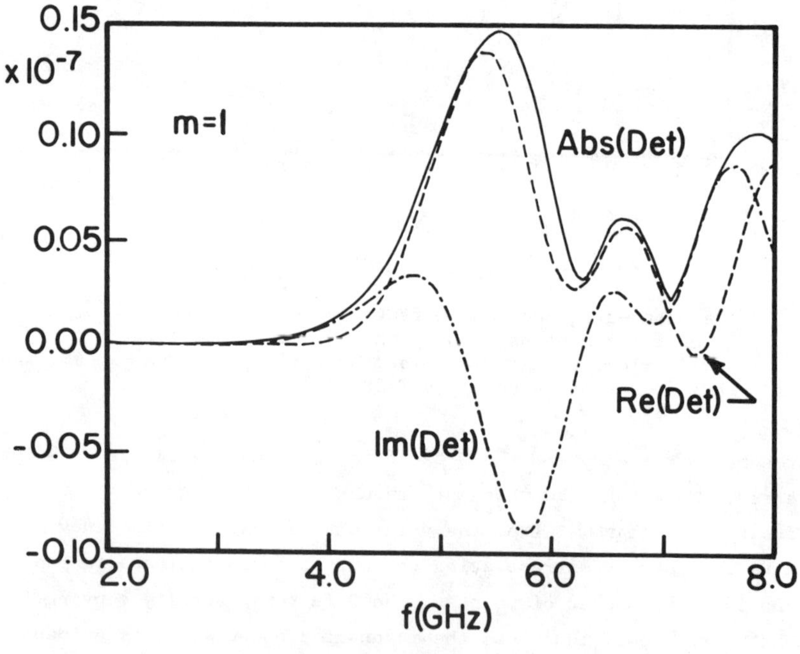

Fig. 6.8 Moment matrix determinant for the Fourier component m = 1 along the imaginary axis ($s = j\omega$) of the complex frequency plane for a cylindrical dielectric resonator with $\epsilon_r = 35$, radius a = 5 mm, and height h = 5 mm (reference [19], ©1983 IEEE)

denoted by HEM_{mnp} in accordance with the notation of Ch. 3. The first, second, and third subscripts in this notation specify the nature of the azimuthal, radial, and axial variations, respectively. Generally, for the modes of interest encountered in dielectric resonators, the third subscript is smaller than unity and is denoted by the symbol δ.

The search for complex roots of (6.29), however, provides the investigator only with the knowledge that a resonant mode exists at the complex frequency $s_{m,\nu}$. Identification of the mode number ν (or the mode numbers n and p, if the resonator is cylindrical in shape) is another problem. Some use may be made of reference materials on similar resonators or, as above, of mode charts for other structures such as the dielectric rod waveguide. In a limited sense, one may also utilize the equivalent surface currents on the resonator or equivalently, the surface fields, to identify the mode corresponding to a particular resonant frequency. For a reliable mode identification, however, it is necessary to compute the detailed field distribution in and around the resonator. Discussions on the computation of the equivalent surface currents and the field distribution, as well as the use of these quantities in the identification of modes, are presented in subsequent sections. Before proceeding to these topics, we present in the next section a comparison between numerical results for resonant frequencies and Q factors obtained via the procedures described here and some experimental results for isolated dielectric resonators.

6.6 Numerical and Experimental Results

Numerical and experimental results have been obtained and compared for a cylindrical "pillbox" dielectric resonator having relative permittivity $\epsilon_r = 38$, radius a = 5.25 mm and height h = 4.6 mm. The measured data were obtained using a network analyzer and the experimental set-up depicted in Fig. 6.9. The transmission method was used to obtain most of the data. The resonator was situated in a box padded with absorbing material to simulate the free-space environment and was electromagnetically coupled to semirigid coaxial cables by a small balanced loop and by a balanced dipole. The balanced arrangement was found to be essential for avoiding external currents on the cable shields, which caused serious difficulties at the beginning of the experimental investigation.

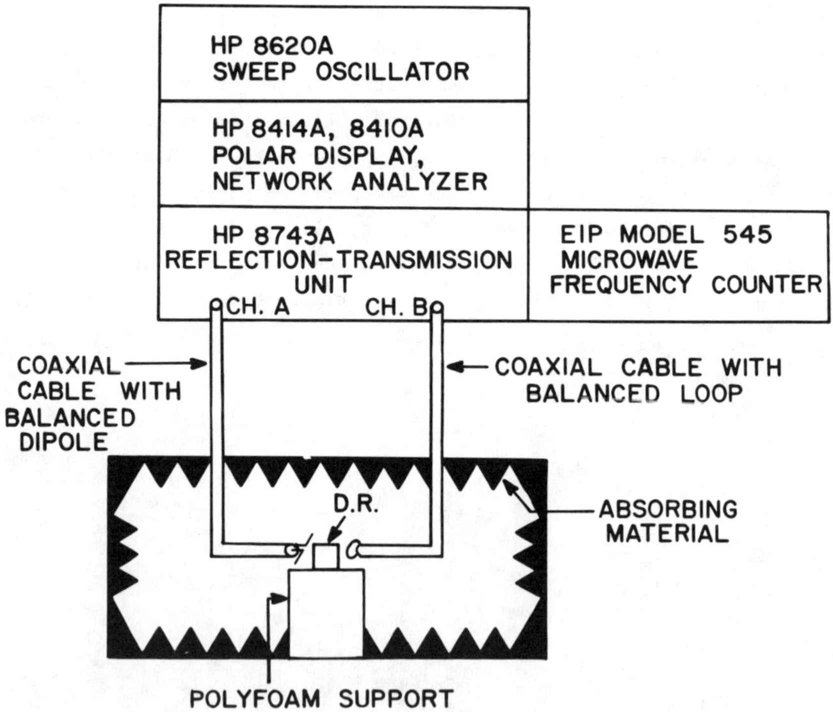

Fig. 6.9 Block diagram of apparatus for measuring resonant frequencies and Q factors of isolated dielectric resonators using the transmission method

Computed values of the resonant frequencies and Q factors for the above resonator are shown in Table 6.1. These values have been obtained for the $TE_{01\delta}$ and the $TM_{01\delta}$ modes with N = 16, which corresponds to an 18-point model of the generating arc and 66 unknowns. The values for the HEM modes have geen obtained using a 27-point model for the resonator (N = 25; 102 unknowns). The computed and measured values of resonant frequencies of various modes agree to within about 0.5 % for the modes for which measured data were obtained. The agreement between the computed and measured values of Q factor is not quite as good. In the worst case (mode $HEM_{21\delta}$), the measured value of Q is about 38 % lower than the computed value when the transmission method is used for the measurement. As the transmission method is not very reliable for the Q measurement, the reflection method was also attempted. It proved to be difficult to obtain sufficient coupling to the coaxial line using this method, especially for the modes with a low Q factor. For the two modes where the reflection measurement was possible, the computed and measured values of Q agreed to within about 3 % for initial calculations using a resonator model with N = 11 (46 unknowns). Subsequent calculations [22] in which the more accurate resonator model was used (N = 25; 102 unknowns), however, showed the higher theoretical Q value of 327 for the $HEM_{21\delta}$ mode indicated in Table 6.1. This value represents a difference of 13 % when compared to the measured value.

Numerical computations show that the resonant frequency of the $HEM_{11\delta}$ mode is 6.333 GHz, which is within the frequency range covered by Table 6.1. The computed Q factor of 30.7 for this mode, however, is significantly lower than that of the other modes. In spite of repeated attempts, accurate experimental observation of the resonance of the $HEM_{11\delta}$ mode was not possible for the resonator in the simulated free-space environment, probably because of the difficulty in coupling to such a low-Q mode. Therefore, measured values for this mode are not included in the chart. It is worth noting here, however, that when the resonator is placed within a shielding enclosure, the $HEM_{11\delta}$ mode is generally well-defined and is easily observed experimentally. Because of the low Q factor (high radiation loss) of this mode for an isolated resonator, one might consider the possibility of using a resonator operating in this mode as a radiating element. This idea is discussed further in a subsequent section.

The numerical procedure has also been used to compute a universal mode chart for isolated cylindrical dielectric resonators with $\epsilon_r = 38$. The chart, which is shown in Fig. 6.10, provides resonant frequency information in terms of k_0a for several modes as a function of the resonator radius to height ratio a/h. In order to economize the computer time, the resonant frequencies for this chart were determined by simply tracking the minimum of the determinant for each mode on the imaginary axis of the complex plane as the geometry is changed. In addition, an attempt was made to keep the number of points used to model the resonator generating arc to a minimum (13 points; N = 11), but it was necessary to increase the number of points for certain values of a/h. Such changes of N represent part of the reason for the slight kinks in the curves shown in Fig. 6.10. The kinks may also result from the difficulty in determining the "minimum" along the imaginary axis, as well as from the fact that the resonant frequency indicated by a minimum

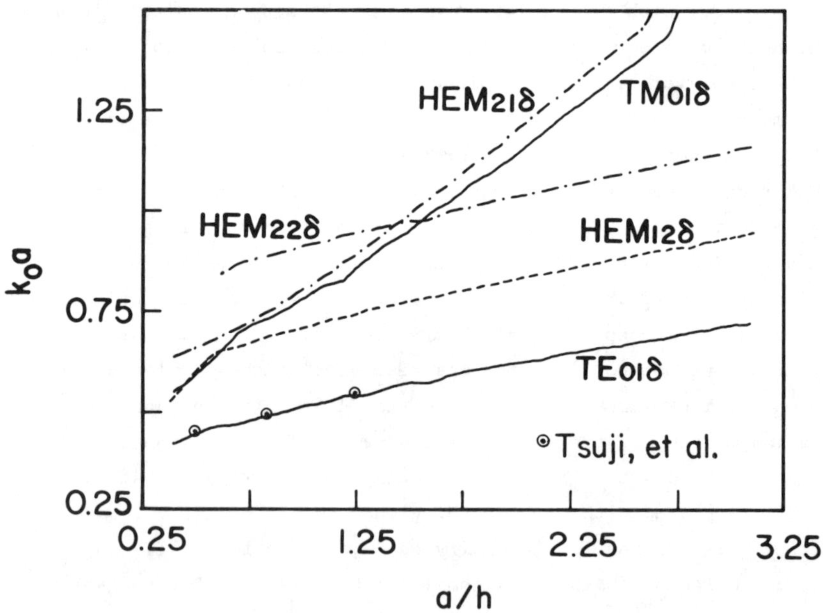

Fig. 6.10 Universal mode chart for isolated cylindrical dielectric resonators with permittivity $\epsilon_r = 38$ (reference [19], ©1983 IEEE)

on the imaginary axis is slightly different from the actual complex plane resonant frequency.

Shown on the mode chart for comparison are three points taken from a curve computed for the $TE_{01\delta}$ mode by Tsuji et al. [20]. The $HEM_{11\delta}$ mode is omitted from this mode chart for two reasons. First, the minimum of the determinant for this mode is poorly defined on the imaginary axis because of its low Q, so that it was often overlooked in our "economized" search process. Second, the experimental investigation did not confirm the existence of this mode.

Table 6.1 COMPARISON OF COMPUTED AND MEASURED RESONANT-FREQUENCY AND Q-FACTOR DATA FOR A CYLINDRICAL DIELECTRIC RESONATOR WITH ϵ_r = 38, RADIUS a = 5.25 mm, AND HEIGHT h = 4.6 mm

Mode	f (GHz)	
	Computed	Measured
$TE_{01\delta}$	4.829	4.85
$HEM_{11\delta}$	6.333	--
$HEM_{12\delta}$	6.638	6.64
$TM_{01\delta}$	7.524	7.60
$HEM_{21\delta}$	7.752	7.81

Mode	Q		
	Computed	Measured (Transmission Method)	Measured (Reflection Method)
$TE_{01\delta}$	45.8	51	47
$HEM_{11\delta}$	30.7	--	--
$HEM_{12\delta}$	52.1	64	--
$TM_{01\delta}$	76.8	86	--
$HEM_{21\delta}$	327.1	204	288

6.7 Computation of Modal Surface Current Distributions

When the surface integral equation approach is used to model a dielectric resonator, computation of the equivalent surface currents is a prerequisite for calculation of the field distribution in the vicinity of the resonator. Knowledge of the equivalent surface current distribution also provides a relatively rapid, but preliminary, means of mode identification when a resonant frequency is found, since the surface currents actually represent the electric and magnetic fields tangential to the body surface. Once a search in the complex frequency plane, as described in Sec. 6.5, has yielded a resonant frequency to a sufficient degree of accuracy, it is relatively easy to calculate the modal surface current distributions. The first step is to compute the moment matrix $\underset{\sim}{Z}_m$ with the frequency at the newly found complex resonant frequency $s_{m,\nu}$. At this frequency, the determinant of the moment matrix is zero and the homogeneous system of equations represented by

$$\underset{\sim}{Z}_m \, |I_m\rangle = |0\rangle \qquad (6.33)$$

must be satisfied, where $|I_m\rangle$ is the vector containing the unknown coefficients of the surface current expansions. Since the moment matrix determinant is zero, there is no unique solution for the coefficient vector $|I_m\rangle$. If the determinant has a simple zero, however, the rank of the matrix will be 1 less than its order. In this case, one may choose the value for one of the coefficients of the vector $|I_m\rangle$ and the remaining coefficients will be uniquely determined [37]. Whether the determinant has only a simple zero or not depends on the system geometry and on the operators from which the matrix is generated. For the dielectric body of revolution and the combined integro-differential operators used in (6.16) and (6.17), one would expect degeneracies in the surface current solution and in the electromagnetic field of the same type as those which exist within cylindrical cavities, i.e., those of the cos(mϕ) and sin(mϕ) type. Such degeneracies would generally lead to multiple zeros of the moment matrix determinant. In the development of the moment method solution procedure in Sec. 6.4, however, we have effectively assumed that only one of these variations exists by specifying the relationship between the modes which vary as exp(jmϕ) and those varying as exp(-jmϕ). Thus, we expect the moment matrix $\underset{\sim}{Z}_m$ to have a simple zero. Inclusion of the remaining variation is easily accomplished after the vector $|I_m\rangle$ is computed, if desired.

Since we assume that the determinant of the moment matrix has a simple zero at the frequency $s_{m,\nu}$, it is convenient to employ a Gaussian elimination procedure to reduce the matrix and obtain a solution for $|I_m\rangle$. The last step in the reduction of a system of simultaneous equations in this manner leads to a scalar equation of the form:

$$a_{MM} i_M = 0 \tag{6.34}$$

where M represents the number of equations in the system (and the total number of unknowns). Thus, i_M is the last element in the vector $|I_m\rangle$ and, when the moment matrix is evaluated at a resonant frequency $s_{m,\nu}$, the elimination process leads to $a_{MM} = 0$. (In practice a_{MM} is usually a very small non-zero value, due to round-off error, that is taken to be zero. Also, the element i_M is not necessarily the "last" element in the original vector $|I_m\rangle$ if pivoting is used in the Gaussian elimination process.) Since $a_{MM} = 0$, (6.34) is simply a statement that i_M may take on any arbitrary value and, hence, that there are an infinite number of solution vectors $|I_m\rangle$ satisfying (6.33), which may be obtained by back substitution. If, however, we choose an arbitrary value for i_M, say $i_M = 1$, and begin the back substitution process, we obtain a unique solution for the vector $|I_m\rangle$ containing the coefficients of the surface current expansion (unique for this particular choice of i_M).

One problem which has been encountered during the computation of the current coefficient vector $|I_m\rangle$, when one directly generates the moment matrix as described in [32-34], is that numerical instabilities may appear when large numbers of unknowns are used and higher-order modes are studied. These instabilities occur because the individual elements in the unknown surface current vector $|I_m\rangle$ (as defined in [32-34]) are expressed in mixed physical units and because the rows of the moment matrix represent mixed physical units (because (6.16) and (6.17) do not have the same units). This mixing of units results in a moment matrix which is not as well-conditioned as it could be. Kajfez et al. have shown in [22] that a straightforward normalization of the moment matrix can significantly improve the condition number defined by Klein and Mittra for the matrix [38] and, consequently, remove or reduce the numerical difficulties associated with the computation of the modal surface currents. It is recommended that such a normalization procedure always be applied to the system of simultaneous equations (6.33).

6.8 Interpretation of Modal Surface Current Distributions

The modal surface current distributions represented by the coefficient vector $|I_m>$ are directly related to the tangential components of the electric and magnetic fields at the surface of the body. Hence, for a particular root found in the complex frequency plane search, these distributions can serve as guides in determining the mode indices n and p, which represent the radial and longitudinal variations of the fields, respectively, for cylindrical resonators. As long as a resonator is "relatively cylindrical" in shape, the currents may be useful in determining indices n and p, even though the indices would begin to lose their correspondence to the radial and longitudinal directions. For non-cylindrical resonators, however, these two mode indices can be replaced by the single index ν, and in such cases the relationship of the surface current distribution to the mode index may not be a priori clear. The reader should recall that the mode index m in, for example, HEM_{mnp} represents the azimuthal variation of the field and is selected via the testing functions defined by (6.27) and (6.28) in Sec. 6.4 before the resonant frequency is found. The surface current distributions should be used only as an indicator for the remaining mode indices n and p, however. A precise determination of the mode indices should be made based on evaluation of the field distribution in the resonator as described in Sec. 6.10.

The current coefficient vector $|I_m>$ determined from the solution of (6.33) at a resonant frequency actually represents four distinct surface current quantities. Thus, it is convenient to partition the vector in four column subvectors before normalization as

$$|I_m> = \begin{pmatrix} |I_t> \\ |J_\phi> \\ |K_t> \\ |M_\phi> \end{pmatrix} \tag{6.35}$$

where the subvectors correspond to the two vector components of quantities related to the electric and magnetic currents on the surface. The subscripts ϕ and t in (6.35) denote the spatial vector components of the currents in the azimuthal direction and in the direction along the generating curve for the body of revolution, respectively. Figure 6.11

illustrates the orientation of the components of the equivalent electric surface current densities J_t and J_ϕ on various parts of a dielectric resonator which is cylindrical in shape. For a more general geometry, one should refer to Fig. 6.3. For the cylindrical resonator of Fig. 6.11, however, note that J_t is parallel to the z-axis for points on the cylindrical surface, whereas on the two flat end faces of the resonator, the orientation of J_t corresponds to either the positive or negative radial direction. The J_ϕ component is directed along the ϕ coordinate lines on any face. Similar comments apply to the equivalent magnetic surface current densities M_t and M_ϕ.

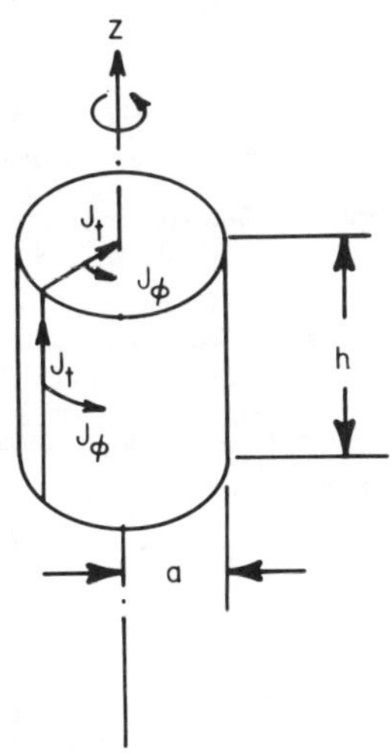

Fig. 6.11 Orientation of components of equivalent currents on surface of a cylindrical dielectric resonator (reference [22], ©1984 IEEE)

The current coefficient vector $|I_m>$ is partitioned before normalization as indicated in (6.35) for consistency with previous work [32-34]. The subvectors $|I_t>$ and $|K_t>$ appearing on the right side of (6.35) do not represent the surface current densities J_t and M_t directly. Instead, they actually represent coefficients of the expansion of the quantities $2\pi\rho J_t$ and $2\pi\rho M_t$, where ρ is the radial distance to the current source in a cylindrical system of coordinates. Expansions for these quantities are used, rather than for the current densities J_t and

M_t, because their use simplifies the evaluation of the moment matrix. In addition, as has been indicated in the preceding section, the current vector $|I_m>$ should be normalized to improve the condition number of the moment matrix. The normalization process, however, changes the physical units of some elements of the current coefficient vector. After computation of the current coefficient vector $|I_m>$, it is, therefore, desirable to rescale the vector into the unnormalized form of (6.35). Before attempting to interpret the results it is also convenient to convert the subvectors $|I_t>$ and $|K_t>$ to vectors representing surface current densities by dividing the ith element in each subvector by $2\pi\rho_i$ for i = 1 to N, where ρ_i is the radial coordinate of the ith point defining the generating arc, to obtain subvectors $|J_t>$ and $|M_t>$.

The sample results displayed in Figs. 6.12 and 6.13 have been computed for a cylindrical dielectric resonator, such as illustrated in Fig. 6.11, with dimensions a = 5.25 mm, h = 4.6 mm, and with permittivity ϵ_r = 38. In these figures the circular data points represent

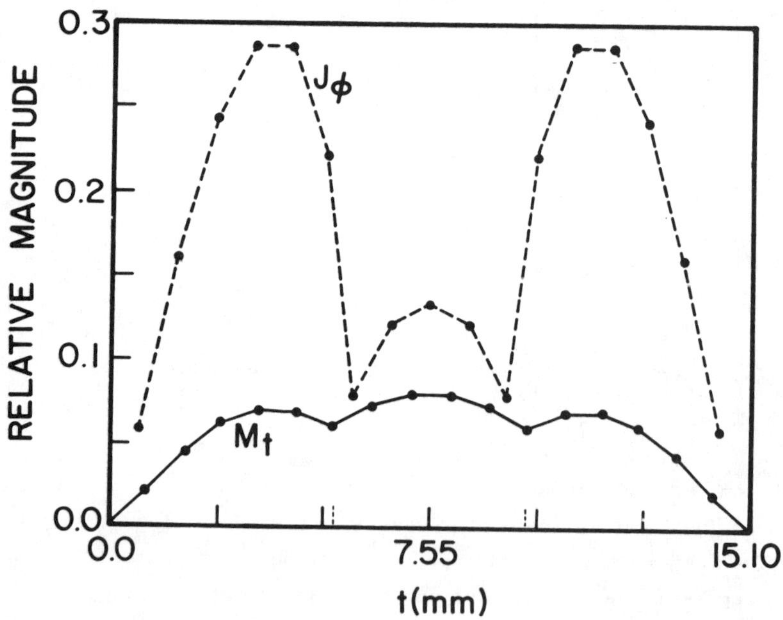

Fig. 6.12 Distributions of magnetic and electric surface current densities for $TE_{01\delta}$ mode of cylindrical dielectric resonator (reference [19], ©1983 IEEE)

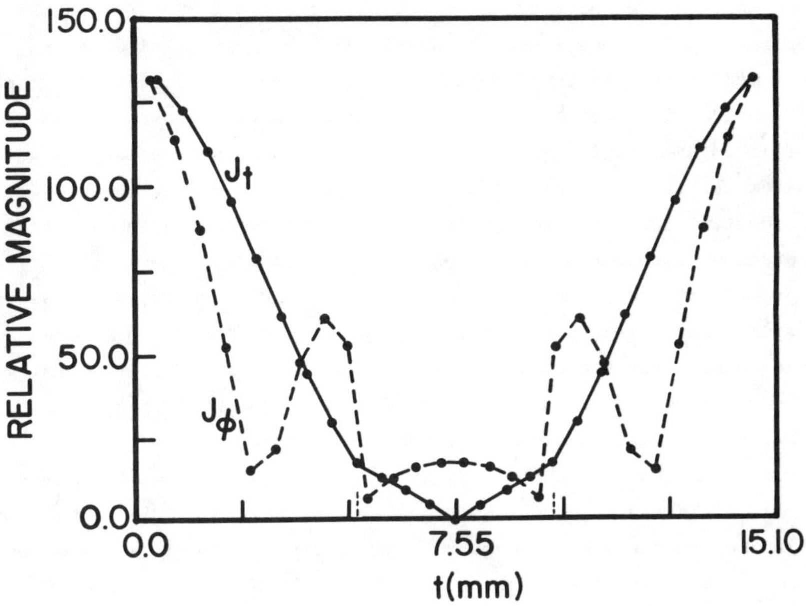

Fig. 6.13 Distribution of electric surface current density for $HEM_{12\delta}$ mode of cylindrical dielectric resonator (reference [19], ©1983 IEEE)

the magnitudes of the elements of one of the current coefficient subvectors $|J_t>$, $|J_\phi>$, $|M_t>$, or $|M_\phi>$. Straight lines are used to connect the data points to provide a better indication of the current distribution. The actual current magnitude distribution, of course, does not exhibit any slope discontinuities, except possibly at the corners of the cylindrical resonator and at points where the current magnitude is zero. The curves are plotted as a function of the arc length variable t along the surface for a constant value of the ϕ-coordinate. The variation in the ϕ direction for each individual current component is either $\cos(m\phi)$ or $\sin(m\phi)$.

For the resonator specified above, the amplitudes of the magnetic surface current density M_t (solid line) and the electric surface current density J_ϕ (dashed line) computed for an azimuthally symmetric mode (m = 0) with a complex resonant frequency located at $s = 2\pi(-0.05272 + j4.829) \times 10^9$ are shown in Fig. 6.12. The currents were computed using an 18-point model of the resonator (66 unknowns). The remaining two current components M_ϕ and J_t are zero at this frequency. From Sec. 6.6

we find that the resonant frequency of this mode is 4.829 GHz and the Q factor is 45.8 (for the 18-point model). The current magnitudes are plotted beginning from t = 0 at the center of one end face of the resonator, progressing radially outward to the resonator corner, up the side (parallel to the z-axis) to the second corner, and then radially inward to the center of the remaining face. The positions of the corners of the resonator along the arc length variable are indicated in the figure by the dashed tick marks. The amplitudes indicated on the vertical scale in the figure are only significant in relative terms, since we are finding the natural response of the system. Since $\underline{M} = \underline{E} \times \hat{\underline{n}}$ at the surface, where $\hat{\underline{n}}$ is the outward-directed surface normal, it is readily seen from Fig. 6.11 that M_t is proportional to the E_ϕ component of electric field everywhere on the resonator surface. Using $\underline{J} = \hat{\underline{n}} \times \underline{H}$, on the other hand, reveals that J_ϕ is proportional to the radial magnetic field H_ρ on the end faces of the resonator, but is proportional to H_z on the cylindrical surface of length h. Similarly, J_t would be proportional to H_ϕ on all surfaces of the resonator, while M_ϕ would be proportional to E_ρ and E_z on the end faces and the resonator side, respectively, if they were non-zero. Because the resonant frequency s was located assuming no azimuthal variation (m = 0), and because $M_\phi = 0$ implies $E_z = 0$ along the side of the resonator, one might surmise that the mode is a TE mode (transverse electric to z). It clearly cannot be a TM mode because J_ϕ, and, hence, H_z, is non-zero along the cylindrical side of the resonator. The relatively constant behavior of M_t (E_ϕ) and the single extremum of J_ϕ (H_z) in the z-direction along the side of the resonator suggest that the axial mode index $p = \delta$ should be used to indicate less than one complete half-wavelength of variation in the z-direction. The single pronounced extremum of J_ϕ (H_ρ) on each end face of the resonator suggests that the radial mode index n = 1 should be used to indicate variation close to one half-wavelength in the radial direction. In this region, M_t also exhibits only one extremum on each face. Thus, based on the surface current distributions, it is established that the azimuthally symmetric mode located at 4.829 GHz is the $TE_{01\delta}$ mode.

If one searches in the complex plane for resonant frequencies of modes having the Fourier index m = 1, i.e., those having an azimuthal variation of $\exp(j\phi)$ (or equivalently, $\cos\phi$), one of the resonances found, using a 27-point model (102 unknowns) for the same resonator as before, corresponds to a frequency of 6.638 GHz and Q factor of 52.1. The magnitudes of the modal electric surface current components evaluated at this resonance are shown in Fig. 6.13. The plot of the modal magnetic current distribution has been omitted for brevity. The distributions shown in Fig. 6.13 are slightly more complicated than in the previous example, but are interpreted similarly. If we investigate the fields on the end faces of the resonator in the same manner as before, we find that the radial magnetic field (J_ϕ) varies more rapidly than does the other component of magnetic field and it clearly exhibits two extrema in the radial direction. This indicates a probable radial mode index of n = 2. Both electric current components have variations along the cylindrical surface of the resonator characteristic of a half-wavelength (or slightly less) resonance, indicating a z-variation mode index of $p = \delta$. One would, therefore, conclude that the mode with azimuthal variation specified by m = 1 and resonant frequency 6.638 GHz is the $HEM_{12\delta}$ mode.

The behavior of the modal currents on the end faces of the resonator may be better understood when compared with actual field patterns. A convenient comparison can often be made with the field patterns in an infinitely long dielectric rod waveguide. The fields for several modes of a dielectric rod waveguide have been computed and displayed via computer graphics by Kajfez [39]. The waveguide fields of several modes have also been displayed in this manner and discussed in Ch. 3.

6.9 Computation of Modal Field Distributions

As indicated previously, it is generally desirable to know the actual field structure in and around the resonator for mode identification and so that more efficient coupling or mode suppression techniques can be developed. While use of the surface currents (i.e., the fields tangential to the resonator surface) for mode identification is illustrated in the preceding section, the interpretation of these currents may be difficult and could lead to a mistake in identification of some modes, particularly in the case of the higher-order hybrid modes. The equivalent surface currents are computed primarily so that we may use them to calculate the field anywhere in space, either inside or outside the resonator.

The electric and magnetic surface current densities are used to compute the fields in the region outside the resonator by allowing them to radiate via the homogeneous-region electric and magnetic vector and scalar potentials in which the medium parameters are those of the region exterior to the resonator, as indicated by equations (6.1) and (6.2) with $i = 1$ and by Fig. 6.2(a). To compute the fields inside the resonator, the same equations would be used, but with the sign of the computed currents changed and with all medium parameters in the equation set to those of the dielectric resonator ($i = 2$ in the equations), so that the currents $(-\underline{J}, -\underline{M})$ radiate in a homogeneous region as in Fig. 6.2(b). For the body of revolution, however, the process can be made somewhat simpler because it is already known that each element of the Fourier component current set $(\underline{J}_m(t), \underline{M}_m(t))$ with index m varies as $\exp(jm\phi)$ and produces fields $(\underline{E}_m(t), \underline{M}_m(t))$, which also vary as $\exp(jm\phi)$. Furthermore, for the present dielectric resonator analysis, we are interested in computing the fields due to only one of these modal current sets. It is, therefore, possible to reduce the equations so that the fields $(\underline{E}_m(t), \underline{M}_m(t))$ are computed along a single cut, for example, $\phi = 0$, and to include the $\exp(jm\phi)$ variation later. Thus, one would simply write a program that accepts the current coefficient vector $|I_m>$ as input data, substitutes these coefficients appropriately into current and charge expansions, such as equations (6.23) and (6.26), and accumulates the vector sum of the fields resulting from each of the terms in these expansions. Computation of the field due to each expansion term involves, of course, a numerical surface integration over the source band for an expansion function. This integration must be performed carefully when the observation point is near the source region.

A somewhat more convenient procedure in which the preceding difficulties have already been resolved can be developed if one recalls that the simultaneous equation system $\underset{\sim}{Z}_m|I_m>$ actually represents equations of the form (cf. equations (6.14) and (6.15))

$$\hat{\underline{n}} \times [\underline{E}^s(\underline{J},\underline{M}) - \underline{E}(-\underline{J},-\underline{M})] = 0 \quad (6.36)$$

$$\hat{\underline{n}} \times [\underline{H}^s(\underline{J},\underline{M}) - \underline{H}(-\underline{J},-\underline{M})] = 0, \quad r \epsilon s \quad (6.37)$$

where $(\underline{E}^s,\underline{H}^s)$ and $(\underline{E},\underline{H})$ are the fields evaluated just outside and just inside the resonator surface, respectively. In particular, a single element of the moment matrix for mode m is related to the difference between one component of a field, for example, E_ϕ, evaluated at point $\underline{r}$ due to a unit current source "band" (e.g., a basis function for M_ϕ such as $\Pi^i_\phi(t')e^{jm\phi'}$ existing over a range $t_{i-1} < t' < t_i$, $-\pi < \phi' < \pi$, and the same field component evaluated at $\underline{r}$ due to the negative of the same unit source, but with the sources radiating in a different homogeneous medium. Because of the additive nature of the fields in (6.36) and (6.37), it is relatively simple to create a modified moment matrix program in which a single matrix element is related directly to the field at a point $\underline{r}$ produced by a source basis function. This can be done by simply eliminating all terms associated with the second homogeneous medium. The original moment matrix program, of course, restricts the observation point to the surface S of the body, as indicated in (6.36) and (6.37). It is again a simple modification to allow the observation points to be selected from a new set of coordinates which could be anywhere in space.

To compute the fields, therefore, we first specify the generating arc for a phantom surface of revolution on which we wish to know the tangential fields. The fields tangential to this surface due to unit sources on the resonator surface are then computed using the modified version of the moment matrix routine which includes only the potentials involving the Green's function of the medium in which the field is to be evaluated. The resulting matrix, which essentially represents a numerical Green's function, is multiplied by the previously computed modal current solution ($|I_m>$ for $\underline{r}$ in region 1 of Fig. 6.1, $-|I_m>$ for $\underline{r}$ in region 2) to obtain the tangential fields on the generating arc of the phantom surface. The fields so computed are actually weighted field values, since (6.36) and (6.37) were "tested" with the functions defined in (6.27) and (6.28) to obtain expressions for moment matrix elements. The appropriate values for the fields are easily obtained, however, by dividing by the length of the phantom surface subdomain at the observation point.

Fields anywhere on the phantom surface are obtained by including the $\exp(jm\phi)$ behavior of the field. It is generally desirable to express the field behavior in terms of $\cos(m\phi)$ or $\sin(m\phi)$ rather than $\exp(jm\phi)$. To do this, it is necessary to determine the current coefficient vector $|I_{-m}\rangle$ for the Fourier mode with index -m. It can be shown in a manner similar to that given in [34] that $|I_{-m}\rangle$ can be expressed in terms $|I_m\rangle$. The variation of the field components can then be readily determined. For the hybrid modes ($m \neq 0$), however, each mode has a degeneracy in the sense that the field can acquire either $\cos(m\phi)$ or $\sin(m\phi)$ variation. If we choose, for example, the mode in which the field component E_t (which could represent either E_ρ, E_z, or a combination thereof) varies as $\cos(m\phi)$, we find that the H_ϕ field also varies as $\cos(m\phi)$, while the remaining two components vary as $\sin(m\phi)$. If any asymmetry is introduced into the system, such as the use of two different types of probes, the modes indicated above are no longer degenerate. This fact has been used to advantage by Guillon and Garault [40] and Fiedziuszko [41] to effectively produce a dual-mode filter using a single resonator (see Sec. 9.8).

The preceding process for computing the fields tangential to the phantom surface may be repeated as often as necessary for new phantom surfaces until the fields have been calculated over the desired region. The major disadvantage of computing the fields in this manner is that, because only the t and ϕ components of the field are computed on the phantom surface, two phantom surfaces are required to compute both the ρ and z components of the field at a particular point in space. The procedure does, however, provide an expedient means for computing near-field distributions without developing complicated new computer subroutines which include integration over the surfaces of the elementary bands and the appropriate vector summation of all contributions from the four surface current components (J_ϕ, J_t, M_ϕ, and M_t).

For the $TE_{01\delta}$ mode, field distributions obtained via this procedure have been compared with the theoretical distribution for a resonant section of a dielectric rod waveguide terminated by two parallel magnetic walls, for which the solution is available in terms of Bessel functions [42]. This comparison is shown in Fig. 6.14 for resonators having radius a = 5.25 mm and ϵ_r = 38. The height of the isolated resonator is h = 4.6 mm. The resonant frequency used for the computations was 4.829 GHz. One observes in Fig. 6.14 that the agreement is quite good inside the resonator (peak values have been normalized). The field of the isolated resonator, however, decays more slowly outside the resonator than does the field for a resonant section of dielectric rod.

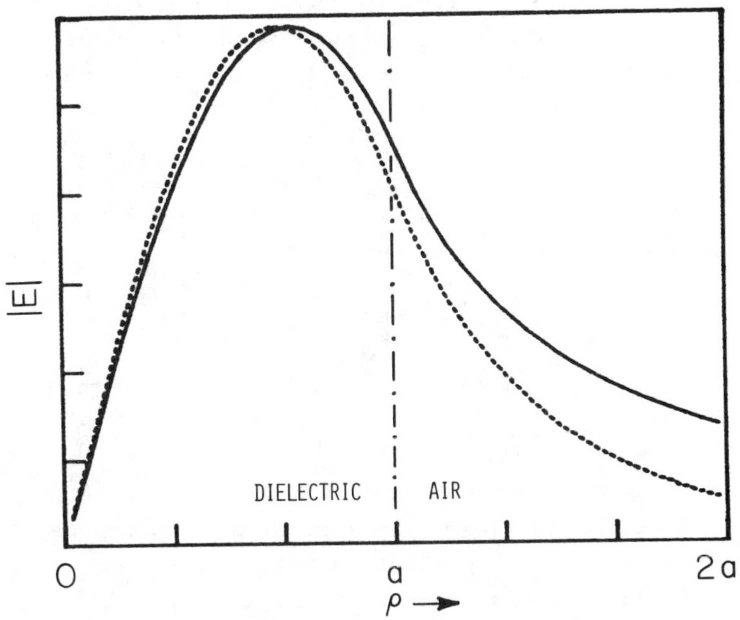

Fig. 6.14 Comparison of $TE_{01\delta}$ mode electric field (solid line) for isolated dielectric resonator and TE_{011} mode electric field (dashed line) for dielectric rod waveguide terminated by parallel magnetic walls (reference [22], ©1984 IEEE)

6.10 Interpretation and Use of Modal Field Distributions

In this section we present a catalog of electric and magnetic field patterns for several low-order resonant modes in isolated dielectric resonators which have been computed with the methods described in the preceding sections. In conjunction with the field plots we also attempt to provide an indication of how these patterns may be useful for designing tuning mechanisms, coupling devices, or mode suppression devices. The electric and magnetic fields for a particular mode in an isolated resonator, which we may represent generically as the vector field $\underline{F}$, oscillate and decay exponentially as functions of time. Even if the decaying nature of the field is ignored, it is difficult to represent graphically the three-dimensional spatial distribution of the magnitude and the phase of the vector field $\underline{F}$. Therefore, we display instead the instantaneous values of the vector:

$$\mathrm{Re}(\underline{F}\, e^{j\omega_{mnp} t})$$

at several instants of time, such as

$$\omega_{mnp} t = 0, \frac{\pi}{4}, \frac{\pi}{2}, \text{ etc.}$$

In the above we use ω_{mnp} to represent the imaginary part of the complex natural frequency of the mode (m,n,p), i.e., we do not include the exponential time decay factor.

A computer-generated graphical display is used to show the orientation of the component of the field tangential to an observation plane at equidistant points within the plane, as well as to provide some relative amplitude information at these points. In this section the fields are displayed in a plane either parallel or perpendicular to the axis of rotation (the z-axis), as indicated in Fig. 6.15. A plane containing the z-axis is referred to as a meridian plane. The plane that passes through the center of the resonator and is perpendicular to the z-axis is referred to as the equatorial plane. In plots showing the field in a meridian plane cut, the resonator cross section appears as a rectangle (side view of the resonator). In equatorial plane cuts, the resonator cross section is a circle (top view of the resonator). In all of the field distribution plots, the double-line arrows are used to indicate a computed field value of less than 3 dB below the maximum computed value,

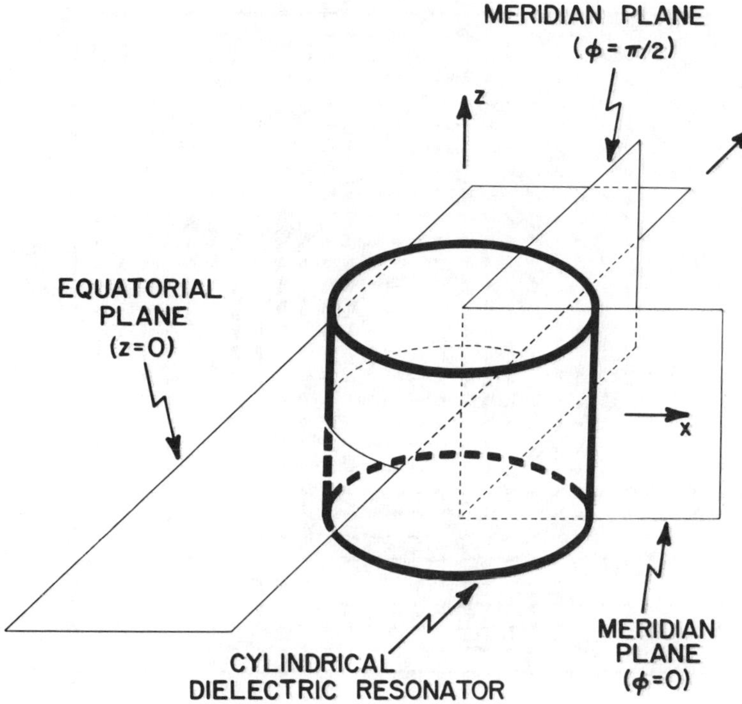

Fig. 6.15 Equatorial and meridian planes for dielectric resonator

while the longer single lines indicate a level between 3 and 10 dB below the maximum, and the shorter lines indicate a level between 10 and 20 dB below the maximum. When the field component tangential to the observation plane is more than 20 dB below the maximum value of the field, the points are left blank. All the plots presented in this section were obtained for a resonator having radius a = 5.25 mm and height h = 4.6 mm, and which is made of material with permittivity $\epsilon_r = 38$. The resonant frequencies and the Q factors (due to radiation) for this resonator are listed in Table 6.1.

Figures 6.16 and 6.17 display the electric and magnetic fields for the $TE_{01\delta}$ mode, respectively. The electric field is shown in the equatorial plane at the moment $\omega_{01\delta}t = 0$. The magnetic field in the same plane is zero. The magnetic field in a meridian plane is shown in Fig. 6.17. Since this is an azimuthally symmetric mode, the plot of the magnetic field would be the same in any meridian plane. The magnetic

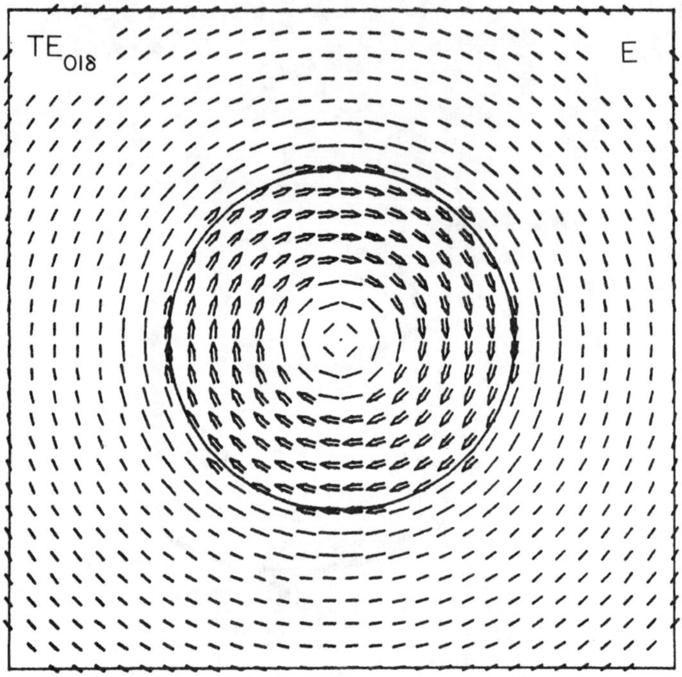

Fig. 6.16 Electric field distribution in equatorial plane for $TE_{01\delta}$ mode (reference [22], ©1984 IEEE)

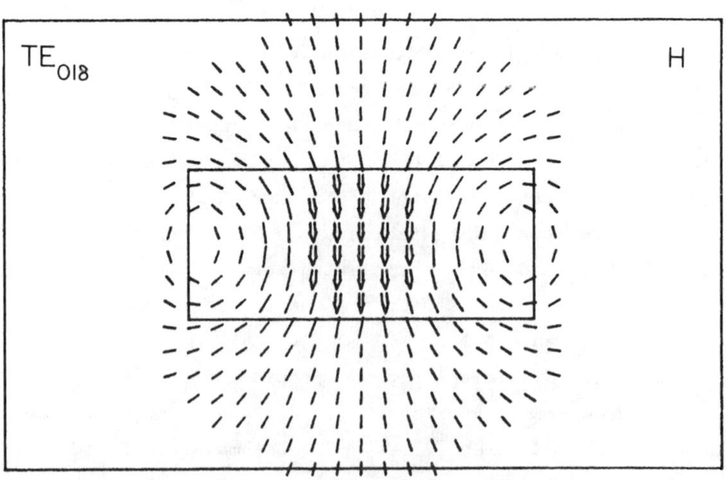

Fig. 6.17 Magnetic field distribution in meridian plane for $TE_{01\delta}$ mode (reference [22], ©1984 IEEE)

field is perpendicular to the electric field of Fig. 6.16, and its maximum value occurs one-quarter period later in time at $\omega_{01\delta}t = \pi/2$. In general, all the magnetic field patterns are in time quadrature with the electric field patterns, so we do not specify the time at which the field is evaluated in subsequent figures.

One observes in Fig. 6.16 that the electric field is quite strong everywhere within the equatorial plane of the resonator, except near the resonator center. A consequence of this is that a cylindrical plug could be removed from the center of the resonator (leaving a rectangular doughnut shape) without disturbing the field, and, hence, the resonant frequency, too much. Such a resonator is called a tubular or ring resonator and is illustrated in Fig. 6.18. Also shown in the vicinity of

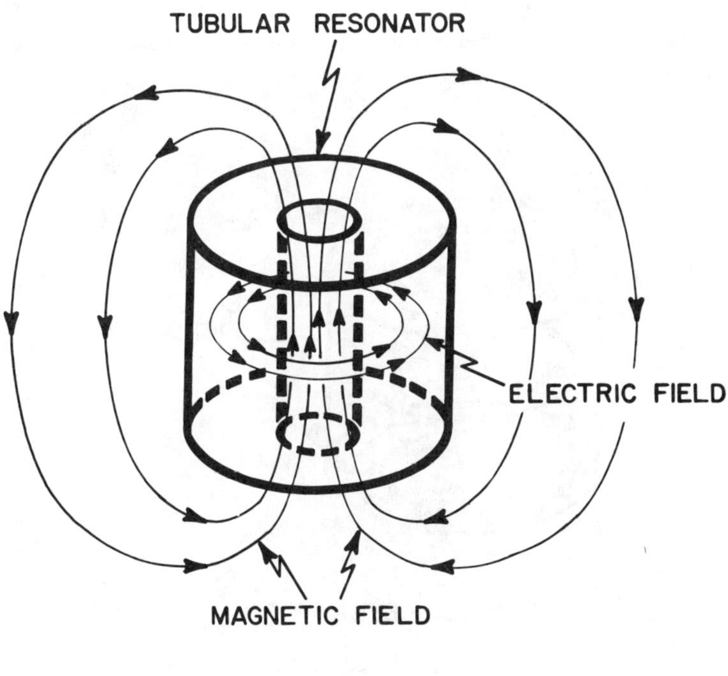

Fig. 6.18 Dielectric resonator with cylindrical plug removed showing strongest electric and magnetic field lines

the resonator are lines indicating the strongest electric and magnetic fields. One notes that the magnetic field is strong down the center of the resonator. The presence or absence of the dielectric plug, however, has little direct effect on the magnetic field. Thus, one cannot effectively tune the $TE_{01\delta}$ mode by inserting, removing, or changing the position of a dielectric plug. However, since the magnetic field is strongest down the center of the resonator, as indicated in Figs. 6.17 and 6.18, one would expect that modification of the magnetic field within the plug region in some manner would change the resonant frequency. Thus, one might insert a plug that would have a significant effect on the magnetic field, such as a ferrite, into the hole through the center of the resonator to tune this mode.

One might also propose a mode suppression device for the $TE_{01\delta}$ mode by observing the electric field in Figs. 6.16 and 6.14. The electric field appears to be strongest at a distance of about three-quarters of the radius away from the axis of the resonator. If this holds true at the end face of the resonator as well (it does; cf. Fig. 6.12), then a thin wire loop with approximate radius 3a/4 placed on the end face, as illustrated in Fig. 6.19, would tend to suppress the $TE_{01\delta}$ mode [43].

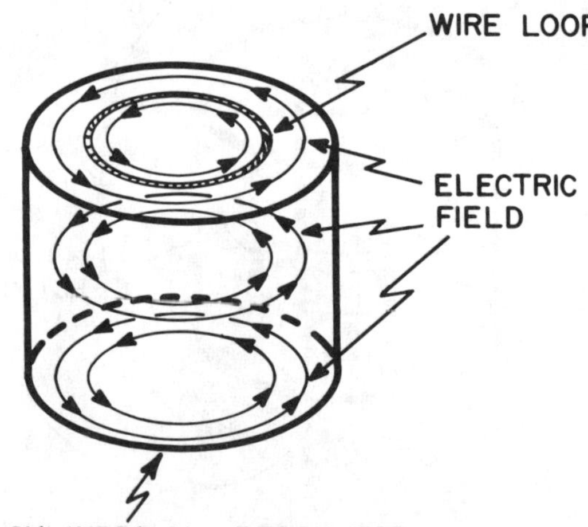

Fig. 6.19 Wire loop placement for suppression of $TE_{01\delta}$ mode

The mode is suppressed because the wire conductor forces the electric field along the line of the loop to zero. If the desired mode of operation is the $TM_{01\delta}$ mode, for example, the loop does not affect the field of the operating mode, whereas it suppresses the undesired $TE_{01\delta}$ mode.

Coupling to the $TE_{01\delta}$ mode is often accomplished through the magnetic field via a small horizontal wire loop placed in the equatorial plane or by placing the resonator end face on a substrate near a microstrip line as indicated in Fig. 6.20, so that the magnetic field lines link with those of the loop or the microstrip. Coupling to a waveguide operating in the TE_{01} mode can be accomplished by placing the resonator on its side (rather than on the flat end face) within the waveguide, as

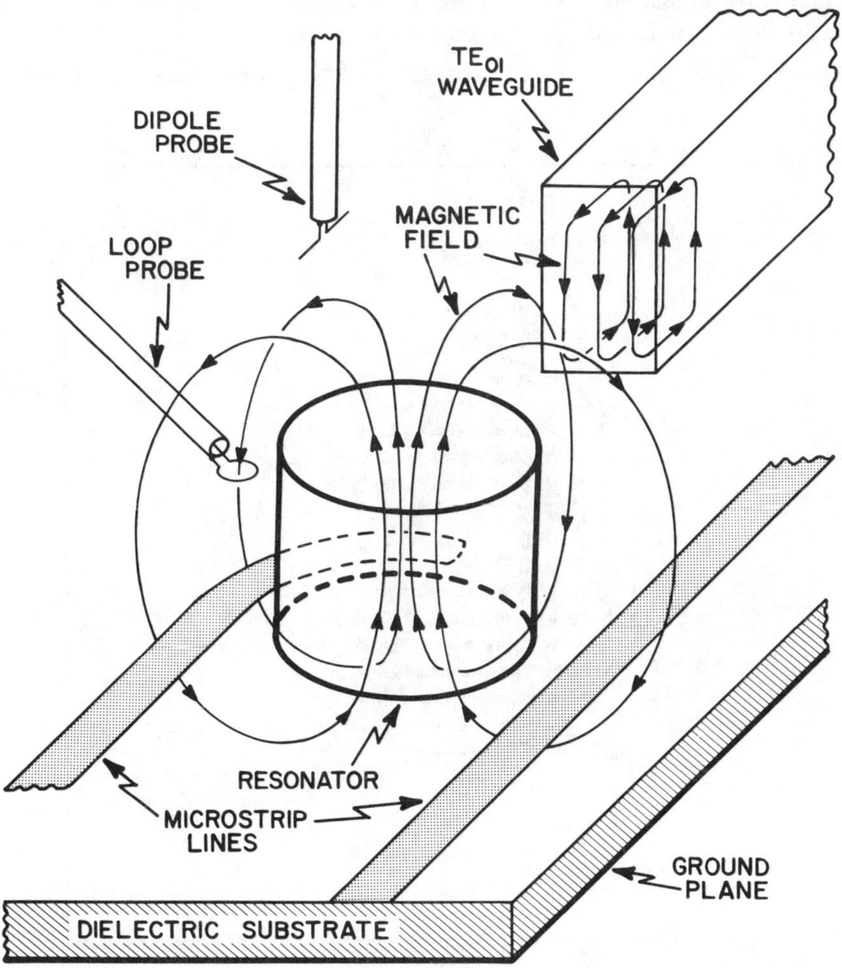

Fig. 6.20 Methods of coupling to $TE_{01\delta}$ dielectric resonator mode

is also implied in Fig. 6.20. This coupling method is useful in cutoff waveguide sections for filtering (see Ch. 8 and 9). Coupling to this mode via the electric field can also be obtained using a small horizontal dipole or a bent monopole as shown in the figure.

The magnetic and electric fields of the $TM_{01\delta}$ mode in the equatorial and meridian planes are shown in Figs. 6.21 and 6.22, respectively. Observe that the magnetic field of this mode is well-contained within the resonator. The outside electric field is not as well-contained and is relatively strong near the top and bottom faces of the resonator. In three-dimensional perspective, the field lines for the $TM_{01\delta}$ mode are similar to those shown in Fig. 6.18 for the $TE_{01\delta}$ mode, if one simply

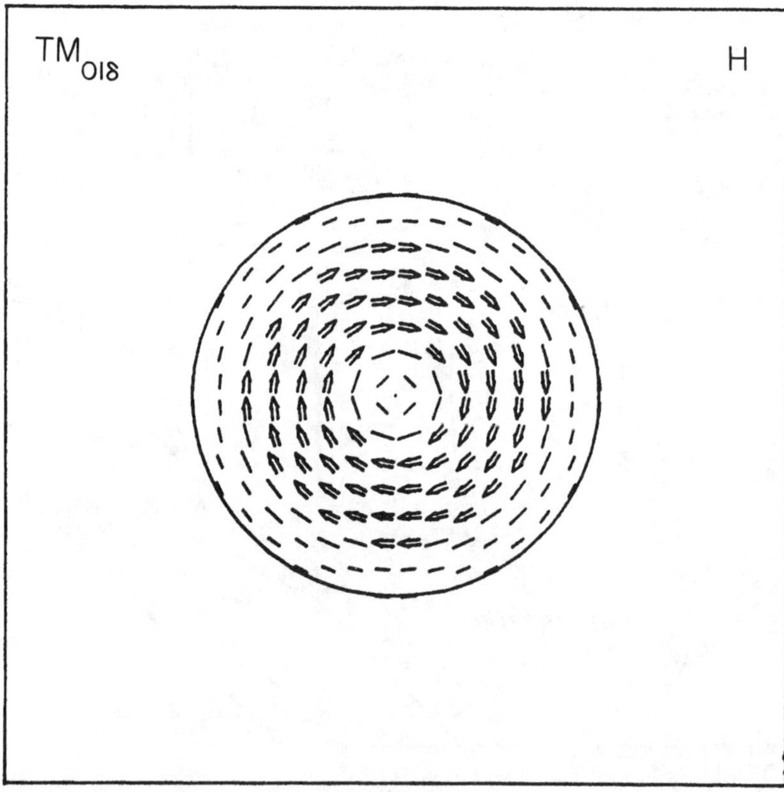

Fig. 6.21 Magnetic field distribution in equatorial plane for $TM_{01\delta}$ mode (reference [22], ©1984 IEEE)

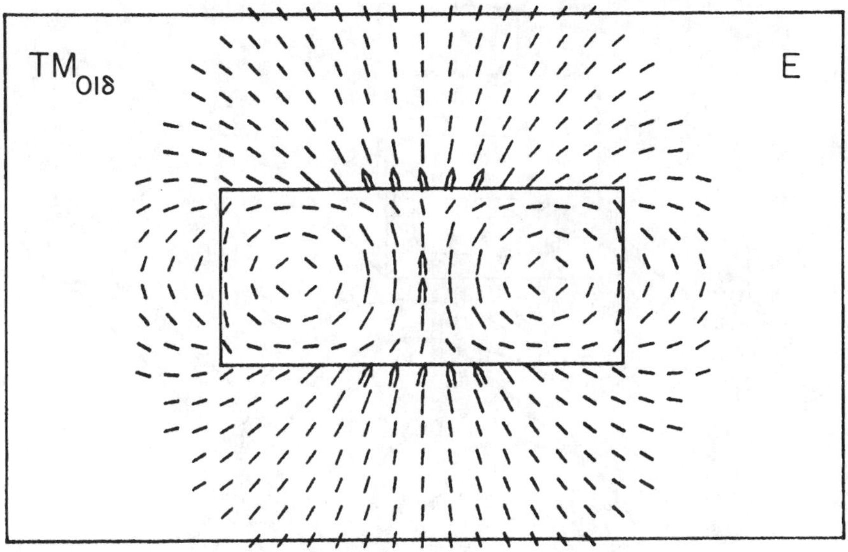

Fig. 6.22 Electric field distribution in meridian plane for $TM_{01\delta}$ mode (reference [22], ©1984 IEEE)

reverses the labels on the electric and magnetic field lines. One notes that, because the electric field is relatively strong along the axis of rotation, it is possible to tune this mode by removing the cylindrical center section (again leaving a rectangular doughnut shape) and replacing it by a movable dielectric rod [44].

Simply removing a cylindrical rod from the center has the effect of increasing the resonant frequency of the $TM_{01\delta}$ mode. Since the $TE_{01\delta}$ mode resonant frequency is not much affected by this action, the tubular shaped resonator was originally introduced to reduce interference from the $TM_{01\delta}$ mode when $TE_{01\delta}$ is the desired mode of operation. Another mode suppressor for the $TM_{01\delta}$ mode might consist of a thin metal rod inserted through the center of the resonator, since this would force the axial electric field to zero.

Several methods of coupling to the $TM_{01\delta}$ mode are illustrated in Fig. 6.23. Because the electric field is not well-contained and is strong along the axis of the resonator, a short capacitive probe, directed along the axis of rotation, should be well-suited for coupling to this mode. The advantage of a capacitive probe for coupling in this case is that it produces no coupling to modes other than the TM_{0np}

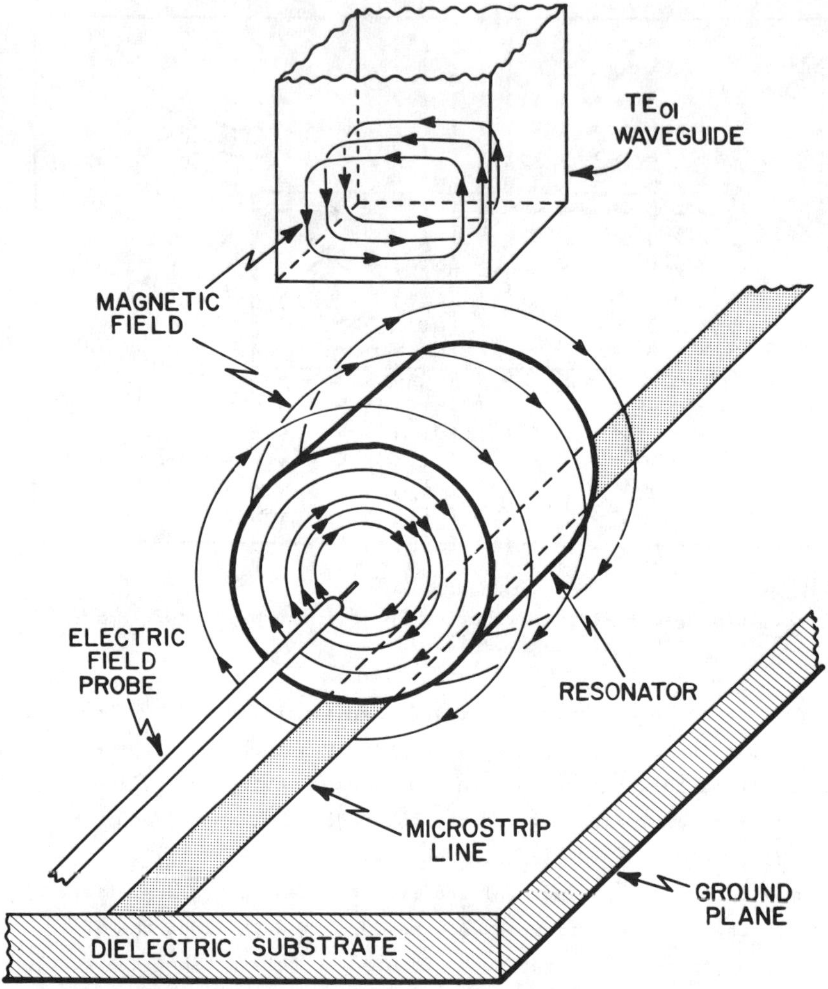

Fig. 6.23 Methods of coupling to $TM_{01\delta}$ dielectric resonator mode

modes, as long as the probe is well-aligned with the axis of the resonator. Coupling to a microstrip line may be accomplished by placing the resonator on its side with its axis of rotation parallel to the microstrip so that the magnetic field lines loop around the microstrip line (see Sec. 8.10). The resonator may be coupled to the TE_{01} waveguide mode by placing the resonator on its end in the waveguide as implied by the orientation of the waveguide in Fig. 6.23.

The remaining three resonant modes displayed here are hybrid electromagnetic modes (HEM) with respect to the axis of rotation. As previously indicated, each of these modes has a degeneracy in the sense that, as a function of the angle ϕ, it can acquire either a $\cos(m\phi)$ or a $\sin(m\phi)$ dependence. For this reason, the angular reference $\phi = 0$ is indicated in the figures to specify the orientation of various field patterns with respect to each other.

The hybrid mode with the lowest resonant frequency is the $HEM_{11\delta}$ mode, which is depicted in Figs. 6.24 through 6.27. As we have previously noted in Sec. 6.6, it was not possible to observe this mode experimentally with our measurement apparatus which simulated the free-space environment. Table 6.1 shows that the Q factor of this mode is

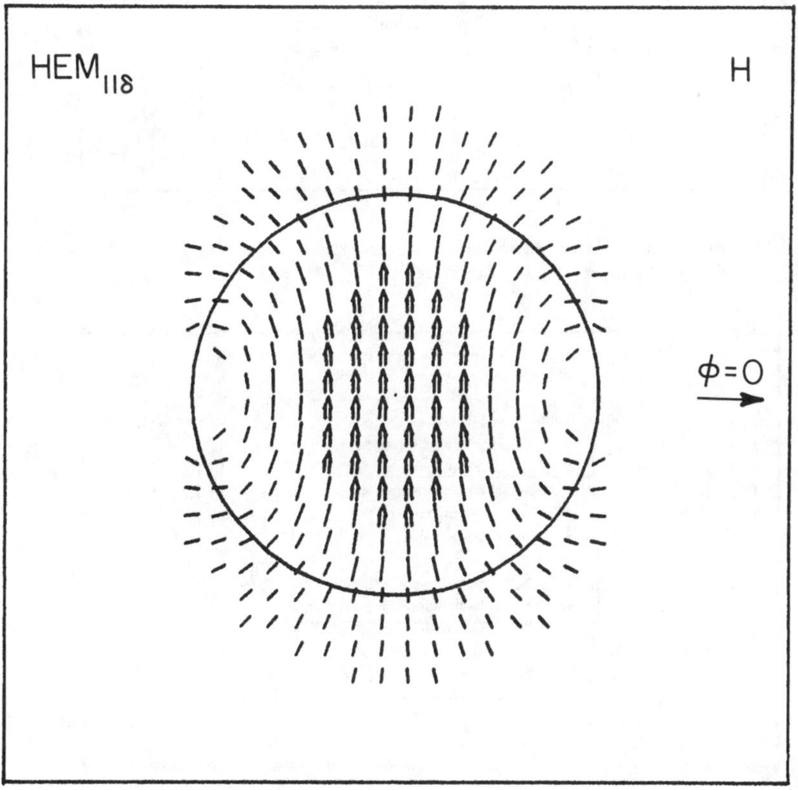

Fig. 6.24 Magnetic field distribution in equatorial plane for $HEM_{11\delta}$ mode (reference [22], ©1984 IEEE)

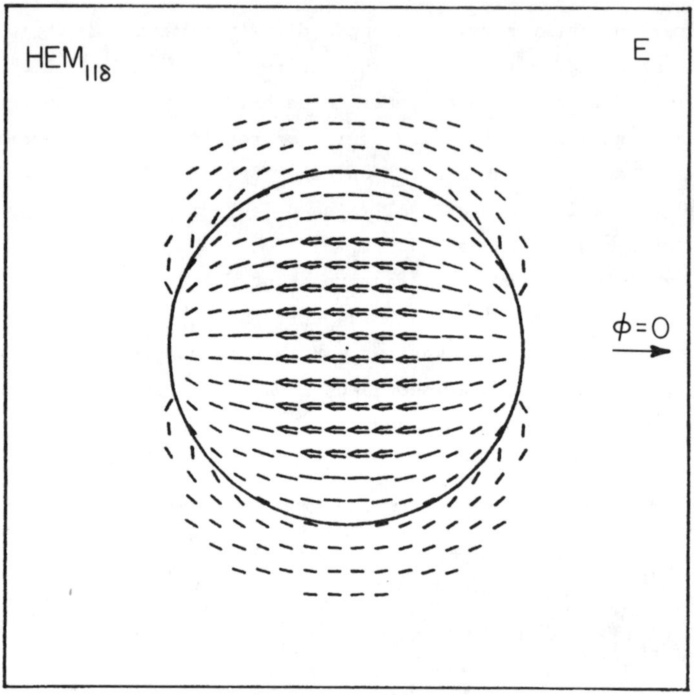

Fig. 6.25 Electric field distribution in plane near resonator end face for $HEM_{11\delta}$ mode (reference [22], ©1984 IEEE)

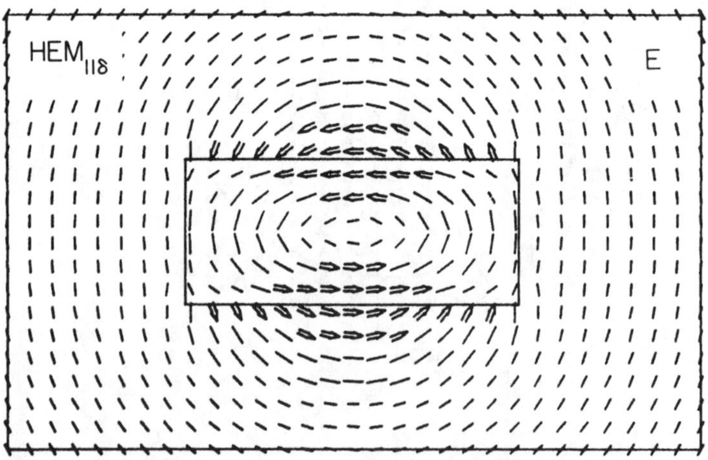

Fig. 6.26 Electric field distribution in meridian plane $\phi = 0$ for $HEM_{11\delta}$ mode (reference [22], ©1984 IEEE)

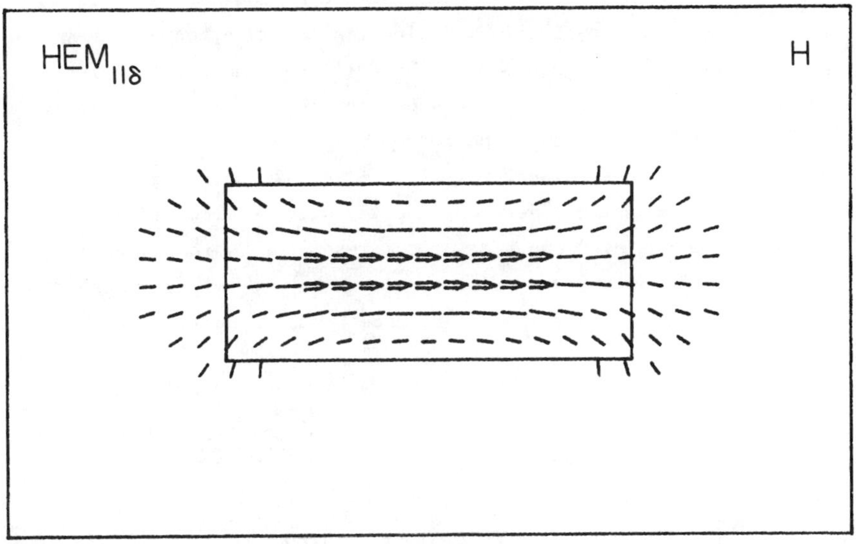

Fig. 6.27 Magnetic field distribution in meridian plane $\phi = \pi/2$ for $HEM_{11\delta}$ mode (reference [22], ©1984 IEEE)

the lowest of all five modes investigated. The low Q factor makes it very difficult to achieve sufficient coupling to the coaxial cable leading to the measurement apparatus.

The magnetic field in the equatorial plane is given in Fig. 6.24. The corresponding equatorial plane components of the electric field are always zero, because the E-field pattern has an odd symmetry with respect to the equatorial plane. The electric field shown in Fig. 6.25 has, therefore, been computed in a plane parallel with the equatorial plane, but displaced by a distance 2.15 mm toward the resonator end face (which is located at $z = 2.30$ mm). The meridian plane in which the maximum electric field intensity occurs is the $\phi = 0$ plane (Fig. 6.26). This mode has $\cos(\phi)$ variation so that in the $\phi = \pi/2$ meridian plane there is only a ϕ-component of electric field, whereas only the ρ-component of the electric field exists in the $\phi = 0$ plane, as can be seen in Fig. 6.25. The maximum magnetic field intensity occurs in the meridian plane $\phi = \pi/2$. Note that the magnetic field is very weak outside each resonator face, while the electric field is strongest there.

The $HEM_{11\delta}$ mode has recently been utilized by Long et al. [45] in the so-called resonant cylindrical dielectric cavity antenna. They excited the resonator on a ground plane by drilling a hole in the resonator near the edge and inserting a short monopole probe as indicated in Fig. 6.28. With image theory, the ground plane corresponds to the equatorial plane of the resonator in Fig. 6.26. In this figure, it can indeed be seen that a small vertical dipole, which is located on the equatorial plane and near the side of the resonator, strongly excites the electric field of the $HEM_{11\delta}$ mode. As we have previously noted, another application of this mode is to create dual-mode filters [40,41] (see also Ch. 8 and 9).

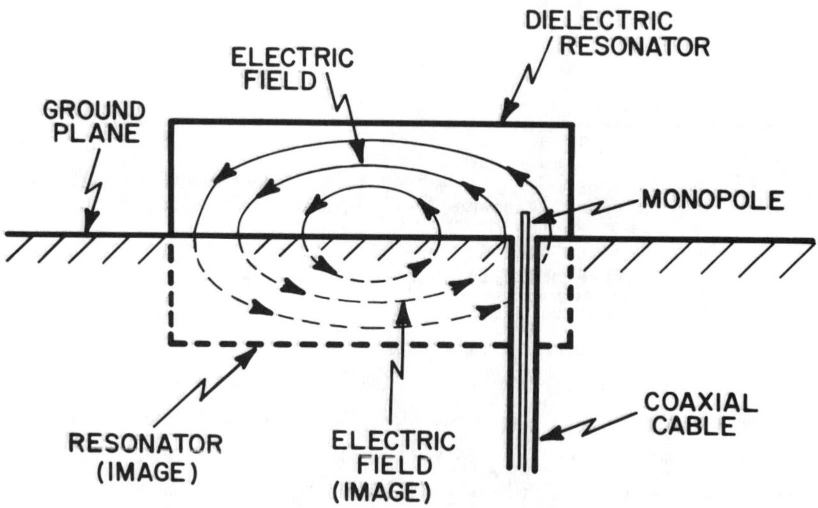

Fig. 6.28 Dielectric resonator operating as radiating element in the $HEM_{11\delta}$ mode when excited by monopole

One notes from Table 6.1 that the $HEM_{11\delta}$ mode has a resonant frequency closer to that of the $TE_{01\delta}$ mode than the other modes. Under circumstances other than the isolated case, the $HEM_{11\delta}$ mode may be in even closer proximity and is likely to cause interference with the $TE_{01\delta}$ mode operation. A well-known technique for suppressing the $HEM_{11\delta}$ mode in filters is illustrated in Fig. 6.29. The mode is suppressed by wrapping two wires around the resonator at right angles in "package" style [46]. It is clear from Figs. 6.25, 6.26, and 6.29 that this mode suppression technique works because one of the wires forces the strong electric field tangential to the top, bottom, and sides of the resonator to zero along the wire length. The second wire, which crosses the resonator at right angles to the first, suppresses the electric field of the degenerate mode. These wires, however, do not significantly affect the $TE_{01\delta}$ mode because the electric field of that mode is always normal to the axes of the wires (see Fig. 6.16). Also indicated in Fig 6.29 is one possible method for external coupling to the $HEM_{11\delta}$ mode. One observes in Fig. 6.26 that the electric field is strong just outside the

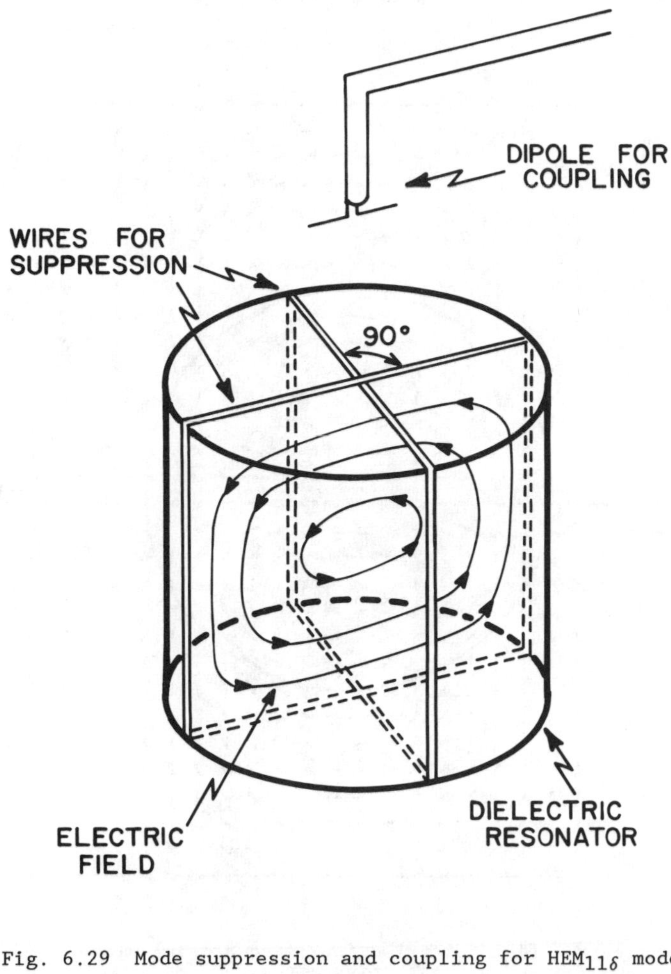

Fig. 6.29 Mode suppression and coupling for $HEM_{11\delta}$ mode

resonator end face and near the center of the resonator. In this region, the field is oriented parallel to the surface of the resonator and primarily in a uniform direction. Thus, a short balanced dipole probe centered on the axis of rotation and oriented along a diameter of the resonator end face should excite this mode well.

Figures 6.30 through 6.32 depict the $HEM_{12\delta}$ mode, which has a resonant frequency only 5 % higher than the $HEM_{11\delta}$ mode (see Table 6.1). One may compare the field distributions shown there with those in Ch. 3 for the HEM_{12} wave guided by a dielectric rod waveguide. The electric field in the equatorial plane is shown in Fig. 6.30. In Fig. 6.31, which shows the electric field in the meridian plane, one notices the strong localized field at the four corners of the resonator. Thus, when

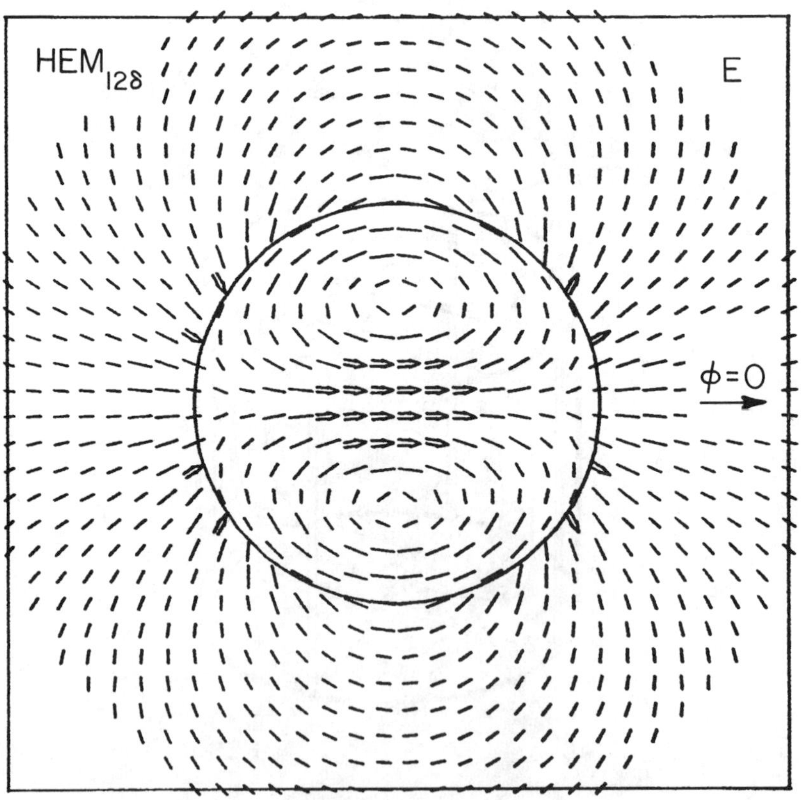

Fig. 6.30 Electric field distribution in equatorial plane for $HEM_{12\delta}$ mode (reference [22], ©1984 IEEE)

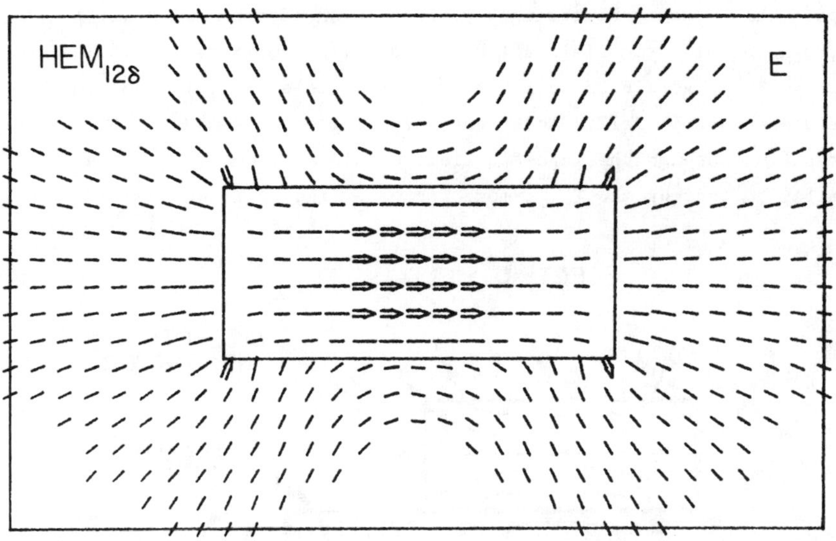

Fig. 6.31 Electric field distribution in meridian plane $\phi = 0$ for $HEM_{12\delta}$ mode (reference [22], ©1984 IEEE)

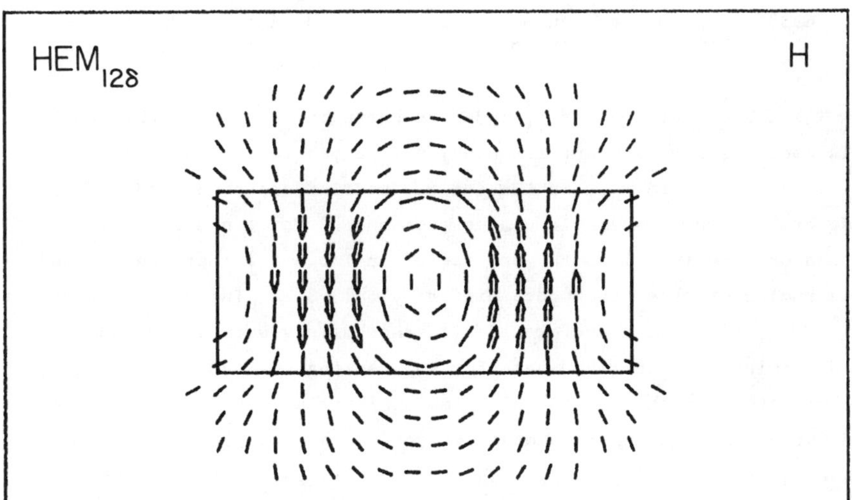

Fig. 6.32 Magnetic field distribution in meridian plane $\phi = \pi/2$ for $HEM_{12\delta}$ mode (reference [22], ©1984 IEEE)

the resonator is placed face down on a microstrip substrate, one might expect this mode to be strongly coupled to an adjacent microstrip line through the electric field action (capacitive coupling), as indicated in Fig. 6.33. Figure 6.32 shows the magnetic field for this mode in the meridian plane $\phi = \pi/2$. Additional coupling of this mode to the microstrip line through the magnetic field should also occur in a manner similar to that of the $TE_{01\delta}$ mode (compare Figs. 6.17 and 6.32).

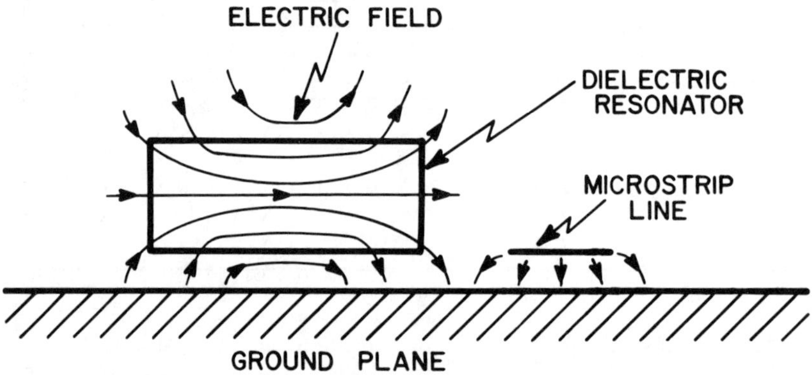

Fig. 6.33 Coupling between microstrip line and $HEM_{12\delta}$ mode of dielectric resonator

The fields for the $HEM_{21\delta}$ mode are shown in Figs. 6.34 through 6.37. This mode has its resonant frequency very close to the resonance of the $TM_{01\delta}$ mode, and if $TM_{01\delta}$ is the desired mode of operation, the $HEM_{21\delta}$ mode creates an undesirable nearby resonance. For the dielectric resonator used in this work, one sees from Table 6.1 that the resonant frequencies of these two modes differ by only 3 %. The field pattern of the $HEM_{21\delta}$ mode is, therefore, of interest mainly because one would like to determine an effective means for suppressing the mode. The magnetic field pattern in the equatorial plane exhibits an octupole character, consisting of two linear quadrupoles rotated by $\pi/2$ with respect to each other, as shown in Fig. 6.34. The octupole is an inefficient radiator and, consequently, the Q factor of this mode is much higher than that of any other mode listed in Table 6.1. The electric field distribution again has an odd symmetry about the equatorial plane. Therefore, the electric field in Fig. 6.36 is shown in a plane close to the end face of

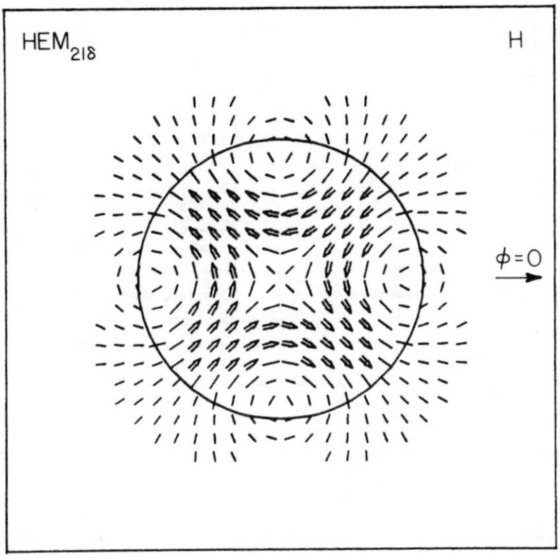

Fig. 6.34 Magnetic field distribution in equatorial plane for $HEM_{21\delta}$ mode (reference [22], ©1984 IEEE)

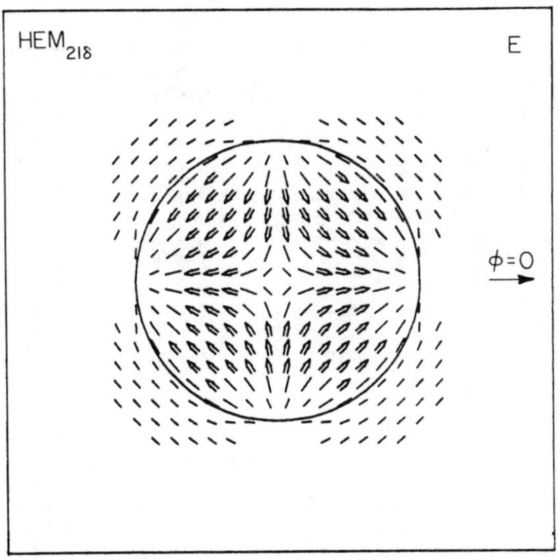

Fig. 6.35 Magnetic field distribution in meridian plane $\phi = \pi/4$ for $HEM_{21\delta}$ mode (reference [22], ©1984 IEEE)

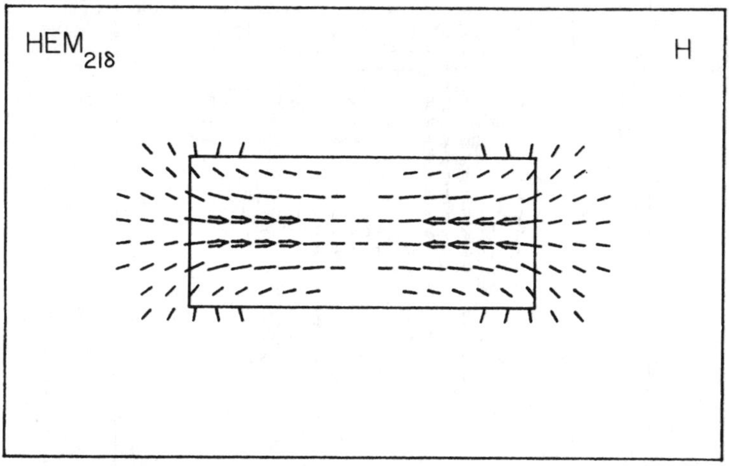

Fig. 6.36 Electric field distribution in plane near resonator end face for $HEM_{21\delta}$ mode (reference [22], ©1984 IEEE)

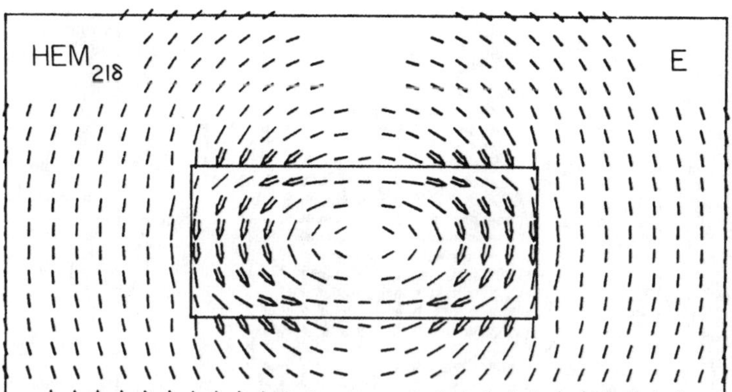

Fig. 6.37 Electric field distribution in meridian plane $\phi = 0$ for $HEM_{21\delta}$ mode (reference [22], ©1984 IEEE)

the resonator because the transverse electric field is always zero at the equatorial plane. The maximum magnetic field occurs in the meridian plane $\phi = 45°$. Figure 6.35 shows that the magnetic field is strongest near the equator. The electric field in the meridian plane $\phi = 0$ is shown in Fig. 6.37.

A distinguishing feature of the $HEM_{21\delta}$ mode, which might permit its suppression without significantly affecting the $TM_{01\delta}$ mode, is that it possesses strong components of the electric field in the ϕ-direction. One observes in Fig. 6.36 that the component of electric field tangent to the observation plane is still quite large near the end face of the resonator. Thus, one might expect that a simple thin wire loop placed on the end face of the resonator would suppress the $HEM_{21\delta}$ mode just as it does for the $TE_{01\delta}$ mode (cf. Fig. 6.19). The radius of such a loop probably should be adjusted to coincide with the maximum value of E_ϕ along the line $\phi = \pi/4$ or $\phi = 3\pi/4$ on the end face of the resonator ($E_\phi = 0$ for $\phi = 0$ and $\phi = \pi/2$). It is not clear from Fig. 6.36 where this value occurs. From the equivalent surface current distribution for this mode [36], however, one finds that the maximum value again occurs at about $\rho = 3a/4$.

6.11 Other Resonator Shapes and Environments

In the preceding sections of this chapter we have described a surface integral equation approach (SIE) for the analysis of dielectric resonators and have applied the method for cases involving isolated, cylindrical dielectric resonators. The SIE approach has several advantages which make its use attractive, particularly for resonators situated in an open space, as we have indicated previously. One of the primary advantages of this approach, however, is that it can be applied to the analysis of dielectric resonators that have shapes which are more complex than the simple cylindrical resonator. It should also be possible to adapt the approach to the analysis of dielectric resonators in the presence of other objects, which may or may not be rotationally symmetric.

The SIE approach can be applied directly in the form presented here to the analysis of dielectric resonators of more complex shape. The only restriction is that the resonator be rotationally symmetric. Cross sections of several body of revolution shapes which may be appropriate for study as dielectric resonators are sketched in Fig. 6.38. The usefulness of the tubular geometry with or without a tuning rod has been indicated previously. For an increased tuning range, the rod and the resonator may be conically shaped, as indicated in Fig. 6.38(d). The multiple resonators shown in Figs. 6.38(e) and 6.38(f) can be used to model a resonator above a ground plane, or, for example, two resonators having different dielectric constants or different sizes. The configuration of Fig. 6.38(f) is also of interest for the study of tuning with dielectric discs, such as described in [47]. Other resonator shapes may be useful for increasing the Q factor, increasing the coupling factor for a specific configuration, eliminating proximity of interfering modes, etc.

The SIE approach can also be used in the manner described in this chapter to treat configurations involving dielectric resonators with metal tuning rods or plates, or resonators in partial or complete metal shields, such as indicated in Fig. 6.39, as long as the metal objects have the same axis of revolution as the resonator. The inclusion of the metal object is possible because the equivalence principle used in Sec. 6.2 can be applied to perfect electric conductors as well. The dielectric resonator within the closed metal cavity (Fig. 6.39(d)) could also be treated by using a modified surface integral equation approach,

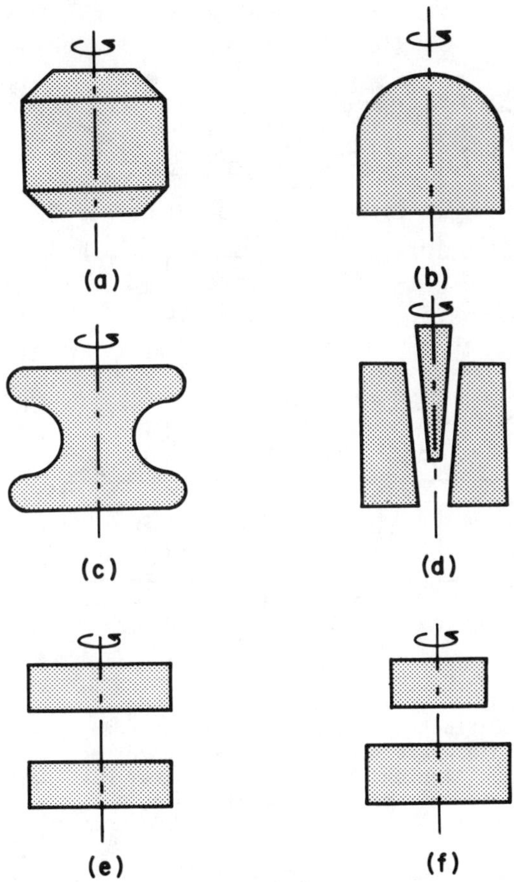

Fig. 6.38 Cross sections of some possible dielectric resonator geometries: (a) cylindrical with corners removed, (b) cylindrical with hemispherical cap, (c) rounded disks of different radii, (d) conical tubular with conical tuning rod, (e),(f) multiple cylinders

in which the resonator is modeled by the SIE, but the presence of the metal cavity is accounted for by use of the Green's function for the cavity in appropriate terms of the SIE, rather than the homogeneous-space Green's function used previously. The latter approach of including the cavity Green's function would reduce the number of unknowns required in the numerical model, but would increase the difficulty and time necessary for computing the required surface integrals. Furthermore, it would be suitable only for cavities for

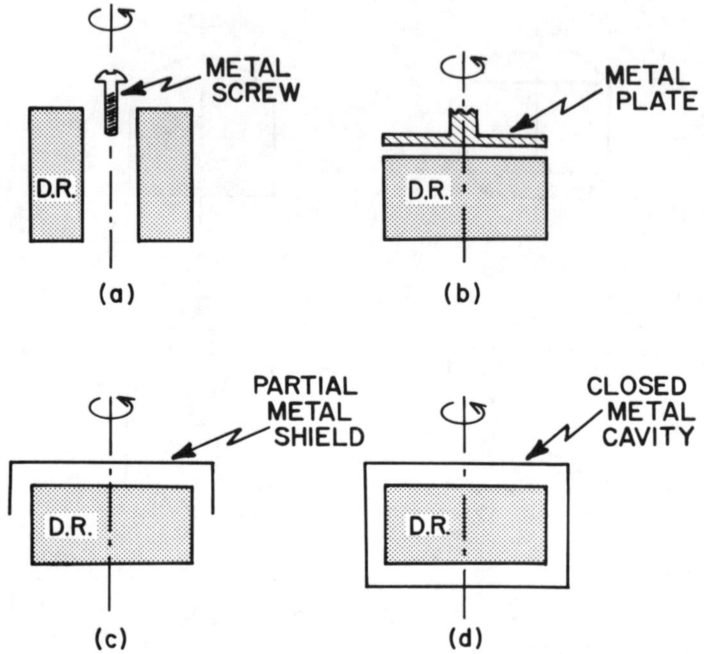

Fig. 6.39 Cross sections of dielectric resonator geometries with metal: (a) tubular resonator with metal tuning screw, (b) resonator with metal tuning plate, (c) resonator with partial metal shield, (d) resonator in metal cavity

which the Green's function can be determined, such as a purely cylindrical cavity. The SIE approach described in this chapter, on the other hand, could be applied for any azimuthally symmetric cavity shape.

Application to more complex configurations involving objects which do not have the same axis of symmetry as the resonator is also possible with some adaptation of the method. For example, one can model the resonant cylindrical dielectric cavity antenna [45] from Fig. 6.28 to any degree of accuracy desired (in principle) by using procedures such as those described in [27]. In the crudest model for this geometry, which may be entirely adequate for computing the radiated fields, one simply ignores the presence of the monopole. Resonant frequencies and field distributions are then the same as those for an isolated resonator of twice the thickness of the original resonator (from image theory) and may be computed as indicated in Sec. 6.10. If more accurate results are required, a point-source or elemental-dipole current distribution may be included as the excitation of the resonator. Inclusion of a source in the model, however, places this situation back into the realm of electromagnetic scattering problems. A response (radiation) would exist for a source of any given frequency. Also, the lack of axial symmetry in the source theoretically would cause an infinite number of modes to be excited in the resonator. In principle, this problem can be solved numerically at any real frequency and for any mode, although some difficulty may be encountered in the inversion of the moment matrix near the complex resonant frequencies of the high-Q modes. Such a model might be quite useful in the determination of the optimal location of the monopole feed, or for determining the extent to which undesired modes are excited in the resonator by the source. If more accuracy is required (for instance, for computation of the input impedance), the coaxial aperture and the monopole (center conductor of the coaxial cable) can also be modeled numerically in a manner similar to that presented in [27].

Several configurations involving objects which do not have the same axis of symmetry, such as coupled dielectric resonators, as well as other factors increasing the complexity of analysis, such as the proximity of a microstrip transmission line and the presence of a dielectric substrate, are indicated in composite form in Fig. 6.40. Approximate methods for modeling such configurations exist. For some configurations these methods may be adequate, while for others they may fail or may not be sufficiently accurate for a particular project. Numerical solutions for geometries involving such complexities via the surface integral equation method appear to be feasible and should provide accurate, reliable results, but it may be impractical in many cases. The practicality of applying the surface integral equation approach in a given situation will depend on the particular geometry to be investigated and on whether simpler approximate methods provide sufficient accuracy. Further research also needs to be performed to investigate the possibility of difficulty in inverting the moment matrix or solving the system of simultaneous equations for frequencies very near complex plane resonances when sources are present in the problem.

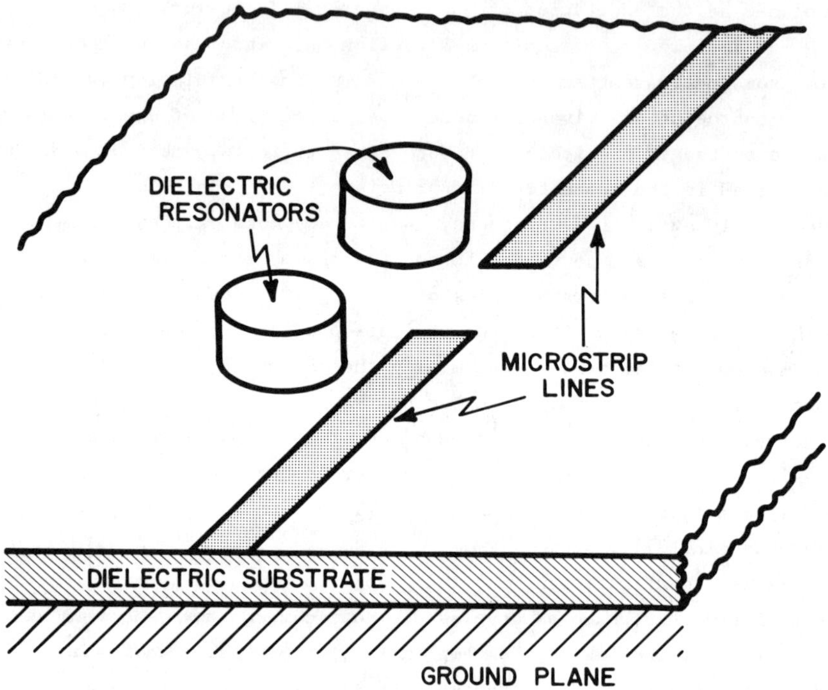

Fig. 6.40 Composite geometry showing coupled dielectric resonators located near a microstrip transmission line

References

[1] H.Y. Yee, "Natural resonant frequencies of microwave dielectric resonators," IEEE Trans. Microwave Theory Tech., vol. MTT-13, p. 256, March 1965.

[2] Y. Garault and P. Guillon, "Best approximation for design of natural resonance frequencies of microwave dielectric disc resonants," Electron. Lett., vol. 10, pp. 505-507, Nov. 1974.

[3] J. Van Bladel, "On the resonances of a dielectric resonator of very high permittivity," IEEE Trans. Microwave Theory Tech., vol. MTT-23, pp. 199-208, Feb. 1975.

[4] M. Verplanken and J. Van Bladel, "The electric dipole resonances of ring resonators of very high permittivity," IEEE Trans. Microwave Theory Tech., vol. MTT-24, pp. 108-112, Feb. 1976.

[5] Y. Konishi, N. Hosino, and Y. Utsumi, "Resonant frequency of a $TE_{01\delta}$ dielectric resonator," IEEE Trans. Microwave Theory Tech., vol. MTT-24, pp. 112-114, Feb. 1976.

[6] T. Itoh and R. Rudokas, "New method for computing the resonant frequency of dielectric resonators," IEEE Trans. Microwave Theory Tech., vol. MTT-25, pp. 52-54, Jan. 1977.

[7] P. Guillon and Y. Garault, "Accurate resonant frequencies of dielectric resonators," IEEE Trans. Microwave Theory Tech., vol. MTT-25, pp. 916-922, Nov. 1977.

[8] M.W. Pospieszalski, "Cylindrical dielectric resonators and their applications in TEM line microwave circuits," IEEE Trans. Microwave Theory Tech., vol. MTT-27, pp. 233-238, March 1979.

[9] M. Jaworski and M.W. Pospieszalski, "An accurate solution to the cylindrical dielectric resonator problem," IEEE Trans. Microwave Theory Tech., vol. MTT-27, pp. 639-642, July 1979.

[10] P. Guillon, J.P. Balabaud, and Y. Garault, "TM_{01p} tubular and cylindrical dielectric resonator mode," IEEE MTT-S Int. Microwave Symp. Dig., pp. 163-166, June 1981.

[11] P. Gelin, S. Toutain, P. Kennis, and J. Citerne, "Scattering of the TE_{01} modes on transverse discontinuities in a rod dielectric waveguide--application to the dielectric resonators," IEEE Trans. Microwave Theory Tech., vol. MTT-29, pp. 712-719, July 1981.

[12] R.R. Bonetti and A.E. Atia, " Design of cylindrical dielectric resonators in inhomogeneous media," IEEE Trans. Microwave Theory Tech., vol. MTT-29, pp. 323-326, April 1981.

[13] Y. Kobayashi, N. Fukuoka, and S. Yoshida, "Resonant modes for a shielded dielectric rod resonator," Electron. Comm. Japan, vol. 64-B, pp. 44-51, Nov. 1981.

[14] M. Tsuji, H. Shigesawa, H. Aoki, and K. Takiyama, "Analytical and experimental considerations on the resonant frequency and the

quality factor of dielectric resonators," IEEE Trans. Microwave Theory Tech., vol. MTT-30, pp. 1952-1957, Nov. 1982.

[15] R. De Smedt, "Dielectric resonator above an electric or magnetic wall," Arch. Elek. Ubertragung, vol. 37, pp. 6-14, Jan. 1983.

[16] M. Tsuji, H. Shigesawa, and K. Takiyama, "On the complex resonant frequency of open dielectric resonators," IEEE Trans. Microwave Theory Tech., vol. MTT-31, pp. 392-396, May 1983.

[17] D. Maystre, P. Vincent, and J.C. Mage, "Theoretical and experimental study of the resonant frequency of a cylindrical dielectric resonator," IEEE Trans. Microwave Theory Tech., vol. MTT-31, pp. 844-848, Oct. 1983.

[18] K.A. Zaki and A.E. Atia, "Modes in dielectric loaded waveguides and resonators," IEEE Trans. Microwave Theory Tech., vol. MTT-31, pp. 1039-1045, Dec. 1983.

[19] A.W. Glisson, D. Kajfez, and J. James, "Evaluation of modes in dielectric resonators using a surface integral equation formulation," IEEE Trans. Microwave Theory Tech., vol. MTT-31, pp. 1023-1029, Dec. 1983.

[20] M. Tsuji, H. Shigesawa, and K. Takiyama, "Analytical and experimental investigations on several resonant modes in open dielectric resonators," IEEE Trans. Microwave Theory Tech., vol. MTT-32, pp. 628-633, June 1984.

[21] R. De Smedt, "Correction due to a finite permittivity for a ring resonator in free space," IEEE Trans. Microwave Theory Tech., vol. MTT-32, pp. 1288-1293, Oct. 1984.

[22] D. Kajfez, A.W. Glisson, and J. James, "Computed modal field distributions for isolated dielectric resonators," IEEE Trans. Microwave Theory Tech., vol. MTT-32, pp. 1609-1616, Dec. 1984.

[23] A.J. Poggio and E.K. Miller, "Integral equation solutions of three-dimensional scattering problems," in Computer Techniques for Electromagnetics, R. Mittra, Ed., New York: Pergamon, 1973, Ch. 3.

[24] R.F. Harrington, Field Computation by Moment Methods. New York: Macmillan, 1968.

[25] S.S.M. Rao, D.R. Wilton, and A.W. Glisson, "Electromagnetic scattering by surfaces of arbitrary shape," IEEE Trans. Antennas Propagat., vol. AP-30, pp. 409-418, May 1982.

[26] D.H. Schaubert, D.R. Wilton, and A.W. Glisson, "A tetrahedral modeling method for electromagnetic scattering by arbitrarily shaped, inhomogeneous dielectric bodies," IEEE Trans. Antennas Propagat., vol. AP-32, pp. 77-85, Jan. 1984.

[27] A.W. Glisson and C.M. Butler, "Analysis of a wire antenna in the presence of a body of revolution," IEEE Trans. Antennas Propagat., vol. AP-28, pp. 604-609, Sept. 1980.

[28] R.F. Harrington, Time-Harmonic Electromagnetic Fields. New York: McGraw-Hill, 1961, Ch. 3.

[29] E.N. Vasil'ev and L.B. Materikova, "Excitation of dielectric bodies of revolution," Soviet Phys. - Tech. Phys., vol. 10, pp. 1401-1406, 1966.

[30] T.K. Wu and L.L. Tsai, "Scattering from arbitrarily-shaped lossy dielectric bodies of revolution," Radio Science, vol. 12, pp. 709-718, Sept. 1977.

[31] J.R. Mautz and R.F. Harrington, "Electromagnetic scattering from a homogeneous material body of revolution," Arch. Elek. Ubertragung, vol. 33, pp. 71-80, Feb. 1979.

[32] A.W. Glisson and D.R. Wilton, "Simple and efficient numerical methods for problems of electromagnetic radiation and scattering from surfaces," IEEE Trans. on Antennas and Propagat., vol. AP-28, pp. 593-603, Sept. 1980.

[33] A.W. Glisson and D.R. Wilton, "Electromagnetic scattering by bodies of revolution," in Applications of the Method of Moments to Electromagnetics Fields, B. J. Strait, Ed., Orlando: SCEEE Press, 1980, Art. 4.

[34] A.W. Glisson, "On the development of numerical techniques for treating arbitrarily-shaped surfaces," Ph.D. Dissertation, University of Mississippi, University, MS, June 1978.

[35] W.H. Hayt, Jr. and J.E. Kemmerly, Engineering Circuit Analysis, 3rd Edition. New York: McGraw-Hill, 1978, Ch. 13.

[36] J. Joseph, "Evaluation of resonant modes in isolated dielectric resonators using a surface integral equation formulation," M.S. Thesis, University of Mississippi, University, MS, July 1984, (also available as an NSF report through NTIS, Accession No. PB85110831).

[37] E. Kreyszig, Advanced Engineering Mathematics, (Fourth Ed.). New York: Wiley, 1979, Ch. 7.

[38] C. Klein and R. Mittra, "Stability of matrix equations arising in electromagnetics," IEEE Trans. Antennas Propagat., vol. AP-21, pp. 902-905, Nov. 1973.

[39] D. Kajfez, "Modal field patterns in dielectric rod waveguide," Microwave Journal, vol. 26, pp. 181-192, May 1983.

[40] P. Guillon and Y. Garault, "Dielectric resonator dual modes filters," Electron. Lett., vol. 16, p. 646, Aug. 1980.

[41] S.J. Fiedziuszko, "Dual-mode dielectric resonator loaded cavity filters," IEEE Trans. Microwave Theory Tech., vol. MTT-30, pp. 1311-1316, Sept. 1982.

[42] C.C. Johnson, Field and Wave Electrodynamics. New York: McGraw-Hill, 1965.

[43] F.C.F. Tan and J. Helszajn, "Suppression of higher order modes in waveguide-junction circulators using coupled open dielectric resonators," IEEE Trans. Microwave Theory Tech., vol. MTT-24, pp. 271-273, May 1976.

[44] E.J. Hwan and D. Kajfez, "Microwave oscillator with $TM_{01\delta}$ dielectric resonator," U. S. Patent 4,521,746, June 4, 1985.

[45] S.A. Long, M. McAllister, and L.C. Shen, "The resonant cylindrical dielectric cavity antenna," IEEE Trans. Antennas Propagat., vol. AP-31, pp. 406-412, May 1983.

[46] A. Karp, H.J. Shaw, and D.K. Winslow, "Circuit properties of microwave dielectric resonators," IEEE Trans. Microwave Theory Tech., vol. MTT-16, pp. 818-828, Oct. 1968.

[47] J. Delaballe, J. Fouillet, Y. Le Nohaic, A. Osias, and D. Osias, "Ultra-high frequency filter with a dielectric resonator tunable in a large bandwidth," U. S. Patent 4,459,570, July 10,1984.

Chapter 7
MATERIAL PROPERTIES
Darko Kajfez

7.1 Introduction

In this chapter, we will take a closer look at the imperfections occurring in dielectric resonators. The two major imperfections are (i) the losses within the dielectric and (ii) the temperature dependence of the mechanical and electrical properties of the dielectric. Each of the two is of great significance because they impose limits on the two most important properties of dielectric resonators: the high Q factor and the high temperature stability.

Inside an isotropic dielectric material which also has a non-vanishing conductivity σ, the Maxwell equation for time-harmonic variation is expressed as

$$\nabla \times \underline{H} = (\sigma + j\omega\epsilon)\underline{E} \tag{7.1}$$

The finite conductivity σ is an obvious reason for losses, namely, for converting electromagnetic energy into heat.

Another mechanism which produces losses in the dielectric material at microwave frequencies is the damping caused by the alternating polarization of material exposed to the time-harmonic electric field. These losses may be expressed by defining a dielectric constant which is a complex number [1,2]:

$$\epsilon = \epsilon' - j\epsilon'' \tag{7.2}$$

In such a notation, (7.1) becomes

$$\nabla \times \underline{H} = j\omega[\epsilon' - j(\epsilon'' + \frac{\sigma}{\omega})]\underline{E} \tag{7.3}$$

It can be seen that ϵ' takes the role of a traditional dielectric constant, and the total losses are caused partially by ϵ'' and partially by σ. The loss tangent is defined as the ratio of the imaginary part to the real part in brackets above:

$$\tan \delta = \frac{\epsilon''}{\epsilon'} + \frac{\sigma}{\omega\epsilon'} \tag{7.4}$$

At microwave frequencies, the first part of expression (7.4) is dominant. Typically, ϵ' is constant and ϵ'' grows with frequency.

It should be noted that the symbol δ used in (7.4) has nothing in common with the skin depth δ of an electric conductor:

$$\delta = \frac{1}{\sqrt{\pi f \mu \sigma}} \tag{7.5}$$

The confusion can be avoided if one uses the dielectric Q factor, Q_d, to denote the dielectric losses. When a resonant cavity is uniformly filled with the lossy dielectric, the dielectric Q factor is an inverse of the loss tangent:

$$Q_d = \frac{1}{\tan \delta} \tag{7.6}$$

The manufacturers of dielectric resonators usually specify the value of Q_d to be inversely proportional to frequency:

$$Q_d = \frac{C}{f} \tag{7.7a}$$

For example, the material D-8512 manufactured by Trans Tech has C = 40000, when f is expressed in GHz [3]. At 4 GHz, this material has Q_d = 10000, and at 8 GHz the same material has Q_d = 5000.

The linear growth of the loss tangent with frequency can be expressed in a more general form as follows:

$$\tan \delta = A + Bf \tag{7.7b}$$

For instance, material $(Zr{\cdot}Sn)TiO_4$, produced by Murata, is quoted [4] to have $A = 0.205 \cdot 10^{-4}$ and $B = 0.170 \cdot 10^{-4}$/GHz. At a frequency of 8 GHz, we compute from (7.7b) and (7.6) that this material has Q_d = 6390.

Sections 7.2 and 7.3 of this chapter treat the theoretical and experimental procedures for evaluating the dielectric losses (as well as other losses) in resonant cavities that are partially filled with various dielectric materials. The rest of the chapter is devoted to the temperature effects pertinent to dielectric resonators.

7.2 Resonant Cavities Filled with Inhomogenous Materials

The Q factor of a resonant cavity is a ratio of the stored energy to the dissipated power, as can be recalled from Ch. 2:

$$Q_L = \frac{2\omega W_e}{P_{tot}} \tag{7.8}$$

In the above, ω is the resonant frequency, W_e is the total stored electric energy in the volume of the cavity, and P_{tot} is the total dissipated power.

In a practical microwave cavity, the power may be dissipated in several different ways. The general situation is shown in Fig. 7.1. The cavity is enclosed within a metal shield. Inside the shield we see three regions, each filled with a different dielectric material. For

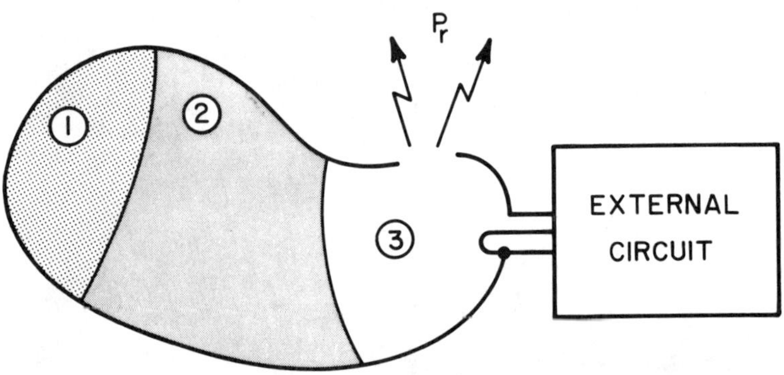

Fig. 7.1 Resonant cavity filled with inhomogeneous dielectric

instance, region 1 may be the dielectric substrate of a microstrip circuit, region 2 may be the dielectric resonator, and region 3 may be air. The total stored electric energy is obtained by integrating over all three regions:

$$W_e = \sum_{i=1}^{3} \frac{\epsilon_i'}{4} \int_{V_i} |\underline{E}_i|^2 dV_i \tag{7.9}$$

The total power dissipated in the dielectric is obtained by a similar integration:

$$P_d = \sum_{i=1}^{3} \frac{\sigma_i + \omega\epsilon_i''}{2} \int_{V_i} |\underline{E}_i^2| \, dV_i \tag{7.10}$$

The total power dissipated in the conducting shield is obtained by integrating over the surface of the shield:

$$P_c = \frac{1}{2} \int_S R_s |J_s|^2 dS \tag{7.11}$$

where J_s is the surface current density on the shielding conductor. If the shield is not completely closed, some power P_r may be lost to radiation. Finally, some power P_{ext} may be coupled to an external microwave circuit, and dissipated there. The total dissipated power is then

$$P_{tot} = P_d + P_c + P_r + P_{ext} \tag{7.12}$$

It is customary to define the partial Q factors according to the manner in which the cavity power is dissipated. The dielectric Q factor is, thus,

$$Q_d = \frac{2\omega W_e}{P_d} \tag{7.13}$$

and the Q factor due to conductor losses is

$$Q_c = \frac{2\omega W_e}{P_c} \tag{7.14}$$

Similarly, the radiation Q factor is

$$Q_r = \frac{2\omega W_e}{P_r} \tag{7.15}$$

and the external Q factor is

$$Q_{ext} = \frac{2\omega W_e}{P_{ext}} \tag{7.16}$$

The total Q factor is, then, from (7.8) and (7.12):

$$\frac{1}{Q_L} = \frac{1}{Q_d} + \frac{1}{Q_c} + \frac{1}{Q_r} + \frac{1}{Q_{ext}} \tag{7.17}$$

The first three parts on the right-hand side of (7.17) comprise what is usually called the unloaded Q factor of the resonant cavity:

$$\frac{1}{Q_u} = \frac{1}{Q_d} + \frac{1}{Q_c} + \frac{1}{Q_r} \tag{7.18}$$

Most often, the shield is entirely closed so that there is no radiation. In this case, the third term in (7.18) can be ignored.

Another useful quantity in dealing with inhomogeneously filled resonant cavities is the electric energy filling factor p_e. It is defined as the ratio of the stored electric energy in a given region to the total stored electric energy in the cavity. For the cavity in Fig. 7.1, the three filling factors are

$$p_{ei} = \frac{W_{ei}}{W_e}, \quad \text{where } i = 1,2,3. \tag{7.19}$$

The integrals appearing on the right-hand sides of (7.9) and (7.10) are identical to each other within each region V_i. Therefore, the dissipated power and the stored energy in each volume are related as

$$\frac{P_{di}}{2W_{ei}} = \omega \left(\frac{\epsilon_i''}{\epsilon_i'} + \frac{\sigma_i}{\omega \epsilon_i'} \right) = \omega \tan \delta_i \tag{7.20}$$

Using the filling factor from (7.19), we obtain

$$P_{di} = 2W_{ei}\omega \tan \delta_i = p_{ei}\, 2\omega W_e \tan \delta_i \tag{7.21}$$

so that the partial Q factor due to dielectric losses in the region i (i = 1,2,3) can be expressed in terms of the filling factor and the loss tangent as follows:

$$Q_{di} = \frac{2\omega W_e}{P_{di}} = \frac{1}{p_{ei} \tan \delta_i} \tag{7.22}$$

If all three regions in Fig. 7.1 are filled with lossy dielectrics, the overall Q factor due to dielectric losses can be computed from the partial dielectric Q factors in the following manner:

$$\frac{1}{Q_d} = \sum_{i=1}^{3} \frac{1}{Q_{di}} \tag{7.23}$$

As an illustration of a resonant cavity filled with inhomogeneous dielectric, we will analyze the Courtney holder shown in Fig. 7.2. For

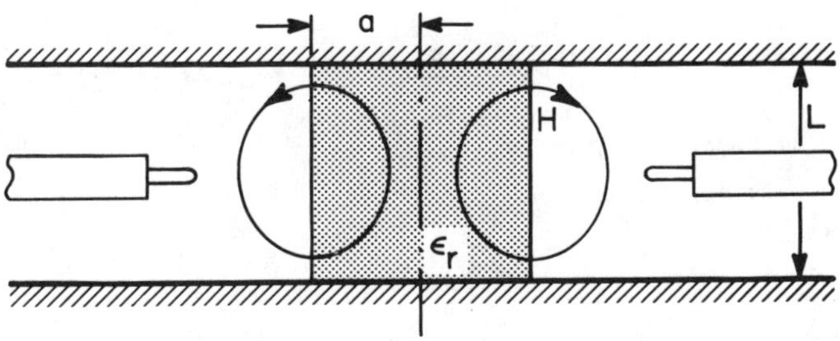

Fig. 7.2 Courtney holder

the resonant mode TE_{01p}, the length L is equal to p half-wavelengths. From (3.76):

$$\beta L = p\pi \quad \text{for } p = 1,2,3,... \tag{7.24}$$

The electric field of this mode, inside dielectric (region 1), is given by

$$E_{\phi 1} = E_0 J_1(k_{\rho 1}\rho) \sin(p\pi \frac{z}{L}) \tag{7.25}$$

Outside the dielectric (region 2), the electric field is

$$E_{\phi 2} = E_0 \frac{J_1(k_{\rho 1}a)}{K_1(k_{\rho 2}a)} K_1(k_{\rho 2}\rho) \sin(p\pi \frac{z}{L}) \tag{7.26}$$

where $k_{\rho 1}$ and $k_{\rho 2}$ are the radial wave numbers in regions 1 and 2. Substituting (7.25) into (7.9), and using the integral (4.53), one finds the electric energy stored in the dielectric rod:

$$W_{e1} = 2\pi \frac{\epsilon_0 \epsilon_r}{4} E_0^2 \frac{a^2 L}{4} [J_0^2(x) - \frac{2}{x} J_0(x) J_1(x) + J_1(x)] \qquad (7.27)$$

The electric energy stored in region 2 is obtained from (7.26), (7.9), and (3.55)

$$W_{e2} = 2\pi \frac{\epsilon_0}{4} E_0^2 \frac{J_1^2(x)}{K_1^2(y)} \frac{a^2 L}{4} [K_0^2(y) + \frac{2}{y} K_0(y) K_1(y) - K_1^2(y)] \qquad (7.28)$$

Arguments x and y stand for the following:

$$x = k_{\rho 1} a \quad \text{and} \quad y = k_{\rho 2} a \qquad (7.29)$$

The ratio of the two energies is

$$\frac{W_{e2}}{W_{e1}} = \frac{W}{\epsilon_r} \qquad (7.30)$$

where the factor W is defined as

$$W = \frac{\dfrac{K_0^2(y)}{K_1^2(y)} + \dfrac{2}{y} \dfrac{K_0(y)}{K_1(y)} - 1}{\dfrac{J_0^2(y)}{J_1^2(y)} - \dfrac{2}{x} \dfrac{J_0(x)}{J_1(x)} + 1} \qquad (7.31)$$

For high-ϵ dielectrics, W is a number typically smaller than unity, as can be seen in Fig. 7.3. For the dimensions of the resonator in Table 3.1 ($\epsilon_r = 38$, $a = 4.25$ mm, $L = 4.25$ mm) and for $p = 1$, we read the value $k_0 a = 0.71$. Thus, from Fig. 7.3, we find that $W = 0.14$.

The filling factor for region 1 of the resonator in Fig. 7.2 is

$$p_{e1} = \frac{W_{e1}}{W_{e1} + W_{e2}} \qquad (7.32)$$

By using (7.30), p_{el} becomes

$$p_{el} = \frac{1}{1 + W/\epsilon_r} \tag{7.33}$$

Substituting the values from the above example, we compute the filling factor for region 1 of the TE_{011} mode to be $p_{el} = 0.996$. In other words, 99.6% of the entire stored electric energy is contained within the dielectric rod. In this case, the filling factor is practically equal to unity.

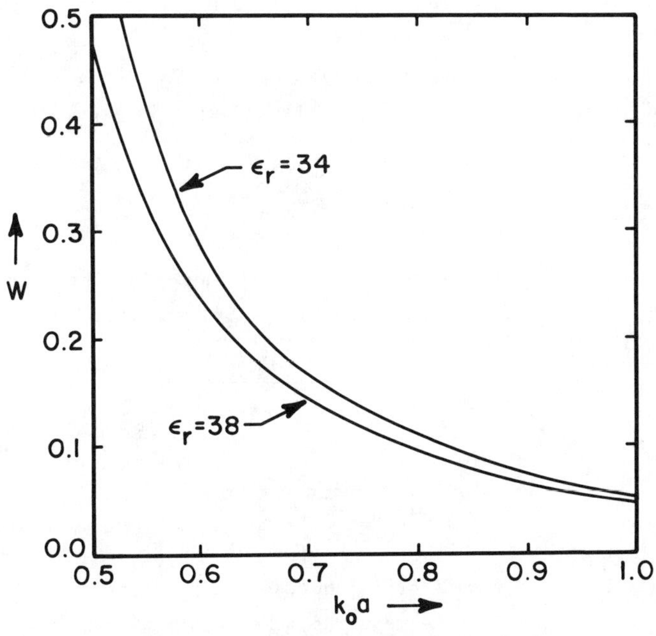

Fig. 7.3 Factor W, utilized to evaluate the ratio of energies for TE_{01p} resonant mode

Region 2 is filled with air. Usually, losses in air can be neglected. Therefore, the Q factor of the TE_{01p} mode due to dielectric losses is

$$Q_d = Q_{d1} = \frac{1 + W/\epsilon_r}{\tan \delta} \tag{7.34}$$

Next, we will compute the Q factor due to conductor losses. This could be done by integrating the square of the surface current over the two parallel metal plates. However, we will utilize an alternative procedure, the so-called incremental frequency rule from (2.142):

$$Q_c = \frac{f_0}{\Delta f_0(\delta)} \tag{7.35}$$

In the above, f_0 is the resonant frequency of the cavity, and Δf_0 is the change in resonant frequency computed for the case where all the metal walls are moved inward for one skin depth δ.

A slight complication in evaluating the differential Δf_0 in (7.35) is caused by the fact that the resonant frequency f_0 of the Courtney holder cannot be expressed explicitly. The resonant frequency is obtained by solving the eigenvalue equation of the TE_{01} mode in the dielectric rod waveguide. From (4.95), this equation is expressed as

$$\frac{J_1(x)}{xJ_0(x)} + \frac{K_1(y)}{yK_0(y)} = 0 \tag{7.36}$$

The resonant frequency is computed from (7.36) and from the resonance condition (7.24). Substituting (7.24) in (4.96) and (4.97), we find

$$x = \sqrt{(k_0a)^2\epsilon_r - (p\pi a/L)^2} \tag{7.37}$$

$$y = \sqrt{(p\pi a/L)^2 - (k_0a)^2} \tag{7.38}$$

If the length L of the resonator is shortened for two skin depths, as the incremental frequency rule requires, the resonant frequency f_0 will change and, accordingly, k_0a will change. Variables x and y in (7.37) and (7.38) can be considered as being dependent on L and k_0a:

$$x = x(L, k_0a) \tag{7.39}$$

$$y = y(L, k_0a) \tag{7.40}$$

Furthermore, (7.36) can be written as a sum of two terms, where one is a function of x only, and the other is a function of y only:

$$F(x) + G(y) = 0 \tag{7.41}$$

The differential of this expression is

$$\frac{\partial F}{\partial x}\,\Delta x + \frac{\partial G}{\partial y}\,\Delta y = 0 \tag{7.42}$$

Using the fact that

$$\Delta L = -2\delta \tag{7.43}$$

and

$$\Delta k_0 a = k_0 a\,\frac{\Delta f_0}{f_0} \tag{7.44}$$

it is possible to evaluate all the needed derivatives and substitute them into (7.35) to obtain

$$Q_c = \frac{\epsilon_r + W}{1 + W} \cdot \left(\frac{k_0 L}{p\pi}\right)^2 \cdot \frac{L}{2\delta} \tag{7.45}$$

In the above, factor W is the same factor, from (7.31), as evaluated earlier in the process of finding the Q factor due to dielectric losses.

In order to get an idea about the magnitude of the Q factor which can be achieved for the Courtney holder, let us evaluate Q_c for the same example of the TE_{011} resonator from Table 3.1. Assuming both plates to be made of copper, and assuming the conductivity is the same as at low frequencies ($\sigma = 5.8 \cdot 10^7$ Siemens/meter), we find from (7.5) the skin depth $\delta = 0.740\ \mu m$. The Q factor due to conductor losses is then

$$Q_c = \frac{38 + 0.14}{1 + 0.14} \cdot \left(\frac{0.71}{\pi}\right)^2 \cdot \frac{4.25 \cdot 10^{-3}}{2 \cdot 0.740 \cdot 10^{-6}}$$

$$Q_c = 4909$$

For the same resonator, assume the manufacturer's listed value for the dielectric Q factor is 10,000 at 4 GHz (see Resomics-U material, Appendix 7.A). By using (7.7a) and (7.34), we find

$$Q_d = 10000 \cdot \frac{4}{7.98} \cdot \frac{1}{0.996} = 5033$$

Therefore, for this Courtney holder, dielectric losses are roughly of the same magnitude as conductor losses. The unloaded Q factor is then $Q_u = 1/(1/4909 + 1/5033) = 2485$.

Finally, we may check the possibility of radiation losses for the above resonator. When the distance L becomes larger than $\lambda_0/2$, a radially outgoing wave may propagate between the parallel plates. For L = 4.25 mm, the cutoff frequency of such a propagating mode is 35.3 GHz. Therefore, there is no danger of radiation losses occuring at 7.98 GHz. However, the possibility of radiation should be kept in mind, because when the separation is larger than a half-wavelength in free space, the Q factor drops abruptly to a value that is an order of magnitude smaller. Experimental results showing the occurrence of radiation in the Courtney holder can be seen in [4].

7.3 Measurement of Loss Tangent

The first step in an experimental verification of the loss tangent of the dielectric material is to measure the unloaded Q factor of the cavity containing this dielectric material. Traditionally, the measurement is performed for the TE_{011} mode in the Courtney holder. When the separation L is small enough that the mode is not radiating, one obtains from (7.18:)

$$\frac{1}{Q_d} = \frac{1}{Q_u} - \frac{1}{Q_c} \tag{7.46}$$

Using (7.34) and (7.45), the above equation may be solved for the loss tangent:

$$\tan\delta = \frac{1 + W/\epsilon_r}{Q_u} - \frac{1 + W}{\epsilon_r}\left(\frac{p\pi}{k_0 L}\right)^2 \frac{2\delta}{L} \tag{7.47}$$

For the TE_{011} mode, the integer p in the above equation is set to unity. The equation is also valid for higher values of p, which will be needed later. The reader may recognize this as equation (3.83), which was originally derived by Hakki and Coleman [5], and also used by Courtney [6].

In (7.47), Q_u and ϵ_r are measured values, as are the spacing L and the resonant frequency. When all of these quantities are substituted into (7.47), one may compute the loss tangent. Unfortunately, the results of the loss tangent evaluated in this way are very inaccurate. Courtney [6] quotes an example where the estimated error of $\tan\delta$ is $\pm$ 40 %. There are two main reasons for the inaccuracy of this method:

(i) The conductivity σ (needed to compute the skin depth δ) at microwave frequencies is not the same as the conductivity at low frequencies. The difference may be as large as 20 %, depending on surface roughness, possible corrosion, and thermal annealing treatment of the metal plates.

(ii) The measurement is based on equation (7.46) in which one positive real quantity is subtracted from another. When the two quantities are of comparable magnitude, their difference acquires a large relative error.

The indeterminacy of the value σ can be removed by clever use of two different resonant modes, namely, TE_{011} and TE_{01p}, as described in a paper [7] by Kobayashi and Katoh. The principle of their procedure can be explained as follows. When a given dielectric resonator of radius a and length L is placed between two parallel plates as shown in Fig. 7.2, the resonance of the mode TE_{011} occurs at some frequency f_0. Now, if a number of identical resonators is stacked on top of each other (total number of resonators is equal to an integer p), the resonant frequency of the TE_{01p} mode between the parallel plates is the same f_0 as before. In Fig. 7.4(a) one can see the distribution of the magnetic field for the resonant mode TE_{011}. When the resonator length is made two times longer (p = 2), the field of the TE_{012} mode is as shown in Fig. 7.4(b). The skin depth for the two different resonant structures is the same

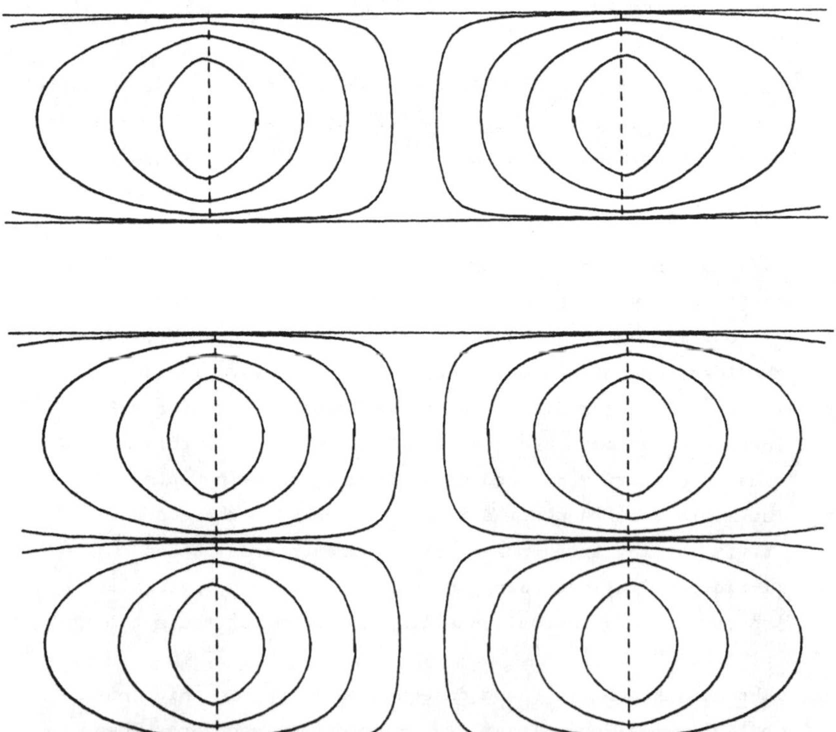

Fig. 7.4 Magnetic field patterns for: (a) mode TE_{011} and (b) mode TE_{012}

(provided the same two metal plates have been used in both measurements). Then, by substituting the measured values twice in (7.47) (once for each measurement), it becomes possible to eliminate δ so that the final formula for the loss tangent becomes [7]:

$$\tan \delta = \frac{1 + W/\epsilon_r}{p - 1} \left(\frac{p}{Q_{up}} - \frac{1}{Q_{u1}} \right) \tag{7.48}$$

In the above, Q_{u1} is the unloaded Q factor of the TE_{011} mode, and Q_{up} is the measured unloaded Q factor of the TE_{01p} mode (for the resonator with p times larger length).

The estimated tolerances on $\tan \delta$ quoted by Kobayashi and Katoh range between ± 1.1 % (for a material with $\tan \delta = 18 \cdot 10^{-4}$), and ± 12 % (for a material with $\tan \delta = 1.59 \cdot 10^{-5}$). Naturally, care must be exercised so that the separation between plates is small enough to prevent the radiation.

Kobayashi and Katoh did not stack several resonators on top of each other, as was implied above, but rather they machined two specimens of the same radius, one of them being four times longer than the other. Also, their paper shows how to account for small inaccuracies in the physical dimensions and the fact that the two resonant frequencies slightly differ from each other.

The second reason for inaccuracy in measuring the loss tangent is caused by the fact that the conductor losses constitute a large part of the total losses. Because dielectric materials are manufactured with larger and larger values of Q_d, the conductor losses in the metal plates make it virtually impossible to measure the dielectric part of the losses by the Courtney method. To reduce the conductor losses, the metal shield should be removed from the immediate vicinity of the dielectric resonator. The magnetic field intensity on the surface of the conductor becomes small, and the power lost in the conductor decays as the square of the field intensity, according to (7.11) (recall that the surface current is proportional to the tangential component of the magnetic field intensity). The question is, how far should one move the metal shield for best results?

DelaBalle, Guillon, and Garault have computed in [8] the influence of a cylindrical metal shield enclosing the resonator. Their results show that for a cylindrical shield of the same aspect ratio as the resonator itself (see Fig. 7.5), namely, for $h/b = L/a$, Q_u becomes essentially equal to Q_d when the ratio b/a is anywhere between 2 and 3.

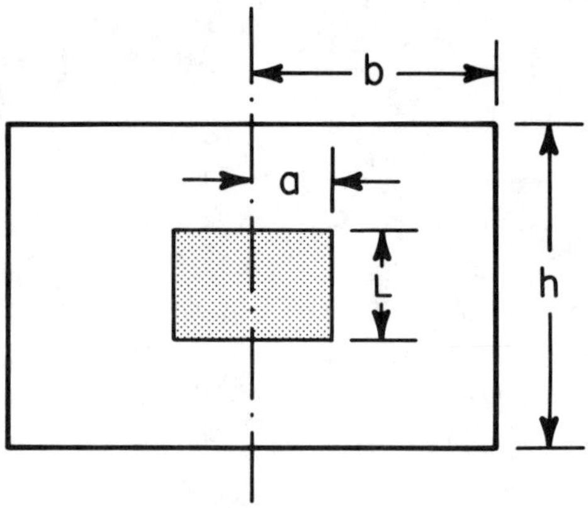

Fig. 7.5 Dielectric resonator in a cylindrical cavity

Therefore, they recommend measuring tan δ by placing the dielectric resonator in such a cylindrical cavity, and by assuming $Q_u = Q_d$. In other words, when $2 < b/a < 3$, they claim that the conductor losses have a negligible effect on the unloaded Q factor of the cavity. Leading manufacturers of dielectric resonators have adopted this method of measuring the loss tangent. Typically, their shielding cavities are about three times larger than the dielectric resonators which are being tested.

An interesting physical interpretation of a very similar phenomenon was given by Imai and Yamamoto [9]. They have studied a semispherical dielectric resonator in a semispherical conductive shield. The radius of the dielectric resonator is $R_1 = 1.2$ mm, and its relative dielectric constant is $\epsilon_r = 36$ (see Fig. 7.6(a)). When the radius R_2 of the conductive shield was gradually increased while R_1 was kept constant, the computed resonant frequency of the cavity behaved in the way indicated in Fig. 7.6(b). When R_2 was only slightly larger than R_1 (region (A) in Fig. 7.6(b)), the frequency changed rapidly as a function of R_2. Then, in region (B), the frequency remained essentially independent of R_2. Finally, in region (C), frequency again changed rapidly as a function of R_2. The rapid change of frequency in region (C) closely follows the

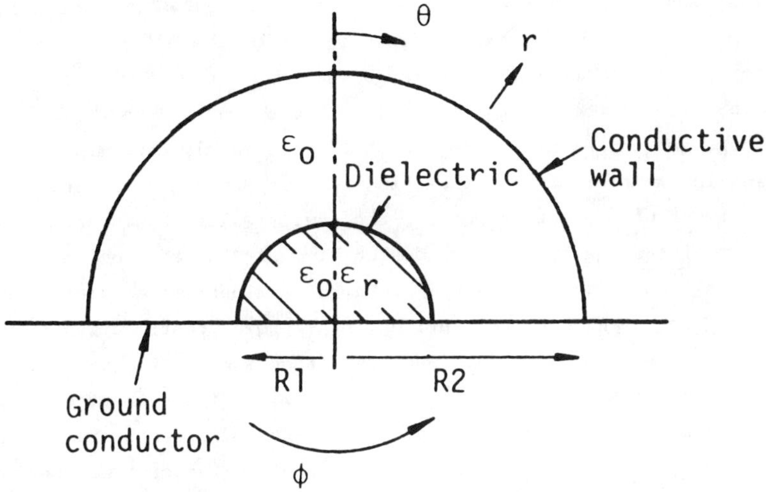

Fig. 7.6(a) Semispherical resonator (reference [9], ©1984 Scripta Publishing Co.)

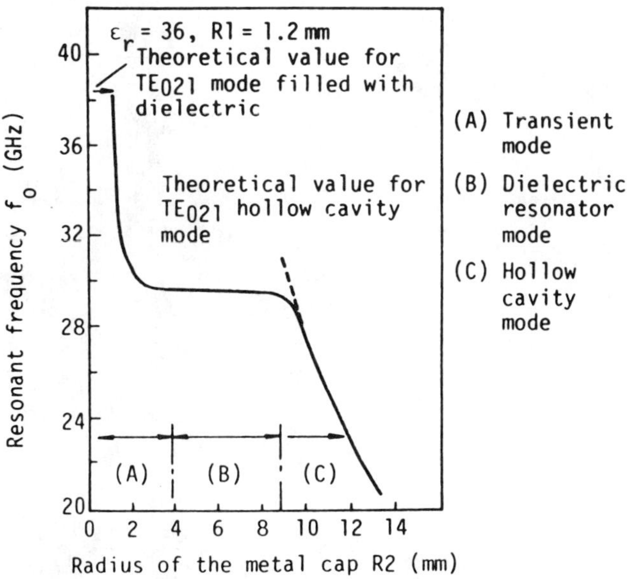

Fig. 7.6(b) Resonant frequency of the semipherical resonator in a semispherical metal cavity (reference [9], ©1984 Scripta Publishing Co.)

resonance of an empty cavity, while the rapid change in region (A) corresponds to the frequency of a cavity fully filled with dielectric.

In view of the incremental frequency rule, the Q factor due to conductor losses in the shield is large when the slope df/dR_2 (frequency vs. the size of the cavity) is small. Therefore, in the range (B), Q_c is large and can be neglected, so that Q_u is mainly determined by the losses in the dielectric. Thus, in order to perform an accurate measurement of dielectric losses, the cavity walls should be placed at a certain distance away from the dielectric. At the time of this writing, an actual measurement procedure along these lines, which would properly quantify the small, but not entirely negligible, contribution from the conductor losses, has not yet been published. However, the analysis and measurements made by Kobayashi and Katoh [7] indicate that, even for a Courtney holder, if the conductor plates are sufficiently removed from the dielectric resonator itself, the unloaded Q factor is approximately equal to the dielectric Q factor.

7.4 Linear Coefficients

The application of dielectric resonators in microwave filters has been attempted in the mid-1960s, but it was soon realized that a widespread use would be impractical unless the materials could be found with better temperature stability [10]. Such materials were developed about ten years later [11], and a temperature stability was achieved which is comparable to that of microwave hollow resonators made out of invar. Widespread use followed immediately [12].

The first temperature effect which comes to mind when talking about a microwave resonator is the expansion of material. It is an experimental fact that most solids expand linearly with an increase in temperature. A rod of length L will expand by ΔL when the temperature increases by ΔT. The constant of proportionality is α, the linear coefficient of expansion:

$$\frac{\Delta L}{L} = \alpha \Delta T \tag{7.49}$$

The relative expansion per degree centigrade is very small, on the order of 10^{-6} or 10^{-5}, as can be seen in Table 7.1, which was compiled from [13, Ch. 21] and [14]. Strictly speaking, the coefficients vary slightly as functions of temperature, and the values given in the table are the averaged results observed in the range between 0°C and 100°C. The abbreviation often used to denote 10^{-6} is ppm (parts per million).

Table 7.1 LINEAR EXPANSION COEFFICIENTS

Material	α (per °C)	Material	α (per °C)
Aluminum	23×10^{-6}	Glass (ordinary)	9×10^{-6}
Brass	19×10^{-6}	Glass (pyrex)	3.2×10^{-6}
Copper	17×10^{-6}	Teflon	90×10^{-6}
Steel	11×10^{-6}	Polyethilene	190×10^{-6}
Invar	0.7×10^{-6}	Fused quartz	0.57×10^{-6}

Consider the simple example of a microwave coaxial half-wavelength TEM resonator shown in Fig. 7.7. The electric field intensity is largest at the midpoint between the two shorted ends. The resonance occurs when the length L is one-half wavelength, λ, of the TEM wave which is, furthermore, equal to the free-space wavelength λ_0. We are

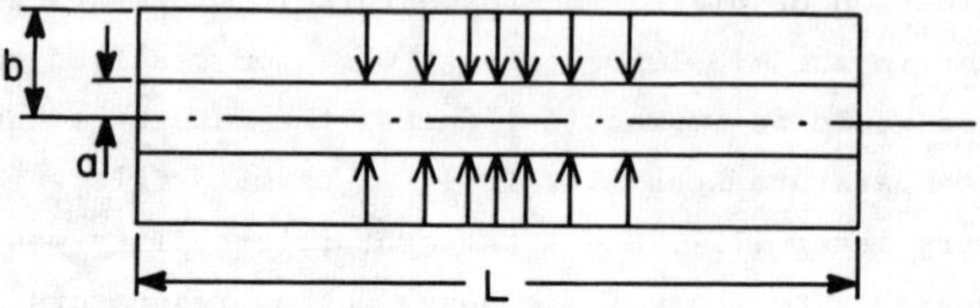

Fig. 7.7 Half-wavelength coaxial resonator

specifically assuming that the resonator is filled with air. The resonant frequency is, therefore, determined by the condition:

$$L = \frac{\lambda}{2} \tag{7.50}$$

and since

$$\lambda = \lambda_0 = \frac{c}{f} \tag{7.51}$$

we have

$$f = \frac{c}{2L} \tag{7.52}$$

where $c = 3 \times 10^8$ m/s.

When temperature T changes by an increment ΔT, the relative change in frequency can be computed by taking the differential of (7.52) as follows:

$$\frac{\Delta f}{f} = \frac{1}{f}\frac{df}{dL}\Delta L$$

By taking the derivative of (7.52) and then using (7.49), one obtains

$$\frac{\Delta f}{f} = -\alpha \Delta T \tag{7.53}$$

As an example of the use of (7.53), we find that an air-filled coaxial resonator which is made of brass should demonstrate a relative change in resonant frequency with temperature of -19 ppm/°C. If the same resonator was made of invar, the temperature sensitivity would be reduced to -0.7 ppm/°C. The sign of the result is negative because an increase in temperature makes the resonator longer, thus, its resonant frequency is decreased.

The linear expansion of solid materials is the same in each direction in space, unless the material has anisotropic expansion properties. For the coaxial resonator in Fig. 7.7, the radii a and b expand with temperature by the same expansion coefficient as the length L does. However, the resonant frequency of the TEM resonator is not influenced by either a or b.

The second effect of temperature on the resonant frequency of microwave resonators comes from the fact that the relative dielectric constant ϵ_r also varies as a function of temperature. As a first approximation, the change is linearly proportional with temperature. The constant of proportionality is denoted by τ_ϵ, the temperature coefficient of the dielectric constant:

$$\frac{\Delta\epsilon_r}{\epsilon_r} = \tau_\epsilon \Delta T \tag{7.54}$$

Suppose the coaxial resonator from Fig. 7.7 has been entirely filled with an insulator of relative dielectric constant ϵ_r. The wavelength λ of the TEM field in the resonator is, thus,

$$\lambda = \frac{\lambda_0}{\sqrt{\epsilon_r}} \tag{7.55}$$

and the resonant frequency of the half-wavelength resonator is, hence, specified by

$$f = \frac{c}{2L\sqrt{\epsilon_r}} \tag{7.56}$$

As a function of temperature, the resonant frequency will shift for two reasons: because of the linear expansion, and because of the change in dielectric constant. This situation is expressed as follows:

$$\frac{\Delta f}{f} = \frac{df}{fdL}\Delta L + \frac{df}{fd\epsilon_r}\Delta\epsilon_r \tag{7.57}$$

The sensitivity of the resonant frequency with temperature is denoted τ_f, the temperature coefficient of the resonant frequency:

$$\frac{\Delta f}{f} = \tau_f \Delta T \tag{7.58}$$

By taking the appropriate derivatives of (7.56), we find

$$\tau_f = -\alpha - \frac{1}{2}\tau_\epsilon \tag{7.59}$$

The above equation suggests an obvious method of achieving a temperature compensated resonator: τ_ϵ must have an opposite sign from α, and have approximately double magnitude. All solid materials expand with rising temperature. Therefore, the dielectric material must exhibit a negative τ_ϵ in order to be suitable for temperature compensation.

Equation (7.59) is not only valid for TEM resonators, but also for any other hollow waveguide resonator which is completely filled with an insulator of relative dielectric constant ϵ_r.

As another example, consider a circular cylindrical hollow resonator of radius a and length L, as shown in Fig. 7.8. For simplicity of

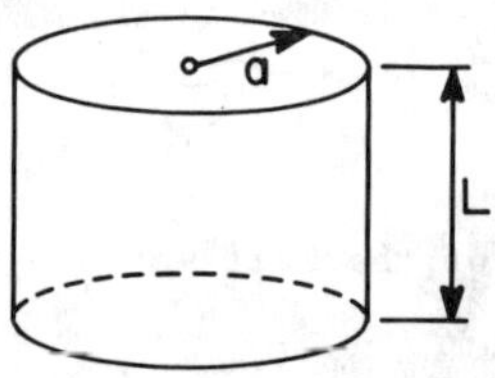

Fig. 7.8 Cylindrical cavity resonator

discussion, assume that the resonator is made of a solid isotropic dielectric material which is plated with a thin layer of good conductor, such as silver or copper. Then, only α and τ_ϵ of the dielectric material will influence the resonant frequency, since the thin metal layer will be forced to expand at the same rate as the bulk dielectric.

The resonant frequency is obtained from (2.89) or (2.96) as follows:

$$f^2 = \left(\frac{c}{2\pi}\right)^2 \frac{1}{\epsilon_r}\left(\left(\frac{p\pi}{L}\right)^2 + \left(\frac{x}{a}\right)^2\right) \tag{7.60}$$

where x is a zero of a Bessel function or its derivative, depending on the type of mode chosen. For example, for the TE_{011} mode, x = 3.832, p = 1, etc. When the expansion properties of the dielectric material are isotropic, we have

$$\frac{\Delta L}{L} = \frac{\Delta a}{a} = \alpha \tag{7.61}$$

In this case, the differential of (7.60) gives the same result (7.59) as in the TEM case. This means that (7.59) is valid for any TE_{mnp} or TM_{mnp} resonant mode of the cylindrical resonator in Fig. 7.8.

For the resonant system which consists of several regions filled with various dielectric materials, (7.59) does not hold exactly true, but it remains, nevertheless, a fairly good approximation. An important example of such a resonator filled with inhomogeneous dielectric is the parallel-plate dielectric resonator used in the Courtney measurement. The resonant frequency of this resonant cavity is given by the set of equations (7.36), (7.37), and (7.38). If the temperature T is an independent variable, the quantities which become dependent variables are: the resonant frequency, the radius and the length of the dielectric rod, and the relative dielectric constant. The analysis starts with (7.42), using

$$\Delta x = \frac{\partial x}{\partial k_0 a} \frac{\partial k_0 a}{\partial T} \Delta T + \frac{\partial x}{\partial \epsilon_r} \frac{\partial \epsilon_r}{\partial T} \Delta T \tag{7.62}$$

and

$$\Delta y = \frac{\partial y}{\partial k_0 a} \frac{\partial k_0 a}{\partial T} \Delta T \tag{7.63}$$

Evaluating all the necessary differentials and utilizing definitions (7.49), (7.54), and (7.58), we obtain an interesting relationship:

$$\tau_f = -\alpha - \frac{1}{2} \frac{\tau_\epsilon}{1 + \frac{W}{\epsilon_r}} \tag{7.64a}$$

The difference between (7.64a) and (7.59) is in the appearance of the filling factor p_e, defined earlier by (7.33):

$$\tau_f = -\alpha - \frac{1}{2} \tau_\epsilon \, p_e \tag{7.64b}$$

It may be recalled from the example treated earlier that p_e is very close to unity. Therefore, at least for the Courtney holder, the exact relationship between τ_f and τ_ϵ is almost the same as for a uniformly filled cavity. For other cavities filled with inhomogeneous matierals, the filling factor may differ considerably more from unity, so that the use of the approximate equation (7.59) may become the cause of significant errors in determining the value of τ_ϵ. This topic will be pursued further in the next section.

7.5 Covered DR on Microstrip Substrate

In a typical microwave integrated circuit application the dielectric resonator is mounted on a dielectric substrate, and the whole unit is closed in a metal housing. Neglecting the presence of the microstrip conductor, the resonant system can be modeled as shown in Fig. 7.9. The dielectric resonator radius is a, and its length is L. The relative dielectric constant of the resonator is ϵ_r, and the relative dielectric constant of the microstrip substrate is ϵ_{r2}. The surrounding material is either air ($\epsilon_{r1} = 1$) or some polyfoam filling. Due to temperature

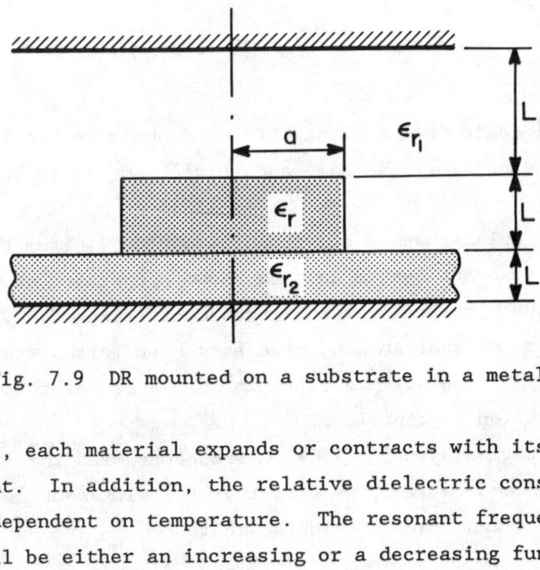

Fig. 7.9 DR mounted on a substrate in a metal housing

variations, each material expands or contracts with its own linear coefficient. In addition, the relative dielectric constants are also linearly dependent on temperature. The resonant frequency of the entire system will be either an increasing or a decreasing function of temperature. In an ideal situation, the proper selection of materials and the suitable combination of dimensions may result in a temperature-compensated system.

The above problem has been studied by Higashi and Makino [15], for the resonator operating in $TE_{01\delta}$ mode. They compute the resonant frequency of the system by an interesting simplification of the Itoh and Rudokas procedure. The computation is essentially the same as the one described in Ch. 4, except that the following approximation is used for computation of the eigenvalue x in place of (4.103):

$$x = 0.951x_{01} + 0.222\sqrt{(k_0a)^2(\epsilon_r - 1) + 0.951x_{01}^2} \tag{7.65}$$

where $x_{01} = 2.405$. Use of (7.65) enables one to compute the resonant frequency without utilizing Bessel functions, which would otherwise complicate the evaluation of differentials needed in the computation of temperature dependence.

When all the differentials are evaluated, the relative change of frequency due to temperature looks as follows:

$$\frac{\Delta f}{f} = C_a\frac{\Delta a}{a} + C_L\frac{\Delta L}{L} + C_{L1}\frac{\Delta L_1}{L_1} + C_{L2}\frac{\Delta L_2}{L_2} + C_{\epsilon r1}\frac{\Delta\epsilon_{r1}}{\epsilon_{r1}} + C_{\epsilon r2}\frac{\Delta\epsilon_{r2}}{\epsilon_{r2}} + C_{\epsilon r}\frac{\Delta\epsilon_r}{\epsilon_r} \tag{7.66}$$

Each of the coefficients C is a straightforward but somewhat lengthy expression, given explicitly in [15], and it will not be reproduced here.

The practical implementation of (7.66) is obvious. From the dimensions of the resonant system and the known properties of materials, one can evaluate the numerical value of each coefficient C. This value indicates the relative importance of each part used in the system. Then, it is possible to select the parts in such a manner that the overall frequency variation is minimized.

So that the notation used in (7.66) corresponds with the linear coefficients introduced earlier, we note that the differential increment of any length L as a function of temperature is

$$\Delta L = \frac{dL}{dT}\Delta T \tag{7.67}$$

Using the linear temperature coefficients defined by (7.49), (7.54), and (7.58), we can rewrite (7.66) as follows:

$$\tau_f = C_a\alpha_a + C_L\alpha_L + C_{L1}\alpha_{L1} + C_{L2}\alpha_{L2} + C_{\epsilon r1}\tau_{\epsilon 1} + C_{\epsilon r2}\tau_{\epsilon 2} + C_{\epsilon r}\tau_{\epsilon} \tag{7.68}$$

In the above, α denotes the temperature expansion coefficient, and τ_ϵ denotes the temperature coefficient of the relative dielectric constant. The subscripts indicate that part of the entire system to which the individual coefficient refers. τ_f is the resulting temperature sensitivity of the entire resonant system, given in ppm/°C.

An example of the values which one may encounter in practice is given in Table 7.2. The example is taken from [15] for a system with resonant frequency f = 10.80 GHz. First of all, one can see that the three dominant coefficients are C_a, C_L, and $C_{\epsilon r}$. Since the linear

Table 7.2 TEMPERATURE COEFFICIENTS OF A COVERED DR ON MICROSTRIP SUBSTRATE

<u>Dimensions:</u>

L = 2.26 mm, L_1 = 2 mm, L_2 = 0.62 mm, a = 2.88 mm

<u>Materials:</u>

$\epsilon_r = 35$, $\epsilon_{r1} = 1$, $\epsilon_{r2} = 2$

<u>Coefficients C:</u>

$C_a = -0.567$, $C_L = -0.363$, $C_{L1} = -0.0132$, $C_{L2} = -0.0574$

$C_{\epsilon r} = -0.493$, $C_{\epsilon r1} = -0.002$, $C_{\epsilon r2} = -0.0008$

expansion of the dielectric resonator is isotropic, $\alpha_a = \alpha_L$, and we have approximately (by ignoring all the other coefficients):

$$\tau_f \simeq (C_a + C_L)\alpha_a + C_{\epsilon r}\tau_\epsilon \tag{7.69}$$

For the values given in the Table, $C_a + C_L = -0.93$ and $C_{\epsilon r} = -0.493$. It can be observed that in the first approximation the relationship specified by (7.59) is still valid. The next closest factor of importance is the expansion of the dielectric substrate material, denoted C_{L2}, and so on.

The dielectric resonator system in Fig. 7.9 fits the simple model discussed in Ch. 4 for which the PC computer programs DRESP and DRESV2 have been developed. Program DRESV2 is based on an improved Itoh and Rudokas procedure. The same procedure in simplifed form has been used by Higashi and Makino in their study of temperature stability. Thus, we should be able to use the computer program DRESV2 to obtain the same results numerically as those derived analytically by Higashi and Makino.

Numerical evaluation of the coefficient C_a is done as follows. The dimensions and the values of dielectric constants as given in Table 7.2

are entered into the program. The resonant frequency computed by DRESV2 is freq(I&R) = 10.91248. Next, the radius a of the resonator is increased by 0.1% (the new value of a is 2.88288 mm), and the progam is executed again. The new resonant frequency for the Itoh and Rudokas model comes out to be 10.90622 GHz. The coefficient C_a is then computed as follows:

$$C_a = 1000(10.90622/10.91248 - 1) = -0.5736$$

This value agrees fairly well with -0.567, the result obtained by Higashi and Makino [15]. The more accurate variational result from DRESV2 gives $C_a = -0.560$.

In general, if the dimension L_i is increased to $L_i + \Delta L_i$ and the new frequency of the resonator is computed to be f_i, any of the coefficients C_{Li} in (7.68) can be obtained numerically as follows:

$$C_{Li} \simeq \frac{L_i}{\Delta L_i} \frac{f_i - f}{f} \tag{7.70}$$

Using an analogous procedure for other dimensions and relative dielectric constants, it is possible to find all the coefficients C in (7.68) by a strictly numerical procedure. A warning is appropriate regarding such a use of the program DRESV2. Because the program is written in BASIC language, it operates in single-precision arithmetic. On the other hand, when very small differences are evaluated, the single precision arithmetic does not always ensure sufficient accuracy of results.

Higashi and Makino do not give any example of designing an actual circuit for temperature stable operation. Other authors, like Ishihara et al. [16] have found, experimentally, that it is indeed possible to achieve a highly temperature stabilized operation of an oscillator which utilizes a DR on a dielectric substrate as a frequency determining element. They found that the combination of the expansion coefficients of the DR and other parts is such that the perfect thermal compensation is possible only at a single position L_1 of the tuning plate (see Fig. 7.9). While the entire tuning range of the oscillator is up to 1.5 GHz in the best case, the high stability is achieved in a tuning range of only 50 MHz.

The stability is indeed impressive: only ± 150 kHz, centered at 11.85 GHz, over the temperature range between -20°C and +60°C. If this change was linear, it would amount to ± 0.16 ppm/°C as quoted by Ishihara et al., but such an interpretation is erroneous. The frequency exhibits a bell-shaped curve as a function of temperature, as can be seen in Fig. 7.10, which is taken from [16]. Therefore, the frequency sensitivity is zero at the center of the stablized range of temperatures, but it is consequently higher at the edges of the temperature range. A graphical estimation of the slopes of Fig. 7.10 yields frequency sensitivites up to 0.55 ppm/°C, but this is still a very good stability.

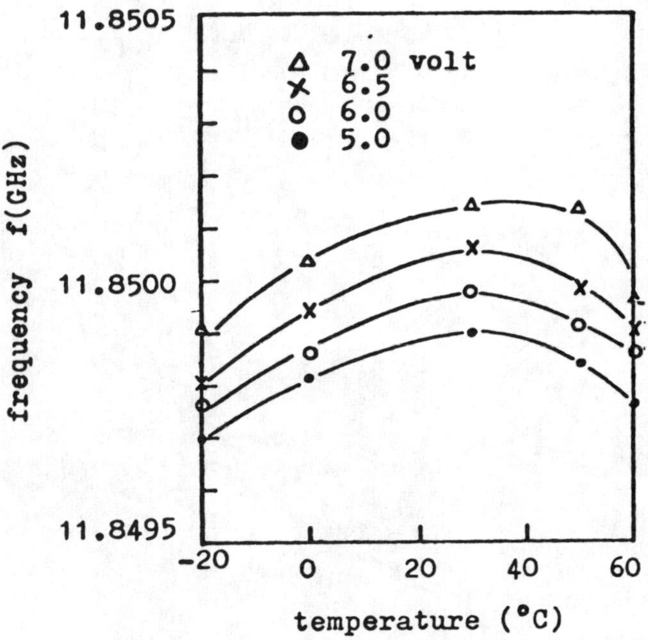

Fig. 7.10 DRO frequency vs. temperature diagram (reference [16], ©1980 IEEE)

It is true that the temperature compensation is not possible in the entire tuning range, but, as Ishihara et al. point out, using a DR with a different τ_ϵ makes it possible to achieve the compensation at another frequency. In this manner, they were able to manufacture the oscillators with the same high stability at any operating frequency from 9 to 14 GHz.

7.6 Measurement of Temperature Coefficients

One of the best understood and analyzed methods for measuring the dielectric constant and its temperature dependence is the one known as the Courtney method [6]. Courtney, actually, only perfected and scrutinized a parallel-plate arrangement introduced earlier by Hakki and Coleman [5].

The side view of the parallel-plate resonator utilized in this measurement is shown in Fig. 7.2. The circular cylinder of relative dielectric constant ϵ_r, the value which is to be measured, is placed between two metal plates. Depending on frequency and the nature of the field excitation, many resonant modes are possible in this truncated dielectric rod waveguide, as described in Ch. 3. The lowest resonant mode is HEM_{111}, and the next higher mode, TE_{011}, is the one which is used in the measurement. The magnetic field, $\underline{H}$, for the latter mode is plotted in Fig. 7.4(a). Suitable coupling to the TE_{011} mode is achieved by the use of coaxial cables terminated in small, horizontally oriented loops. The actual choice of the coupling mechanism is not critical: Courtney used small, horizontally bent monopoles and coaxial cables, whereas Hakki and Coleman utilized an iris in the rectangular waveguide which was placed outside of the lower metal plate. The resonant frequency of the TE_{011} mode is then measured in the transmission method. Because the resonant frequency of this configuration is known exactly, it is possible to compute the value of ϵ_r which corresponds to the measured frequency and to the measured dimensions a and L.

Courtney has measured values of ϵ_r for various materials and found the results to be very consistent. His estimate is that by measuring the dimensions with an accuracy of ± 0.5 mils (± 127 μm), and by measuring frequency with an accuracy of ± 1 MHz, the error in ϵ_r is less than 0.3 %. He has also placed the resonator in an oven and measured ϵ_r as a function of temperature in a range between 20°C and 120°C. The measured results are shown in Fig. 7.11, which is taken from [6]. Notice a highly linear dependence of ϵ_r with temperature. After correcting the results for the fact that the resonator dimensions have also expanded due to temperature increase, Courtney was able to obtain very accurate values of the temperature coefficients τ_ϵ.

Unfortunately, the present-day materials commonly used in dielectric resonators were not available at the time when Courtney performed his measurements. Consequently, the highly reliable data in the Courtney

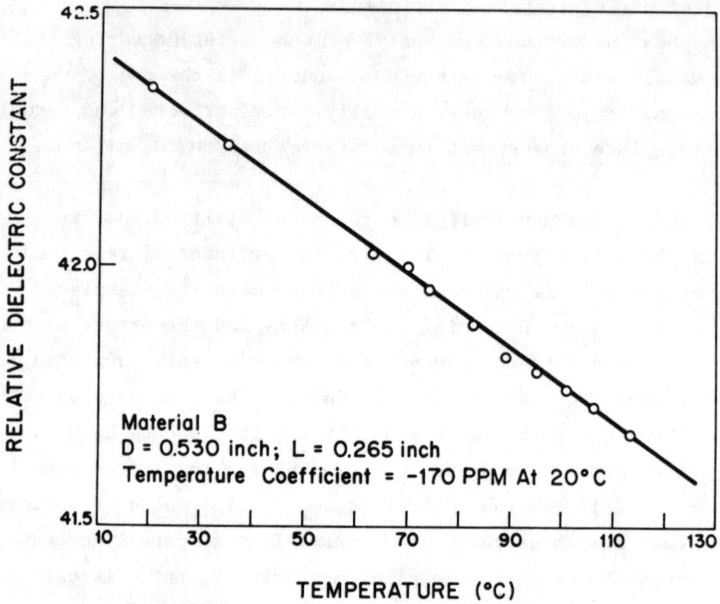

Fig. 7.11 Measured ϵ_r vs. temperature (reference [6], ©1970 IEEE)

paper [6] do not include those materials which are commercially available today. Later researchers selected other, indirect, methods for measuring the temperature variation of dielectric constant, as will be described in what follows. For this reason, the values τ_ϵ of the present-day materials used in dielectric resonators are largely unkown.

It can be recalled from Sec. 7.3 that the Courtney method is not well suited for the measurement of the loss tangent of the dielectric material. When the materials are studied from the point of view of their possible use in dielectric resonators, the loss tangent of the material is very important and should be closely monitored. For this reason, it is convenient to remove the dielectric resonator from the immediate vicinity of the metal walls, and thus reduce the surface currents in the walls. In that case, the conductor losses become a minor part of the loss mechanism, and the measured unloaded Q factor is a good indicator of the quality of the dielectric material itself.

Plourde et al. [11] have placed the resonator in the center of a rectangular waveguide, the size of which is selected so that it is in the propagating region for the dominant TE_{10} mode. The measurement of the unloaded Q factor was then performed by the reflection-type method (see Ch. 2). Such a measurement is faster and more indicative of the loss tangent of the resonator itself.

Besides the unloaded Q factor, the temperature stability of the dielectric resonator is a very important property to monitor. Plourde et al. simply monitored the resonant frequency of the resonator mounted in a waveguide and observed the change of frequency as a function of temperature. This measurement provided τ_f, the temperature coefficient of the resonant frequency, as defined in (7.58). The temperature coefficient of expansion, α, has been measured independently. Finally, the temperature coefficient of the dielectric constant was computed from (7.59) as follows:

$$\tau_\epsilon = -2(\tau_f + \alpha) \tag{7.71}$$

It has shown in the previous section that the above formula is not accurate for resonators filled with inhomogeneous materials. The use of this formula in determining the approximate value of τ_ϵ is justified in such a systematic testing of a large number of samples, as performed by Plourde et al., in order to select the best material. They have found that by varying the percentage of TiO_2 and BaO, the two basic components in their experiment, it was possible to observe a distinct minimum in τ_f at 81.8 mol % of TiO_2. The resulting compound, $Ba_2Ti_9O_{20}$, had $\epsilon_r = 39.8$, $Q = 8000$, and $\tau_f = 2 \pm 1$ ppm/°C. The thermal stability of this material was two orders of magnitude better than the previously used rutile, TiO_2, which had $\tau_f \simeq 400$ ppm/°C [10].

The discovery of Plourde et al. started the proliferation of commercially available temperature-stable dielectric resonators made of this and other similar ceramic materials. Unfortunately, even today, the manufacturers of dielectric resonators determine the temperature properties of their materials in the same approximate way as Plourde et al. did by measuring only τ_f, and not τ_ϵ.

As we have seen in the previous section, τ_f is the temperature coefficient of the inhomogeneous resonant system, such as the one in Fig. 7.5. Granted, if the user utilizes the test cavity of the same size and shape as the manufacturer's test cavity, he will get the same τ_f as the manufacturer. However, the typical user wishes to miniaturize his resonant system. When the dimensions of the shield in Fig. 7.5 are modified, the resulting τ_f may come out to be very different from the one specified by the manufacturer. In order to make an intelligent design, one has to perform an analysis which yields an equation like (7.68), and select the value of τ_ϵ which will minimize τ_f.

Unfortunately, the manufacturers do not know accurately the τ_ϵ of their materials because they do not measure it. If pressed, the manufacturers provide the values of α and suggest that the user, himself, compute τ_ϵ from (7.71). As we know, this is an unreliable procedure.

A numerical example, which illustrates the magnitude of error involved in this procedure, will be taken from [3], The test cavity specified by the manufacturer is such as shown in Fig. 7.5. The resonator has $\epsilon_r = 38.6$, a = 6.35 mm, L = 5.08 mm. The shielding cavity is made of gold-plated aluminum with radius b = 19.05 mm and height h = 26.67 mm. The thermal expansion coefficient for the dielectric resonator is $\alpha_r = 9.4 \cdot 10^{-6}/°C$, and for aluminum $\alpha_c = 23 \cdot 10^{-6}/°C$. The temperature coefficient of the resonant frequency for this cavity is measured to be $\tau_f = 4 \cdot 10^{-6}/°C$. We want to find the value of the temperature coefficient of the dielectric constant τ_ϵ.

Using (7.71), we compute

$$\tau_\epsilon = -2(4 \cdot 10^{-6} + 9.4 \cdot 10^{-6}) = -26.8 \cdot 10^{-6}/°C$$

An accurate expression should take into account the expansion coefficients of different materials. In principle, resonant frequency is a function of the mechanical dimensions and of the relative dielectric constant:

$$f = f(a,L,b,h,\epsilon_r) \tag{7.72}$$

The temperature coefficient of resonant frequency is then

$$\frac{1}{f}\frac{df}{dT} = \frac{1}{f}\left(\frac{df}{da}\frac{da}{dT} + \frac{df}{dL}\frac{dL}{dT} + \frac{df}{db}\frac{db}{dT} + \frac{df}{dh}\frac{dh}{dT} + \frac{df}{d\epsilon_r}\frac{d\epsilon_r}{dT}\right) \tag{7.73}$$

This can be written as

$$\tau_f = C_{\alpha r}\alpha_r + C_{\alpha c}\alpha_c + C_{\epsilon r}\tau_\epsilon \tag{7.74}$$

The linear coefficients C_α are defined in the same way as in (7.68).

The resonant frequency of the shielded resonator cannot be expressed analytically in a form which would permit the evaluation of derivatives in (7.73). In order to evaluate C_α one must evaluate the numerical derivatives according to (7.70) by using a computer program of the type described in Ch. 5. For the above example, such a numerical evaluation gives:

$$C_{\epsilon r} = -\ 0.486$$

$$C_{\alpha r} = -\ 0.999$$

$$C_{\alpha c} = -\ 0.0017$$

From (7.74), we express $\tau_{\epsilon r}$ as follows:

$$\tau_{\epsilon r} = \frac{1}{C_{\epsilon r}}\left(\tau_f - C_{\alpha r}\alpha_r - C_{\alpha c}\alpha_c\right) \qquad (7.75)$$

For the above example, (7.75) gives $\tau_\epsilon = -\ 27.6 \cdot 10^{-6}/°C$. The difference between this value and the previously computed approximate value is 3.0 %.

In the example shown, the manufacturer has used a test cavity which was three times as wide, and more than five times as high as the dielectric resonator. Because of this, the thermal expansion of the metal cavity did not play a great role in the overall temperature coefficient of the resonant frequency. For a smaller cavity, the influence would become more significant. For example, if the same resonator was inserted in a cavity which was only two times as wide and two times as high as the DR itself, the discrepancy between (7.75) and (7.71) would increase to 26 %.

What really should be done is to measure τ_ϵ directly by the Courtney method and determine the slope of the straight line, such as that shown in Fig. 7.11. Then, the user would know a number which specifies the material itself, and this number would be valid regardless of whether the resonator is used in the $TE_{01\delta}$, $HEM_{11\delta}$, $TM_{01\delta}$, or any other mode, and regardless of the size of the air gap between the resonator and the nearest metal wall.

7.7 Temperature Stabilization of DR Oscillators

When a dielectric resonator is used as a frequency-determining element of a free-running microwave oscillator, a highly stable source can be obtained. In order to achieve a good temperature compensation of the oscillator, it is not enough simply to select the dielectric resonator which has a temperature coefficient of resonant frequency equal to zero. An oscillator is specifically a system which consists of a passive circuit (typically DR mounted near a microstrip transmission line), and of an active device, which produces the oscillation. The oscillations occur at the frequency where the susceptance of the active device is equal and opposite to the susceptance of the passive circuit. In order to ensure that the oscillation frequency does not depend on temperature, the temperature coefficient of the passive circuit susceptance must be equal and opposite to the temperature coefficient of the active device susceptance.

Consequently, the general approach of designing a temperature-stable oscillator consists of measuring the temperature coefficient of the active device, and then selecting an appropriate resonator and its associated circuitry to compensate for the temperature dependence of the active device.

A dielectric resonator oscillator (DRO) with very high temperature stability of 1 ppm/°C, operating at 11 GHz, was designed by Komatsu et al. [17]. The entire circuit is built on a microstrip substrate, the essential part of which is shown in Fig. 7.12 (bias lines are not shown). The active device is a GaAs FET transistor. A special stack-shaped dielectric resonator has been developed for this purpose, consisting of two different materials A and B as indicated in Fig. 7.13.

The purpose of fabricating the "stack" DR is to achieve a highly linear temperature coefficient of resonant frequency in a wide range of temperature. Ordinary dielectric resonators made of homogeneous materials exhibit a slight nonlinearity of resonant frequency as a function of temperature. Figure 7.14 shows the values of frequency deviation with temperature, which Komatsu et al. have measured for two DR's made of different zirconate ceramics. The curve denoted by A is measured for an orthorombic structure, and curve B for a cubic structure. The nonlinearity of each curve is quite obvious, but note that one curve is convex, and the other concave. When the two materials are glued together in the stacked DR, the measured relative frequency variation becomes an almost perfectly linear function of temperature (curve C).

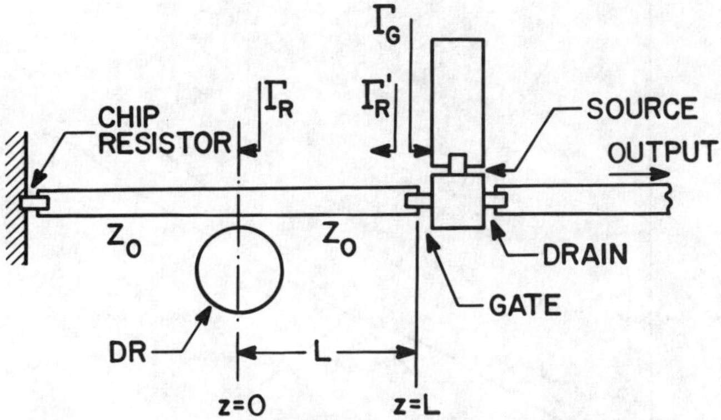

Fig. 7.12 GaAs FET DRO (reference [17], ©1981 IEEE)

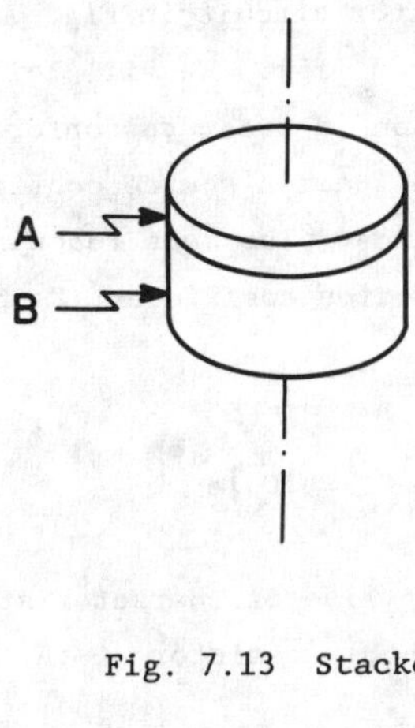

Fig. 7.13 Stacked DR

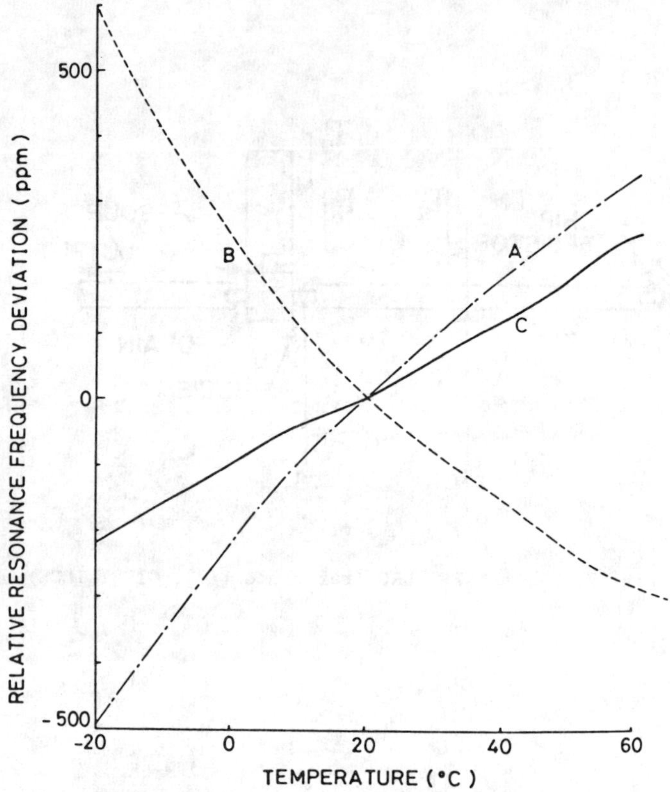

Fig. 7.14 Temperature behavior of the stacked DR (reference [17], ©1981 IEEE)

An analysis of the oscillator circuit in Fig. 7.12 has been presented by Tsironis and Pauker [18]. We will follow their analysis with a slight change in notation in order to conform with the notation used earlier in this book. The common-source configuration of the FET constitutes a negative resistance (plus some reactance) at the gate port. The corresponding reflection coefficient Γ_G has a magnitude larger than unity:

$$\Gamma_G = |\Gamma_G| e^{j\phi_G} \qquad (7.76)$$

The microstrip transmission line of characteristic impedance Z_0 is terminated at its far end by a chip resistor of the same value in order to prevent any possible parasitic oscillation at an unwanted frequency. Looking from the gate port toward the left, the impedance presented by the passive circuit is equal to the characteristic impedance (50 ohms), except at resonant frequencies of the natural modes in the dielectric resonator where some sharp maxima may occur.

The mode of operation is, of course, $TE_{01\delta}$. A simplified equivalent circuit for the position z = 0 (see Fig. 7.12) on the microstrip line is shown in Fig. 7.15. The coupling coefficient is denoted by κ:

$$\kappa = \frac{R_0}{2Z_0} \tag{7.76}$$

so that the impedance in the vicinity of resonance is expressed by

$$Z_R = Z_0 + \frac{R_0}{1 + j2Q_u\delta} \tag{7.77}$$

where the relative frequency is denoted by δ:

$$\delta = \frac{f - f_0}{f_0} \tag{7.78}$$

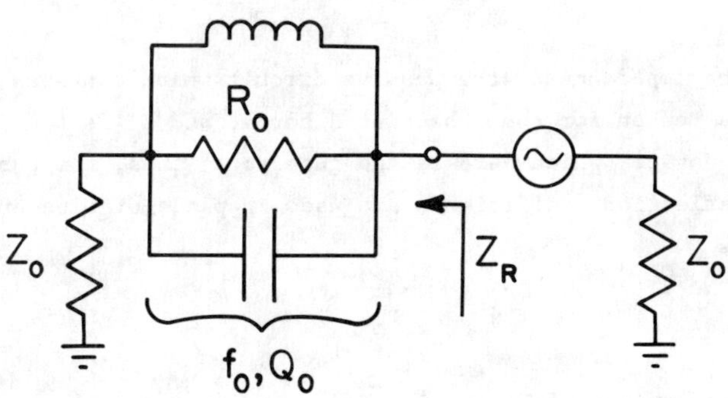

Fig. 7.15 Equivalent circuit of the DRO from Fig. 7.8

The unloaded Q factor of the DR mounted on the substrate is denoted by Q_u, and the resonant frequency of the mode $TE_{01\delta}$ is denoted by f_0. The reflection coefficient at z = 0 is computed as

$$\Gamma_R = \frac{Z_R - Z_0}{Z_R + Z_0} \tag{7.79}$$

from which we obtain

$$\Gamma_R = \frac{\kappa}{1 + \kappa + j2Q_u\delta} \tag{7.80}$$

The operating frequency occurs not far from the resonant frequency of the unloaded resonator so that the imaginary part in the denominator of (7.80) is considerably smaller than the real part:

$$\frac{2Q_u\delta}{1 + \kappa} \ll 1 \tag{7.81}$$

The reflection coefficient Γ_R is transformed along the microstrip line of the length L into the following value:

$$\Gamma'_R = \Gamma_R e^{-j2\theta} \tag{7.82}$$

where θ is the electric length between the resonator and the gate. The oscillation condition of the negative-resistance oscillator is usually expressed in the form

$$Z'_R + Z_G = 0 \tag{7.83}$$

where Z'_R is the impedance of the passive circuit which contains the resonator (the reason for the subscript R here), and Z_G is the impedance of the active device (which here is the gate port; thus, the subscript G). If the reflection coefficients are used in place of impedances, (7.83) becomes

$$\Gamma'_R \Gamma_G = 1 \tag{7.84}$$

Substituting the values of Γ'_R and Γ_G, found earlier, results in

$$\frac{\kappa e^{-j2\theta}}{1 + \kappa + j2Q_u\delta} \cdot |\Gamma_G| e^{j\phi_G} = 1 \tag{7.85}$$

The absolute value of (7.85) provides the oscillation amplitude, and the phase angle provides the oscillation frequency. To analyze the frequency stability, we need the phase of (7.85):

$$-2\theta - \tan^{-1}\left(\frac{2Q_u\delta}{1+\kappa}\right) + \phi_G = 0 \tag{7.86}$$

It is convenient to introduce the loaded Q factor, which is defined as

$$Q_L = \frac{Q_u}{1+\kappa} \tag{7.87}$$

In view of inequality (7.81), it is possible to simplify (7.86) as follows:

$$\phi_G \simeq 2\theta + 2Q_L\delta \tag{7.88}$$

which is the equation for determining the operation frequency of the system. Equation (7.88) will be the starting point for the study of temperature dependence of the system.

When the temperature of the circuit varies, each term in (7.88) has a distinct physical interpretation. For instance, the term on the left-hand side, ϕ_G, is the phase of the reflection coefficient of the active device. As a function of temperature, ϕ_G is going to change considerably, and this change is an important factor in the overall stability of the operating frequency f of the oscillator. ϕ_G is a quantity which can be measured as a function of temperature by observing the active device alone. Tsironis and Pauker have measured a number of points ϕ_G vs. temperature, and then plotted a best-fit straight line through these points in order to determine the slope $d\phi_G/dT$. The measured values reported by Tsironis and Pauker are between $2000 \cdot 10^{-6}$ and $3000 \cdot 10^{-6}$ rad/°C.

The first term on the right-hand side of (7.88) is twice the phase shift in the microstrip line of physical length L. The second term on the right-hand side also has a clear physcial interpretation. Relative frequency δ is defined by (7.78). When the temperature varies, both f and f_0 are influenced (recall that f is the operating frequency and f_0 is the resonant frequency of the unloaded resonator). Therefore,

$$\frac{d\delta}{dT} = \frac{1}{f_0}\frac{df}{dT} - \frac{f}{f_0^2}\frac{df_0}{dT} \tag{7.89}$$

Since f is nearly equal to f_0, we can write

$$\frac{d\delta}{dT} \simeq \tau_f - \tau_{f0} \tag{7.90}$$

where the temperature coefficients of frequency are defined as

$$\tau_f = \frac{1}{f}\frac{df}{dT} \tag{7.91}$$

$$\tau_{f0} = \frac{1}{f_0}\frac{df_0}{dT} \tag{7.92}$$

The value of the loaded Q factor which multiplies δ in (7.88) can be measured with a network analyzer by observing the reflection coefficient Γ_R in Fig. 7.12 as a function of frequency. This measurement is known as the reaction method, which was described in Ch. 2.

As seen in (7.87), Q_L depends on Q_u and on the coupling coefficient κ. The value of κ may be adjusted by changing the lateral distance d between the resonator and the microstrip (see Fig. 7.16). Another convenient adjustment is to lift the resonator away from the surface of the microstrip substrate. As a consequence, the losses in the ground plane are reduced, and the unloaded Q factor is increased. By placing the

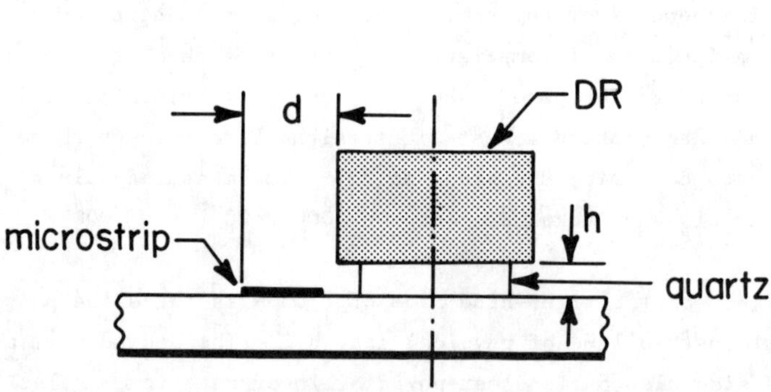

Fig. 7.16 DR mounted on microstrip for oscillator application

stacked resonator on a fused quartz spacer of height h = 0.5 mm rather than directly on the aluminum substrate, Q_u was increased from 600 to 1000 (one of the examples quoted in [18]). Table 7.3 lists several values of the loaded Q factor as a function of distance d. The values were read off the diagram of measured values in [18], and for that reason they may be somewhat inaccurate.

Table 7.3 LOADED Q FACTOR AS A FUNCTION OF DISTANCE d IN FIG. 7.16

d	0 mm	0.4 mm	0.8 mm	1.2 mm
Q_L for h = 0	91	161	304	512
Q_L for h = 0.5 mm	205	293	482	745

To find the temperature variation of the operating frequency, one has to evaluate the differential of (7.88):

$$\frac{d\phi_G}{dT}\,\Delta T \simeq 2\,\frac{d\theta}{dT}\,\Delta T + 2\left(\frac{dQ_L}{dT}\,\delta + Q_L\,\frac{d\delta}{dT}\,\Delta T\right) \tag{7.93}$$

Tsironis and Pauker have found that θ and Q_L change very little with temperature, so that the terms $d\theta/dT$ and dQ_L/dT can be ignored. Equation (7.93) thus simplifies, using (7.90), to the following:

$$\tau_f \simeq \tau_{f0} + \frac{1}{2Q_L}\,\frac{d\phi_G}{dT} \tag{7.94}$$

For temperature-stable operation, one requires $\tau_f = 0$, which leads to the requirement:

$$\tau_{f0} = -\,\frac{1}{2Q_L}\,\frac{d\phi_G}{dT} \tag{7.95}$$

We have obtained an equation which tells us how to make the oscillation frequency independent of temperature. For a given device, the value of $d\phi_G/dT$ is fixed, but Q_L may be selected, almost at will, by adjusting the coupling coefficient between the DR and microstrip line. Different combinations of Q_L and τ_{f0} may be tried with the idea of improving other properties of the oscillator, like constant output power or better tunability.

As an example, for the measured phase drift $d\phi_G/dT = -\,2600 \cdot 10^{-6}$ rad/°C, and the lateral distance d = 0.8 mm for a DR without a quartz spacer, Tsironis and Pauker quote the required temperature sensitivity of the DR to be 4.4 ppm/°C. If we take the value $Q_L = 304$ from Table 7.3 and substitute in (7.95), we obtain $\tau_{f0} = 4.27$ ppm/°C. The discrepancy of 3 % is probably caused by our incorrect reading of Q_L from the Tsironis and Pauker diagram.

When we try to apply the above principles to a particular DRO design, we may run into the same difficulty of interpreting the manufacturer's data as described in the previous section. Namely, we need to know τ_{f0} of the DR when it is mounted on a particular microstrip substrate, and possibly on top of a special quartz spacer. The manufacturer's data on τ_{f0} have been obtained in a different test cavity, the shape of which is usually not even known to us. Then, how useful is such τ_{f0} data?

If one had the information on the true value of τ_ϵ, one could compute τ_{f0} for the DR mounted on the microstrip by using the methods described in Sec. 7.5. In the absence of such information, the only practical solution left to each user is to measure τ_{f0} of the DR in his particular environment. After all, such a measurement may not necessitate an extra expenditure of time or equipment, since the measurement in the temperature chamber must be performed anyway because of the need to measure $d\phi_G/dT$.

Appendix 7.A COMMERCIALLY AVAILABLE MATERIALS

Listed below are tables with detailed technical specifications of some materials which are currently available in the United States. The list of materials is not intended to be comprehensive. The main purpose of these tables is to give the reader an idea of what is typically known about properties of materials used in dielectric resonators. The dimensions of the actual resonators available from stock are not listed here because this would take considerable additional space.

When certain data are not specified by the manufacturer, the corresponding slot in the tables is listed as "unspec.".

****

Manufacturer: Murata Erie North America, Inc.
Brand: Resomics-K series
Composition: $(Ba,Pb)Nd_2Ti_5O_{14}$
Q_d = 5000 at 1 GHz
Frequency range: 0.8 to 5 GHz
$\alpha = 8.5 \pm 0.42$ ppm/°C

Type	ϵ_r	τ_ϵ (ppm/°C)	τ_f (ppm/°C)
unspec.	90 ± 2.7	-29 ± 4	0 ± 2
unspec.	90 ± 2.7	-17 ± 4	+6 ± 2

Manufacturer: Murata Erie North America, Inc.

Brand: Resomics-M series

Composition: $(Zr,Sn)TiO_4$

Q_d = 15000 at 4 GHz

Frequency range: 1 to 12 GHz

$\alpha = 6.5 \pm 0.3$ ppm/°C

Type	ϵ_r	τ_ϵ (ppm/°C)	τ_f (ppm/°C)
C	37.3 ± 0.5	−13 ± 2	0 ± 2*
D	37.7 ± 0.5	−17 ± 4	+2 ± 2*
E	38.0 ± 0.5	−21 ± 4	+4 ± 2*
F	38.4 ± 0.5	−25 ± 4	+6 ± 2*
G	38.8 ± 0.5	−29 ± 4	+8 ± 2*

*Note: Available also in τ_f tolerances of ± 1 and ± 0.5 ppm/°C

****

Manufacturer: Murata Erie North America, Inc.

Brand: Resomics-U series

Composition: $(Zr,Sn)TiO_4$

Q_d = 10000 at 4 GHz

Frequency range: 1 to 12 GHz

$\alpha = 6.5 \pm 0.3$ ppm/°C

Type	ϵ_r	τ_ϵ (ppm/°C)	τ_f (ppm/°C)
A	36.5 ± 0.5	−5 ± 4	−4 ± 2*
B	36.9 ± 0.5	−9 ± 4	−2 ± 2*
C	37.3 ± 0.5	−13 ± 4	0 ± 2*
D	37.7 ± 0.5	−17 ± 4	+2 ± 2*
E	38.0 ± 0.5	−21 ± 4	+4 ± 2*
F	38.4 ± 0.5	−25 ± 4	+6 ± 2*
G	38.8 ± 0.5	−29 ± 4	+8 ± 2*
H	39.2 ± 0.5	−33 ± 4	+10 ± 2*

*Note: Available also in τ_f tolerances of ± 1 and ± 0.5 ppm/°C

Manufacturer: Murata Eire North America, Inc.

Brand: Resomics-S series

Composition: $Ba(Zr,Zn,Ta)O_3$

Q_d = 10000 at 10 GHz

Frequency range: 5 to 25 GHz

α = 10.2 ± 0.5 ppm/°C

Type	ϵ_r	τ_ϵ (ppm/°C)	τ_f (ppm/°C)
C	28.6 ± 0.5	-20.4 ± 4	0 ± 2*
D	29.2 ± 0.5	-24.4 ± 4	+2 ± 2*
E	29.7 ± 0.5	-28.4 ± 4	+4 ± 2*
F	30.1 ± 0.5	-32.4 ± 4	+6 ± 2*
G	30.5 ± 0.5	-36.4 ± 4	+8 ± 2*

*Note: Available also in τ_f tolerances of ± 1 and ± 0.5 ppm/°C

****

Manufacturer: Murata Erie North America, Inc.

Brand: Resomics-X series

Composition: $Ba(Zr,Zn,Ta)O_3$

Q_d = 14000 at 10 GHz

Frequency range: 5 to 25 GHz

α = 10.2 ± 0.5 ppm/°C

Type	ϵ_r	τ_ϵ (ppm/°C)	τ_f (ppm/°C)
D	30.3 ± 0.5	-24.4 ± 4	+2 ± 2*
E	30.6 ± 0.5	-28.4 ± 4	+4 ± 2*
F	31.0 ± 0.5	-32.4 ± 4	+6 ± 2*

*Note: Available also in τ_ϵ tolerances of ± 1 and ± 0.5 ppm/°C

Manufacturer: Trans-Tech, Inc.

Brand: Trans-Tech

Composition: Zr/Sn Titanate

Q_d > 10000 at 4 GHz

Frequency range: 2 to 45 GHz

α = 5.6 ± unspec. ppm/°C

Type	ϵ_r	τ_ϵ (ppm/°C)	τ_f (ppm/°C)
D-8516	35.9 ± 0.5	+1.6 ± unspec.	-3 ± 0.5
D-8515	36.0 ± 0.5	-6.9 ± unspec.	0 ± 0.5
D-8514	36.0 ± 0.5	-9.7 ± unspec.	+3 ± 0.5
D-8513	36.9 ± 0.5	-13.1 ± unspec.	+6 ± 0.5
D-8517	36.4 ± 0.5	-19.0 ± unspec.	+9 ± 0.5

****

Manufacturer: Trans-Tech, Inc.

Brand: Trans-Tech

Composition: Barium Tetratitanate

Q_d > 10000 at 4 GHz

Frequency range: 2 to 45 GHz

α = 9.5 ± unspec. ppm/°C

Type	ϵ_r	τ_ϵ (ppm/°C)	τ_f (ppm/°C)
D-8512	38.6 ± 0.6	-10.4 ± unspec.	+4 ± 0.5

Manufacturer: Thomson-CSF Components Corp.

Brand: E36

Composition: $(Zr,Sn)TiO_4$

Q_d = 4000 at 10 GHz

Frequency range: 2 to 100 GHz

$\alpha = 5 \pm 1$ ppm/°C

Type	ϵ_r	τ_ϵ (ppm/°C)	τ_f (ppm/°C)
E-1336	36.9 ± 0.4	-4 ± 1	-3 ± 1
E-2036	37.0 ± 0.4	-10 ± 1	0 ± 1
E-2336	37.1 ± 0.4	-16 ± 1	+3 ± 1
E-2636	37.2 ± 0.4	-22 ± 1	+6 ± 1
E-2936	37.3 ± 0.4	-28 ± 1	+9 ± 1

References

[1] R.E. Collin, Foundations for Microwave Engineering, New York: McGraw-Hill, pp. 17-20, 1966.

[2] R.F. Harrington, Time-Harmonic Electromagnetic Fields, New York: McGraw-Hill, pp. 23-26, 1961.

[3] *** , "Microwave Dielectric Technical Briefs," No. 831, Trans-Tech, Inc., Adamstown, MD, undated.

[4] Y. Kobayashi, T. Aoki, and Y. Kabe, "Influence of conductor shields on the Q-factors of a TE_0 dielectric resonator," IEEE MTT-S Int. Microwave Symp. Dig., pp. 281-284, St. Louis, June 1985.

[5] B.W. Hakki and P.D. Coleman, "A dielectric resonator method of measuring inductive capacities in the millimeter range," IRE Trans. Microwave Theory Tech., vol. MTT-8, pp. 402-410, July 1960.

[6] W.E. Courtney, "Analysis and evaluation of a method of measuring the complex permittivity and permeability of microwave insulators," IEEE Trans. Microwave Theory Tech., vol. MTT-18, pp. 476-485, Aug. 1970.

[7] Y. Kobayashi and M. Katoh, "Microwave measurement of dielectric properties of low-loss materials by the dielectric rod resonator method," IEEE Trans. Microwave Theory Tech., vol. MTT-33, pp. 586-592, July 1985.

[8] J. DelaBalle, P. Guillon, and Y. Garault, "Local complex permittivity measurement of MIC substrates," AEU, Electronics and Communication, vol. 35, pp. 80-83, Feb. 1981.

[9] N. Imai and K. Yamamoto, "A design of high-Q dielectric resonators for MIC applications," Electronics and Communications in Japan, vol. 67-B, pp. 59-67, Dec. 1984.

[10] S.B. Cohn, "Microwave band-pass filters containing high-Q dielectric resonators," IEEE Trans. Microwave Theory Tech., vol. MTT-16, pp. 218-227, April 1968.

[11] J.K. Plourde, H.M. O'Bryan, Jr., D.F. Linn, and J. Thomson, Jr., "$Ba_2Ti_9O_{20}$ as a microwave dielectric resonator," Jour. Amer. Ceramic Soc., vol. 58, pp. 418-420, Sept. - Oct. 1975.

[12] J.K. Plourde and C.L. Ren, "Application of dielectric resonators in microwave components," IEEE Trans. Microwave Theory Tech., vol. MTT-29, pp. 754-660, Aug. 1981.

[13] D. Halliday and R. Resnick, Physics for Students of Science and Engineering, New York: John Wiley & Sons, 1960.

[14] *** , Reference Data for Radio Engineers, New York: International Telephone and Telegraph Co., 1949.

[15] T. Higashi and T. Makino, "Resonant frequency stability of the dielectric resonator on a dielectric substrate," IEEE Trans. Microwave Theory Tech., vol. MTT-29, pp. 1048-1052, Oct. 1981.

[16] O. Ishihara, Y. Murakami, T. Yamaguchi, T. Otobe, and M. Hirabayashi, "A highly stabilized GaAs FET oscillator using a dielectric resonator feedback circuit in 9-14 GHz," IEEE Trans. Microwave Theory Tech., vol. MTT-28, pp. 817-824, Aug. 1980.

[17] Y. Komatsu, T. Mori, H. Sawano, and M. Nakatani, "A frequency-stabilized MIC oscillator using a newly-developed dielectric resonator," IEEE Int. Microwave Symp. Dig., pp. 313-315, Los Angeles, 1981.

[18] C. Tsironis and V. Pauker, "Temperature stabilization of GaAs MESFET oscillators using dielectric resonators," IEEE Trans. Microwave Theory Tech., vol. MTT-31, pp. 312-314, March 1983.

Chapter 8
COUPLING
Pierre Guillon

8.1 DR Mounted on Microstrip

The use of dielectric resonators in microwave circuits necessitates accurate knowledge of the coupling between the resonator and the microwave circuits (lines, waveguides, loops, etc.) and also of the mutual coupling between two adjacent dielectric resonators. In this chapter we study these two kinds of coupling successively.

We start with a discussion of coupling of the $TE_{01\delta}$ mode in the dielectric cylindrical resonator with a microstrip line. It is well known that the $TE_{01\delta}$ mode in the dielectric resonator can be approximated by a magnetic dipole of moment M [1]. The coupling between the line and the resonator is accomplished by orienting the magnetic moment of the resonator perpendicular to the microstrip plane so that the magnetic lines of the resonator link with those of the microstrip line [2,3,4], as shown in Fig. 8.1.

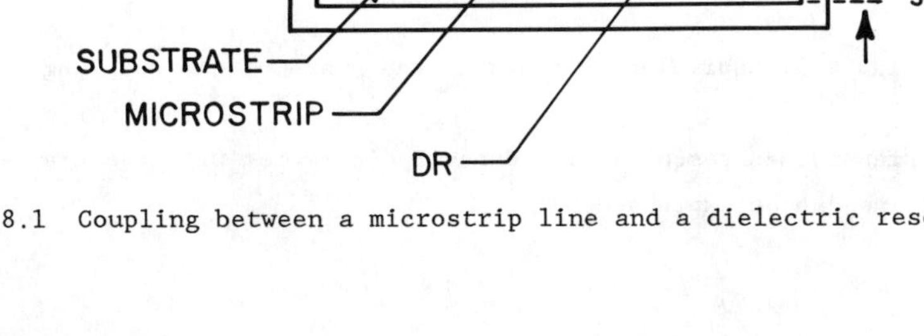

Fig. 8.1 Coupling between a microstrip line and a dielectric resonator

The dielectric resonator placed adjacent to the microstrip line operates like a reaction cavity which reflects the RF energy at the resonant frequency. The equivalent circuit of the resonator coupled to the microstrip line is shown in Fig. 8.2, [5], [6]. In this figure, L_r, C_r, R_r are the equivalent parameters of the dielectric resonator, L_1, C_1, R_1 those of the line. The magnetic coupling is characterized by the mutual inductance L_m.

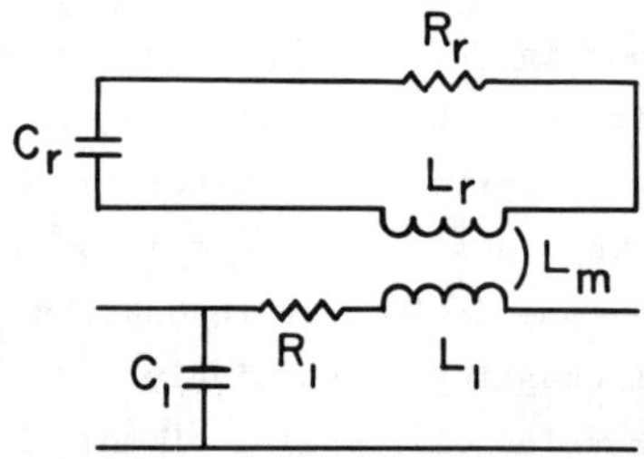

Fig. 8.2 Equivalent circuit of the dielectric resonator coupled with a line

In the coupling plane, the network of Fig. 8.2 can be simplified to that of Fig. 8.3 (the line is assumed to be without losses). Furthermore, the equivalent circuit in Fig. 8.3 can be put in the form shown in Fig. 8.4.

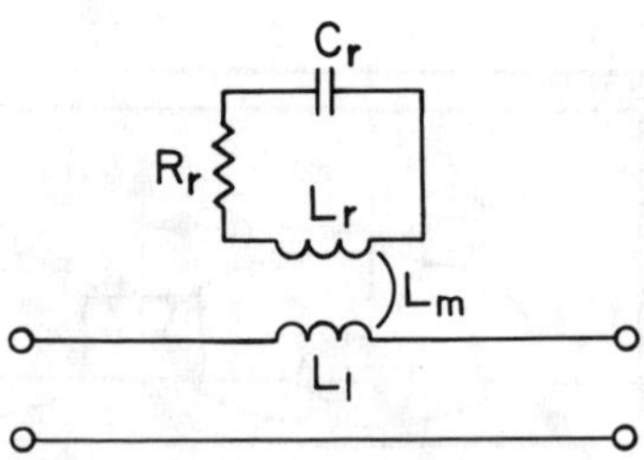

Fig. 8.3 Equivalent circuit in the plane of the coupling

The transformed resonator impedance Z_t in series with the transmission line can be expressed by

$$Z_t = j\omega L_1 + \frac{\omega^2 L_m^2}{R_r + j\left(\omega L_r - \frac{1}{\omega C_r}\right)} \tag{8.1}$$

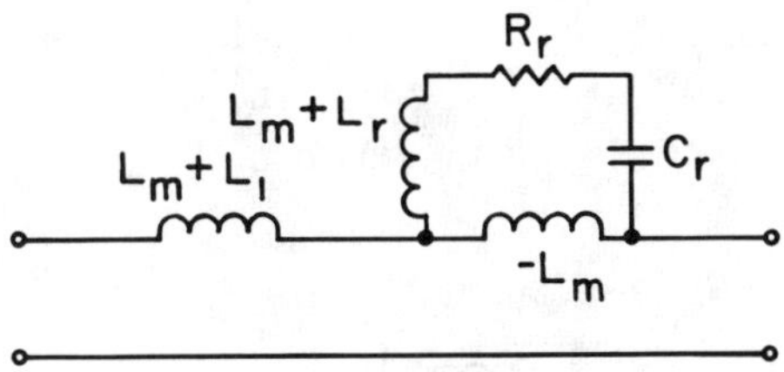

Fig. 8.4 Simplified equivalent circuit

Around the resonant frequency, ωL_1 can be neglected and Z_t becomes

$$Z_t = \omega Q_0 \frac{L_m^2}{L_r} \frac{1}{1 + jX} \tag{8.2}$$

where the following notation has been used:

$$X = 2 Q_0 \frac{\Delta\omega}{\omega} \tag{8.3}$$

$$Q_0 = \frac{L_r \omega_0}{R_r} \quad \text{and} \quad \omega_0 = \frac{1}{\sqrt{L_r C_r}}$$

At resonant frequency X = 0 and the transformed impedance becomes real:

$$Z_t = R = Q_0 \omega_0 \frac{L_m^2}{L_r} \tag{8.4}$$

Equation (8.2) indicates that Fig. 8.4 can be represented by the simple parallel tuned circuit, such as shown in Fig. 8.5. L, R, C satisfy the following equations:

$$L = \frac{L_m^2}{L_r} \tag{8.5}$$

$$C = \frac{L_r}{\omega_0^2 \; L_m^2} \tag{8.6}$$

$$R = Q_0 \; \omega_0 \; \frac{L_m^2}{L_r} \tag{8.7}$$

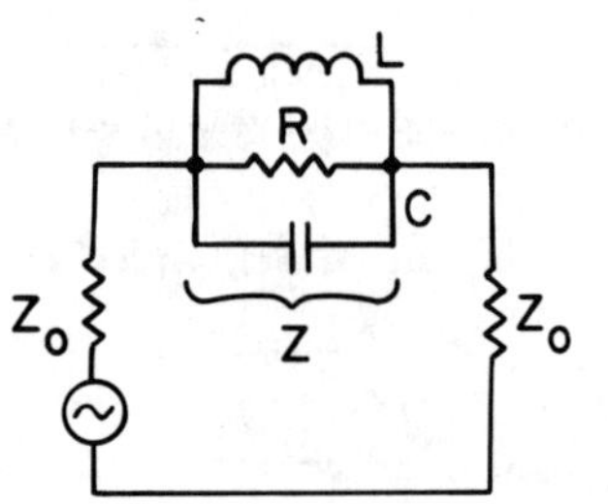

Fig. 8.5 Final equivalent of a dielectric resonator coupled with a microstrip line

The coupling coefficient κ is defined by

$$\kappa = \frac{Z_t(\omega_0)}{R_{ext}} \tag{8.8}$$

Using $R_{ext} = 2\,Z_0$, we obtain

$$\kappa = \frac{\omega_0 Q_0}{2Z_0} \; \frac{L_m^2}{L_r} \tag{8.9}$$

Let Q_e be the external quality factor which characterizes the coupling of the resonator with the microstrip line, and let Z_0 be characteristic impedance of the line. Then,

$$Q_e = \frac{Q_0}{\kappa} = \frac{2Z_0 L_r}{\omega_0 L_m^2} \tag{8.10}$$

To determine the coupling coefficient completely it is necessary to evaluate the Q_e factor as a function of distance d between the line and the resonator. The factor which characterizes this variation in (8.10) is L_r/L_m^2 .

Two methods will be presented for evaluation of Q_e. The first of them is called H/I method [4].

Let I be the current flowing in the microstrip line. The induced voltage e_r in the resonator is

$$e_r = j\omega\, L_m\, I \tag{8.11}$$

The magnetic field H due to the current I flowing in the microstrip line satisfies

$$e_r = \int \underline{E} \cdot d\underline{l} = j\omega\, \mu_0 \int_S \underline{H} \cdot d\underline{S} \tag{8.12}$$

$\underline{E}$ and $\underline{H}$ are respectively the electric and magnetic fields calculated at the center of the loop of area S. For a small loop the value of the magnetic field can be approximately taken to be constant over the area of the loop. From (8.11) and (8.12), we can write L_m in the following form:

$$L_m = \frac{\mu_0}{I}\, H\, S \tag{8.13}$$

Let M be the magnetic moment of the loop. If the resonator current is I_r, the magnetic moment is

$$M = S\, I_r \tag{8.14}$$

Let W be the stored energy in the dielectric resonator:

$$W = \frac{1}{2}\, L_r\, I_r^2 \tag{8.15}$$

Substituting (8.14) and (8.15) into (8.13), one obtains

$$\frac{L_m^2}{L_r} = \mu_0^2\, \frac{M^2}{2W} \left(\frac{H}{I}\right)^2 \tag{8.16}$$

The H/I ratio permits one to obtain a relation between L_m^2/L_r (and so Q_e) and the distance separating the line and the resonator.

Two procedures can be used to evaluate the H/I ratio.

Procedure (a). A first, simplified, procedure assumes that the magnetic field value results from two currents: the current flowing in the strip (width w) and the current flowing in the ground plane supposed uniformly distributed over a width 3w [4]. Using these assumptions we can determine an approximate relation between H/I and the distance d as follows:

$$\frac{H}{I} = \ln \frac{\left(d + \frac{w}{2}\right)^2 + \left(\frac{H}{2}\right)^2}{\left(d - \frac{w}{2}\right)^2 + \left(\frac{H}{2}\right)^2} - \frac{1}{3} \ln \frac{\left(d + \frac{3w}{2}\right)^2 + \left(h_S + \frac{H}{2}\right)^2}{\left(d - \frac{3w}{2}\right)^2 + \left(h_S + \frac{H}{2}\right)^2} \tag{8.17}$$

Procedure (b). Another expression for H/I can be derived by using the finite element method. Using the quasi-TEM approximation for the mode propagation in the microstrip line we can evaluate the current density in the strip and in the lower and upper plane of the metallic structure. From these values, we deduce the amplitude of the magnetic field H at any point P of the structure due to the current densities. The value of the ratio H/I depends on substrate permittivity and thickness, and on the distance d_1 between any point P of the structure and the line. An example of computed values (H/I) obtained by using the finite element method is given in Fig. 8.6.

The second method of evaluating Q_e is called the magnetic field flux method. Again, the dielectric resonator acting in the $TE_{01\delta}$ mode is modelled by a magnetic dipole of moment M.

Let I_r be the current flowing in the dielectric resonator. The voltage induced in the microstrip line by this current I_r is

$$e = j\omega L_m I_r \tag{8.18}$$

This induced voltage can also be computed from the magnetic flux through the cross section of the microstrip substrate (see Fig. 8.7):

$$e = j\omega \mu_0 \int_S \underline{H} \cdot \underline{dS} \tag{8.19}$$

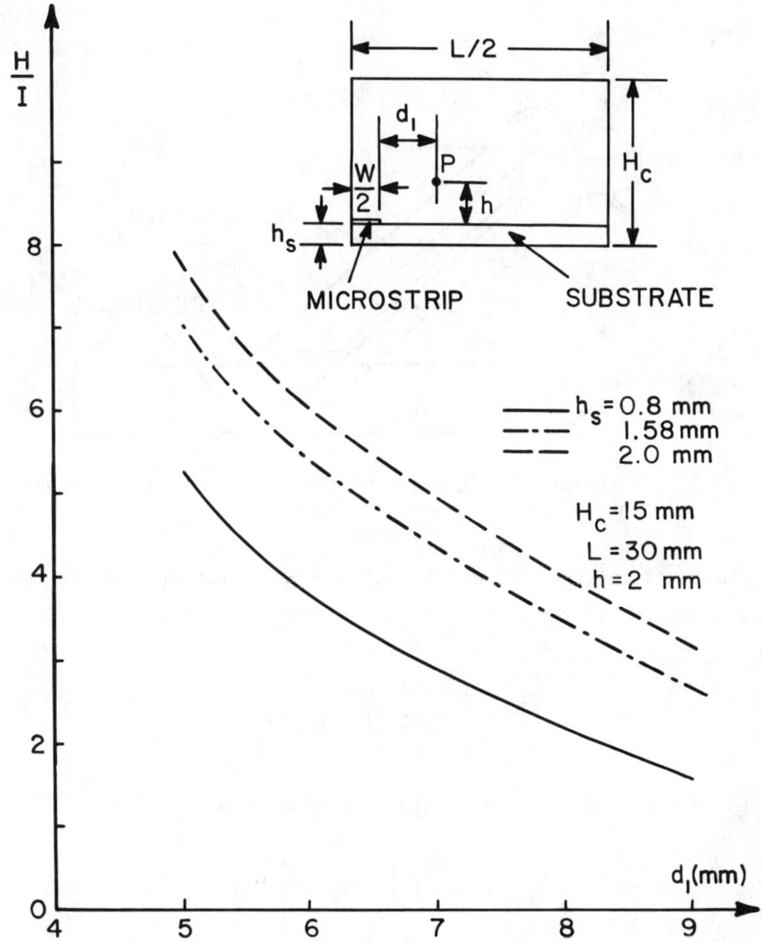

Fig. 8.6 H/I as a function of d_1 (reference [5], ©1981 IEEE)

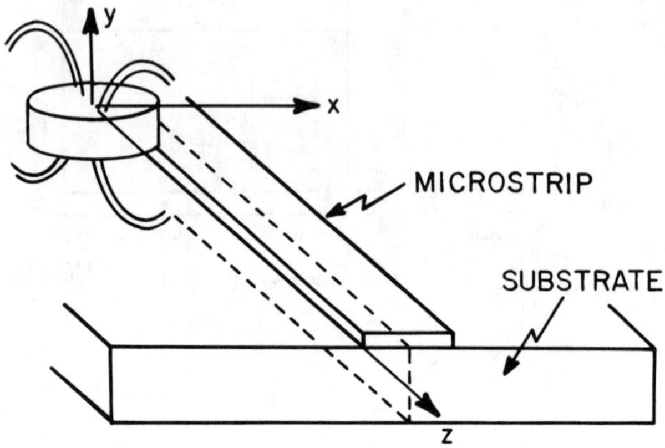

Fig. 8.7 Magnetic flux linkage of the resonator field into the microstrip line

Combining (8.18), (8.19) and (8.15) and substituting into (8.10) the coupling is expressed by

$$Q_e = \frac{2Z_0}{\omega_0 \; \mu_0^2} \frac{W}{\left(\int_S \underline{H} \; d\underline{S} \right)^2} \tag{8.20}$$

Note: In each method W and $\underline{M}$ should be evaluated as follows:

$$W = \frac{1}{2} \epsilon_0 \epsilon_i \int_{V_i} \underline{E}_i \cdot \underline{E}_i^* \, dV_i \tag{8.21}$$

$$\underline{M} = \frac{1}{2} j\omega \, \epsilon_0 \, \epsilon_i \int_{V_i} \underline{R} \times \underline{E}_i \, dV_i \tag{8.22}$$

where

ϵ_i = relative permittivity of the medium i,

$\underline{E}_i$ = electric field vector in the medium i,

$\underline{R}$ = distance vector,

V_i = volume of the medium i.

$\underline{M}$ and W can be computed by using one of the methods presented in the previous chapters to evaluate the electromagnetic parameters (frequencies, fields, etc.) of shielded dielectric resonators.

The magnetic field flux method is easier to use than the H/I method, since with the same program we can obtain the electromagnetic parameters and the external coupling values successively.

Figures 8.8 and 8.9 present the Q_e variation as a function of the distance between the line and the resonator computed by the H/I and the field flux method respectively. $\underline{M}$ and W have been computed here by using the numerical finite difference method.

The dielectric resonator coupled with the microstrip line is identical to a parallel resonant circuit placed in series with the line (Fig. 8.5). Such a circuit can be conveniently described in terms of the scattering parameters. Let θ be the electrical length of the microstrip line, as shown in Fig. 8.10. Then, the parameters of the scattering matrix are [5]:

$$S_{11} = \frac{Z_t/Z_0}{2 + \frac{Z_t}{Z_0}} e^{-2j\theta}$$

$$S_{12} = \frac{2}{2 + \frac{Z_t}{Z_0}} e^{-2j\theta}$$

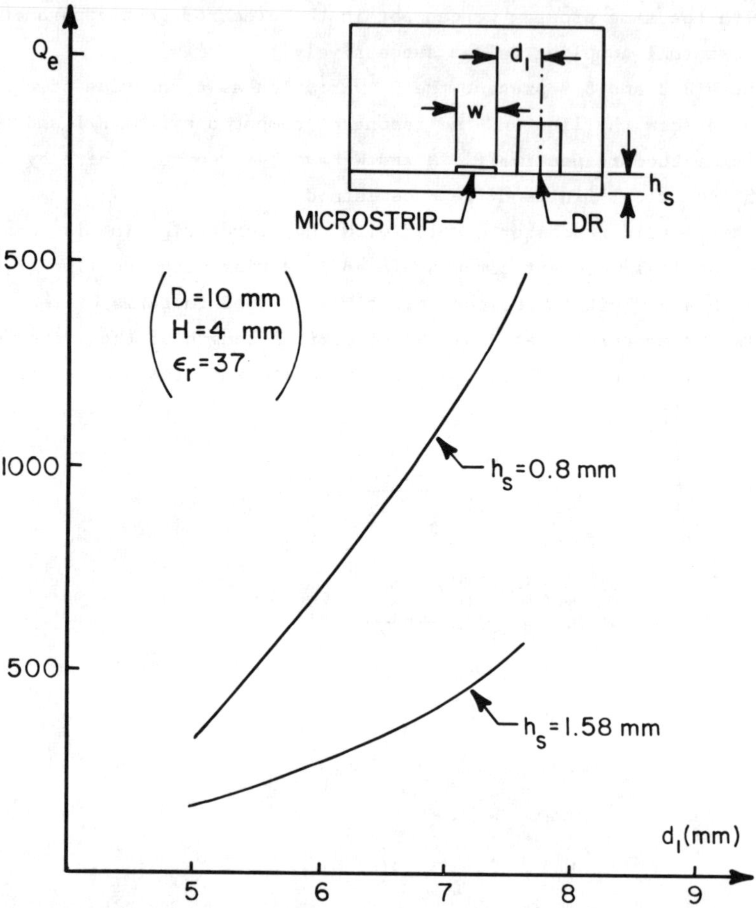

Fig. 8.8 External Q factor as a function of the distance between the line and the resonator, computed by the (H/I) method (reference [5], ©1981 IEEE)

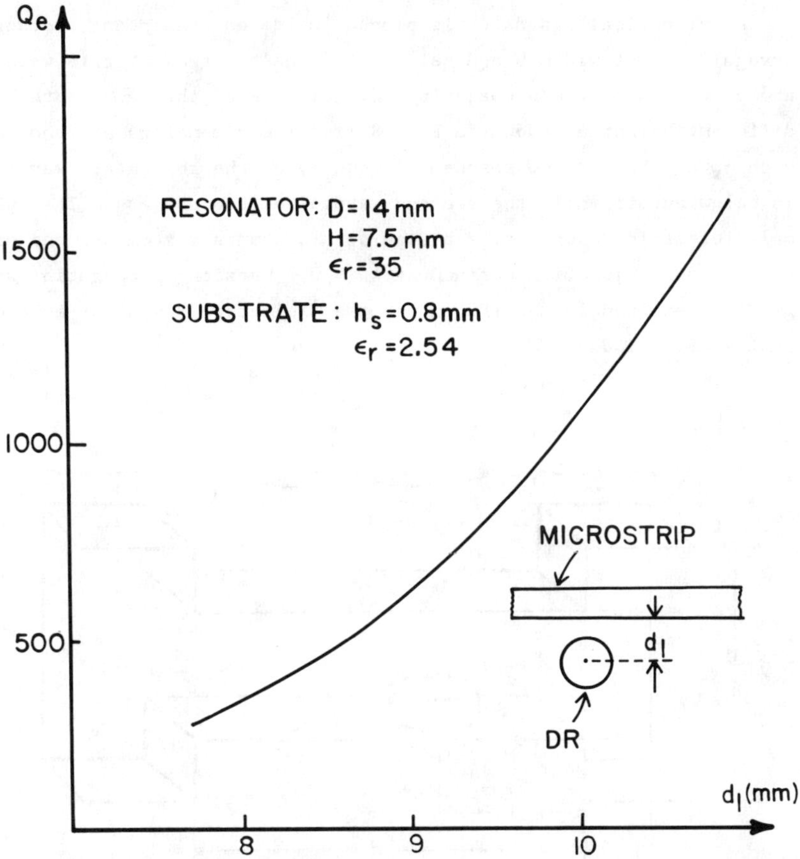

Fig. 8.9 External Q factor as a function of the distance between the line and the resonator, computed by the magnetic flux method (reference [16], ©1985 IEEE)

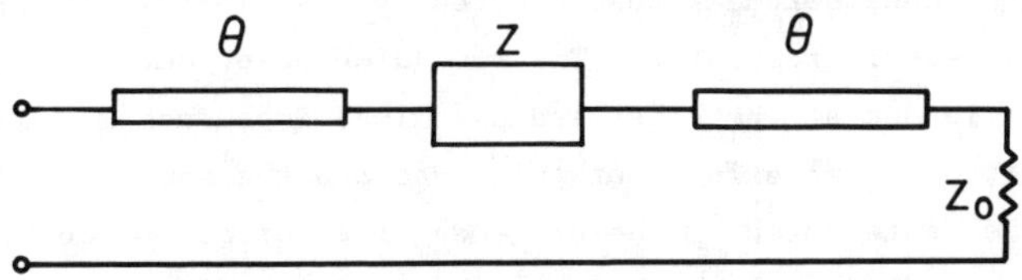

Fig. 8.10 Distributed equivalent circuit (reference [5])

8.2 DR in a Waveguide Below Cutoff

A cylindrical resonator is placed inside an evanescent rectangular waveguide A, of width b and height a'. The two ends of this waveguide are connected to two propagating waveguides B of the same width but of different height a, shown in Fig. 8.11. The dimensions are chosen in such a way that at the resonant frequency of the resonator, waveguide A is below cutoff while the two end waveguides B propagate. The cylindrical dielectric resonator is placed at a distance ℓ from one end of the discontinuity junction defined at $z' = 0$. When the propagating waveguide is excited in the TE_{10} mode, the dielectric resonator is excited in the $TE_{01\delta}$ mode.

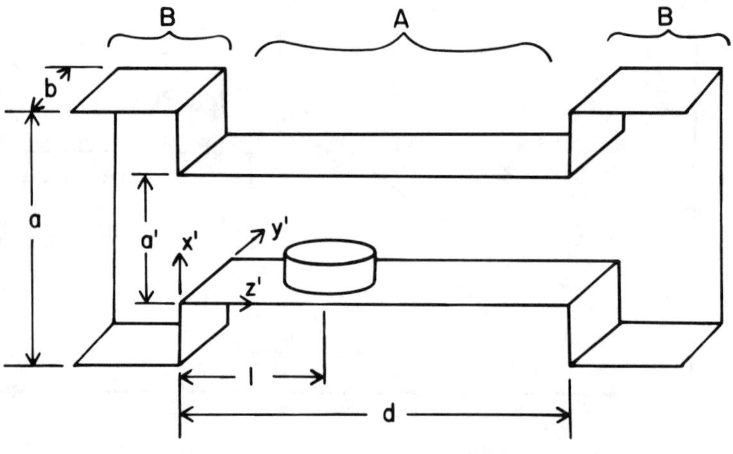

Fig. 8.11 Coupling between the dielectric resonator inserted into a cutoff waveguide and a propagating waveguide

By the principle of reciprocity, the fields in the waveguide below cutoff can be considered as that excited by the electric current density $\underline{J}$ in the dielectric resonator. The waveguide below cutoff is excited at one end and loaded at the other end by a load impedance Z. Under mismatch conditions, the effects of this load and the waveguide discontinuity can be represented by the presence of a reflection coefficient at the two waveguide junctions. Due to the symmetry of the waveguide structure considered, we can represent it by the model shown in Fig. 8.12.

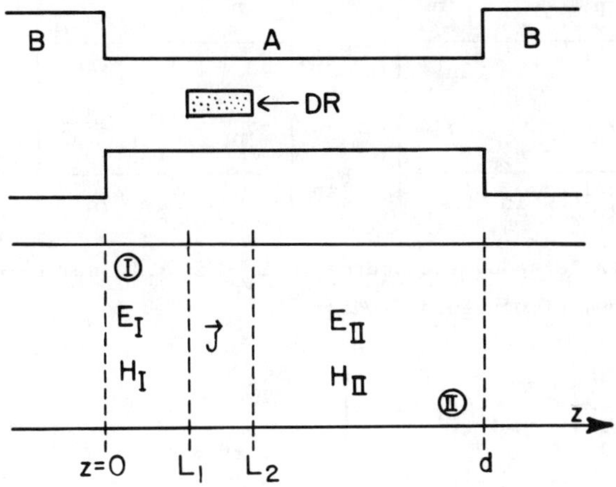

Fig. 8.12 Distribution of sources in the waveguide

In this figure, ρ_{1n} and ρ_{2n} are the reflection coefficients, respectively, at z' = 0 and z' = d when the waveguide is excited at z' > d. Taking into account the multiple reflections at the waveguide junctions, the nth normal mode electric and magnetic fields in region I are

$$\left.\begin{aligned} \underline{e}_{nI} &= \underline{e}_{tn} \left(\frac{e^{\gamma_n z} + \rho_{1n} e^{-\gamma_n z}}{1 - \rho_{1n} \rho'_{2n}} \right) - \underline{e}_{zn} \left(\frac{e^{\gamma_n z} - \rho_{1n} e^{-\gamma_n z}}{1 - \rho_{1n} \rho_{2n}} \right) \\ \underline{h}_{nI} &= - \underline{h}_{tn} \left(\frac{e^{\gamma_n z} + \rho_{1n} e^{-\gamma_n z}}{1 - \rho_{1n} \rho'_{2n}} \right) - \underline{h}_{zn} \left(\frac{e^{\gamma_n z} - \rho_{1n} e^{-\gamma_n z}}{1 - \rho_{1n} \rho'_{2n}} \right) \end{aligned} \right\} \tag{8.23}$$

The vectors $\underline{e}_{tn}$, $\underline{e}_{zn}$, $\underline{h}_{tn}$, $\underline{h}_{zn}$ are the transverse components and z components of the nth normal mode electric and magnetic fields, q_n is the propagation constant of the mode n, and

$$\rho'_{2n} = \rho_{2n} \, e^{-2\gamma_n d}$$

Similarly, when the waveguide is excited at z' < 0 in region II, the modal fields can be expressed by

$$\left.\begin{aligned} \underline{e}_{nII} &= \underline{e}_{tn}\left(\frac{e^{-\gamma_n z} + \rho'_{2n}\, e^{\gamma_n n}}{1 - \rho_{1n}\,\rho'_{2n}}\right) + \underline{e}_{zn}\left(\frac{e^{-\gamma_n z} - \rho'_{2n}\, e^{\gamma_n z}}{1 - \rho_{1n}\,\rho'_{2n}}\right) \\ \underline{h}_{nII} &= \underline{h}_{tn}\left(\frac{e^{-\gamma_n z} - \rho'_{2n}\, e^{\gamma_n n}}{1 - \rho_{1n}\,\rho'_{2n}}\right) + \underline{h}_{zn}\left(\frac{e^{-\gamma_n z} + \rho'_{2n}\, e^{\gamma_n z}}{1 - \rho_{1n}\,\rho'_{2n}}\right) \end{aligned}\right\} \tag{8.24}$$

The fields radiated by the source in $L_1 < z' < L_2$ may be expressed as an infinite sum of waveguide modes:

$$\left.\begin{aligned} \underline{E}_I &= \sum_{n=1}^{\infty} a_{nI}\, \underline{e}_{nI} \\ \underline{H}_I &= \sum_{n=1}^{\infty} a_{nI}\, \underline{h}_{nI} \end{aligned}\right\} \quad 0 < z' < L_1 \tag{8.25}$$

$$\left.\begin{aligned} \underline{E}_{II} &= \sum_{n=1}^{\infty} a_{nII}\, \underline{e}_{nII} \\ \underline{H}_{II} &= \sum_{n=1}^{\infty} a_{nII}\, \underline{h}_{nII} \end{aligned}\right\} \quad L_2 < z' < d \tag{8.26}$$

$$\left.\begin{aligned} a_{nI} &= \frac{1}{2}\,(1 - \rho_{1n}\,\rho'_{2n}) \int_V \underline{e}_{nII} \cdot \underline{J}\, dV \\ a_{nII} &= -\frac{1}{2}\,(1 - \rho_{1n}\,\rho'_{2n}) \int_V \underline{e}_{n1} \cdot \underline{J}\, dV \end{aligned}\right\} \tag{8.27}$$

V is the volume containing the source $\underline{J}$.

If we assume that only the dominant mode is excited, all higher-order excited modes are rapidly attenuated in the waveguide below cut-off, and the resonator is far away from the second sectional plane $\rho_{2n} = 0$, then the field waves radiated by the resonator are very much attenuated, and none of them is reflected from $z' = d$.

From (8.23) to (8.26), the electric field distribution at the sectional plane $z' = 0$ generated by the current source $\underline{J}$ is expressed as follows:

$$E_{pc}\Big|_{z=0} = a_{1I}\, e_{1I}\ \ (0) \tag{8.28}$$

To determine the external Q factor of the dielectric resonator in the evanescent rectangular waveguide, we need to calculate the power radiated from the resonator into the waveguide below cutoff. We simplify the problem by expressing the field excited in the waveguide in terms of an electric field distribution at the cross section S_0 located at $z' = 0$, as shown in Fig. 8.13. The equivalence principle is then applied. The cross section S_0 can be replaced by a perfectly conducting plane supporting the magnetic current source $\underline{M}_S$. This current source

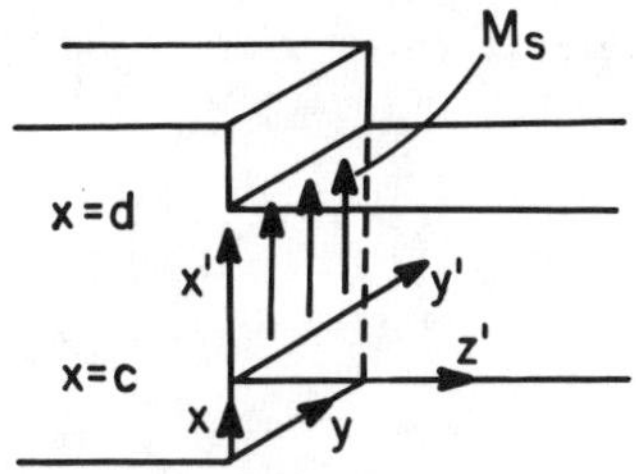

Fig. 8.13 Equivalent surface current at the junction of two waveguides

helps to create the electric field $\underline{E}_{pc}$, thus

$$\underline{M}_S = \underline{E}_{pc}\Big|_{z=0} \times \hat{n} \tag{8.29}$$

In the above, $\hat{n} = -\hat{z}$ is the unit vector, normal to S_0, pointing out from the waveguide below cutoff.

The field $(\underline{E}_r, \underline{H}_r)$ radiated into the waveguide below cutoff is given by

$$\left.\begin{aligned} \underline{E}_r &= C_1\, \underline{e}_{1II} \\ \underline{H}_r &= C_1\, \underline{h}_{1II} \end{aligned}\right\} \tag{8.30}$$

$$C_1 = \frac{1}{2}\int_V \underline{h}_{1I} \cdot \underline{M}_S \, dV \tag{8.31}$$

$$P_r = \int_S (\underline{E}_r \times \underline{H}_r^*) \cdot \hat{z}' \, dS \tag{8.32}$$

The external Q factor Q_e of the dielectric resonator in the waveguide below cut-off is then expressed as

$$Q_e = \omega_0 \frac{W}{P_r} \tag{8.33}$$

Substituting into (8.33) the value of P_r from (8.32) and W from (8.21), we can evaluate Q_e as a function of the distance ℓ between plane S_0 and the resonator. The values of Q_e, shown in Figs. 8.14 and 8.15, have been obtained by using the approximate approach for electromagnetic parameters which permits definition of W and P_r.

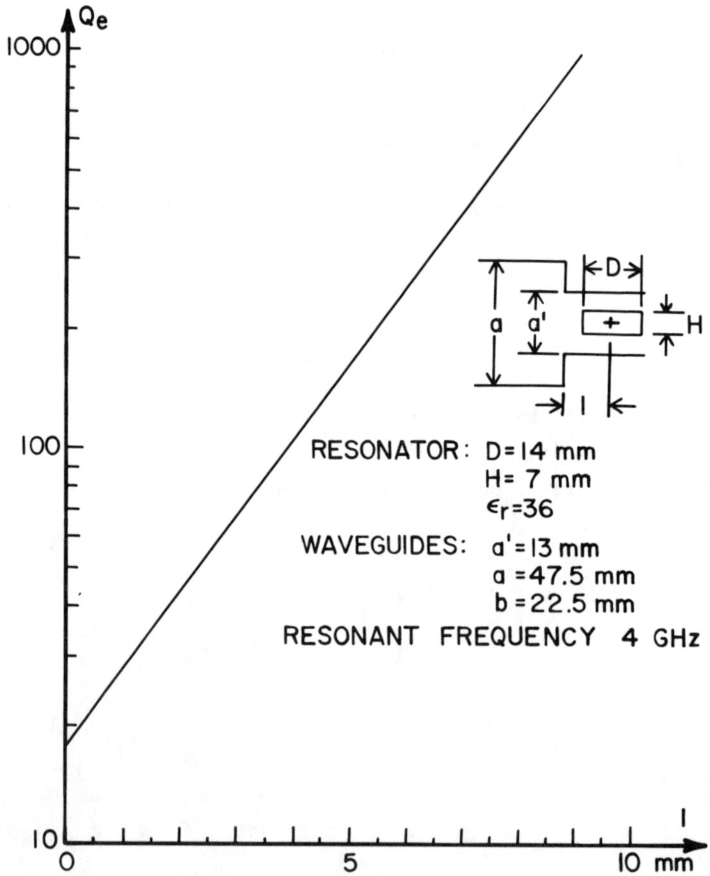

Fig. 8.14 External Q factor of the dielectric resonator inserted into the evanescent waveguide (reference [10], ©1982, IEEE)

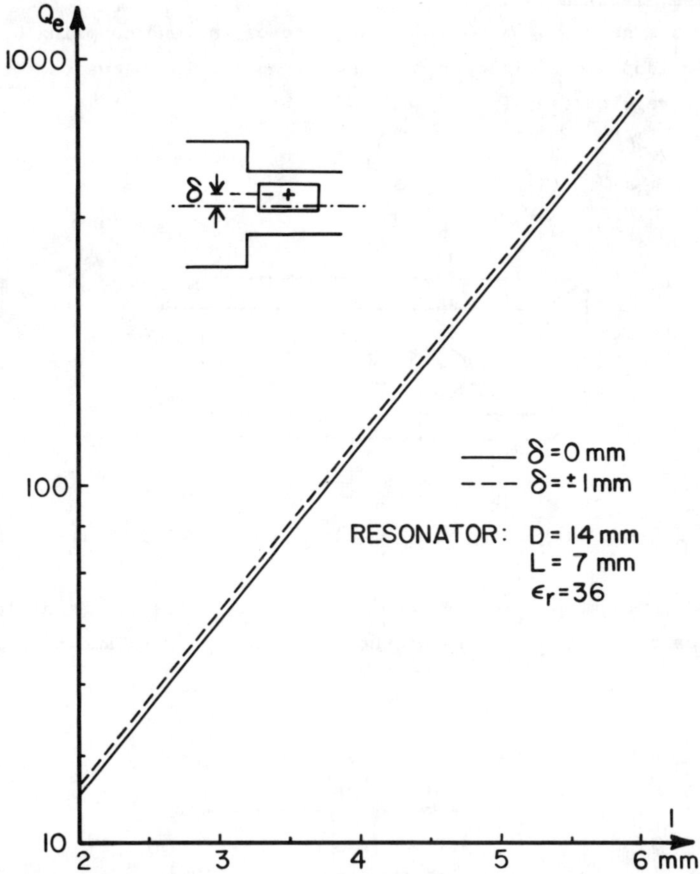

Fig. 8.15 Influence of the off-axis displacement of the DR in the cut-off waveguide (reference [10], ©1982 IEEE)

8.3 Loop Coupling [11]

In this section, we consider the case of the end coupling between a magnetic loop and a dielectric resonator housed in an evanescent metallic waveguide (see Fig. 8.16).

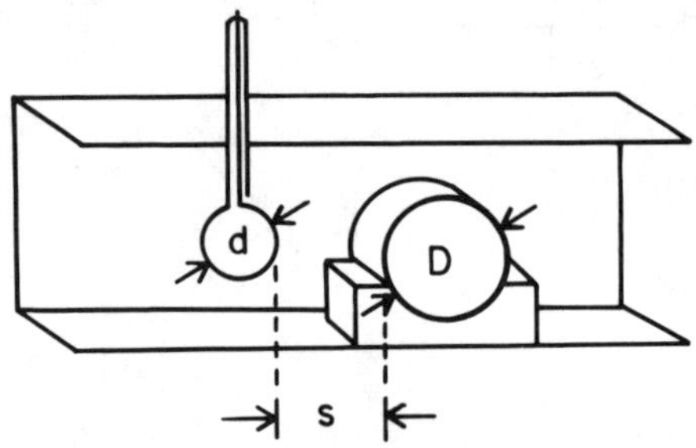

Fig. 8.16 Coupling between a magnetic loop and a dielectric resonator

The coupling between the loop and the $TE_{01\delta}$ mode of the dielectric resonator can be represented by the equivalent network shown in Fig. 8.17.

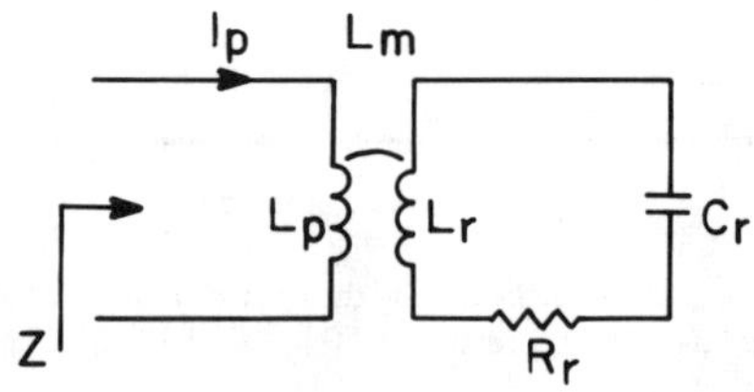

Fig. 8.17 Equivalent circuit of the magnetic loop coupling

L_p is the self inductance of the magnetic loop of diameter d in which flows a current I_p. The input impedance at the resonance is given by

$$Z = \omega_0 Q_0 \frac{L_m^2}{L_r} \tag{8.34}$$

The external Q factor is given by

$$Q_e = \frac{2\ Z_0 L_r}{\omega_0 L_M^2} \tag{8.35}$$

For evaluation of the factor (L_r/L_m^2) in terms of field H, we can consider the field induced in the magnetic loop by the evanescent field of the dielectric resonator (in free space), and so evaluate the magnetic flux through the cross section S of the loop. With this assumption, the factor Q_e is given by

$$Q_e = \frac{2Z_0 W}{\omega_0 \mu_0^2 \left(\iint \underline{H} \cdot d\underline{S}\right)^2} \tag{8.36}$$

Z_0 is the characteristic impedance of the coaxial cable leading to the loop. Alternatively, we can compute the magnetic field at the center of the loop H_p due to the waveguide evanescent modes excited by the magnetic dipole located at the center of the resonator.

For example, in the case of the transverse orientation of the dielectric resonator in a rectangular waveguide, and considering only the mode which has the lowest cut-off frequency, we obtain the following expression for Q_e:

$$Q_e = \frac{2\ W\ Z_0}{\omega_0\ \mu_0\ H_x\ S^2} \tag{8.37}$$

where

$$H_p = H_x = a_{t10}\ h_x\ e^{-\alpha_{10} s}$$

and

$$a_{t10} = \frac{j\omega\mu_0}{2} \sqrt{\frac{2}{a.b\ Z_{10}}}\ M \tag{8.38}$$

In the above, h_x is normalized x component of the magnetic field within the rectangular waveguide, and s is distance between the resonator and the loop. Furthermore,

$$Z_{10} = \frac{\mu_0}{\epsilon_0} \sqrt{\frac{1}{(f_c/f)^2} - 1}$$

where f_c is the cutoff frequency of the TE_{10} rectangular waveguide mode. Finally,

$$h_x = \sqrt{\frac{2}{a.b\ Z_{10}}} \sin \frac{\pi x}{a} \tag{8.39}$$

$$\alpha_{10} = \frac{\pi}{a} \sqrt{1 - \left(\frac{f}{f_c}\right)^2} \tag{8.40}$$

where α_{10} is the attenuation constant of the $TE_{1,0}$ mode in the evanescent waveguide of the rectangular cross section with sides a and b.

Figure 8.18 presents the values of the coupling coefficient between the loop and the resonator.

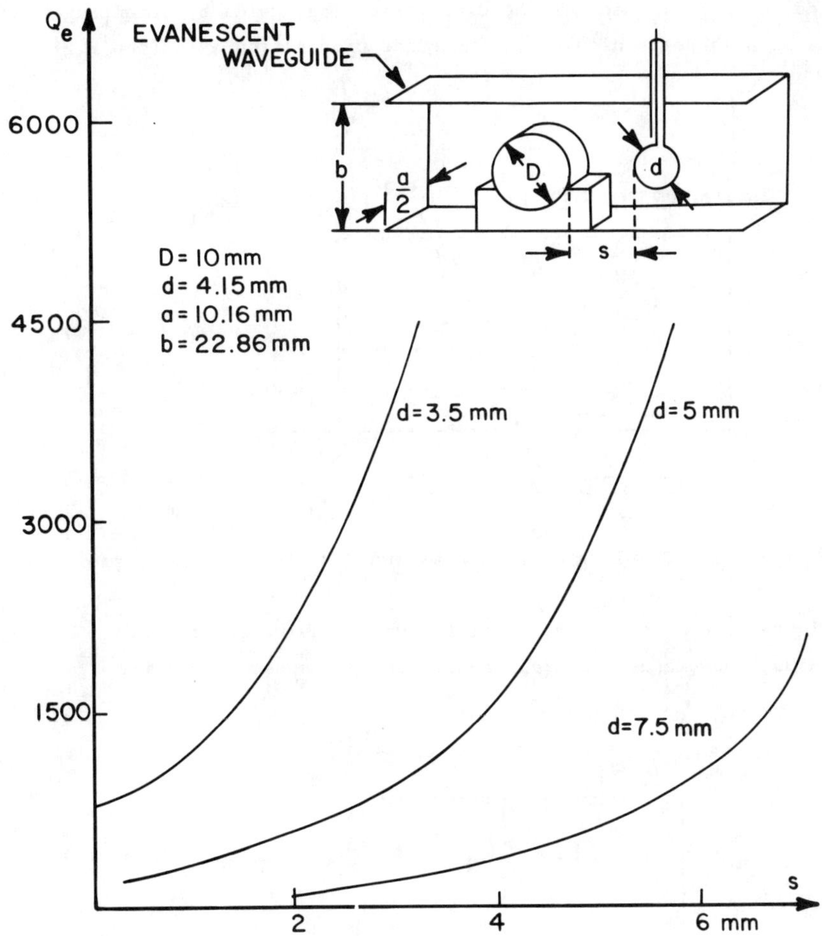

Fig. 8.18 External Q factor as a function of the distance between the loop and the resonator (reference [11], ©1982 IEEE)

8.4 DR in a Dielectric Image Guide [12]

Dielectric guides are used to realize microwave components at millimeter wave frequencies [25]. Inserting the dielectric resonators in such guides will result in integrated dielectric components at these frequencies.

The cylindrical dielectric resonator operating in the $TE_{01\delta}$ mode is coupled to the EH_{11} mode of the dielectric image guide by orienting the resonator axis perpendicular to the guide propagation axis (see Fig. 8.19).

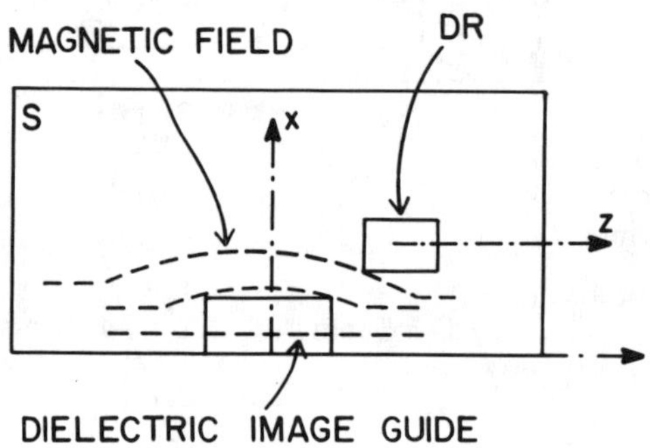

Fig. 8.19 Coupling between a DR and a dielectric image guide

Figure 8.20 gives the appropriate lumped equivalent circuit. L_m is the mutual inductance which characterizes the magnetic coupling.

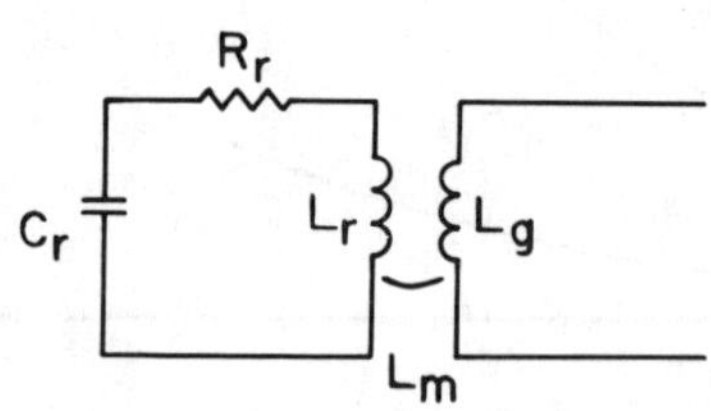

Fig. 8.20 Equivalent circuit of the coupling to the image guide

At the resonant frequency, the input impedance calculated in the coupling plane is given by (8.34). The voltage induced in the dielectric guide by the current I_r flowing in the dielectric resonator is given by (8.18). This voltage can also be calculated by evaluating the magnetic flux linkage between the guide and the resonator:

$$e = j\omega \mu_0 C \int \underline{H}_r \cdot d\underline{S} \tag{8.41}$$

The coefficient C takes into account the misalignment between the magnetic field vector $\underline{H}_r$ of the $TE_{01\delta}$ mode of the dielectric resonator and that of the dielectric guide $\underline{H}_g$ [12]. Thus,

$$C = \frac{\int_S \underline{H}_r \cdot \underline{H}_g^* \, dS}{\left[\int_S \underline{H}_r \cdot \underline{H}_r^* \, dS \int_S \underline{H}_g \cdot \underline{H}_g^* \, dS\right]^{1/2}} \tag{8.42}$$

where S is the cross section shown in Fig. 8.19.

The external quality factor defined by (8.10) has been evaluated as a function of the distance between the guide and the resonator. In this case Z_0 is the characteristic impedance of the dielectric image guide defined by the following expressions:

$$Z_0 = \frac{2P}{I_z^2} \tag{8.43}$$

$$P = \int_0^{\infty} \int_{-\infty}^{+\infty} (E_x H_y - E_y H_x) dx\, dy \tag{8.44}$$

$$I_z = 2 \int_0^{\infty} \hat{y} \times \hat{x} \; H_x \, dx \tag{8.45}$$

The symbols E_x, H_y, H_x denote the field components of the EH_{11} dielectric guide mode. Quantities C and Q_e as functions of the distance between the guide and the dielectric resonator are presented in Figs. 8.21 and 8.22, respectively.

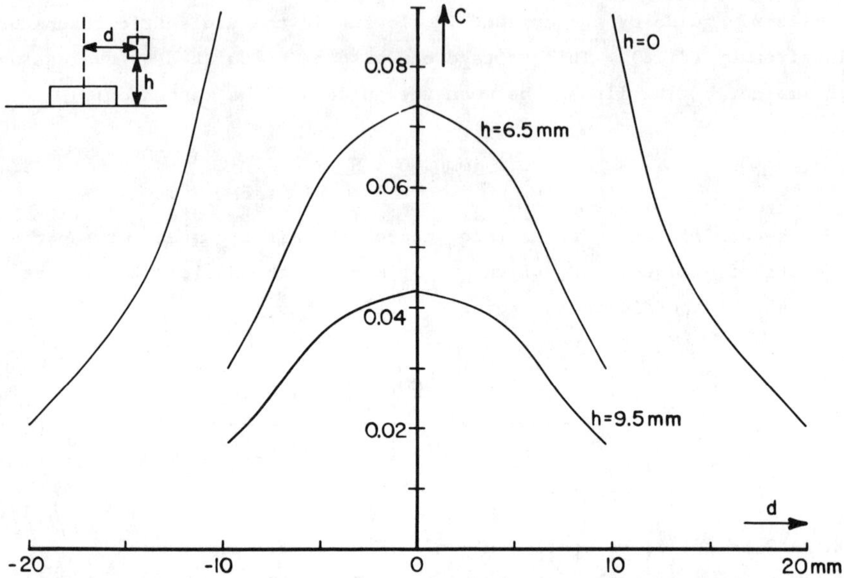

Fig. 8.21 Misalignment coefficient as a function of the position of the DR in the guide (reference [12], ©1984 IEEE)

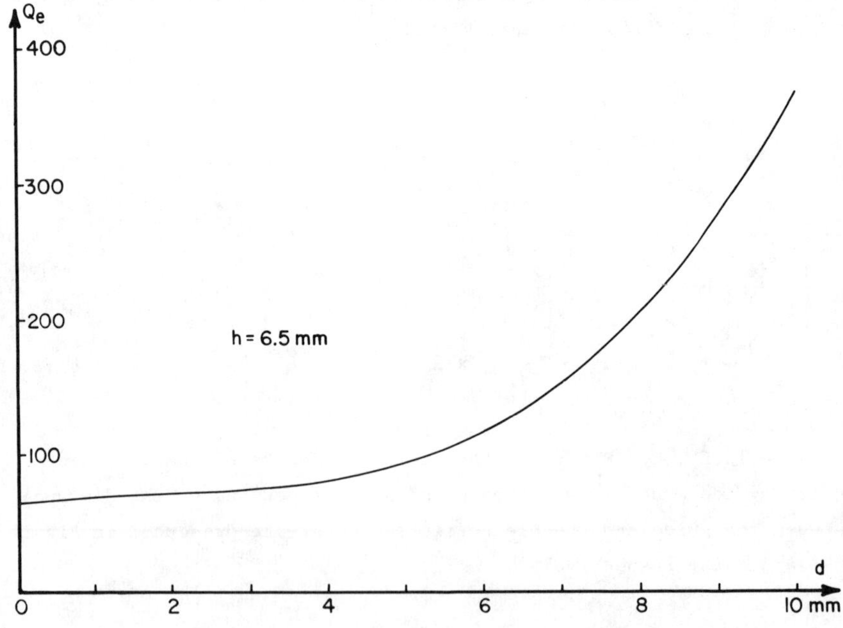

Fig. 8.22 Q_e as a function of the distance between the dielectric guide and the dielectric resonator (reference [12], ©1984 IEEE)

8.5 DR on a Finline

Finline is used for millimeter frequencies (around 90 GHz) [26]. At these frequencies, the dimensions of cylindrical resonators are very small, and so we replace the cylindrical resonator by a spherical one. A chart of the several TE and TM frequency modes for a shielded spherical dielectric resonator is given in Fig. 8.23.

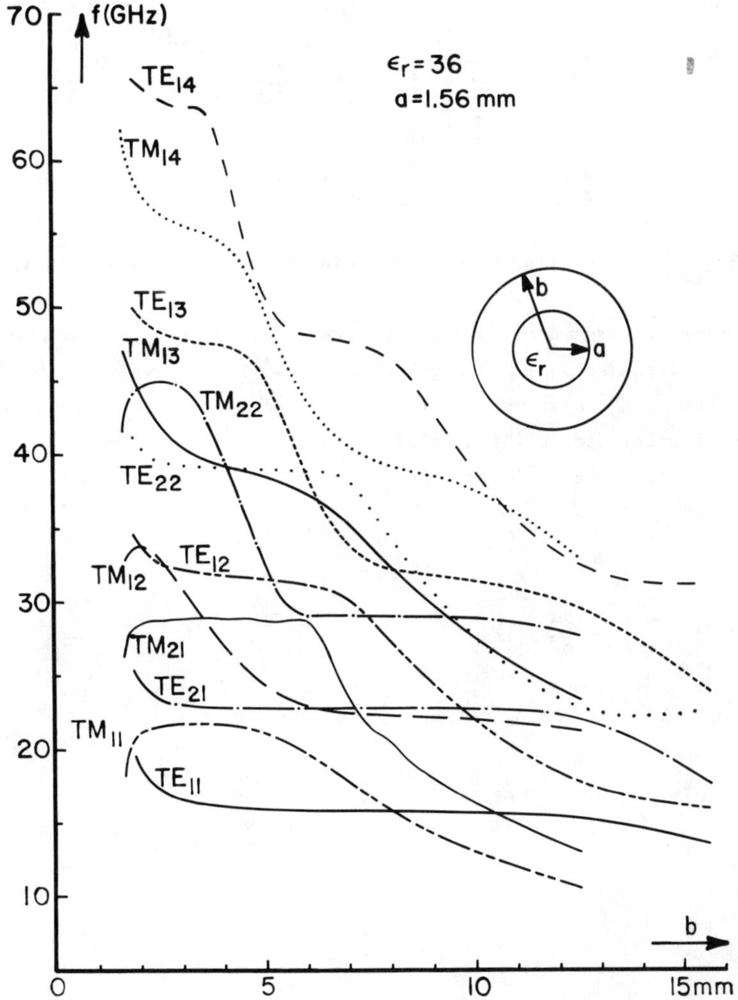

Fig. 8.23 Mode chart of a shielded dielectric sphere

Figure 8.24 shows the position of the spherical dielectric resonator in the finline structure.

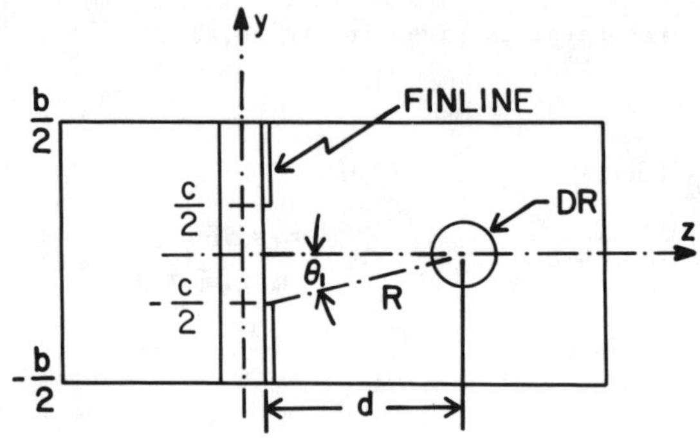

Fig. 8.24 Coupling between the finline and the dielectric sphere

Using the same procedure as the one described earlier for microstrip line, we first establish an equivalent network. We evaluate the input impedance of the system so that we can deduce a relation for the external quality factor which is identical to that given in (8.20).

In this particular case, the magnetic flux calculated over the angle θ_1 in Fig. 8.25 is

$$\phi = \int_S \underline{H} \cdot d\underline{S} = \int_0^{2\pi} \int_0^{\theta_1} H_r R^2 \sin\theta \, d\theta \, d\phi \qquad (8.46)$$

In (8.20), we use Z_0 defined as follows:

$$Z_0 = \frac{V}{I} = \frac{\int_{-c/2}^{+c/2} E_x \, dx}{\int_{-b/2}^{+b/2} H_x \, dx} \qquad (8.47)$$

An example of the coupling coefficient κ as a function of d is given in Fig. 8.25.

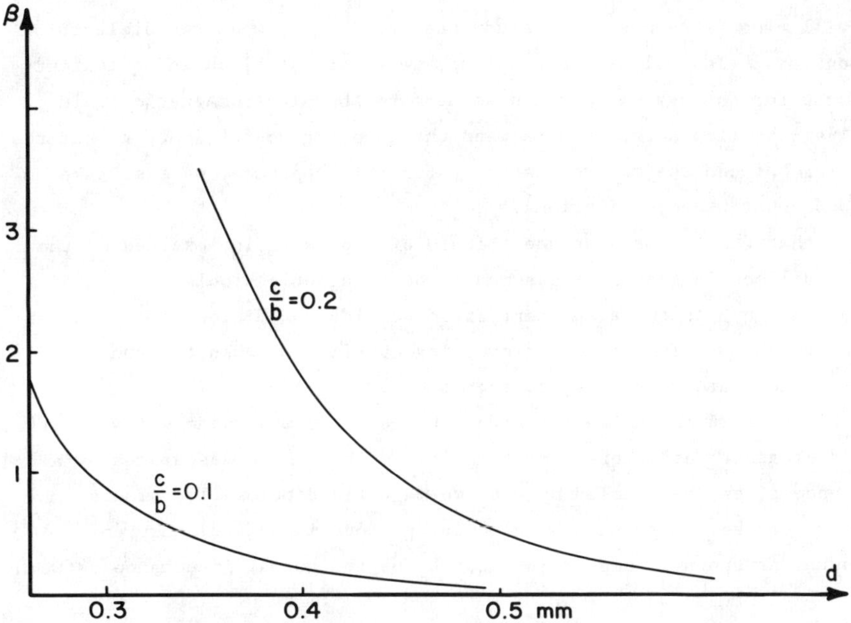

Fig. 8.25 Coupling coefficient as a function of the distance between the dielectric sphere and the finline

8.6 Mutual Coupling Between Two DRs in a Waveguide Below Cutoff

For the applications of dielectric resonators in microwave filters, it will also be necessary to study the coupling between two dielectric resonators. To analyze the coupling, we first establish an equivalent circuit for the system and then we perform the electromagnetic field analysis to find a relation between the coupling coefficient, k, and the spacing between the two resonators. For this approach, we must take into account two contributions:

- that due to the evanescent field of the waveguide excited by the dielectric resonator assumed to be a magnetic dipole, and
- that due to the evanescent field outside the dielectric resonator which contributes to the coupling coefficient when the two resonators are very close to each other.

Figure 8.26 shows two resonators placed in a waveguide below cutoff. Two dielectric resonators operating in the $TE_{01\delta}$ mode, separated by a distance s, can be modeled by the two magnetic dipoles [1]. These dipoles can be represented by conducting loops in an arbitrary enclosure, as shown in Fig. 8.26. Let L_m be the mutual inductance between

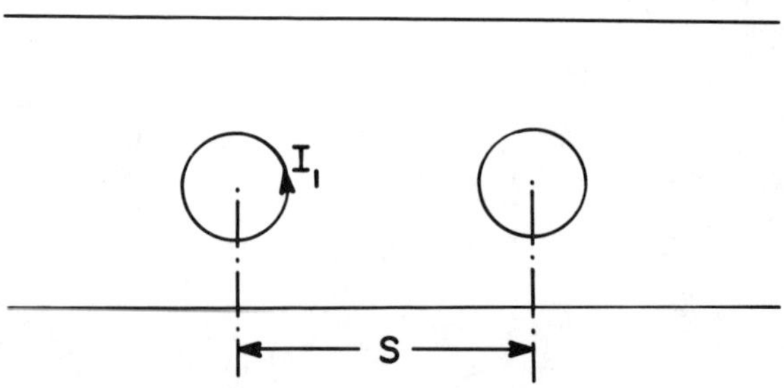

Fig. 8.26 Two dielectric resonators inserted into an evanescent waveguide

the adjacent loops and V_2 is the induced voltage in loop 2 due to i_1 in loop 1:

$$V_2 = j\omega L_m i_1 \tag{8.48}$$

The induced voltage V_2 is also given by the following fundamental integral relationship:

$$V_2 = j\omega \int_S \underline{H}_2 \cdot d\underline{S} \tag{8.49}$$

where $\underline{H}_2$ is the magnetic field value in loop 2 due to the current in loop 1.

Combining (8.48) and (8.49) and taking into account (8.14) and (8.15), we define the coupling coefficient, k, between two identical resonators by

$$k = \frac{L_m}{L_r} = \mu_0 \frac{M}{2W} H_2 \tag{8.50}$$

The E and H fields of a given waveguide mode are obtained from equations (8.25) and (8.26). The amplitude for the waves of type and order p, excited by the magnetic dipole, can be expressed easily by [27]

$$a_p = \frac{j\omega \mu_0}{2} H_p^+ M$$
$$b_p = \frac{j\omega \mu_0}{2} H_p^- M \tag{8.51}$$

H_p^-, H_p^+ are the normalized magnetic fields of mode p traveling in +z and -z directions, respectively. The different cases are now presented.

<u>Transverse orientation of the resonator (Fig. 8.27).</u> The waveguide modes which have their $\underline{H}$ components in the direction of $\underline{M}$ (magnetic dipole) are the TE_{mn} modes with m odd and n even. The following formula for the coupling coefficient between a pair of identical x-directed magnetic dipoles on the central longitudinal axis is obtained by application of (8.50) and (8.51):

$$k = \frac{\mu_0 M^2}{2W} \sum_m \alpha_{m0} e^{-\alpha_{m0} s} + 2 \sum_{m,n} \frac{\alpha_{m0}}{\alpha_{mn}} e^{-\alpha_{m0} s} \tag{8.52}$$

where α_{mn} is the attenuation constant of the m,n mode. Of all the TE_{mn} modes with m odd and n even, TE_{10} is the one which has the lowest cutoff frequency and therefore the lowest attenuation constant. Thus, at a sufficient longitudinal distance from the magnetic dipole, the total

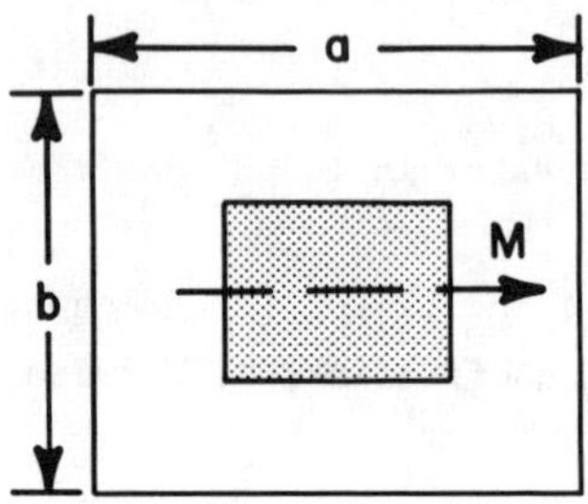

Fig 8.27 Transverse orientation of the dielectric resonator in a rectangular waveguide

field may be represented with a good accuracy by the TE_{10} field alone, so that (8.52) simplifies to

$$k = \frac{\mu_0 M^2}{2W} \alpha_{10} \frac{e^{-\alpha_{10} s}}{ab} \tag{8.53}$$

where a and b are the dimensions of the rectangular waveguide as shown in Fig. 8.27.

For this transverse orientation the coupling coefficient k as a function of the distance between adjacent resonators is presented in Fig. 8.28.

<u>Axial orientation in a rectangular waveguide (Fig. 8.29).</u> Only the modes having $h_z \neq 0$ (i.e., TE_{mn} modes) will contribute to coupling. In the case of disks centered in the cross-section TE_{mn} modes of odd order are not excited. However, if the disks are off center all order modes except m = n = 0 are significant.

<u>Axial orientation in a circular waveguide (Fig. 8.30).</u> Because of the cylindrical symmetry only circular electric modes designated TE_{0n} are excited by the equivalent magnetic dipoles.

<u>Transverse orientation in a circular waveguide.</u> It can be directly concluded that the transverse fields contributing to coupling with the resonators located at waveguide center will be H_r for TE_{mn} modes. The modes which can be taken into account are the TE_{1n} modes where n = 1,2,3... (In general, we can limit ourselves to m unity for r = 0).

An alternative approach can be used to study the coupling between adjacent dielectric resonators. For this purpose, the dielectric resonators are represented by a distribution of displacement current <u>J</u> and

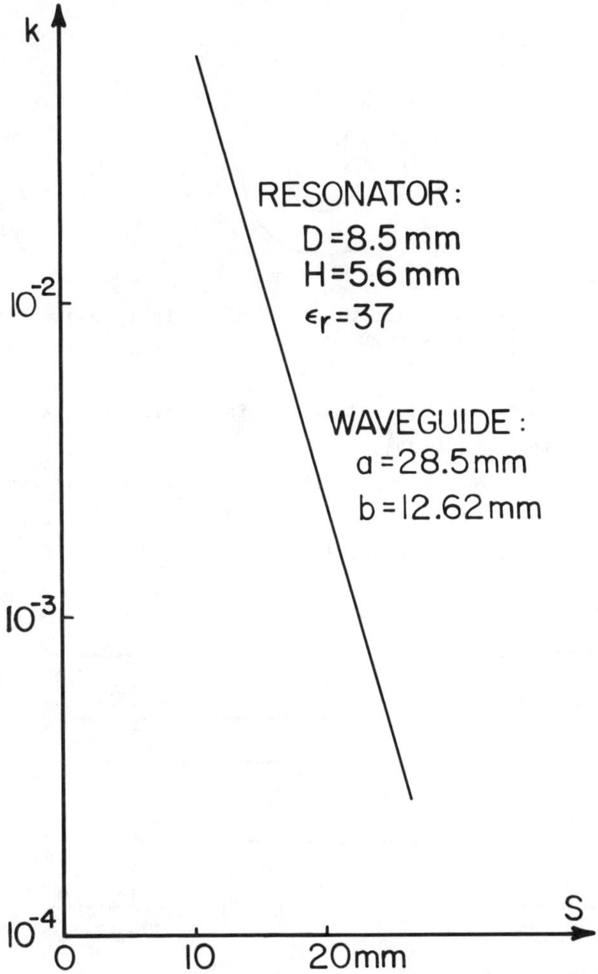

Fig. 8.28 k as a function of the distance between two resonators (transverse orientation in a rectangular waveguide)

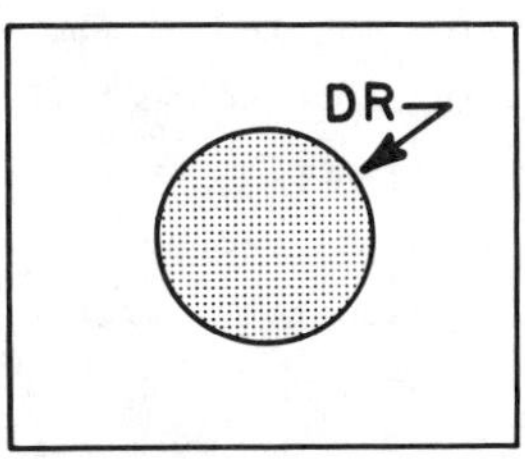

Fig. 8.29 Axial orientation of the dielectric resonator in a rectangular waveguide

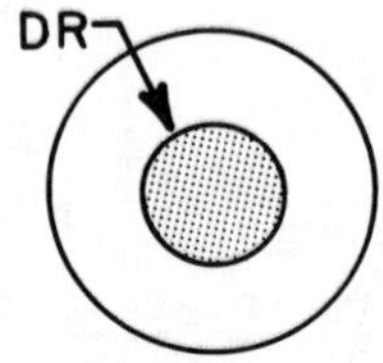

Fig. 8.30 Axial orientation of the dielectric resonator in a cylindrical waveguide

the method used to calculate the coupling coefficient k is based on the perturbation theory [10,14].

Let $\underline{E}_i$ and $\underline{H}_i$ be the unperturbed fields inside the waveguide below cutoff, defined in the region $0 < z' < d$ (see Fig. 8.31), where d is the length of the evanescent waveguide.

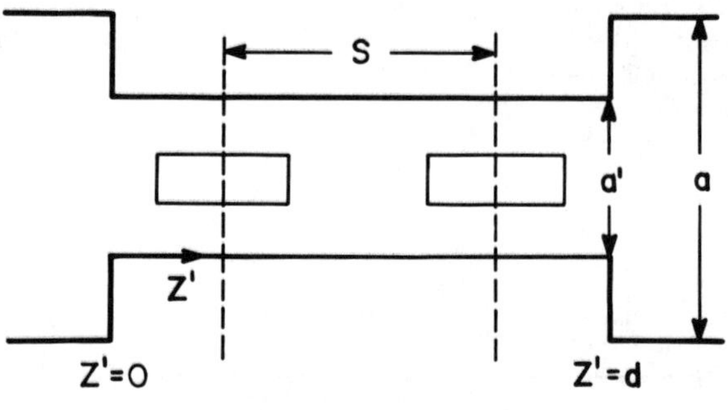

Fig. 8.31 Two dielectric resonators inserted into an evanescent waveguide excited by a propagating waveguide

$\underline{E}_i$ and $\underline{H}_i$ are the fields excited by the equivalent magnetic surface current $\underline{M}_S$ at $z' = 0$, taking into account the reflection coefficient at the waveguide discontinuity junction. Let $\underline{E}$, $\underline{H}$ be the total perturbed fields of the waveguide when the second resonator is placed at a distance s from the first resonator. Writing Maxwell's equations for the unperturbed fields, we have

$$\left.\begin{aligned} -\nabla \times \underline{E}_i &= j\,\omega_i\,\mu\,\underline{H}_i \\ \nabla \times \underline{H}_i &= j\,\omega_i\,\epsilon\,\underline{E}_i \end{aligned}\right\} \qquad (8.54)$$

where ω_i is the unperturbed resonant frequency.

When the second resonator is introduced, the fields are perturbed. Let ω be the new resonant frequency:

$$-\nabla \times \underline{E} = j\omega\,\mu\,\underline{H} \qquad (8.55a)$$

$$\nabla \times \underline{H} = \underline{J} + j\omega\,\epsilon\,\underline{E} \qquad (8.55b)$$

After some calculations, we derive the small change in frequency due to the presence of the second resonator:

$$\omega - \omega_i = \frac{\int_V \underline{J} \cdot \underline{E}_i^* \, dV}{j \int (\mu\,\underline{H}_i \cdot \underline{H}_i^* + \epsilon\,\underline{E}_i \cdot \underline{E}_i^*)\, dV} \qquad (8.56)$$

If a weak coupling is assumed, the field distribution inside the waveguide varies very slightly from the unperturbed fields, and so we can conclude that:

$$\left.\begin{aligned} E_i &\simeq E \\ H_i &\simeq H \end{aligned}\right. \qquad (8.57)$$

The interstage coupling k is defined by

$$k = \left| \frac{\omega - \omega_0}{\omega} \right| \qquad (8.58)$$

Taking into account the field expressed by (8.25) and (8.26), we have drawn in Fig. 8.32 the variation of k as a function of distance s.

Another contribution must be taken into account when evaluating the coupling between two adjacent dielectric resonators. It characterizes the direct coupling between the resonators, namely, the coupling achieved by the evanescent field outside the resonators [15]. When this computation is performed, both resonators are assumed to be situated in an open space.

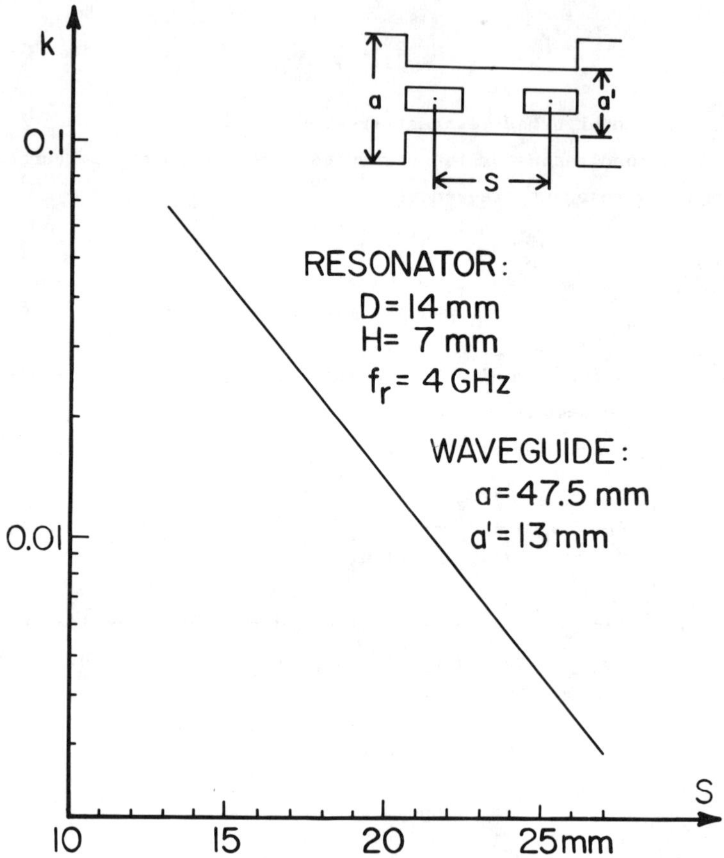

Fig. 8.32 k as a function of the distance between two resonators calculated by using the perturbational method

The expression for the coupling coefficient k is derived from (8.48) and (8.49):

$$k = \frac{V_2 \, I_1}{2j\omega \, W} \tag{8.59}$$

In the above, V_2 is the induced voltage in loop 2 due to the current I_1 flowing in loop 1. It can be computed as follows:

$$V_2 = j\omega \int \underline{B}_2 \cdot d\underline{S} = \int \underline{E}_2 \cdot d\underline{l}$$

If we replace the current loop with a dielectric resonator, we can easily state that the following is true:

$$I_1 \int \underline{E}_2 \cdot d\underline{l} = \int_V \underline{E}_2 \cdot \underline{J} \;\; dV \tag{8.60}$$

$\underline{E}_2$ is the electric field in place of the second resonator due to the first resonator. $\underline{J}$ is the current density in its volume V. Thus, the final expression for the coupling coefficient could be written as

$$k = \frac{j \int \underline{E}_2 \cdot \underline{J} \; dV}{2\omega \; W} \tag{8.61}$$

Quantities W, $\underline{J}$, $\underline{E}_2$ can be defined by many methods. Using the approximate electromagnetic approach [8], we have evaluated k as a function of S. It can be seen in Fig. 8.33 that generally the coupling between adjacent resonators is due to two contributions: one is due to the evanescent dielectric resonator field, and the other is due to the evanescent field in the waveguide below cutoff.

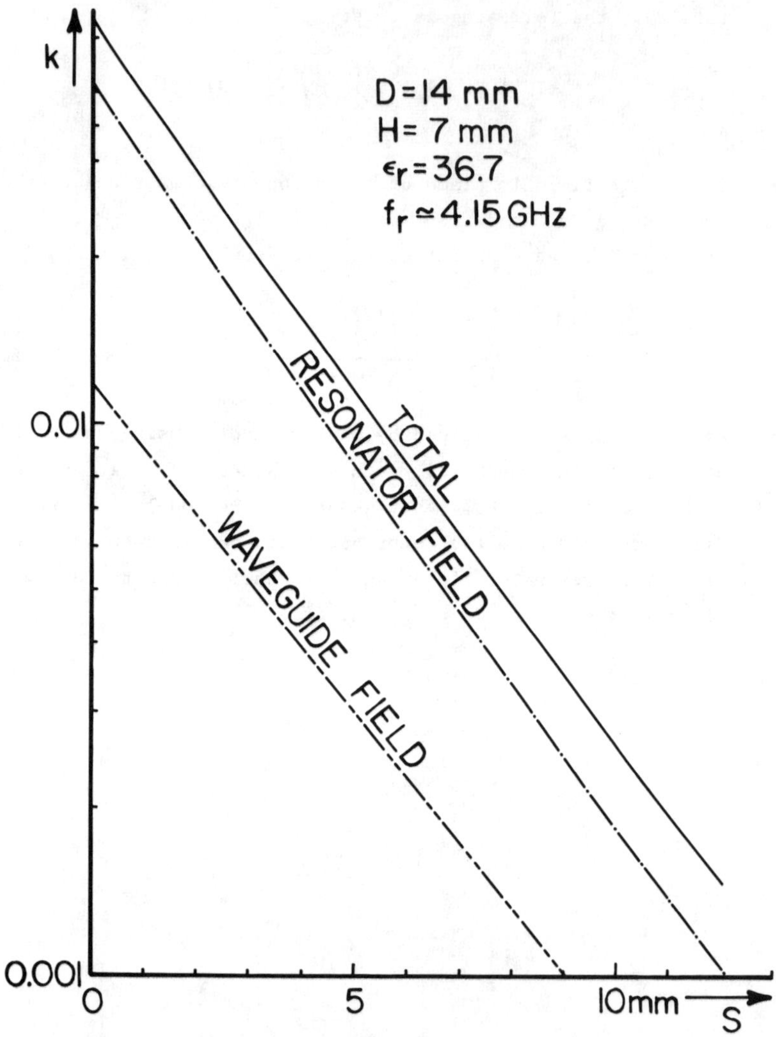

Fig. 8.33 Contributions of the different fields to the coupling between two adjacent dielectric resonators

8.7 Mutual Coupling Between Two DRs Via a Section of Microstrip Line

The determination of the coupling between resonators mounted as shown in Fig. 8.34 can be deduced from a knowledge of coupling to microstrip (recall Fig. 8.1). The details are given in References [6] and [16]. The equivalent circuit of this system is presented in Fig. 8.35.

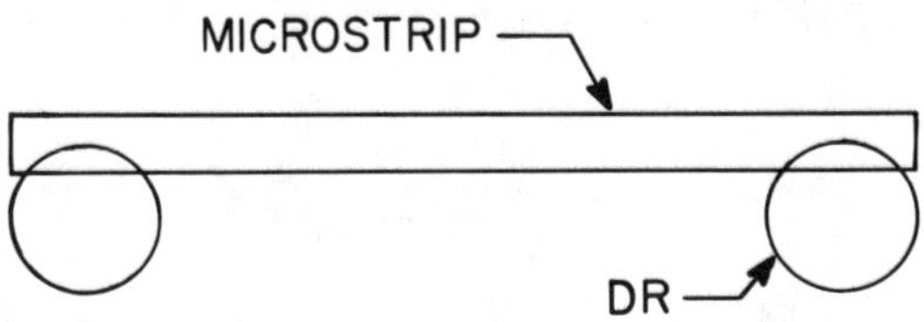

Fig. 8.34 Dielectric resonators coupled by means of a microstrip transmission line

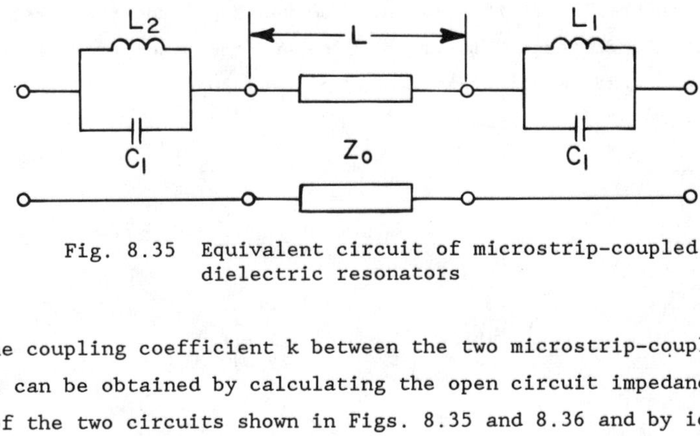

Fig. 8.35 Equivalent circuit of microstrip-coupled dielectric resonators

The coupling coefficient k between the two microstrip-coupled resonators can be obtained by calculating the open circuit impedance parameters of the two circuits shown in Figs. 8.35 and 8.36 and by identifying the corresponding elements.

The condition for equivalence between the two circuits can be easily determined to be

$$\cot \beta L = 0 \quad \text{or} \quad L = \left(\frac{2i+1}{4}\right) \lambda_g \; ; \quad i = 0,1,2\ldots \tag{8.62}$$

where λ_g is the wavelength in the microstrip.

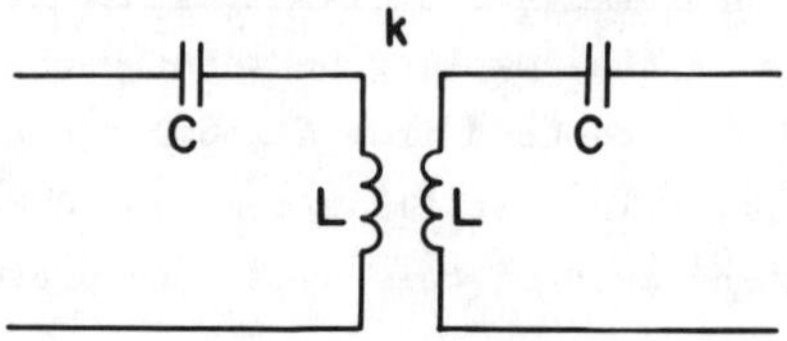

Fig. 8.36 Lumped equivalent circuit

For the values of L given by (8.62), the coupling between the two resonators is

$$k = \frac{1}{\sqrt{Q_{e1}\, Q_{e2}}} (-1)^{i} \tag{8.63}$$

where Q_{e1} and Q_{e2} are the external Q-factors of the resonators. Thus, for a positive coupling coefficient, the line length ℓ must be one-quarter wavelength long, and for a negative coupling it must be three-quarters of a wavelength long. The magnitude of the coupling is controlled by the height of the resonator above the microstrip substrate and also by the distance between the line and the resonator. Figure 8.37 provides some values of k for various spacings.

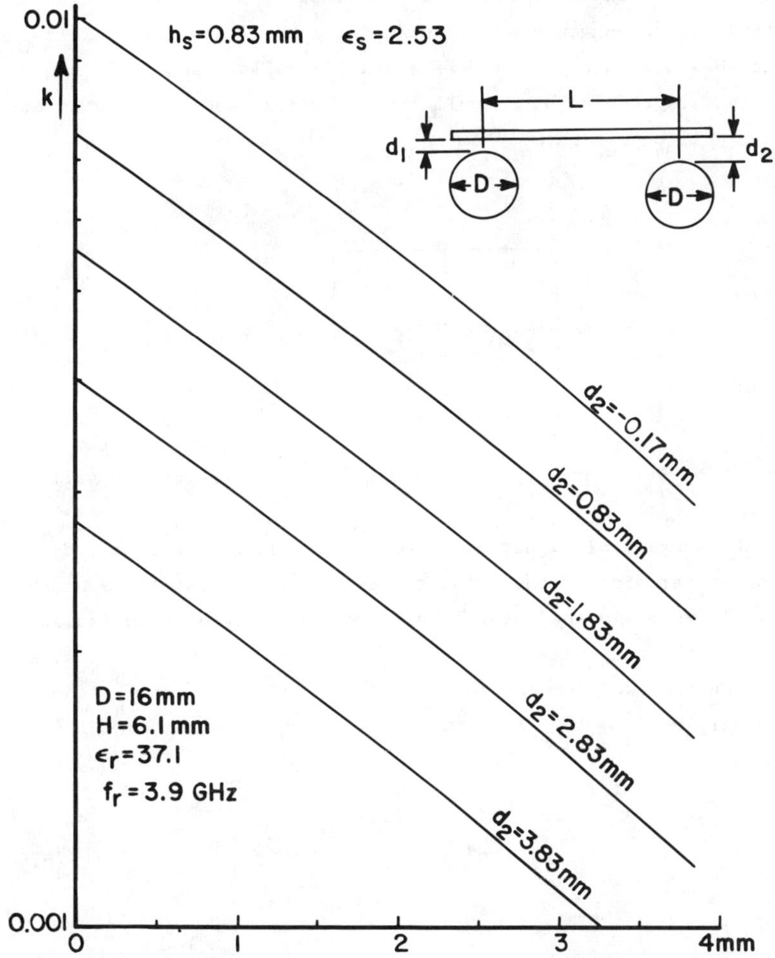

Fig. 8.37 k as a function of the distances of the resonators from the line (reference [16], ©1985 IEEE)

8.8 Mutual Coupling Between Two DRs Through an Iris

Figure 8.38 shows two symmetrically oriented identical cavities, A and B, having a common thin wall containing a small aperture. This configuration has been studied in references [17], [18], and [19]. When the cavities are coupled by a small iris, the resonant frequencies of the system will be ω_α and $\omega_\alpha - \Delta\omega$.

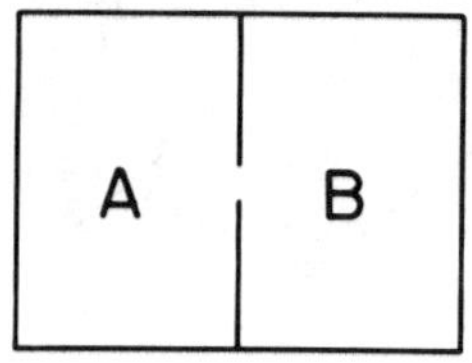

Fig. 8.38 Cavities coupled by a small hole in the common wall

When the tangential magnetic fields are pointing in the same direction on either side of the iris, the cavities will oscillate at the frequency ω_α, which is the natural resonant frequency of each isolated cavity. When the tangential magnetic fields are pointing in opposite directions on the two sides of the coupling iris, the natural resonant frequency will be $\omega_\alpha - \Delta\omega$. When $\Delta\omega$ is small, the coupling coefficient is defined as

$$k = \frac{\Delta\omega}{\omega_\alpha} = \frac{M_a \ \mu_0^2 \ H_t^2}{\epsilon_0 \int_V E^2 \, dV} \tag{8.64}$$

where M_a is the magnetic moment of the coupling aperture, H is the tangential magnetic field at the center of the hole, and E is the normal mode electric field in the cavity. Both cavities are assumed to be filled with air.

The two identical cavities are represented by parallel L-C resonant circuits coupled by an iris which is represented by a reactance X_c. The reactance X_c may be capacitive or inductive, depending upon the type of coupling between the two cavities. For the magnetic coupling, (8.64) becomes

$$k = \frac{P_m H_{\alpha t} H_{\alpha t}}{\int_V H_\alpha H_\alpha \, dV} \tag{8.65}$$

and for the inductive coupling, (8.64) gives

$$k = \frac{P_e E_{\alpha n} E_{\alpha n}}{\int_V E_\alpha E_\alpha \, dV} \tag{8.66}$$

Symbols P_m and P_e denote the magnetic and the electric polarizabilities of the aperture; $E_{\alpha n}$ and $H_{\alpha t}$ are the electric and magnetic field values evaluated in the aperture.

Figures 8.39 and 8.40 present the coupling of two dielectric resonators using an iris. The resonators operate in the $TE_{01\delta}$ and $HEM_{11\delta}$ modes, respectively. A thickness correction and a large-aperture effect can be introduced for the polarizability of an aperture connecting two cavities, which will not be done here.

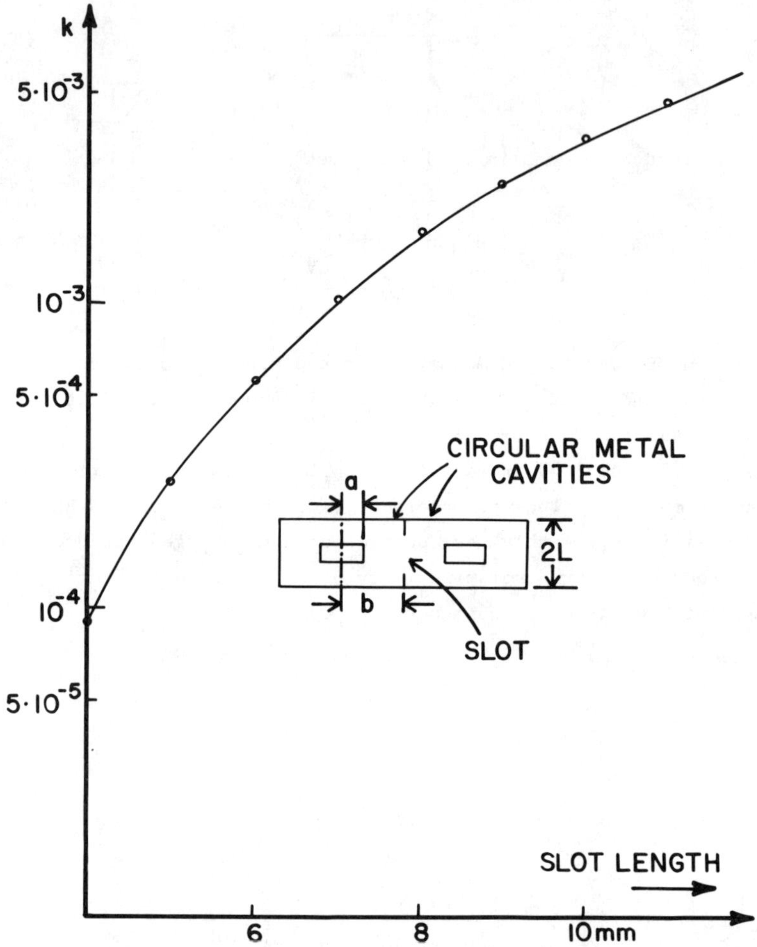

Fig. 8.39 k as a function of the slot length for the $TE_{01\delta}$ mode (reference [19], ©1980 AEU)

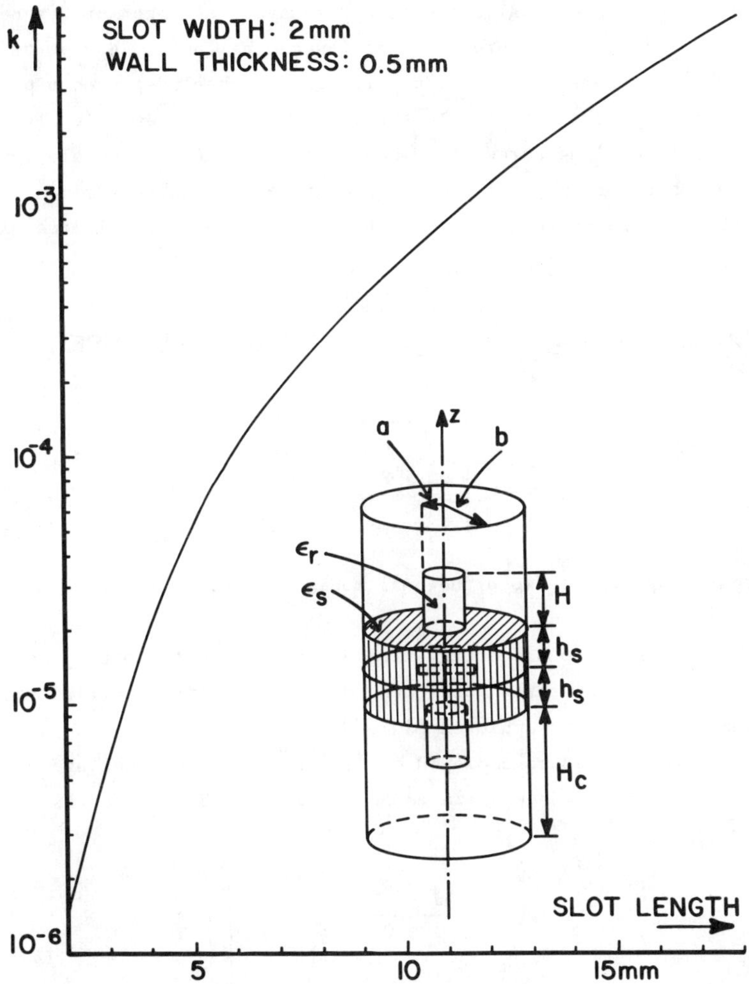

Fig. 8.40 k as a function of the slot length for the $HEM_{11\delta}$ mode (reference [24], ©1985 Eur. Microwave Conf.)

8.9 Dual Modes

In a shielded cylindrical dielectric resonator, a number of degenerate modes with identical natural frequencies can be found. Any perturbation of the electromagnetic fields will destroy the independence of some of the modes and act to couple the energy between them. In Fig. 8.41, two metal coupling screws have been inserted into the shield containing the resonator in such a position as to cause the energy contained within the resonator to split into a pair of orthogonal modes.

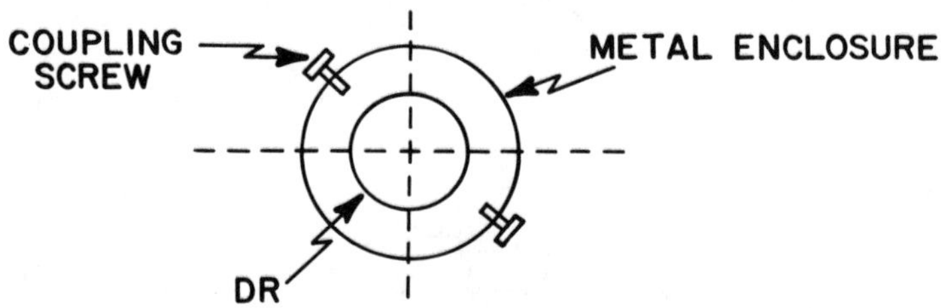

Fig. 8.41 Coupling of two orthogonal modes in a dielectric resonator

It is possible to couple energy from one mode into the other one, and adjust the coupling screws until the desired coupling coefficient between resonant modes is achieved [20].

Following the procedure employed by Slater [22], the coupling coefficient k_{ij} between the degenerate modes i and j can be derived as follows:

$$\left.\begin{aligned} k_{ij} &= \frac{a_{ij} - b_{ij}}{1 + C} \\ a_{ij} &= \int_V H_i H_j \, dV \end{aligned}\right\} \qquad (8.67)$$

$$\left.\begin{aligned} b_{ij} &= \int_V E_i E_j \, dV \\ C &= \int_V (H_i^2 - E_i^2) \, dV \end{aligned}\right\} \qquad (8.68)$$

where V is the volume of the perturbation. E_i, E_j, H_i, H_j are the electromagnetic fields of i,j modes of the original structure. For a small aperture, the perturbed field is very nearly equal to the original one.

Figure 8.42 shows the variation of the factor k_{ij} as a function of the distance between the screw and the resonator. The mode used here is the hybrid mode $HEM_{11\delta}$. Such a coupling permits realization of a two-pole filter with only one resonator [20].

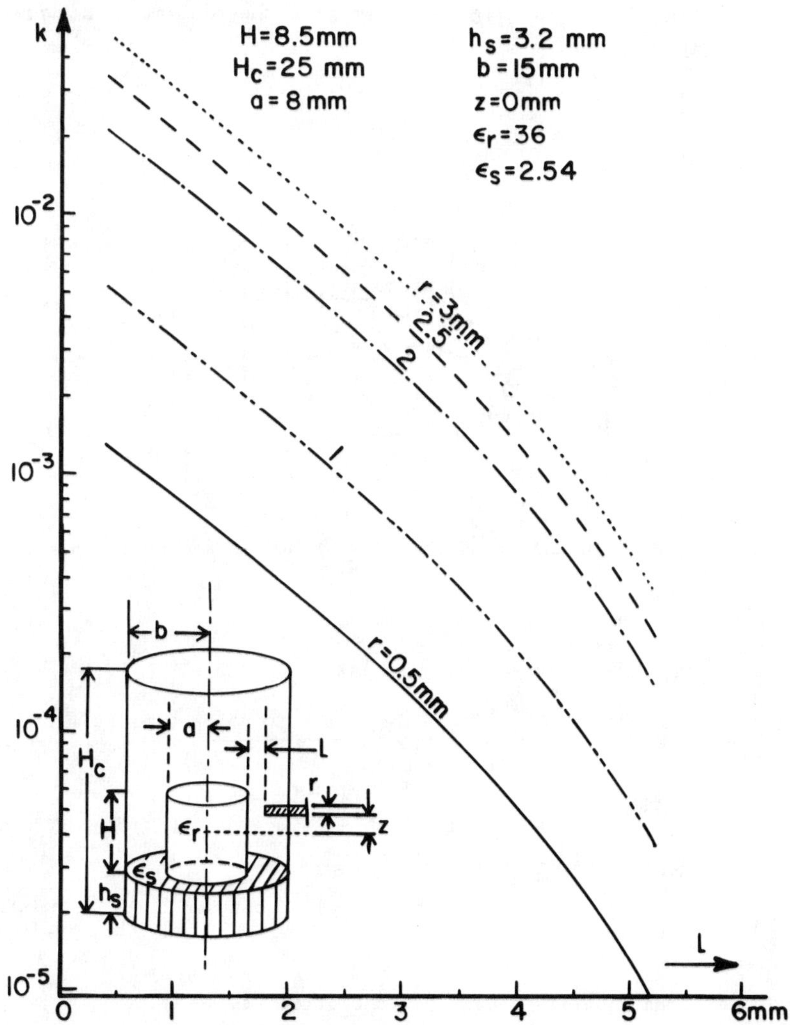

Fig. 8.42 Coupling coefficient for dual modes in a dielectric resonator (reference [24], ©1985 Eur. Microwave Conf.)

8.10 Other Configurations

In this chapter on coupling we have considered essentially the $TE_{01\delta}$ mode of the dielectric resonator. Some considerations concerning hybrid mode $HEM_{11\delta}$ have also been presented.

In addition, the $TM_{01\delta}$ mode can also be used in microwave applications [23]. This mode can be modeled by an electrical dipole moment $\underline{P}$. The excitation of the $TM_{01\delta}$ mode can be achieved via a microstrip line (see Fig. 8.43). In Fig. 8.44, we give the equivalent network of the coupling system.

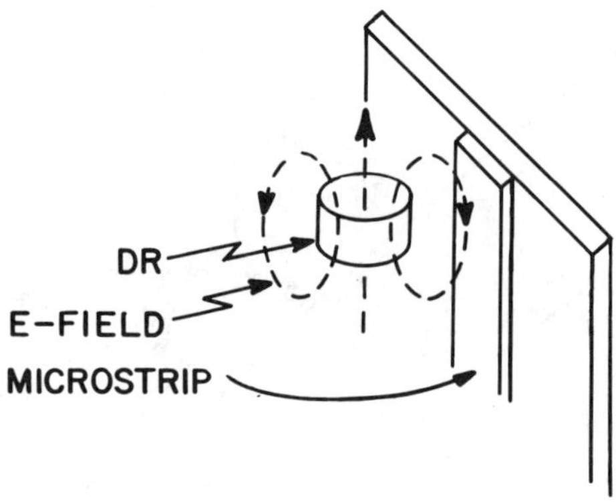

Fig. 8.43 Coupling of the $TM_{01\delta}$ mode and a microstrip line

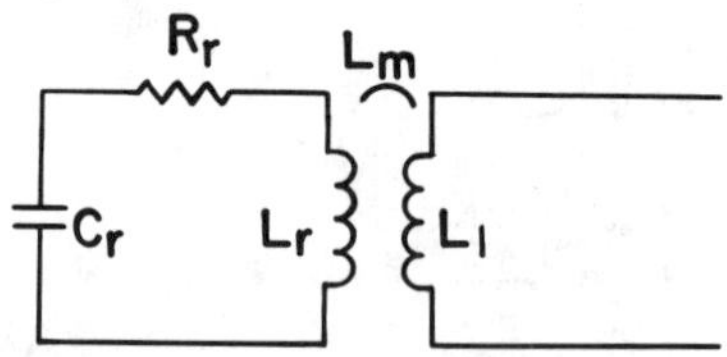

Fig. 8.44 Equivalent circuit of the coupling between the $TM_{01\delta}$ mode and the microstrip line

The external quality factor now satisfies [23]:

$$Q_e = \omega_0 \frac{4 \ Z_0 \ W}{\left[\int_{-H/2}^{+H/2} \underline{E} \cdot d\underline{\ell}\right]^2} \tag{8.69}$$

where W and $\underline{E}$ are the stored energy and the electric field of the $TM_{01\delta}$ mode, respectively.

The variation of Q_e as a function of the distance between the micro-strip line and the resonator is presented in Fig. 8.45.

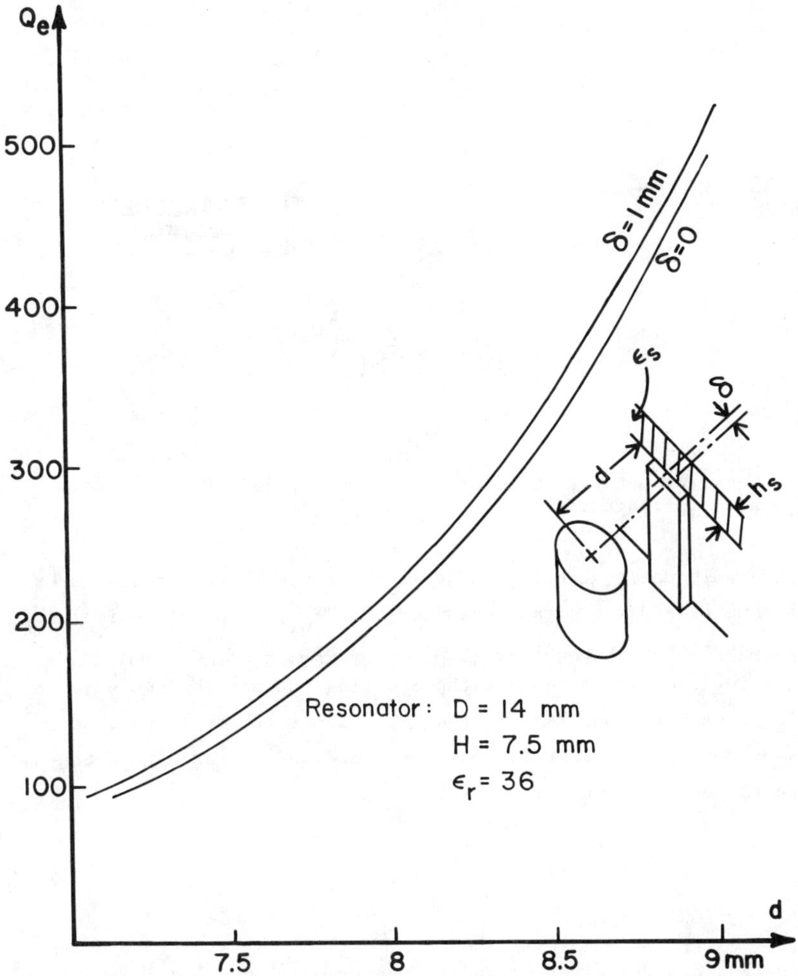

Fig. 8.45 Q_e as a function of the distance between the line and the $TM_{01\delta}$ resonator (reference [23], ©1985 IEEE)

The $TM_{01\delta}$ mode in a dielectric resonator inserted into an evanescent waveguide can be excited by the TE_{10} mode of a propagating waveguide in the manner shown in Fig. 8.46.

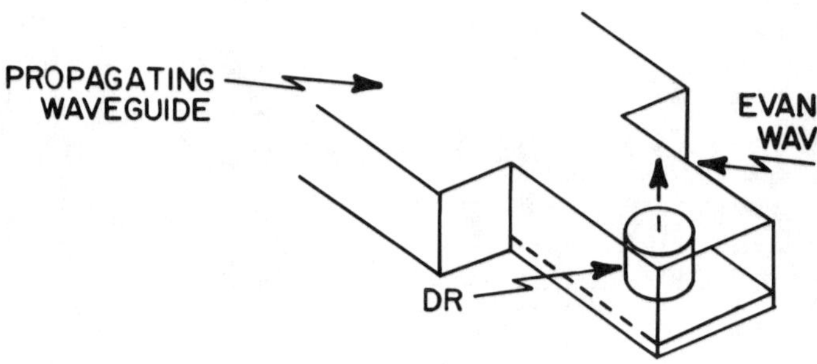

Fig. 8.46 Coupling between the $TM_{01\delta}$ mode in the dielectric resonator and a propagating waveguide

The external quality factor of such a structure can be evaluated by using the same procedure as that used for the $TE_{01\delta}$ mode. An example of Q_e values encountered in practice is given in Fig. 8.47.

Finally, we can also calculate the coupling between the two adjacent dielectric resonators operating in the $TM_{01\delta}$ mode. We can show that in this case only the evanescent fields of the dielectric resonators contribute to the coupling coefficient k:

$$k = \frac{\epsilon_0 (\epsilon_r - 1)}{2W} \int_V E_{z1} E_{z2} \, dV \tag{8.70}$$

E_{z1} and E_{z2} are the exterior and interior longitudinal components of the electric field in the dielectric resonator.

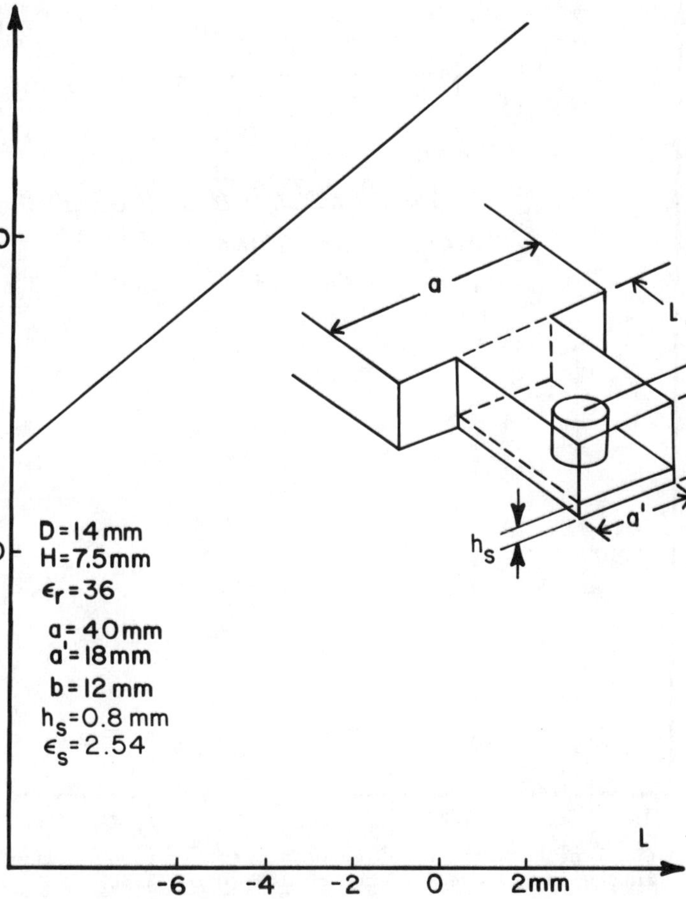

Fig. 8.47 External Q factor as a function of the distance between the dielectric resonator and the junction plane of the two waveguides (reference [23], ©1985 IEEE)

The coefficient k is a function of the distance S between the two resonators, as can be seen in Fig. 8.48.

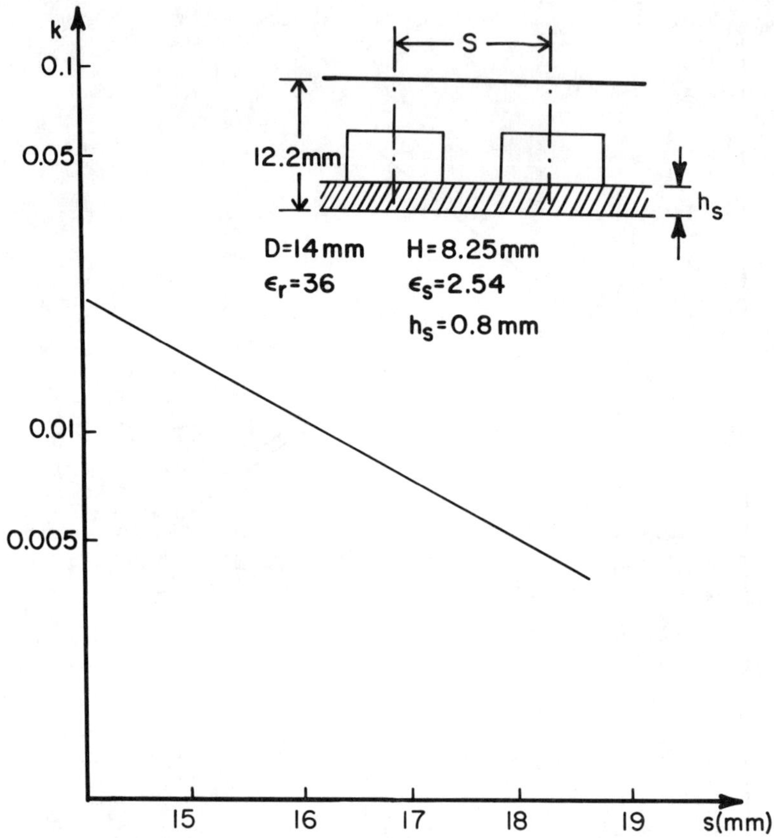

Fig. 8.48 Coupling between two adjacent dielectric resonators operating in the $TM_{01\delta}$ mode (reference [23], ©1985 IEEE)

REFERENCES

[1] S.B. Cohn, "A miniature high Q bandpass filter employing dielectric resonators," IEEE Trans. Microwave Theory Tech., vol. MTT-16, pp. 210-218, April 1968.

[2] T.D. Iveland, "Microstrip bandpass filters with dielectric resonators," IEEE Trans. Microwave Theory and Tech., vol. MTT-19, pp. 643-652, July 1971.

[3] M.W. Pospieszalski, "Cylindrical dielectric resonators and their applications in TEM line microwave circuits," IEEE Trans. Microwave Theory and Tech., vol. MTT-28, p. 434, April 1980.

[4] P. Guillon and Y. Garault, "Coupling between a microstrip transmission line and a dielectric resonator," IEEE MTT-S Int. Microwave Symp. Dig., p. 200, June 1976.

[5] P. Guillon and S. Mekerta, "A bandstop dielectric resonator filter," IEEE MTT-S Int. Microwave Symp. Dig., p. 170, June 1981.

[6] A.E. Atia and R.R. Bonetti, "Generalized dielectric resonator filters," Comsat Technical Review, vol. 11, pp. 321-343, Fall 1981.

[7] S. Mekerta, "Dielectric resonator bandstop filter using $TE_{01\delta}$ and $TM_{01\delta}$ modes," Ph.D. Thesis, Limoges, 1981.

[8] P. Guillon and Y. Garault, "Accurate resonant frequency of dielectric resonators," IEEE Trans. Microwave Theory and Tech., vol. MTT-25, p. 916, Nov. 1977.

[9] Y. Konishi, "External Q of a $TE_{01\delta}$ dielectric resonator in a $TE_{1.0}$ waveguide bandpass filter," Electron. and Commun. in Japan, p. 12, Jan. 1976.

[10] P. Guillon and M.P. Chong, "A bandpass dielectric resonator filter with a high attenuation cut off rate," IEEE MTT-S Int. Microwave Symp. Dig., p. 240, June 1982.

[11] P. Guillon and Y. Garault, "Coupling between a magnetic loop and a dielectric resonator inserted into an evanescent waveguide," IEEE MTT-S Int. Microwave Symp. Dig., p. 392, 1982.

[12] P. Guillon and F. Farzaneh, "Coupling between a dielectric resonator and a dielectric image guide," IEEE MTT-S Int. Microwave Symp. Dig., p. 115, 1984.

[13] T. Itoh and R. Rudokas, "New method for computing the resonant frequencies of dielectric resonators," IEEE Trans. Microwave Theory and Tech., vol. MTT-25, p. 52, Jan. 1977.

[14] L. Pettersson, "On the theory of coupling between finite dielectric resonators," IEEE Trans. Microwave Theory and Tech., vol. MTT-24, p. 615, Sept. 1976.

[15] P. Skalicky, "Direct coupling between two dielectric resonators," Electron. Lett., vol. 18, p. 332, April 1982.

[16] P. Guillon, B. Byzery, and M. Chaubet, "Coupling parameters between a dielectric resonator and a microstrip line," IEEE Trans. Microwave Theory and Tech., vol. MTT-33, pp. 222-226, March 1985.

[17] A.E. Atia and A.E. Williams, "New type of waveguide band-pass filters for satellite transponders," Comsat Technical Review, vol. 1, pp. 402, Fall 1971.

[18] S.J. Fiedziuszko, "Dual mode dielectric resonator loaded cavity filters," IEEE Trans. Microwave Theory and Tech., vol. MTT-30, pp. 1311-1316, Sept. 1982.

[19] H.N. Shama Sundara and P. Guillon, "A new type of dielectric resonator filters," A.E.U., vol. 34, no. 32, p. 63, March 1980.

[20] P. Guillon and Y. Garault, "Dielectric resonator dual modes filters," Electron. Lett., vol. 16, p. 646, August 14 1980.

[21] D. Kajfez, A.W. Glisson, and J. James, "Computed modal field distributions of isolated dielectric resonators," IEEE MTT-S Int. Microwave Symp. Dig., p. 193, June 1984.

[22] J.C. Slater, Microwave Electronics, ch. IV. New York: Van Nostrand, 1950.

[23] B. Byzery and P. Guillon, "$TM_{01\delta}$ dielectric resonator mode uses in microwave circuits," IEEE MTT-S Int. Microwave Symp. Dig., p. 515, 1985.

[24] J.P. Astier and P. Guillon, "Elliptic microwave filter using dual modes of dielectric resonators," Eur. Microwave Conf., Paris, Sept. 1985.

[25] T. Itanami and S. Shindo, "Channel dropping filter for millimeter-wave integrated circuits," IEEE Trans. Microwave Theory and Tech., vol. MTT-26, p. 759, Oct. 1978.

[26] A.K. Sharma and W.J.R. Hoefer, "Empirical expressions for fine line design," IEEE Trans. Microwave Theory and Tech., vol. MTT-31, p. 350, April 1983.

Chapter 9
FILTERS
Pierre Guillon

9.1 Introduction

In the past four decades, tremendous advances have been made in microwave filter technology. There are filter types for almost every usage that can be realized in waveguide and in microstrip technology. Dielectric resonators hold the promise of further enrichment of microwave filter designs. In fact, dielectric resonator filters have also been designed and used in radio systems as low as 1 GHz. In this chapter, the filter design principles, practical design considerations, and examples of dielectric resonator filters are presented.

The procedure for filter design is described as follows. Usually, we have to realize a microwave filter for which we know the type, the bandwidth, and the center frequency. Following the well established design principles, the external Q_e (which defines the input and output coupling) and the coupling coefficient, $k_{j,j+1}$, between two adjacent resonators are computed according to the filter specifications. To relate these values to the physical dimensions of the microwave dielectric filter, we have to compare these theoretical values of Q_e and $k_{j,j+1}$ to those obtained earlier in Ch. 8.

First, we will briefly review the filter design procedure such as described in [1]. Consider a typical low-pass prototype filter shown in Fig. 9.1 having a Chebyshev response. The attenuation is given by

$$L_A(\omega') = 10 \, \mathrm{Log}_{10}\left[1 + \epsilon \cos^2\left(n \cos^{-1} \frac{\omega'}{\omega_1'}\right)\right] \tag{9.1}$$

$$\text{for} \quad \omega' < \omega_1'$$

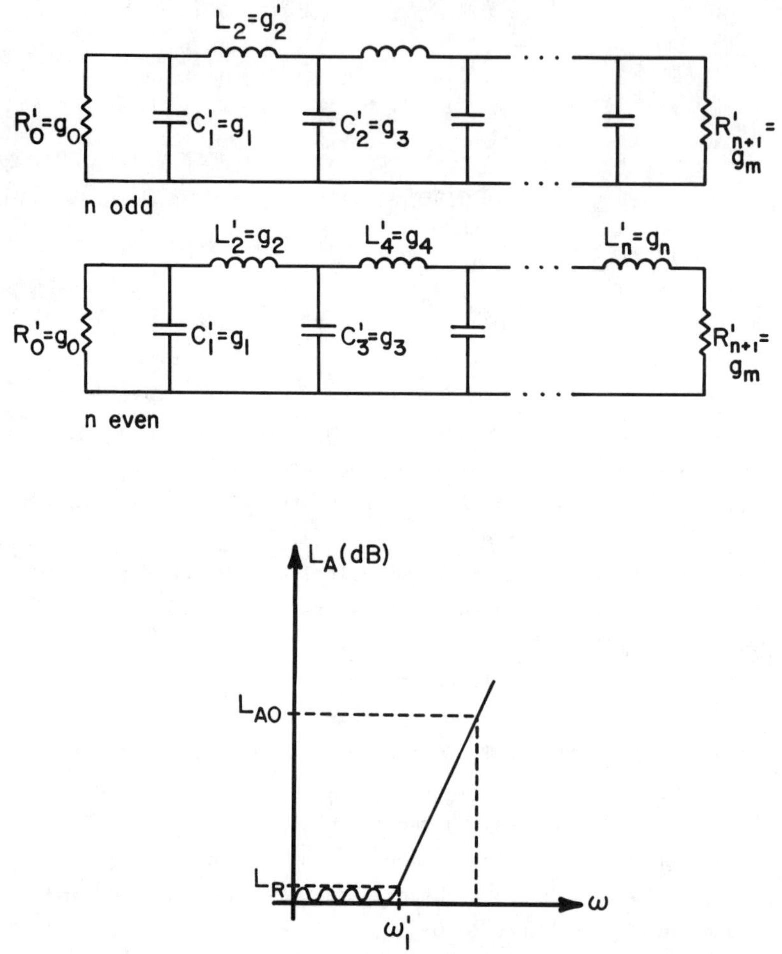

Fig. 9.1 Low-pass filter and its low-pass prototype Chebyshev response

$$L_A(\omega') = 10\ \mathrm{Log}_{10}\left[1 + \epsilon\ \cosh^2\left(n\ \cosh^{-1}\frac{\omega'}{\omega_1'}\right)\right] \tag{9.2}$$

$$\text{for} \quad \omega' > \omega_1'$$

where

$$\epsilon = 10^{L_{Ar}/10} - 1 \tag{9.3}$$

L_{Ar} = pass-band ripple in dB
ω' = operating frequency of the low-pass prototype
ω'_1 = equal ripple cutoff frequency
n = number of reactive elements.

Band-pass filters can be obtained from the low-pass prototype by a frequency transformation:

$$\frac{\omega'}{\omega'_1} = \frac{1}{W}\left[\frac{\omega}{\omega_0} - \frac{\omega_0}{\omega}\right] \tag{9.4}$$

where W is the fractional bandwidth:

$$W = \frac{\omega_2 - \omega_1}{\omega_0} \tag{9.5}$$

and ω_0 is the center frequency:

$$\omega_0 = \sqrt{\omega_1\omega_2} \tag{9.6}$$

while ω_1, ω_2 are the frequency band limits of the band-pass filter.

Figure 9.2(a) shows the band-pass filter obtained by a low-pass to band-pass transformation and Fig. 9.2(b) shows the corresponding band-pass filter response. Introducing the impedance inverters into the circuit, the ladder-type band-pass can be converted to the type which contains only series-tuned circuits $X_i(\omega)$, as shown in Fig. 9.3.

An ideal impedance inverter is essentially a quarter-wave transformer of characteristic impedance K. It has an interesting property such that when it is terminated in an impedance Z_B at one end, the impedance looking in at the other end is Z_A:

$$Z_A = \frac{K^2}{Z_B} \tag{9.7}$$

Although the inverters used in practice are frequency sensitive and the resonators $X_i(\omega)$ are generally not lumped, the equivalent network in Fig. 9.3 gives an approximation which is correct for narrow bandwidths.

The external Q's and coupling coefficients of the band-pass filter having series type resonances are related to the prototype elements by

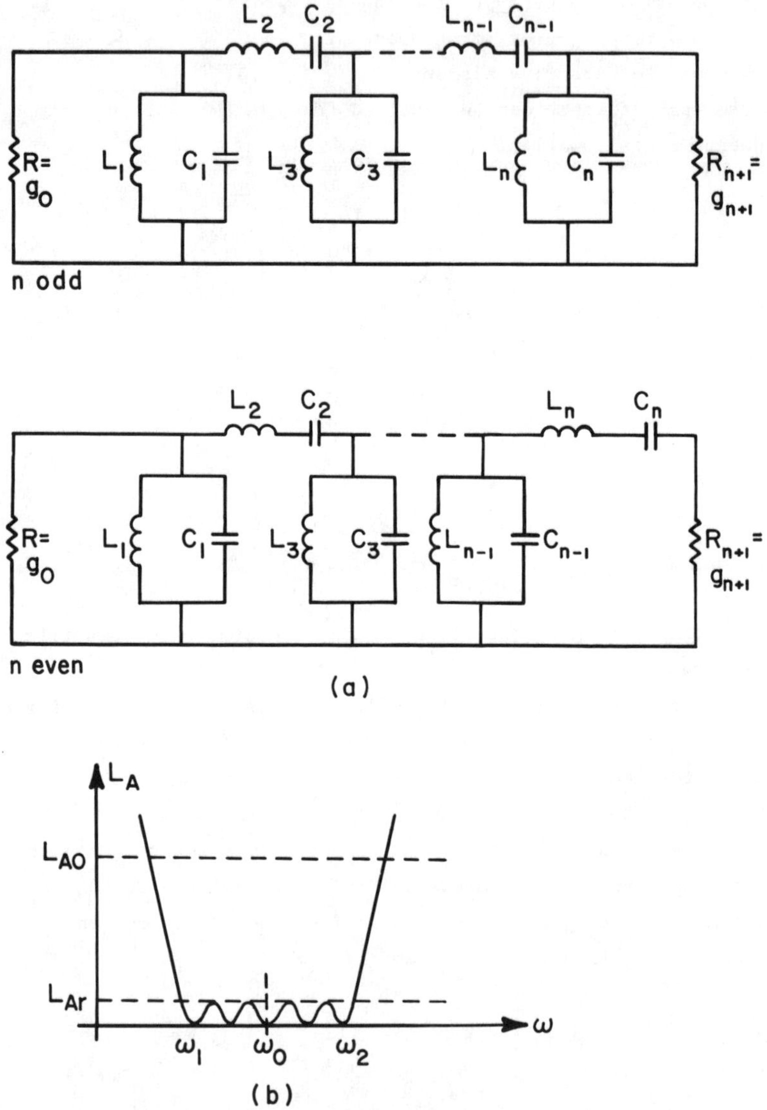

Fig. 9.2 Band-pass filter and its band-pass filter response

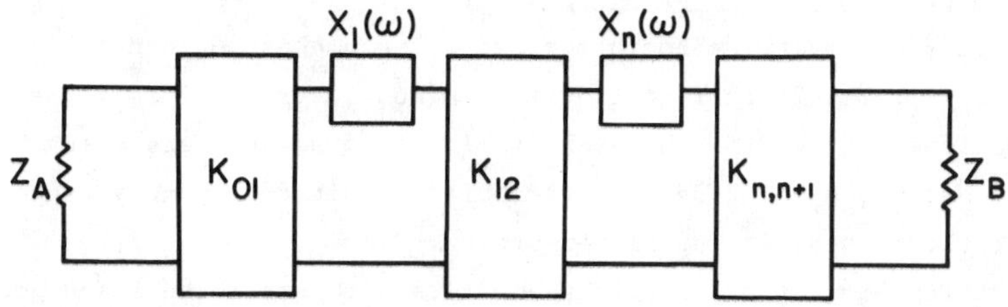

Fig. 9.3 Band-pass filter using impedance inverters

$$(Qe)_{in} = \frac{X_2}{K_{01}^2/R_A} = \frac{g_0 g_2}{W} \omega_2' \qquad (9.8)$$

$$(Qe)_{out} = \frac{X_n}{K_{n,n+1}^2/R_B} = \frac{g_n g_{n+1}}{W} \omega_1' \qquad (9.9)$$

$(Qe)_{in}$ and $(Qe)_{out}$ are the external quality factor at the input and at the output, respectively.

The coupling coefficients are given by

$$k_{j,j+1} = \frac{K_{j \cdot j+1}}{\sqrt{X_j X_{j+1}}} = \frac{W}{\omega_1' \sqrt{g_j g_{j+1}}} \qquad (9.10)$$

X_j is the reactance slope parameters of the jth resonator.

The insertion losses at the center frequency are given by

$$(L_A)_0 = \frac{4.343}{Q_{ui}} \frac{1}{L_{Ar}} \sum_{i=1}^{N} g_i \qquad (9.11)$$

where n is the number of elements, and Q_{ui} is the unloaded Q factor of the ith resonator.

9.2 Waveguide Below Cutoff Band-pass DR Filter

Waveguides or cavity resonators possess the highest unloaded Q's and, hence, will result in filters with minimum insertion loss for a given fractional bandwidth. However, their disadvantages are being relatively bulky and being useful over only a limited frequency range because of the possibility of higher-order modes.

The availability of dielectric resonators with small tan δ and good temperature stability makes them suitable for microwave filters. The band-pass filters realized with DRs are typically about one-third shorter than conventional waveguide filters.

The band-pass filter to be described first is composed of a rectangular waveguide below cutoff containing dielectric resonators. The structure is then connected at both ends to propagating rectangular waveguides as shown in Fig. 9.4.

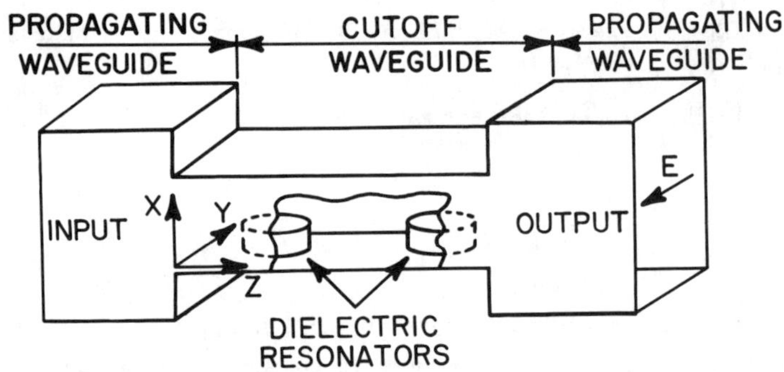

Fig. 9.4 Dielectric-resonator band-pass filter

The input propagating waveguide excites the electromagnetic fields, which decay exponentially with the distance from the junction of the cutoff waveguide. These evanescent fields excite the nearest dielectric resonator in the cutoff waveguide, and the signal is then transmitted through the next dielectric resonator, etc., to the output propagating waveguide.

When the propagating waveguides are excited on the $TE_{1,0}$ mode and if the center line of dielectric resonators is oriented along the x-axis (see Fig. 9.4), the fundamental $TE_{01\delta}$ mode in the dielectric resonator is excited. The transverse orientation of resonators is preferred because they can be tuned with concentric screws so as to avoid spurious mode excitations.

As an example, consider the band-pass filter with dielectric resonators designed to have the following specifications:

4-pole Chebyshev response

center frequency = 4 GHz

bandwidth = 40 MHz

ripple in the passband = 0.2 dB

The dielectric resonators to be employed have a diameter D = 14 mm and a height H = 7 mm. (The aspect ratio of the resonator H/D is chosen in the 0.3 to 0.5 range to minimize the interference of spurious modes.) The unloaded Q factor Q_u is 3800 for the materials at hand.

The side view of the band-pass filter with four dielectric resonators is shown in Fig. 9.5. The distances L_1 and L_4 are measured from

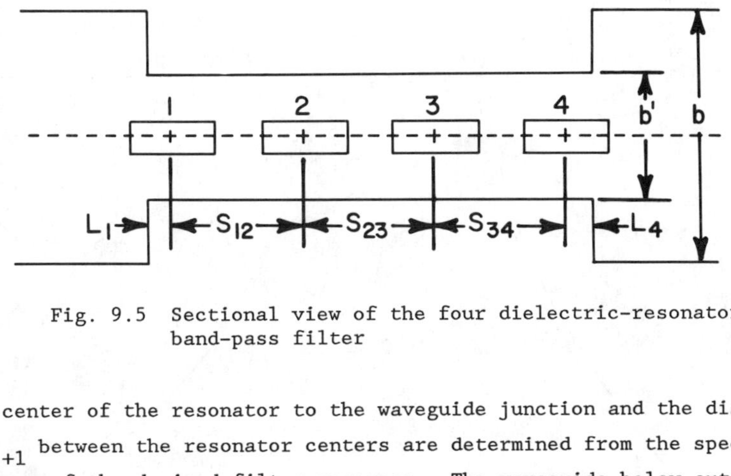

Fig. 9.5 Sectional view of the four dielectric-resonator band-pass filter

the center of the resonator to the waveguide junction and the distances $S_{j,j+1}$ between the resonator centers are determined from the specifications of the desired filter response. The waveguide below cutoff has the following dimensions:

height b' = 13.1 mm

width a' = 22.5 mm

The propagating waveguide has the following dimensions:

height b = 47.5 mm

width a = 22.5 mm

From (9.8), (9.9), and (9.10) and using the element values for the 0.2 dB ripple prototype, we obtain

$Q_{e,in} = Q_{e,out} = 130.3$
$k_{12} = k_{34} = 0.0077$
$k_{23} = 0.00628$

Using the diagrams presented in Ch. 8, we evaluate the distances L_i and S_{ij}, respectively. The insertion losses are computed from (9.11). Figures 9.6 and 9.7 give the theoretical and experimental transmission and reflection characteristics of the band-pass filter. Dielectric resonators in the waveguide below cutoff can also be excited by a magnetic loop located at the end of the coaxial line, such as shown in Fig. 9.8 [5].

The analysis of such a band-pass filter is the same as that developed in the previous sections. Only the couplings of the first and the last resonators are modified and are obtained now by using a magnetic loop. Figure 9.9 presents the response of a four-pole Chebyschev band-pass filter using such a coupling arrangement.

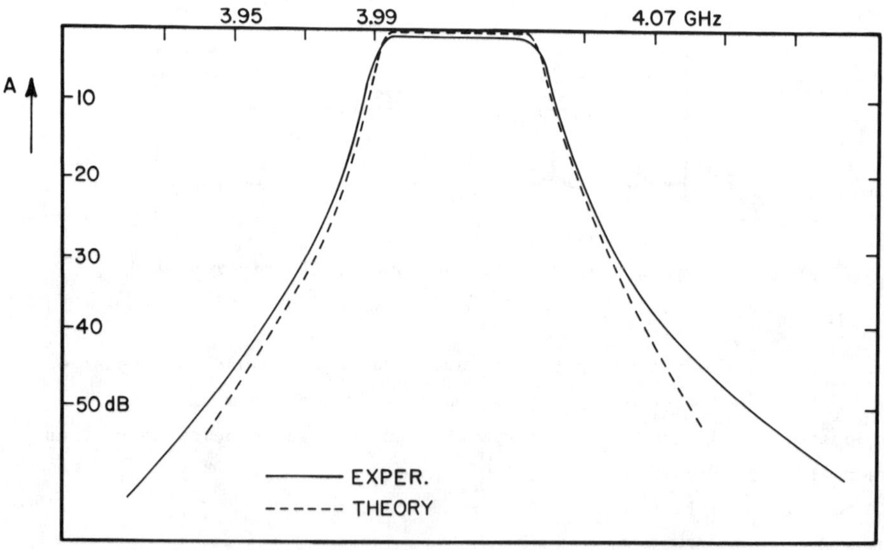

Fig. 9.6 Theoretical and experimental transmission characteristics of the band-pass filter (reference [26], ©1984 IEEE)

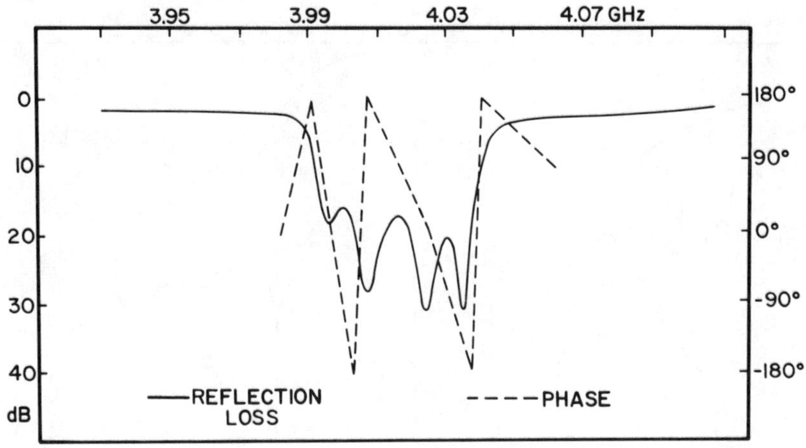

Fig. 9.7 Theoretical and experimental reflection characteristics of the band-pass filter (reference [26], ©1984 IEEE)

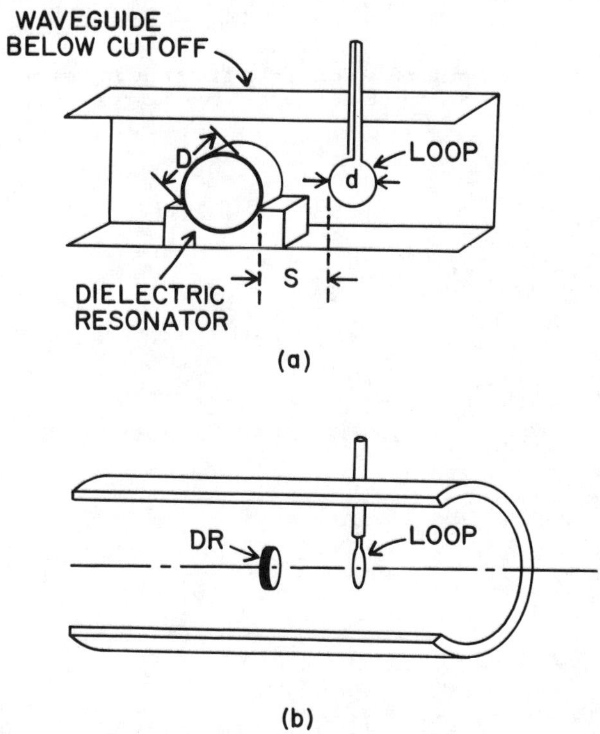

Fig. 9.8 Dielectric resonators in an evanescent waveguide excited by a magnetic loop: a) rectangular waveguide, b) circular waveguide

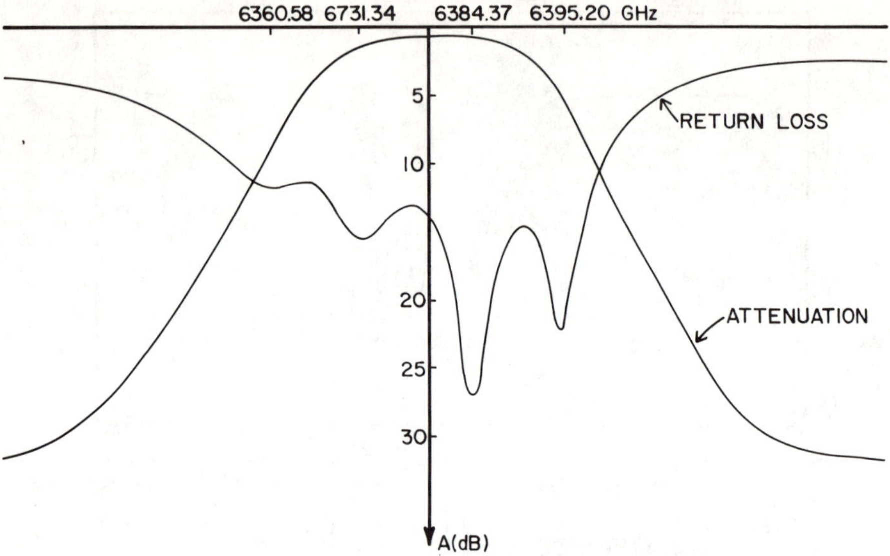

Fig. 9.9 Response of a four-pole filter using magnetic loop for the input and output couplings

9.3 Microstrip Band-pass DR Filter

The coupling between the end resonators is now achieved by using microstrip lines. The design of these filters has been reported in references [6-9]. The filter synthesis is based on the well-known low-pass prototype elements and a low-pass to band-pass mapping. The design is, in this case, completely determined by (9.8), (9.9), (9.10), and (9.11).

A microstrip filter using four dielectric resonators has been realized around 4 GHz. The coupling to the external circuit was performed by two 50 Ω microstrip lines, as shown in Fig. 9.10. The pass-band

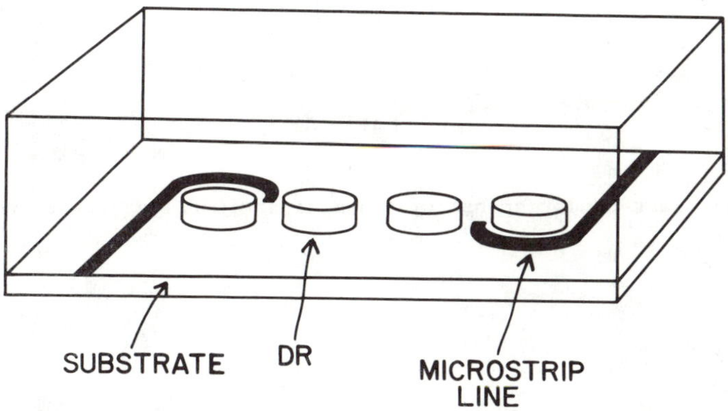

Fig. 9.10 Microstrip dielectric-resonator band-pass filter

response of the filter is shown in Fig. 9.11. The insertion loss is about 1 dB, the VSWR is 1.16. In this realization, the resonators are simply glued to the microstrip substrate.

Using the same method of analysis, a suspended microstrip filter has been studied and realized. The experimental response of this three-dielectric-resonator filter, centered around 16 GHz, is presented in Fig. 9.12.

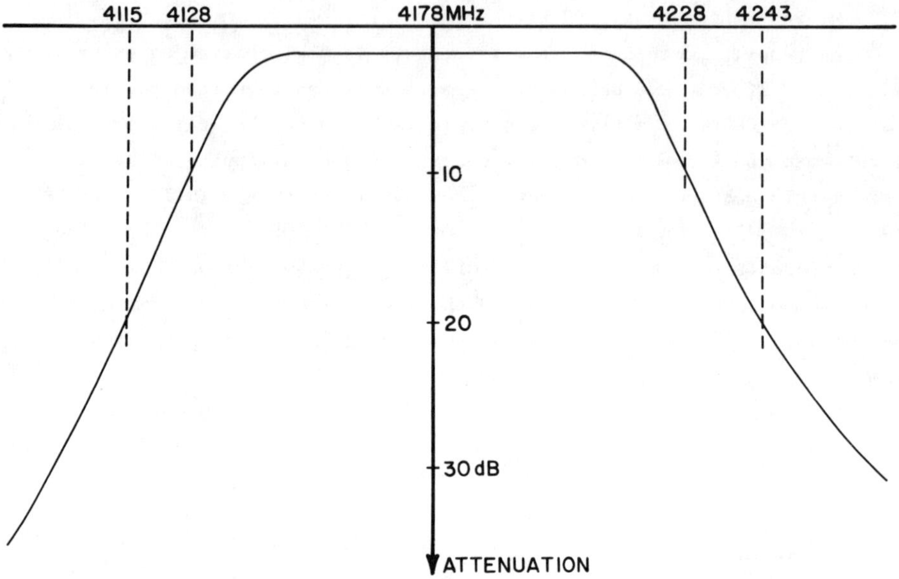

Fig. 9.11 Pass-band response of the microstrip filter (reference [27])

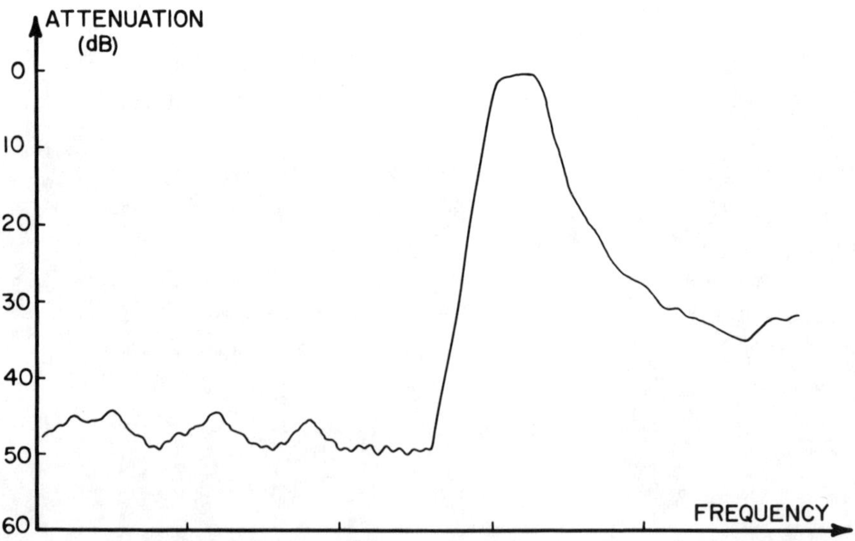

Fig. 9.12 Experimental response of the three dielectric-resonator suspended microstrip filter

9.4 Waveguide Band-Stop DR Filter

When a dielectric resonator is situated in a propagating waveguide, it absorbs power at its resonant frequency and thus offers band-stop properties [10-12]. A band-stop filter can be realized by coupling the dielectric resonator to a propagating waveguide. The resonators are located in their own metal enclosures, as shown in Fig. 9.13. The coupling to the main waveguide is achieved through apertures in the waveguide walls.

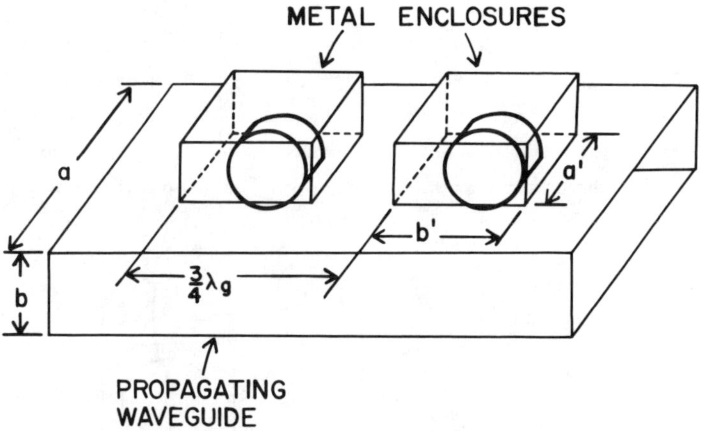

Fig. 9.13 Dielectric-resonator band-stop filter (reference [26], ©1984 IEEE)

The length and the width of the metal enclosure is preferably chosen to be the same as that of the waveguide below cutoff for the band-pass filter. The height b of the propagating waveguide must be small enough to ensure isolation between the dielectric resonators. This isolation is further assured through the separation of the apertures by $n\lambda g/4$, so that no inter-resonator coupling takes place.

When the $TE_{1,0}$ mode is excited in the propagating waveguide, the magnetic field of that mode penetrates into the cutoff waveguide. After decaying exponentially near the aperture, this field couples with the magnetic field of the fundamental $TE_{01\delta}$ mode in the dielectric resonator (see Fig. 9.14).

The H_x coupling configuration is chosen because the orientation of the dielectric resonator is symmetrical with respect to the electric

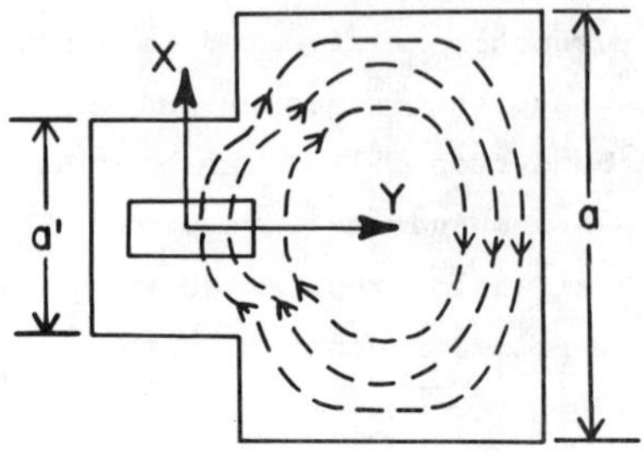

Fig. 9.14 Excitation of the dielectric resonator

field. In this way, the next higher mode to be excited is the $TE_{3,0}$ mode. The height of the metal cavity can be varied by means of a movable shorting plunger placed at the other end, as shown in Fig. 9.15. Its distance from the dielectric resonator modifies the resonant frequency of the resonator and, hence, can be used for tuning purposes.

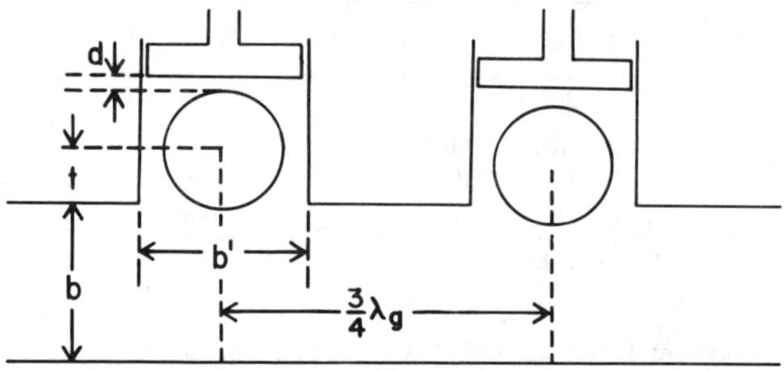

Fig. 9.15 Band-stop filter with two dielectric resonators

As an example, a band-stop filter, such as shown in Fig. 9.13, is to be designed to satisfy the following specifications:

2-pole Butterworth response

center frequency = 3.88 GHz

bandwidth = 20 MHz

The dielectric resonators are supported by teflon stands which are firmly housed in each rectangular cavity. The separation between cavities is three-quarter guide wavelengths. Figure 9.15 shows the side view of the filter. The distance t is measured from the center of the dielectric resonator to the side wall of the propagating waveguide, and d is the distance from the shorting plunger to the dielectric resonator.

Figures 9.16 and 9.17 give the measured values of the external Q factor as a function of the distances d and t, respectively. Figures 9.18 and 9.19 show the influence of the distances t and d on the resonant frequency of the resonator.

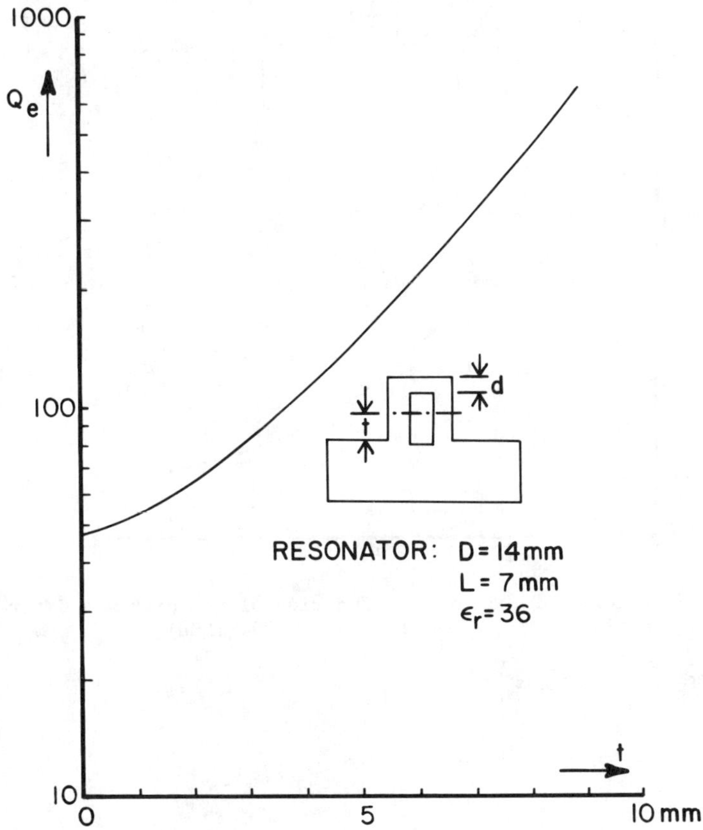

Fig. 9.16 External Q factor as a function of the distance t (valid for d > 5 mm) (reference [26], ©1984 IEEE)

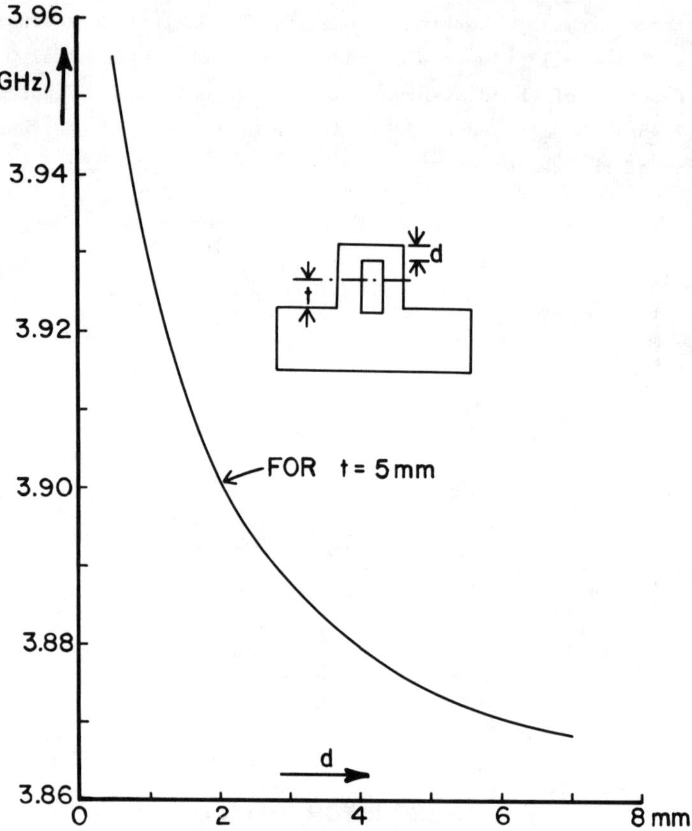

Fig. 9.17 External Q factor as a function of the distance d for different t (reference [26], ©1984 IEEE)

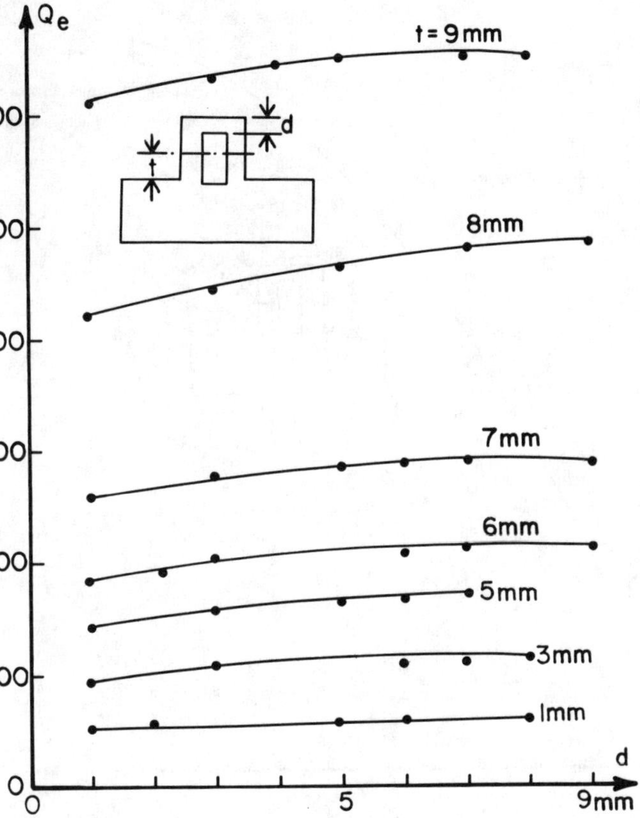

Fig. 9.18 Resonant frequency as a function of d (reference [26], ©1984 IEEE)

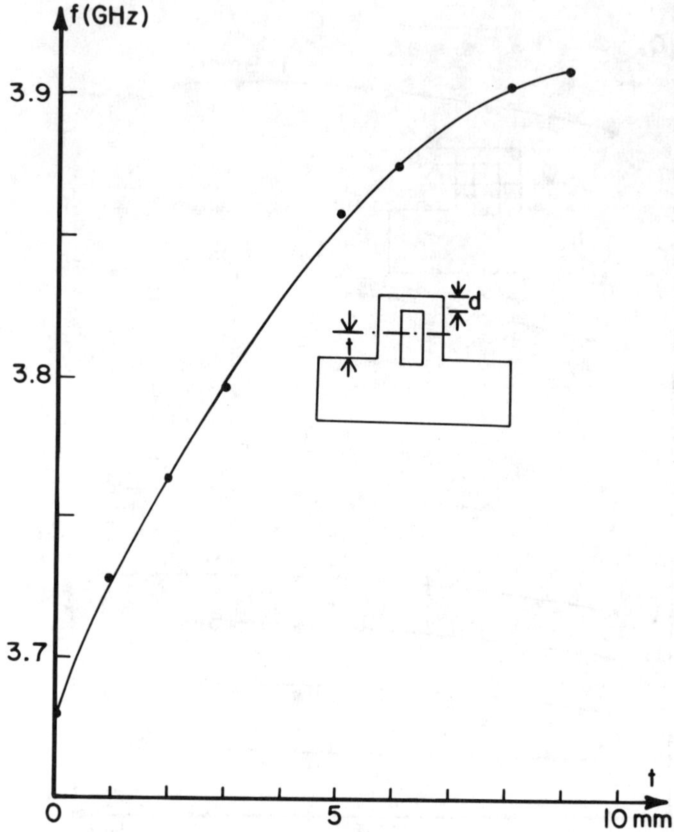

Fig. 9.19 Resonant frequency as a function of t (reference [26], ©1984 IEEE)

The procedure used for the design of the band-stop filter is the same as that which has been presented for the design of the band-pass filter. The theoretical and the experimental responses of the two-resonator Butterworth band-stop filter are given in Fig. 9.20.

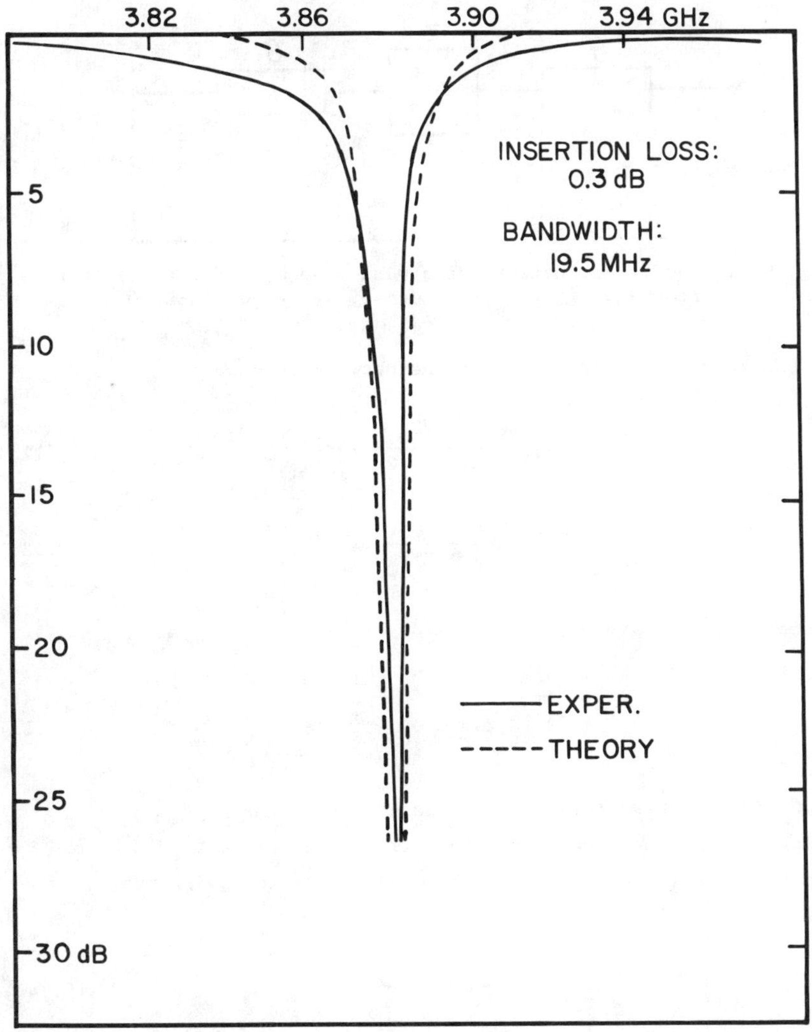

Fig. 9.20 Theoretical and experimental response of the band-stop filter (reference [26], ©1984 IEEE)

9.5 Microstrip Band-stop DR Filter

A dielectric resonator coupled with a microstrip line is identical to a parallel resonant circuit placed in series with the line. It is a localized element for which we have established earlier the equivalent network shown in Fig. 9.21. Denoting the electrical length of the line

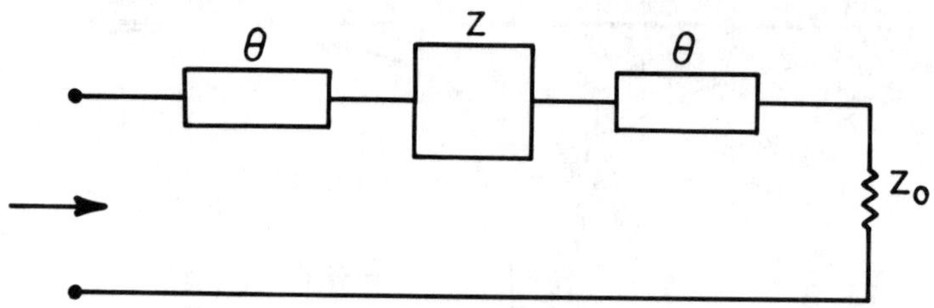

Fig. 9.21 Equivalent network of a band-stop microstrip dielectric-resonator filter (reference [13], ©1981 IEEE)

by θ, the scattering parameters of the network become

$$S_{11} = \frac{2\ Z/Z_0}{Z/Z_0 + 2}\ e^{-2j\theta} \tag{9.12a}$$

$$S_{12} = \frac{2}{2 + Z/Z_0}\ e^{-2j\theta} \tag{9.12b}$$

At resonance, Z is given by

$$Z = \omega_0 Q_0 \frac{L_m^2}{L_r} \tag{9.13}$$

where Z_0 is the characteristic impedance of the line.

Richards [14] has shown that the distributed circuits can be obtained from the lumped RLC circuits by using the following frequency transformation:

$$\Omega = \tan \frac{\pi\omega}{2\omega_0} \tag{9.14a}$$

Using this transformed frequency, Richards introduces a new complex frequency $p = j\Omega$, so that

$$p = j \tan \frac{\pi\omega}{2\omega_0} \tag{9.14b}$$

We use the symbols f_0 and f to denote the frequencies in the microwave domain, and Ω is the frequency of the corresponding low-pass prototype.

If s and p signify, respectively, the microwave and the low frequency domain, we obtain from (9.14b):

$$p = \tan \frac{\pi}{2\omega_0} s \tag{9.15}$$

Using (9.15), a low-pass filter of frequency Ω will be transformed into a band-stop filter centered around ω_0 in the manner illustrated in Fig. 9.22.

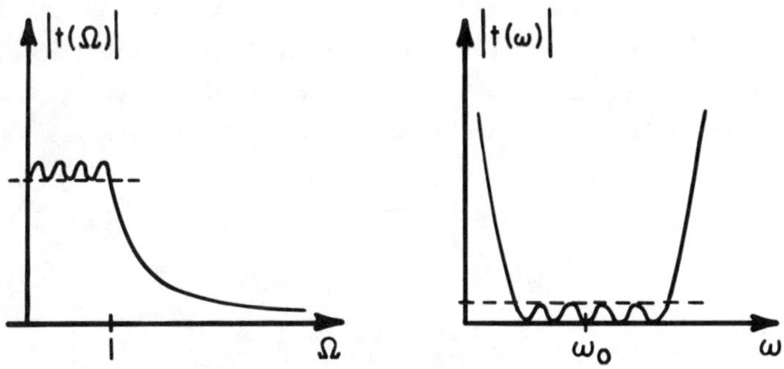

Fig. 9.22 Frequency transformation from low-pass to band-stop

If we consider a quarter-wavelength line of characteristic impedance Z_0, we can establish an equivalence table between the s and p planes (see Table 9.1).

The unit element is a quarter-wavelength transmission line of characteristic impedance Z_0. Its chain matrix is given by

$$\underset{\sim}{T} = \frac{1}{\sqrt{1 - p^2}} \begin{bmatrix} 1 & Z_0 p \\ \frac{p}{Z_0} & 1 \end{bmatrix} \tag{9.16}$$

We first study the equivalent circuit of a microstrip coupled dielectric resonator. We have shown in Ch. 8 that the equivalent circuit

Table 9.1 RICHARDS TRANSFORMATION EQUIVALENCES

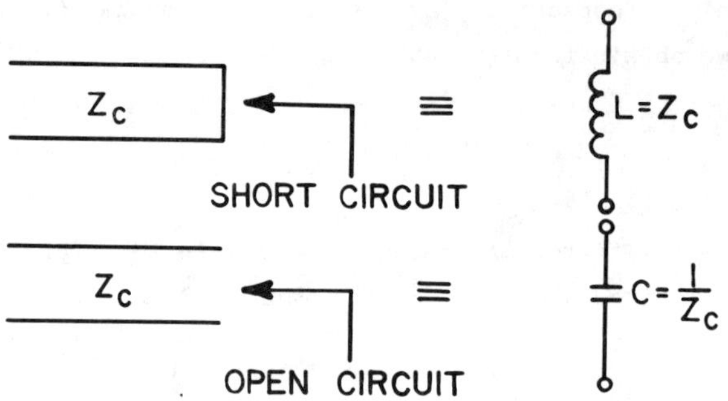

of a low-loss dielectric resonator coupled with a microstrip line is a resonant circuit placed in series with the line, such as shown in Fig. 9.23(a). The equivalent impedance Z expressed in the s plane is

$$Z = \frac{Ls}{1 + LCs^2} \tag{9.17}$$

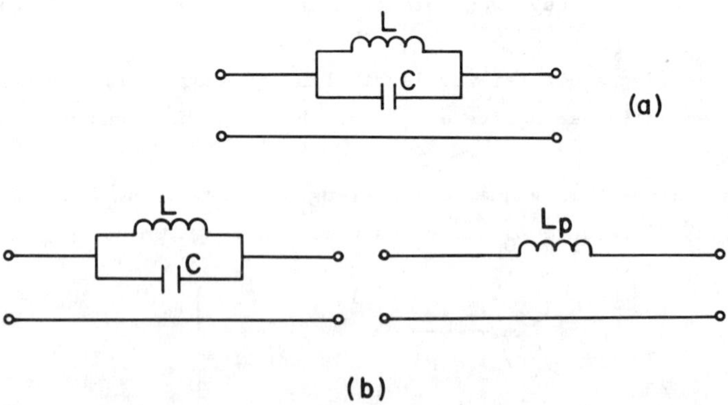

Fig. 9.23 Equivalent circuit of the dielectric resonator coupled to a microstrip line (reference [13], ©1981 IEEE)

Near the resonant frequency $\omega = \omega_0 + \Delta\omega$, so that (9.17) simplifies as follows:

$$Z(s) = \frac{1}{2C(s - s_0)} \tag{9.18}$$

Z is now expressed in the p plane by using (9.14):

$$Z(p) = L_p p \tag{9.19}$$

with $L_p = \frac{\pi}{4} L \omega_0$. The equivalence of the two circuits in the s and p plane is indicated in Fig. 9.23(b).

The three-element low-pass prototype filter is presented in Fig. 9.24. Using the chain matrices of individual elements and the relation

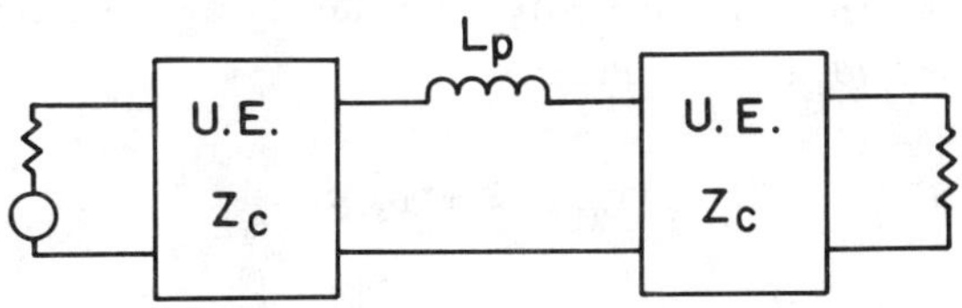

Fig. 9.24 Low-pass prototype filter with unit elements (reference [13], ©1981 IEEE)

between scattering parameters S_{ij} and chain parameters T_{ij}, we obtain

$$|S_{21}|^2 = \frac{1}{1 + |T_{12}|^2} \tag{9.20}$$

The S_{21} coefficient gives the amplitude function t(p) of the filter as follows:

$$|t(p)|^2 = S_{21}(p)S_{21}(-p) \tag{9.21}$$

Therefore,

$$|t(p)|^2 = \frac{1}{1 - \frac{L_p^2}{4Z_0^2}} \tag{9.22}$$

From this expression we can find the external quality factor:

$$Q_e = \frac{\pi Z_0}{2L_p} \tag{9.23}$$

In the paragraphs to follow, we present the design procedure of the band-stop Chebyshev filter with the following characteristics:

number of poles = 3

ripple = 0,2 dB

bandwidth = 120 MHz

center frequency = 6 GHz

The transfer function of such a filter is

$$|t(p)|^2 = \frac{1}{1 + \epsilon^2 T_3^2(p)} \tag{9.24}$$

where $T_3(p)$ is the Chebyshev polynomial of first kind, order three. From (9.20) and (9.24), we derive

$$|T_{21}|^2 = \epsilon^2 T_3^2(p) \tag{9.25}$$

$$\rho(p)^2 = 1 - |t(p)|^2 \tag{9.26}$$

Using

$$Z_i(p) = \frac{1 + \rho(p)}{1 - \rho(p)} \tag{9.27}$$

we obtain

$$Z_i(p) = \frac{p^2 + 1.7159p + 1}{4.9410^{-3} p^3 + p^2 + 2.3311p + 1} \tag{9.28}$$

$Z_i(p)$ is a positive real function associated with the equivalent network presented in Fig. 9.25. Using Richards' theorem we can derive the

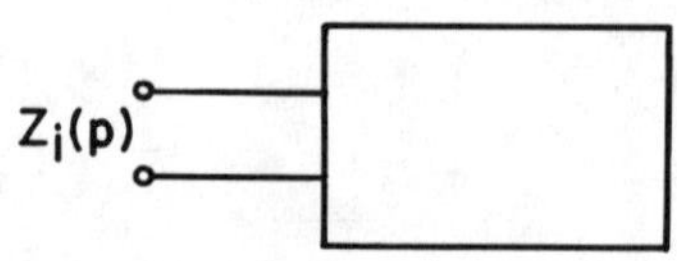

Fig. 9.25 Equivalent network of the band-stop filter

low-pass prototype of the filter which is shown in Fig. 9.26.

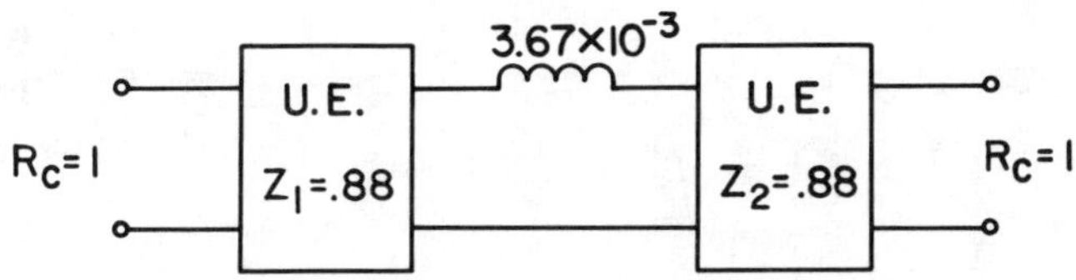

Fig. 9.26 Low-pass prototype of the filter

From the value of L_p we can determine the external quality factor Q_e by using (9.23), which in turn determines the distance between the line and the resonator. The unit elements are represented by quarter-wavelength transmission line sections.

The structure of the filter is represented in Fig. 9.27. The theoretical and the experimental response of this filter are both given in Fig. 9.28. The method of analysis presented in this section is quite general so that it can be extended to the case of a filter having N elements.

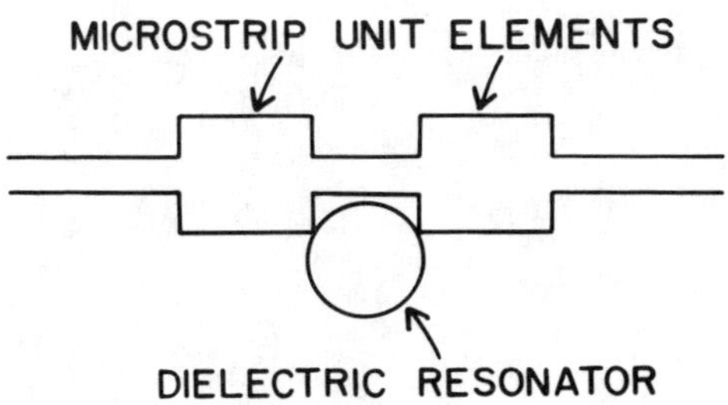

Fig. 9.27 Structure of the microstrip dielectric resonator filter

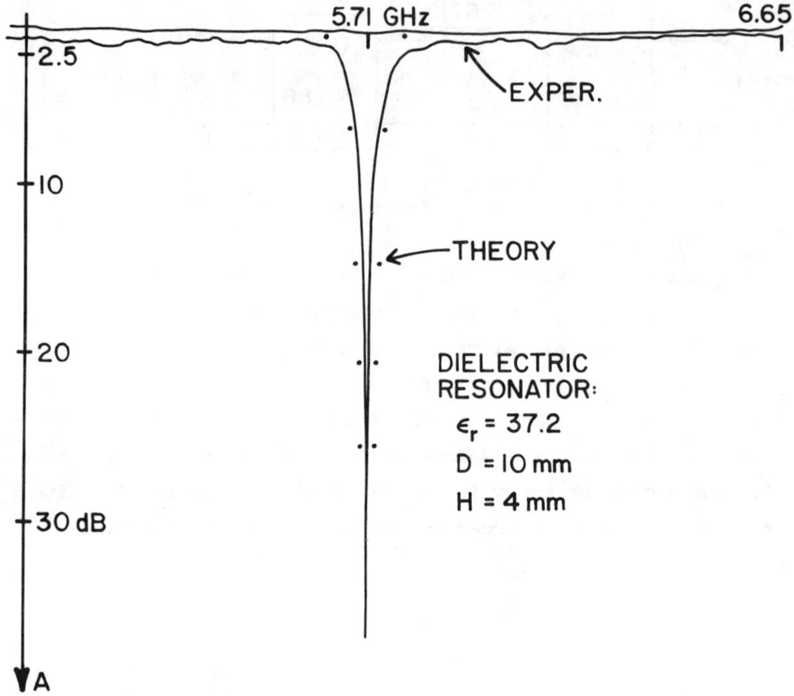

Fig. 9.28 Response of the band-pass filter (reference [13], ©1981 IEEE)

9.6 Other Band-pass DR Filters

An interesting feature occurs when a band-pass filter is cascaded at both ends by two band-stop filters. Their presence introduces two poles of finite attenuation located in the stop band of the band-pass filter response [5,15]. The filter response is shown in Fig. 9.29. The new filter has a higher cutoff attenuation rate (26 dB/10 MHz) compared to the original band-pass filter (14 dB/10 MHz). However, we can note that with the introduction of the band-stop filters, the insertion loss increases.

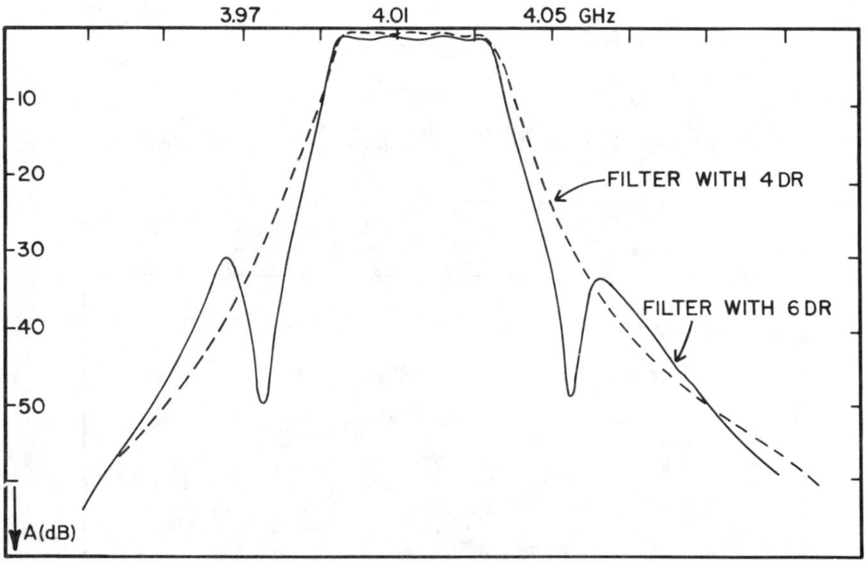

Fig. 9.29 Transmission response of the band-pass filter with a high attenuation cutoff rate (reference [26], ©1984 IEEE)

Using the results obtained on waveguide and microstrip structures, we have realized a filter at 22 GHz for which the input is a waveguide and the output is a microstrip line (see Fig. 9.30). The response of this filter is presented in Fig. 9.31.

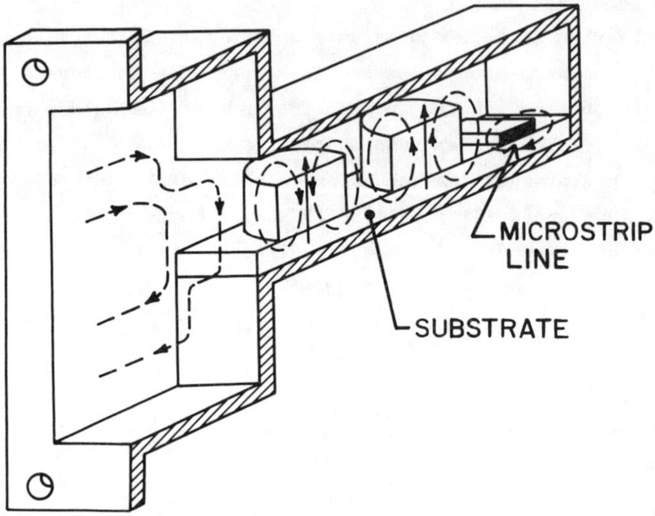

Fig. 9.30 Band-pass filter combining a waveguide and a microstrip line

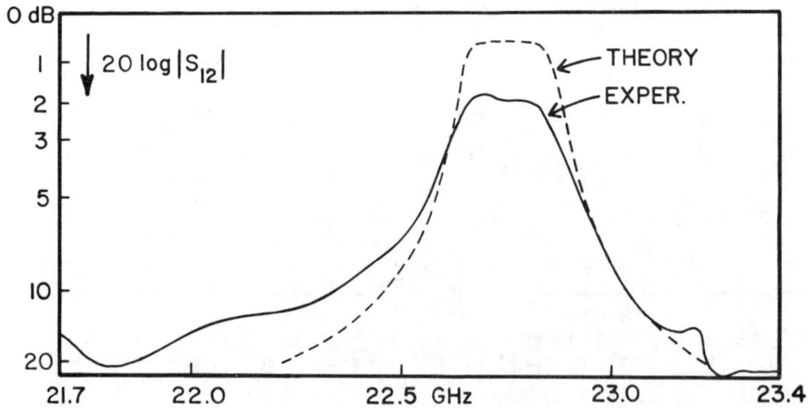

Fig 9.31 Transmission response of the filter centered around 22.5 GHz

9.7 Elliptic Band-pass DR Filters

Consider a coupled-cavity system shown in Fig. 9.32 consisting of four resonant cavities. In a simple cascaded configuration, there are only three coupling coefficients: k_{12}, k_{23}, and k_{34}, but if a new coupling k_{14} is introduced between the first and the fourth cavities which has opposite signs compared to these of k_{12}, k_{23}, and k_{34}, then it is possible to have a transmission zero at a finite frequency [16]. We can accomplish that coupling by using $TE_{01\delta}$ mode of a dielectric resonator in a metallic shield. We have demonstrated in Ch. 8 that such a

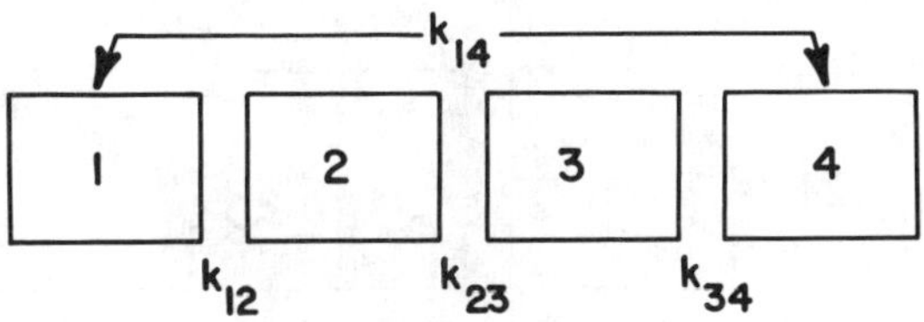

Fig. 9.32 System of four coupled cavities

coupling can be achieved by introducing a hole in the common wall of the metallic shield [17] (see Fig. 9.33). The negative coupling can be obtained by off setting the lower cavity axis with respect to the upper cavity axis. It is evident that an elliptical function filter can be realized with four circular cavities.

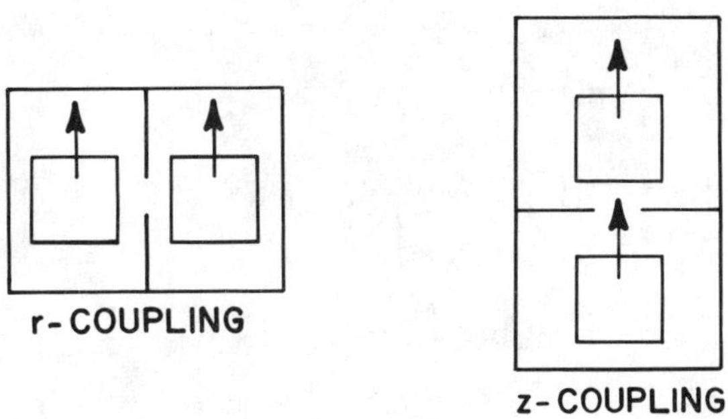

Fig. 9.33 Coupling between two cavities containing dielectric resonators

The synthesis of a filter is done on the basis of an equivalent circuit of four coupled cavities [18]. The equivalent circuit for each cavity will be a parallel circuit.

The dielectric resonators are operating in the $TE_{01\delta}$ mode. Therefore, the coupling is essentially due to magnetic field. The coupling holes between the cavities can be modeled by inductive susceptances. The equivalent circuit of the whole network is shown in Fig. 9.34. The coefficients M_{ij} characterize the coupling between the cavities. The filter elements are normalized to unity bandwidth.

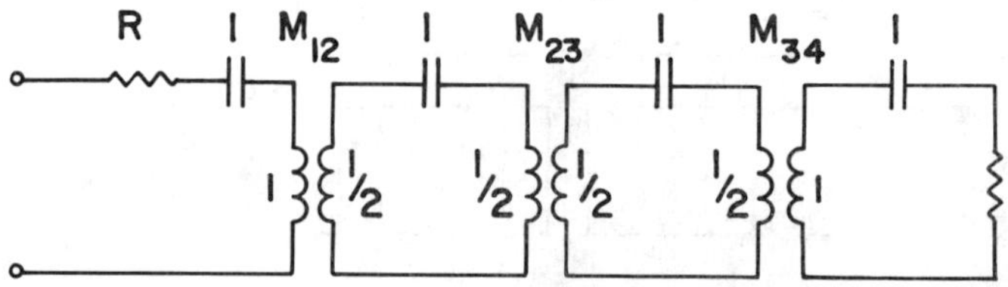

Fig. 9.34 Equivalent network of an elliptic filter

Analyzing the equivalent circuit with the aid of the impedance matrix [16], one obtains the values M_{ij}:

$$M_{12}^2 = RM_{23} + M_{14}\, M_{23} \tag{9.29}$$

$$M_{23} = \frac{\Delta\omega^2 \omega_p}{R\,\sqrt{2\epsilon}}$$

$$M_{14} = \frac{\Delta\omega^2}{2\epsilon R} \tag{9.31}$$

$$R = -\frac{\Delta\omega^2}{2}\left(\omega_z^2 - \frac{2\omega_p}{\sqrt{2\epsilon}}\right) \pm \frac{\Delta\omega^2}{2}\left[\left(\omega_z^2 - \frac{2\omega_p}{\sqrt{2\epsilon}}\right)^2 + 4\left(\frac{\omega_p^2}{2\epsilon} + \frac{1}{4\epsilon^2} - \frac{\omega_p}{\epsilon\sqrt{2\epsilon}}\right)\right]^{1/2} \tag{9.32}$$

The following symbols have been used above:

ω_p = pole frequency

ω_z = zero frequency

ϵ = ripple constant

$\Delta\omega$ = bandwidth

Relations (9.29) to (9.32) permit one to determine the coupling coefficients as functions of poles and zeros. The normalized transfer function of an elliptic filter is presented in Fig. 9.35.

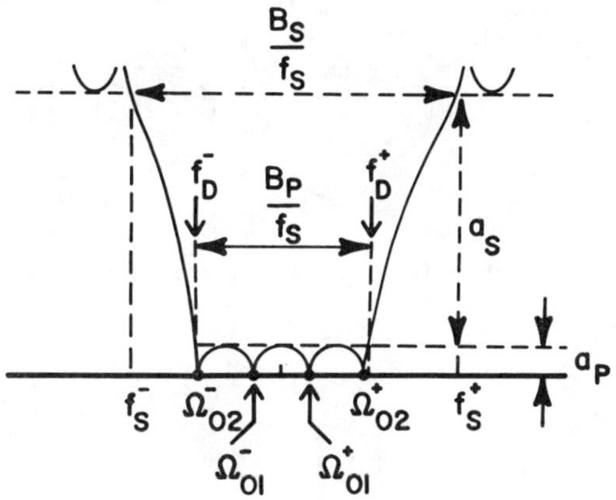

Fig. 9.35 Normalized transfer function of the elliptic filter

An elliptic band-pass filter will now be designed having the following specifications:

Pass-band edges: $f_D^- = 5640$ MHz ,

$f_D^+ = 5680$ MHz

Stop-band edges: $f_S^- = 5590$ MHz ,

$f_S^+ = 5730$ MHz

Reflection coefficient: $\rho = 20$ %

Maximum attenuation in the stop-band: $a_s = 40$ dB

Using these parameters we can evaluate the denormalized pole frequencies, which are, respectively, $f_\infty^+ = 5736.7$ MHz and $f_\infty^- = 5584.23$ MHz. Afterwards, the coefficients M_{ij} are computed.

The appearance of the filter is sketched in Fig. 9.36. Two of the cavities are coupled laterally with $M_{12} = M_{34}$ and the remaining two are coupled longitudinally (positive M_{23} and negative M_{14}). Figure 9.37 gives both the theoretical and experimental responses of the elliptic filter. The VSWR is about 1.2, whereas the insertion loss is quite high ($\simeq$ 1.2 dB).

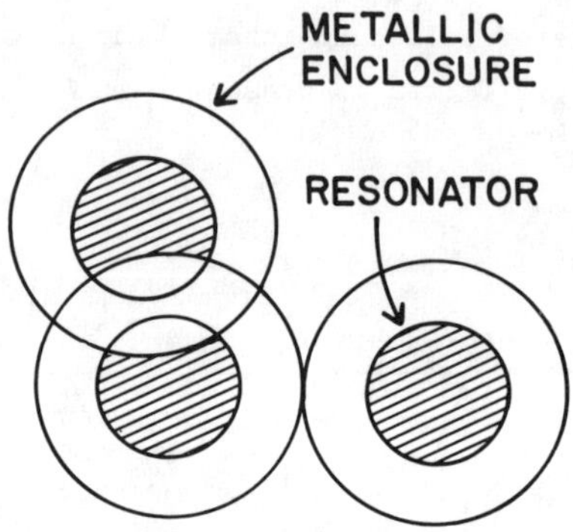

Fig. 9.36 Structure of a $TE_{0\lambda\delta}$ dielectric resonator mode elliptic filter

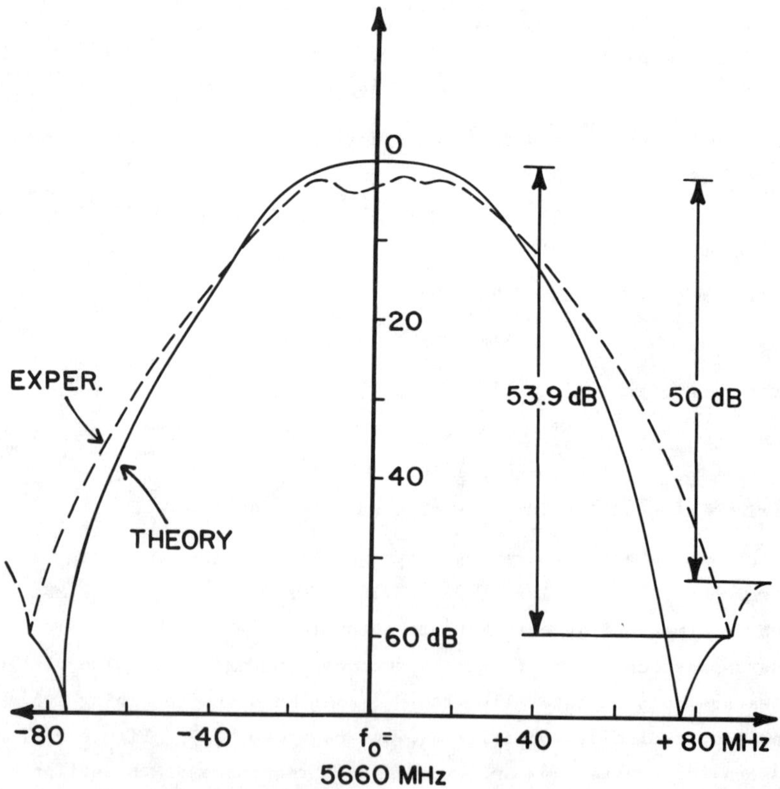

Fig. 9.37 Transmission response of the elliptic filter (reference [17], ©1980 Hirzel Verlag)

9.8 Dual-Mode DR Filters

Dual-mode operation is preferred in satellite applications. It allows the realization of high performance elliptic filters [19,20]. In a shielded cylindrical dielectric resonator, a number of degenerate modes with identical natural frequencies can be found. These degenerate modes can be mutually coupled by perturbing the rotational symmetry of the structure.

For that purpose, we use the hybrid mode $HEM_{11\delta}$, the field pattern of which is presented in Fig. 9.38. As discussed in Ch. 8, any perturbation of the cavity shape will destroy the independence of the modes and couple the energy between them.

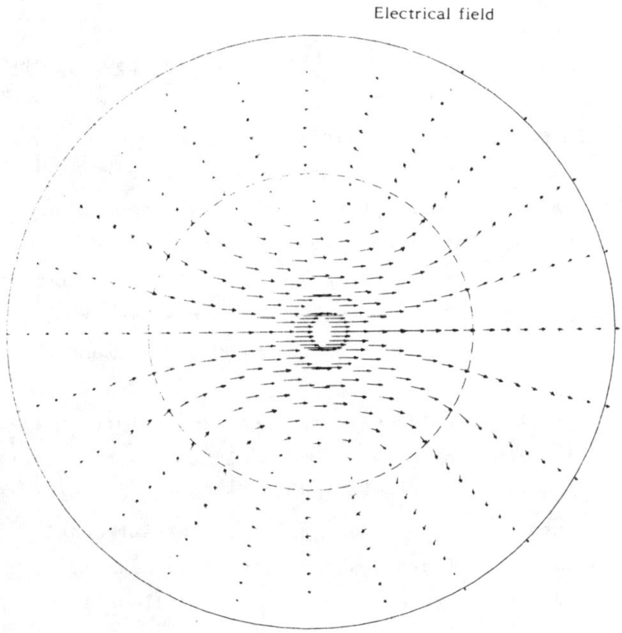

Fig. 9.38 Electrical field of the $HEM_{11\delta}$ mode of the shielded dielectric resonator

A two-pole band-pass filter using only one resonator has been realized. The coupling screw has been inserted into the cavity wall with an angular location of 45° with respect to the input probe, causing the energy within the resonator to split into a pair of orthogonal modes $HEM_{11\delta}$ (see Fig. 9.39).

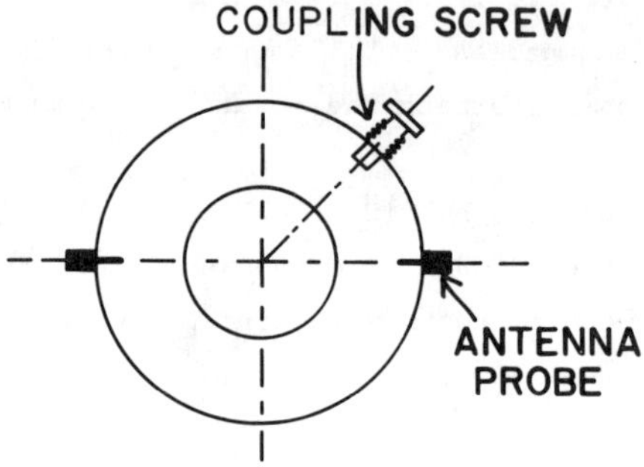

Fig. 9.39 Coupling of the dual $HEM_{11\delta}$ mode of the dielectric resonator

With the use of element values for the Butterworth low-pass prototype, (9.8), (9.9), and (9.10) provide the coupling coefficient between modes and between the input and output ports and the resonator. The comparison of these values with those obtained in Ch. 8 from the electromagnetic field analysis permit one to choose the diameter and depth combination for the coupling screw inserted in the shield.

The measured performance of the dual Butterworth filter is shown in Fig. 9.40. The resonator used for this realization has diameter D = 10 mm and height H = 30 mm. The insertion loss is about 1.5 dB, and the VSWR is about 1.3.

The dual-mode DR filters have also been designed to exhibit an elliptic behavior [20]. The coupling between modes within a single cavity is achieved via a mode coupling screw situated at 45° with respect to the antenna, as in the case of the dual-mode filter. The intercavity couplings are realized by means of cross slots, as can be seen in Fig. 9.41. The arrangement is similar to that used in the metal-wall cavity filter. The dielectric resonators are mounted axially in the center of each evanescent circular cavity.

A four-pole filter has been analyzed using the procedure described in Sec. 9.7. The results concerning the coupling coefficient of the mode $HEM_{11\delta}$ can be found in Ch. 8. In Fig. 9.42, we give the experimental response of the filter. The characteristics of the dielectric resonators are:

diameter = 16 mm

height = 8.5 mm

permittivity ϵ_r = 36

The insertion loss is about 0.8 dB, and the VSWR is 1.5.

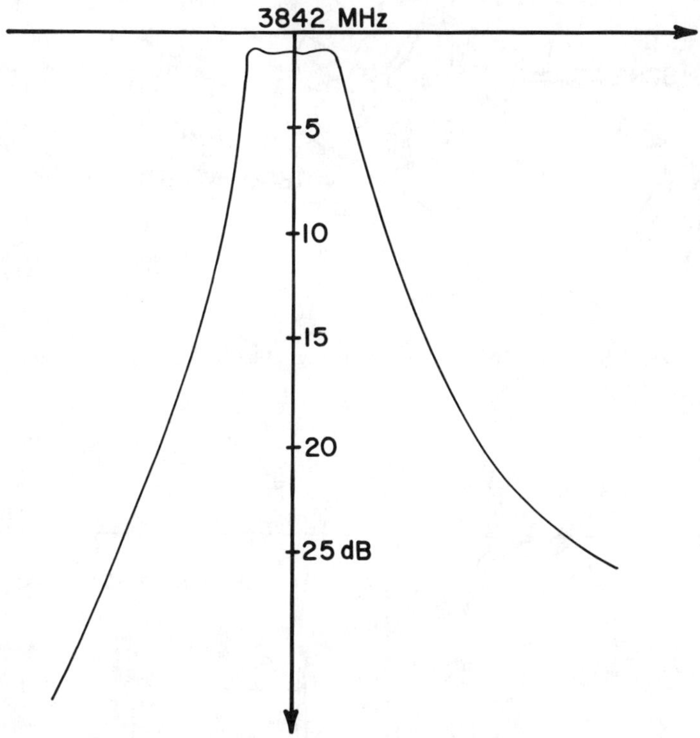

Fig. 9.40 Transmission response of a dual-mode filter (reference [19], ©1980 IEE)

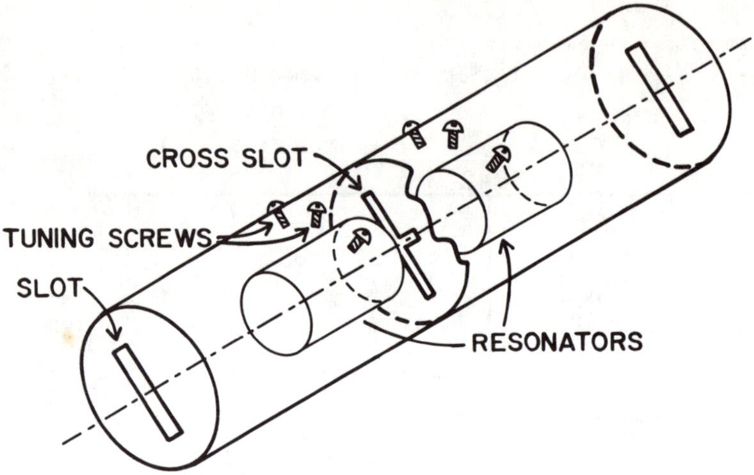

Fig. 9.41 Four-pole elliptic filter using the $HEM_{11\delta}$ mode of the dielectric resonator

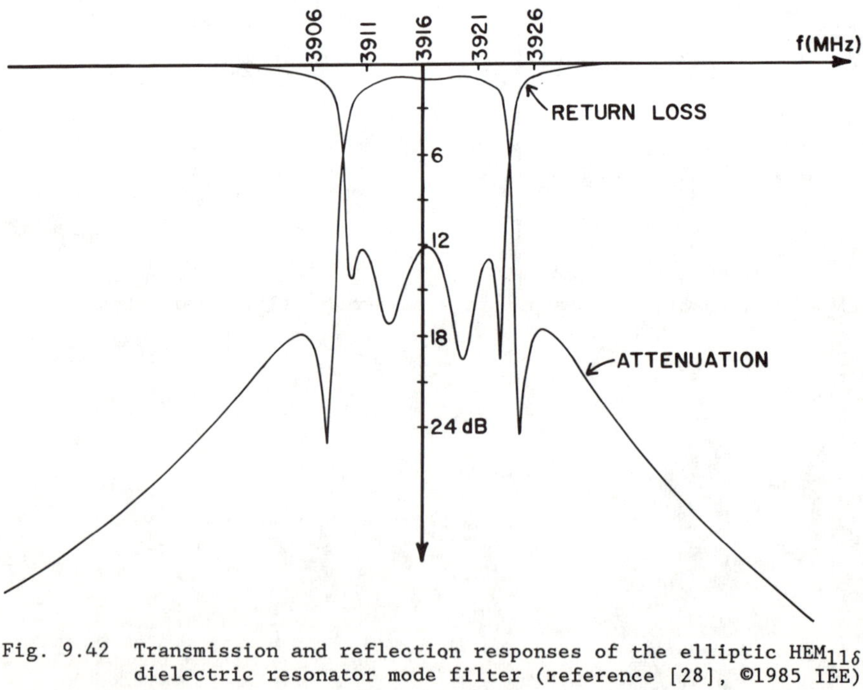

Fig. 9.42 Transmission and reflection responses of the elliptic $HEM_{11\delta}$ dielectric resonator mode filter (reference [28], ©1985 IEE)

9.9 Band-pass DR Filter for Mobile Communications

Dielectric resonator filters have already replaced the helical resonator filters traditionally used in mobile communications in the 800 MHz band. The cylindrical geometry has often been used to obtain the TEM mode in dielectric resonators [21,22]. However, we can also utilize dielectric resonators of rectangular cross section.

Figure 9.43 gives the sketch of a rectangular dielectric resonator utilized for this purpose. Faces 1, 2, and 3 are metallized, whereas face 4 may or may not be metallized. If face 4 is metallized, the

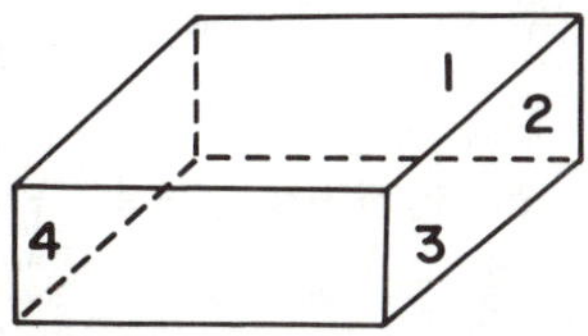

Fig. 9.43 Metallized rectangular dielectric resonator

length of the resonator is $\lambda g/2$; and if not metallized, the length is $\lambda g/4$. As shown in Fig. 9.44, several metallized dielectric resonators excited by a magnetic loop are placed in cascade to realize this filter.

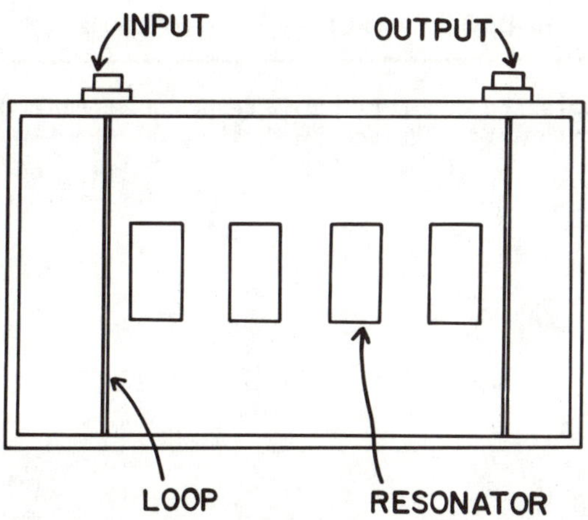

Fig. 9.44 TEM dielectric mode filter structure (reference [29])

In Fig. 9.45, we present the response of a filter centered around 855 MHz. The filter uses dielectric resonators of a square cross section 10 × 10 mm and of the length $\lambda g/4$.

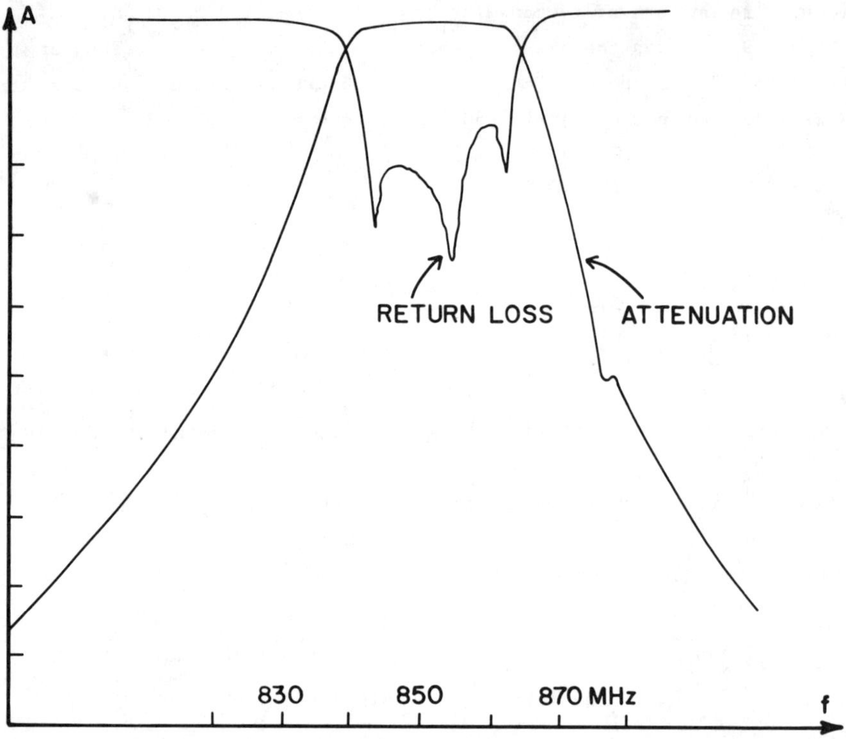

Fig. 9.45 Transmission and reflection response of the TEM dielectric resonator mode filter

9.10 Spurious Responses

The presence of higher resonant modes close to the resonant frequency of the principal mode interferes with the filter's performance [24]. The spurious modes whose resonant frequencies are close to the $TE_{01\delta}$ mode resonant frequency are either HEM_{mnp} or $TM_{01\delta}$ modes. When hybrid modes are used to realize a filter, other hybrid modes as well as $TE_{01\delta}$ and $TM_{01\delta}$ modes can interfere. Some resonant modes of the metallic enclosure (loaded with the dielectric resonator of high permittivity) can also be excited; these modes also perturb the response of the filter. Finally, we can also note that the presence of tuning screws, which are necessary to tune the resonant frequencies, can also excite spurious modes.

In Fig. 9.46, we present an example of the spurious modes that we encounter in a microstrip band-pass filter using the $TE_{01\delta}$ mode.

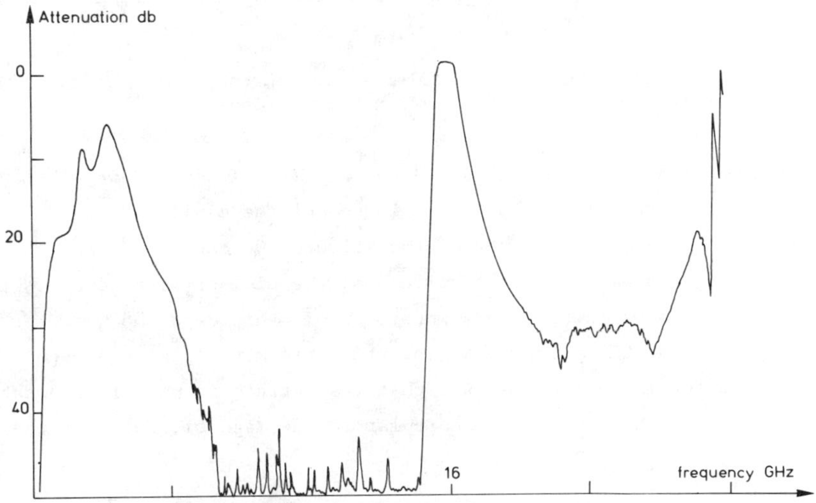

Fig. 9.46 Spurious modes in a $TE_{01\delta}$ dielectric resonator mode filter

Likewise, Fig. 9.47 depicts the spurious responses found in an elliptic filter using the mode $HEM_{11\delta}$.

To eliminate the interferences, the aspect ratio of the dielectric resonators (diameter to height) should be properly chosen to place the resonance of spurious modes outside the operating frequency band. The

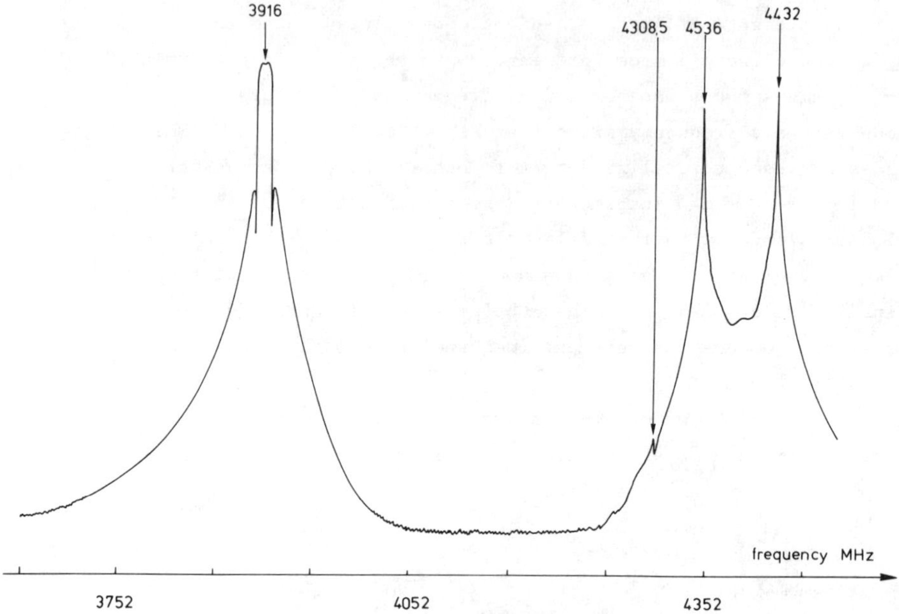

Fig. 9.47 Spurious modes in a hybrid dielectric resonator mode filter (reference [28], ©1985 Eur. Microwave Conf.)

dimensions of the metallic enclosure must be fixed in such a way that the cutoff frequency of the fundamental mode of the cavity be much higher than the center frequency of the filter. We can also place the mode-suppressing elements on the surface of the dielectric resonators to prevent resonance of the spurious modes, but in this case the Q factor is degraded. Ren [25] proposed to insert a thin metal plate between the adjacent dielectric resonators such that its surface is parallel to the coupling magnetic field and perpendicular to the magnetic field of the spurious modes.

References

[1] G.L. Matthaei, L. Young, and E. Jones, Microwave Filters, Impedance Matching Networks and Coupling Structures. New York: McGraw-Hill, 1965.

[2] Y. Konishi, "External Q of a $TE_{01\delta}$ dielectric resonator in a $TE_{1.0}$ waveguide," Electron. and Commun. Japan, vol. E59, p. 143, Jan. 1976.

[3] S.B. Cohn, "Microwave bandpass filters containing high Q dielectric resonators," IEEE Trans. Microwave Theory Tech., vol. MTT-16, no. 4, p. 218, April 1968.

[4] J.K. Ploude and C.L. Ren, "Applications of dielectric resonators in microwave components," IEEE Trans. Microwave Theory Tech., vol. MTT-29, no. 8, p. 754, Aug. 1981.

[5] P. Guillon and Y. Garault, "Coupling between a magnetic loop and a dielectric resonator housed in an evanescent waveguide," Proc. of the Int. Microwave Symp., p. 392, June 1982.

[6] T.D. Iveland, "Dielectric resonator filters for the application in microwave integrated circuits", IEEE Trans. Microwave Theory Tech., vol. MTT-19, no. 7, p. 643, July 1971.

[7] P.L. Clar, "The application of dielectric resonators to microwave integrated circuits," Proc. of the 1970 Int. Microwave Symp., p. 19, May 1970.

[8] A. Karp, H.J. Shaw, and D.K. Winslow, "Circuit properties of microwave dielectric resonators", IEEE Trans. Microwave Theory Tech., vol. MTT-16, no. 10, p. 810, Oct. 1968.

[9] W. Ren, "Dielectric resonator filters for communications system," Proc. of the 1981 Int. Microwave Symp., June 1981.

[10] J.K. Plourde and D.F. Linn, "Microwave dielectric resonator filters using the Ba_2 Ti_9 O_{20} ceramics," Proc. of the 1978 Int. Microwave Symp., p. 227, June 1978.

[11] C.L. Ren, "Waveguides bandstop filter utilizing Ba_2 Ti_9 O_{20} resonators," Proc. of the 1978 Int. Microwave Symp., p. 227, June 1978.

[13] S. Mekerta and P. Guillon, "Dielectric resonator bandstop filters," Proc. of the 1981 Int. Microwave Symp., p. 170, June 1981.

[14] P.I. Richards, "Resistor transmission line circuits," Proc. IRE, vol. 36, p. 217, Feb. 1948.

[15] R.C. Wang and C.L. Ren, "Dielectric resonator filters for communications systems," Proc. of the 1981 Nat'l Telecomm. Conf., New Orleans, p. 6.2.1, Dec. 1981.

[16] A.E. Atia and R.W. Newcomb, "Narrow band multiple coupled cavity synthesis," IEEE Trans. Circuits and Systems, vol. CAS-21, no. 5, p. 649-655, Sept. 1974.

[17] H.N.S. Sundara and P. Guillon, "Dielectric resonator filters," AEU, vol. 34, p. 63, March 1980.

[18] A.E. Atia and A.E. Williams, "General TE_{011} mode waveguide bandpass filters," IEEE Trans. Microwave Theory Tech., vol. 24, no. 10, p. 640, Oct. 1976.

[19] P. Guillon and Y. Garault, "Dielectric resonator dual modes filters," Electron. Lett., vol. 16, p. 646, August 1980.

[20] S.J. Fiedziusko, "Dual mode dielectric resonator loaded cavity filters", IEEE Trans. Microwave Theory Tech., vol. MTT-30, no. 9, p. 1311, Sept. 1982.

[21] K. Wakino, "Miniaturized bandpass filters using half wave dielectric resonators with improved spurious response," Proc. of the 1979 Int. Microwave Symp., p. 230, May 1979.

[22] Y. Kobayashi and S. Yoshida, "Bandpass filters using TM_{010} dielectric rod resonators," Proc. of the 1978 Int. Microwave Symp., p. 233, June 1978.

[23] K. Wakino, "Quarter wave dielectric transmission line diplexer for land mobile communications," Proc. of the 1975 Int. Microwave Symp., p. 278, June 1979.

[24] K. Wakino, "Microwave bandpass filters containing dielectric resonators with improved temperature stability and spurious response," Proc. of the 1975 Int. Microwave Symp., p. 63, June 1975.

[25] L. Ren, "Mode suppressors for dielectric resonator filters," Proc. of the 1982 Int. Microwave Symp., p. 389, June 1982.

[26] M. Chong and P. Guillon, "Dielectric resonators filter with a high attenuation cut-off rate," Proc. of the 1984 Int. Microwave Symp., p. 240, June 1984.

[27] M. Chaubet, "Elliptic bandpass filter using dielectric resonators in a microstrip structure," private communication, Limoges, France, Dec. 1984.

[28] J.P. Astier and P. Guillon, "Elliptic microwave filter using dual modes of dielectric resonators," Proc. of the European Microwave Conf., p. 335, Sept. 1985.

[29] J.C. Mage, "Dielectric resonator filter for mobile communications," private communication, L.C.R. Thomson CSF, Orsai, France, June 1983.

Chapter 10
OSCILLATORS
A.P.S. Khanna

10.1 Introduction

Microwave oscillators form an important part of all microwave systems such as those used in radar, communication links, navigation, and electronic warfare (EW). With the rapid advancement of technology there has been an increasing need for better performance of oscillators. The emphasis has been on low noise, small size, low cost, high efficiency, high temperature stability, and reliability. The transistor dielectric resonator oscillator (TDRO) presents an interesting solution as a quality oscillator for fixed frequency or narrowband tunable oscillators. The present commercially available gallium arsenide field effect transistor (GaAs FET) can be used as the active device in oscillators for the entire microwave frequency range from 1 to 30 GHz. With the advent of temperature stable materials, the dielectric resonator has emerged as a high Q, low loss, and conveniently sized element for applications in various microwave integrated circuits (MICs) for the entire microwave frequency range [1].

This chapter starts with the characterization of the dielectric resonator and the transistor. The oscillation conditions and the design of a TDRO using 3-port scattering matrix parameters (S-parameters) are then described. Temperature stability and the tunability of the TDRO are presented, followed by the measurement of TDRO characteristics.

10.2 S-parameter Characterization of a Microstrip-Coupled Dielectric Resonator

The dielectric resonator is coupled to a microstrip line for applications in different MIC's like filters, discriminators, oscillators, etc. The commonly used configuration is shown in Fig. 10.1. The dielectric resonator placed near the microstrip line on the substrate is enclosed in a metallic shielding box. The shielding conditions, as described in earlier chapters, affect the frequency and Q factor of the

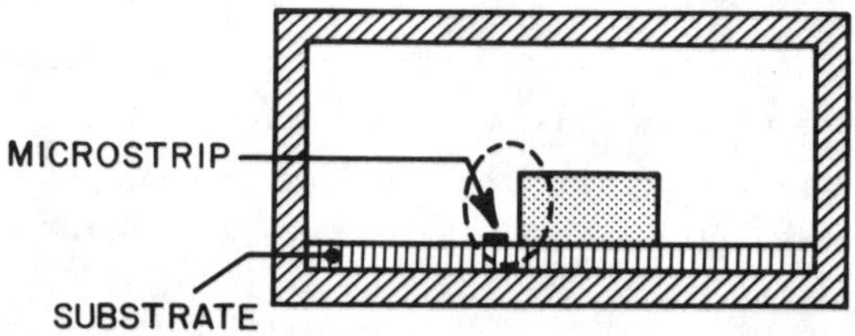

Fig. 10.1 DR in a typical MIC configuration

resonator. Figure 10.2 shows the equivalent circuit of the microstrip-coupled DR under given shielding conditions [2].

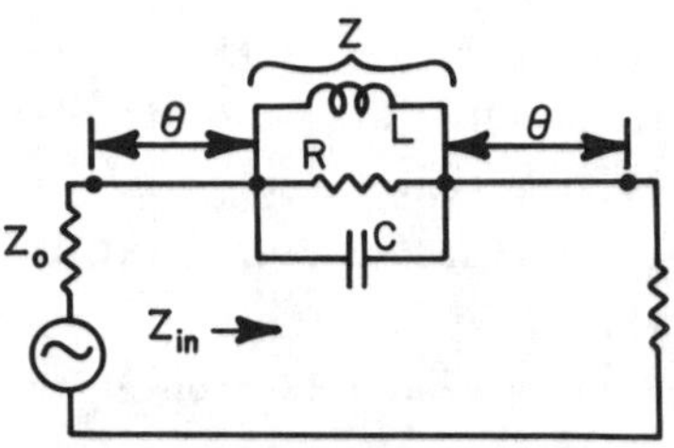

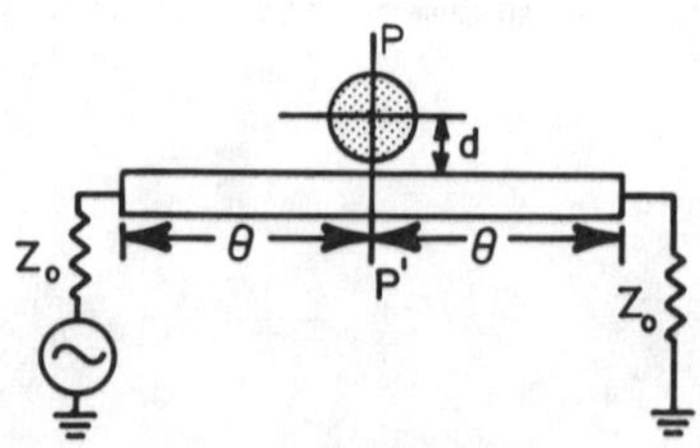

Fig. 10.2 DR coupled to microstrip line and its equivalent circuit (reference [3], ©1983 IEEE)

The coupling coefficient between the resonator and the transmission line is a function of the distance between the resonator and the line, and is given by

$$\kappa = \frac{R}{R_{ext}} = \frac{R}{2Z_0} = \frac{S_{110}}{S_{210}} \tag{10.1}$$

where S_{110} and S_{210} are the reflection and transmission coefficients, respectively, in the resonator plane PP′ at the resonant frequency. The normalized induced input impedance z can be given by

$$z = \frac{2\kappa}{1 + 2jQ_u\delta} + 1 \tag{10.2}$$

where

$$\delta = (f - f_0)/f_0 \tag{10.3}$$

Knowing

$$S_{11} = \frac{z - 1}{z + 1} \tag{10.4}$$

and

$$S_{11} + S_{21} = 1 \tag{10.5}$$

the S-parameters of the DR coupled to a microstrip line in the resonator plane can be written as

$$\underset{\sim}{S}_R = \begin{bmatrix} \dfrac{\kappa}{\kappa + 1 + j2Q_u\delta} & \dfrac{1 + j2Q_u\delta}{\kappa + 1 + j2Q_u\delta} \\ \\ \dfrac{1 + j2Q_u\delta}{\kappa + 1 + j2Q_u\delta} & \dfrac{\kappa}{\kappa + 1 + j2Q_u\delta} \end{bmatrix} \tag{10.6}$$

At resonance ($\delta = 0$), the scattering matrix (S-matrix) becomes

$$\underset{\sim}{S}_{R0} = \begin{bmatrix} \kappa(\kappa + 1)^{-1} & (1 + \kappa)^{-1} \\ \\ (\kappa + 1)^{-1} & \kappa(1 + \kappa)^{-1} \end{bmatrix} \tag{10.7}$$

The effect of the transmission line length (in the input and output planes, Fig. 10.2) on the S-parameter can be given as

$$S_{ij} = S_{ij}e^{-2j\theta} \tag{10.8}$$

It may be noted that a DR coupled to a microstrip line represents a band-stop filter configuration. The insertion loss, which is an increasing function of κ, is given by

$$L_0 = 20 \log (1 + \kappa) \tag{10.9}$$

The coupling coefficient relates the different quality factors in the following way:

$$Q_u = Q_L(1 + \kappa) = \kappa Q_e \tag{10.10}$$

where Q_u, Q_L, and Q_e are the unloaded, loaded, and external quality factors of the DR coupled to the microstrip line under given shielding conditions. These quality factors as well as the coupling coefficient κ can be determined from the network analyzer display of S_{11} or S_{21}. Figure 10.3 represents theoretical curves of S_{11} and S_{21} for different values of the coupling coefficient κ. While κ can be calculated using (1), the loci for the determination of the various quality factors (shown as dashed curves in Fig. 10.3) can be drawn using the following approach [3]. The normalized input impedance z_{in} from Fig. 10.2(b) can be written as

$$z_{in} = 1 + \frac{2\kappa}{1 + 2jQ_u\delta} = 1 + \frac{2\kappa}{1 + 2Q_L(1 + \kappa)\delta} = 1 + \frac{2\kappa}{1 + 2jQ_e\kappa\delta} \tag{10.11}$$

The frequency deviations corresponding to different quality factors are given by

$$\delta_u = \pm \frac{1}{2Q_u}, \quad \delta_L = \pm \frac{1}{2Q_L}, \quad \delta_e = \pm \frac{1}{2Q_e} \tag{10.12}$$

Using (10.12) in (10.11), the Q_u impedance locus is given by

$$z_{in}(u) = 1 + \frac{2\kappa}{1 \pm j} \tag{10.13}$$

The corresponding relations in the S_{11} and S_{21} plane can be obtained using (10.4) and (10.5). The relations for Q_L and Q_e can be calculated using the same approach. Reference [3] provides the necessary formulas

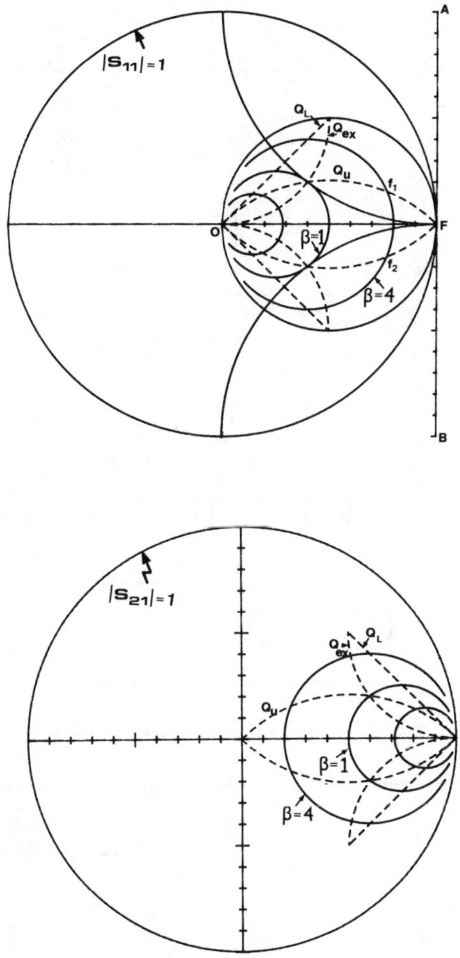

Fig. 10.3 Quality factor determination from a) S_{11} and b) S_{21} measurement (reference [3], ©1983 IEEE)

to draw the loci for Q_u, Q_L, and Q_e on the S_{11} and S_{21} planes, as shown in Fig. 10.3. These loci can be directly used to measure the desired quality factor. The loaded and unloaded Q factors can also be found using the magnitude of the transmission coefficient S_{21} around the resonant frequency, measured with a scalar network analyzer. Using the magnitude relations for S_{21u} and S_{21L} [3], and knowing that the insertion loss is

$$L_0(\text{dB}) = 20 \log S_{210}$$

it becomes possible to determine the quantity x (see Fig. 10.4(a)), needed for the measurement of Q_u and Q_L, in the following way [4]:

$$x(\text{dB}) = 3 - 10 \log \left(1 + 10^{-0.1L_0}\right) \tag{10.14}$$

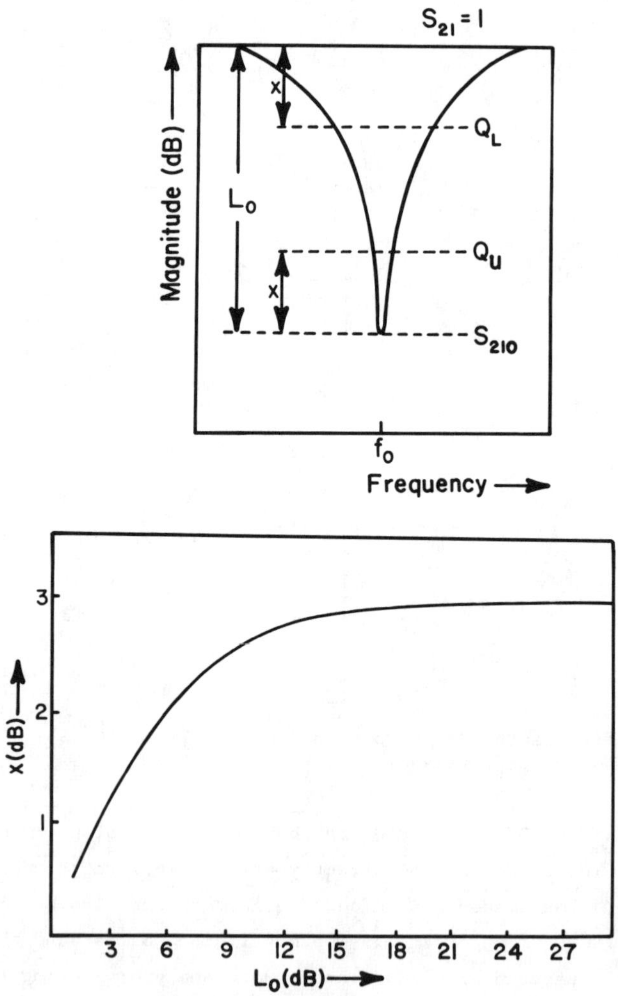

Fig. 10.4 Q_u determination from scalar measurement of S_{21} (reference [4], ©1984 Microwaves & RF)

Figure 10.4b illustrates the above relationship and shows that for a high value of L_0 (> 17 dB) the Q_u and Q_L measuring points approach a 3 dB separation from S_{210} and $S_{21} = 1$, respectively. The coupling coefficient κ can also be determined from the measurement of L_0 by

$$\kappa = 10^{L_0/20} - 1 \tag{10.15}$$

A DR shown in Fig. 10.5 is coupled simultaneously to two microstrip lines. Such a configuration represents a band-pass filter [5]. This network will be used in realizing transistor DRO's of the parallel feedback type, discussed later in this chapter. The equivalent circuit is presented in Fig. 10.6. The distances ℓ_1 and ℓ_2 (Fig. 10.5) are kept equal to one-quarter guided wavelength long in order to create an effective short circuit at the plane AA', necessary for magnetic coupling of the resonator to the microstrip line in the $TE_{01\delta}$ mode.

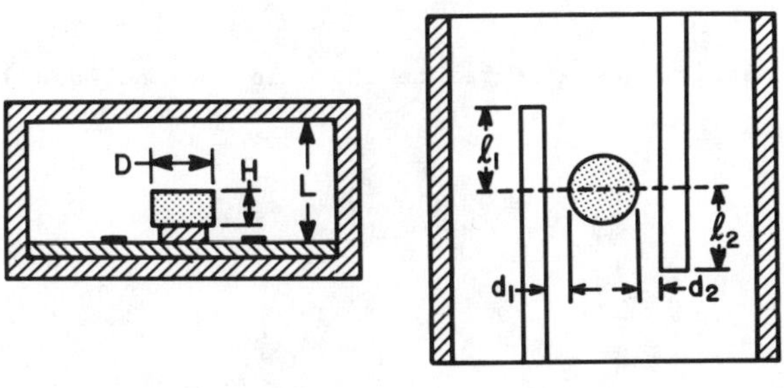

Fig. 10.5 DR coupled simultaneously to two microstrip lines (reference [7], ©1983 Artech House)

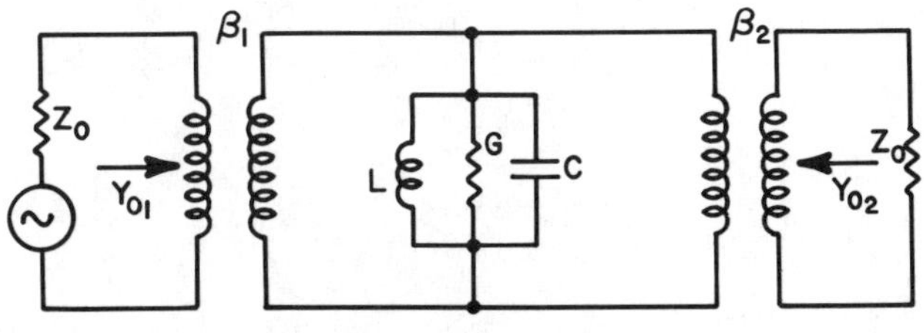

Fig. 10.6 Equivalent circuit of a DR coupled to two microstrip lines (reference [7], ©1983 Artech House)

The S-matrix of this configuration at the resonator plane can be represented by

$$\underset{\sim}{S}_R = \begin{pmatrix} \dfrac{\kappa_1 - \kappa_2 - 1 - j2Q_u\delta}{1 + \kappa_1 + \kappa_2 + j2Q_u\delta} & \dfrac{2\kappa_1\kappa_2}{1 + \kappa_1 + \kappa_2 + j2Q_u\delta} \\ \dfrac{2\kappa_1\kappa_2}{1 + \kappa_1 + \kappa_2 + j2Q_u\delta} & \dfrac{\kappa_2 - \kappa_1 - 1 - j2Q_u\delta}{1 + \kappa_1 + \kappa_2 + j2Q_u\delta} \end{pmatrix} \qquad (10.16)$$

where κ_1 and κ_2 represent the coupling coefficients of the DR with the input and output microstrip lines, respectively, and are given by

$$\kappa_1 = \frac{n_1^2 R}{Z_{01}} \qquad (10.17)$$

$$\kappa_2 = \frac{n_2^2 R}{Z_{02}} \qquad (10.18)$$

The coupling coefficients also relate the unloaded and loaded quality factors by

$$Q_0 = Q_L\,(1 + \kappa_1 + \kappa_2) \qquad (10.19)$$

10.3 3-Port S-parameter Characterization of Transistors

Transistor, a 3-port device, is generally characterized by its 2-port S-parameters with one of its ports grounded. The resulting three different configurations in the case of a GaAs FET are shown in Fig. 10.7. Each configuration has its own advantages (e.g., the common source configuration is used more often for amplifiers, common gate for wideband oscillators, and common drain for medium power oscillators).

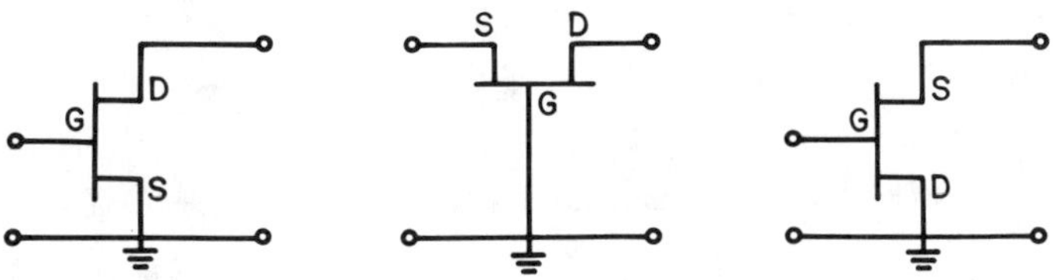

Fig. 10.7 Three configurations of the transistor: (a) common source, (b) common gate, and (c) common drain

The use of 3-port S-parameters, although introduced quite sometime back [6], has not often been used due to the complexity of analysis involved. The availability of desktop computers and computer assisted design (CAD) have now made their use practical. The use of 3-port S-parameters eliminates the otherwise necessary conversion to and from Z and Y parameters to analyze the series and parallel feedback effect as shown later in this chapter. The 3-port indefinite S-matrix of the transistor holds the property of having the sums of the rows and columns to be equal to 1 which helps in determining and eventually correcting the systematic errors in the measurement or analysis. The use of the 3-port S-parameters of the transistor sometimes become essential in the design of oscillators [5].

The transistor as a 3-port device is shown in Fig. 10.8 and the S-matrix of the incident and reflected waves is given by

$$\begin{bmatrix} b_1 \\ b_2 \\ b_3 \end{bmatrix} = \begin{bmatrix} S_{11} & S_{12} & S_{13} \\ S_{21} & S_{22} & S_{23} \\ S_{31} & S_{32} & S_{33} \end{bmatrix} \begin{bmatrix} a_1 \\ a_2 \\ a_3 \end{bmatrix} \tag{10.20}$$

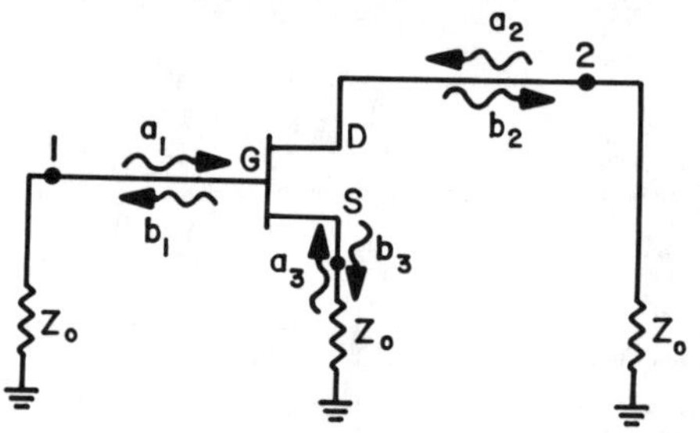

Fig. 10.8 Transistor as a three-port device

The ports 1, 2, and 3 represent gate, drain, and source, respectively. This indefinite 3-port S-matrix satisfies the following conditions:

$$\sum_{j=1}^{3} S_{ij} = 1, \quad \text{for } i = 1,2,3 \tag{10.21}$$

$$\sum_{i=1}^{3} S_{ij} = 1, \quad \text{for } j = 1,2,3 \tag{10.22}$$

The 3-port S-parameters of the transistor can be directly measured or obtained analytically from the commonly supplied 2-port S-parameters using the relations derived in Appendix 10.A. As an example, the 2-port S-parameters of NEC70000 at 10 GHz given by the manufacturer and the 3-port S-parameters calculated using these relations are given below:

$$\underset{\sim}{S}_{2P} = \begin{bmatrix} .73\underline{/-102°} & .1\underline{/42°} \\ 2.23\underline{/96°} & .54\underline{/-49°} \end{bmatrix}$$

$$\underset{\sim}{S}_{3P} = \begin{bmatrix} .85\underline{/-56.6^\circ} & .29\underline{/66.4^\circ} & .61\underline{/47.3^\circ} \\ 1.26\underline{/104.7^\circ} & .83\underline{/-37.1^\circ} & .97\underline{/-47.7^\circ} \\ .99\underline{/-30.9^\circ} & .32\underline{/47.2^\circ} & .28\underline{/104.2^\circ} \end{bmatrix}$$

10.4 Oscillation and Stability Conditions [7]

Any oscillator can be represented in an arbitrary plane on the output line by a nonlinear impedance Z_{NL}, having a negative real part, in series with a load impedance Z_L (Fig. 10.9). We assume that the circuit has a sufficiently high Q factor to suppress the harmonic currents.

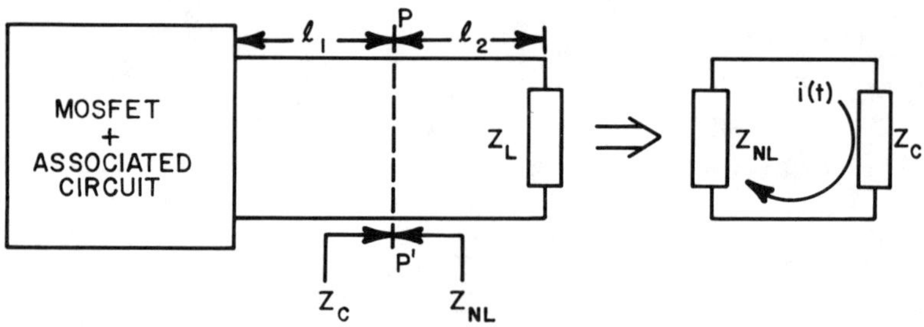

Fig. 10.9 Nonlinear microwave oscillator (reference [7], ©1983 Artech House)

Supposing that a current

$$i(t) = I_o \cos \omega t$$

exists in the circuit shown in Fig. 10.9, we can apply the Kirchhoff voltage law and write in the plane PP':

$$\left[Z_{NL}\ (I_o, \omega_o) + Z_L\ (\omega_o)\right] I_o = 0 \tag{10.23}$$

Let

$$Z_{NL} + Z_L = Z_T = R_T + jX_T \tag{10.24}$$

Since I_o is not equal to zero, (10.22) is satisfied by

$$R_T\ (I_o, \omega_o) = 0 \qquad (10.25$$

and

$$X_T\ (I_o, \omega_o) = 0 \tag{10.26}$$

Since $\mathrm{Re}(Z_L) > 0$, (10.25) implies that $\mathrm{Re}(Z_{NL}) < 0$. Hence, the device needs to present a negative resistance in order to be able to oscillate.

The frequency of oscillation is determined by (10.26), i.e., by the requirement that the load reactance be equal and opposite to the device reactance.

Oscillators can also be represented by a nonlinear admittance Y_L. The oscillation conditions in this case can be determined in the same way to be

$$G_T\ (V_o,\ \omega_o) = 0 \tag{10.27}$$

and

$$B_T\ (V_o,\ \omega_o) = 0 \tag{10.28}$$

At microwave frequencies, it is more convenient to express (10.25) to (10.28) in terms of the corresponding reflection coefficient Γ_{NL} and Γ_L as:

$$|\Gamma_{NL}| \cdot |\Gamma_L| = 1 \tag{10.29}$$

$$\underline{/\Gamma_{NL}} + \underline{/\Gamma_L} = 2\pi \tag{10.30}$$

Relation (10.29) implies that the device reflection coefficient Γ_{NL} modulus should be greater than unity.

An oscillator can be considered as a combination of an active multiport and a passive multiport (the embedding circuit), as shown in Fig. 10.10. With the active device and the embedding circuit characterized

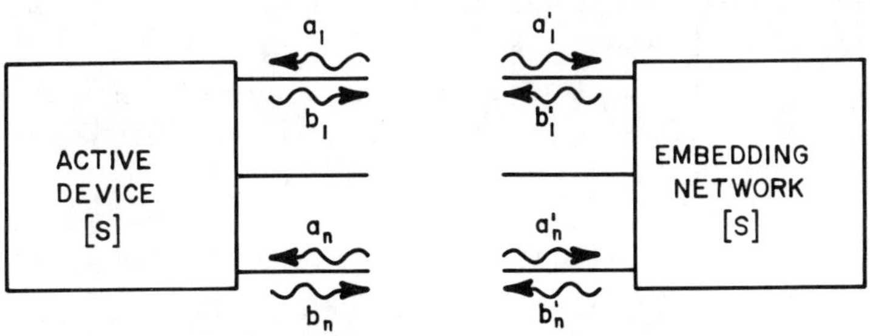

Fig. 10.10 Generalized oscillator configuration (reference [7], ©1983 Artech House)

by their scattering matrix, we have, for the active device [8]:

$$|b\rangle = \underset{\sim}{S}\ |a\rangle \tag{10.31}$$

and for the embedding circuit:

$$|b'\rangle = \underset{\sim}{S}'\ |a'\rangle \tag{10.32}$$

When the active device and the embedding network are connected together, we have, for the oscillation conditions:

$$|b'\rangle = |a\rangle \tag{10.33}$$

and

$$|b\rangle = |a'\rangle \tag{10.34}$$

From (10.31) to (10.34), we can write:

$$|a'\rangle = \underset{\sim}{S}\ \underset{\sim}{S}'\ |a'\rangle \tag{10.35}$$

or

$$(\underset{\sim}{S}\ \underset{\sim}{S}' - \underset{\sim}{1})\ |a\rangle = |0\rangle \tag{10.36}$$

where $\underset{\sim}{1}$ is an identity matrix. Since $|a'\rangle \neq |0\rangle$, it follows that

$$\underset{\sim}{M} = \underset{\sim}{S}\ \underset{\sim}{S}' - \underset{\sim}{1}$$

is a singular matrix, or

$$\det \underset{\sim}{M} = 0 \tag{10.37}$$

Equation (10.37) represents the generalized large-signal oscillation condition for an n-port oscillator.

In fact, the scattering matrix of the active device being defined at small signal level, the n-port oscillation condition at small signal can be represented by

$$\left|\det (\underset{\sim}{S}\ \underset{\sim}{S}' - \underset{\sim}{1})\right| > 0 \tag{10.38}$$

and

$$\operatorname{Arg} \det (\underset{\sim}{S}\ \underset{\sim}{S}' - \underset{\sim}{1}) = 0 \tag{10.39}$$

The oscillations can start as soon as the above relations are satisfied and go on building up until the device nonlinearities cause a steady state to be reached. As an example, consider an active 2-port loaded by two passive impedances, as shown in Fig. 10.11. The active device is

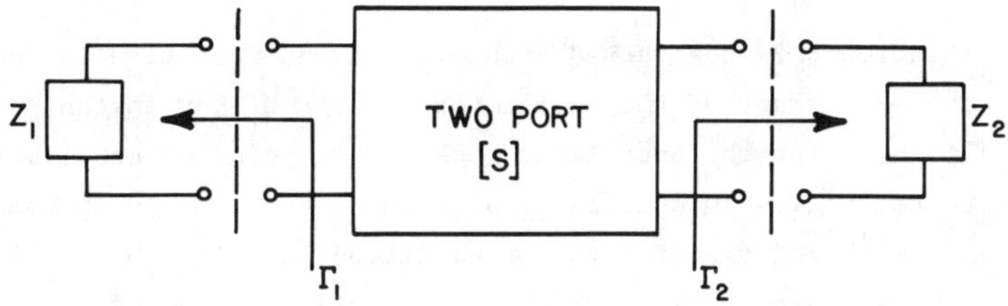

Fig. 10.11 Two-port loaded with two impedances (reference [8], ©1981 IEEE)

described by the scattering matrix:

$$\underset{\sim}{S} = \begin{bmatrix} S_{11} & S_{12} \\ S_{21} & S_{22} \end{bmatrix} \tag{10.40}$$

and the embedding circuit by

$$\underset{\sim}{S}' = \begin{bmatrix} \Gamma_1 & 0 \\ 0 & \Gamma_2 \end{bmatrix} \tag{10.41}$$

The oscillation condition from (10.37) is

$$\det \underset{\sim}{M} = \det \begin{bmatrix} S_{11}\Gamma_1 - 1 & S_{12}\Gamma_2 \\ S_{21}\Gamma_1 & S_{22}\Gamma_2 - 1 \end{bmatrix} = 0 \tag{10.42}$$

which gives

$$(S_{11}\Gamma_1 - 1)(S_{22}\Gamma_2 - 1) - S_{12}S_{21}\Gamma_1\Gamma_2 = 0 \tag{10.43}$$

From the above, we obtain the following two well-known conditions, which are simultaneously satisfied for realizing oscillations [9]:

$$S_{11} + \frac{S_{12}S_{21}\Gamma_2}{1 - S_{22}\Gamma_2} = \frac{1}{\Gamma_1} \tag{10.44}$$

$$S_{22} + \frac{S_{12}S_{21}\Gamma_1}{1 - S_{11}\Gamma_1} = \frac{1}{\Gamma_2} \tag{10.45}$$

The oscillations are considered stable if any perturbation in the RF voltage or the RF current of the oscillator circuit at any instant decays itself, bringing the oscillator back to its point of equilibrium. The oscillator stability is analyzed [10] using a quasistatic approach by applying a small perturbation to the amplitude I_0 (Fig. 10.9). The impedance Z_T defined in (10.24), which is a function of I_0 and the complex frequency p, is developed in a Taylor series about I_0, $j\omega_0$. Since the perturbed current is non-zero, i.e., the oscillations continue to exist after the perturbation, we should have in the plane PP':

$$Z_T\,(I_0,\ \omega_0) + \frac{\partial Z_T}{\partial p}\,\delta p + \frac{\partial Z_T}{\partial I_0}\,\delta I_0 = 0 \tag{10.46}$$

and since $Z_T(I_0,\ j\omega_0) = 0$, we get

$$\delta p = -j\,\frac{\dfrac{\partial Z_T}{\partial I_0} \cdot \dfrac{\partial Z_T}{\partial \omega}^{*}}{\left|\dfrac{\partial Z_T}{\partial \omega}\right|^2} \cdot \delta I_0 \tag{10.47}$$

The increment δp can be decomposed into its real and imaginary parts:

$$\delta p = \alpha + j\delta\omega$$

The oscillator will be stable if, for a positive variation of the current amplitude, the real part of the variation of the complex frequency is negative, i.e., if α is negative, indicating a decreasing wave, returning to its point of equilibrium I_o.

From the above expression for δp, this condition is realized for

$$\frac{\partial R_T}{\partial I_0} \cdot \frac{\partial X_T}{\partial \omega} - \frac{\partial X_T}{\partial I_0} \cdot \frac{\partial R_T}{\partial \omega} > 0 \tag{10.48}$$

This relation represents the stability condition of an oscillator around an amplitude I_0 and angular frequency $j\omega_0$. Moreover, from (10.47), the imaginary part $\delta\omega$ vanishes if the condition:

$$\frac{\partial R_T}{\partial I_0} \cdot \frac{\partial R_T}{\partial \omega} + \frac{\partial X_T}{\partial I_0} \cdot \frac{\partial X_T}{\partial \omega} = 0 \qquad (10.49)$$

is satisfied. This indicates that a variation of amplitude δI_0 will not result into the variation of the oscillator's real angular frequency ω_0. From (10.49) it can also be deduced that, for maximum stability, the device line $Z_T(I_0)$ and load impedance $Z_L(\omega)$ should intersect at right angles at the oscillation equilibrium point I_0, ω_0.

10.5 Stabilized Transistor DRO Design

The origin of solid-state microwave oscillators using Gunn and impact avalanche transit time (IMPATT) diodes dates back to the late 1960s, before which microwave sources used to be massive klystron or magnetron tubes requiring huge power supplies. In less than two decades, solid state oscillators have come a long way. The extension of bipolar transistor oscillators to microwave frequencies and the development of GaAs MESFET devices in the early 1970s has made available to us today, highly cost effective, miniature, reliable, and low noise sources for use right up to the millimeter wave frequency range.

Dielectric resonators due to their high Q, small size, and excellent integrability in microwave integrated circuits (MICs), can be directly used as a frequency-determining element for realizing a stable MIC transistor oscillator. With the recent advent of temperature stable material for dielectric resonators, TDRO is fast becoming an automatic choice for a vast number of applications.

The transistor oscillators can be realized using either bipolar or GaAs FET devices. Bipolar oscillators have a maximum oscillation frequency lower than that of GaAs FET oscillators, while the latter are noisier than the former. GaAs FET oscillators have been reported up to 60 GHz, while oscillators using bipolar transistors have not been reported beyond X-band. Typically, a bipolar oscillator has 6 to 10 dB less frequency modulation (FM) noise close to the carrier as compared to the GaAs FET oscillator. The oscillator design approach presented in this chapter is applicable to oscillators using either type of transistor, although the examples shown are only of GaAs FETs.

A dielectric resonator can be used in two different ways to realize a stable MIC source:

i) As a passive stabilization element coupled appropriately to a free-running transistor oscillator. The oscillator thus obtained can be called a "dielectrically stabilized oscillator" [11].

ii) As a circuit element (e.g., in the feedback or matching network) in the transistor oscillator circuit to determine the oscillation frequency. The oscillator thus obtained can be called "stable transistor dielectric resonator oscillator" [12-15].

The passive stabilization is possible only for those free-running oscillators whose oscillation frequency is sensitive to the variation in the load impedance, in other words, the oscillators with a poor pulling figure. The dielectric resonator is placed in a particular plane with respect to the oscillator plane such that the effective Q factor of the oscillator is increased in such a way as to increase the frequency stability at the cost of higher output power. Once the oscillator is locked onto the resonator frequency, any frequency perturbation will entail a change in the load impedance, which will readjust the oscillation frequency due to the pulling effect.

The dielectric resonator is generally used in the reaction mode to realize a stabilized TDRO, as shown in Fig. 10.12(a). The dielectric resonator coupled to the microstrip line is placed at one-quarter or one-half wavelength from the free running oscillator output plane, depending on the output circuit requirements for the active device, i.e., needing a $dX/d\omega$ positive or $dB/d\omega$ positive, respectively. The analysis of the two cases being similar, we will consider the case for $\theta = \pi$ (half-wavelength).

Using (10.2), the normalized input admittance of the stabilization circuit in Fig. 10.2 can be written as

$$y_{in} = \frac{1}{1 + 2\kappa/(1 + j\Delta)} \tag{10.50}$$

where $\Delta = 2Q_u\delta$. The admittance may be divided into its real and imaginary parts:

$$\mathrm{Re}(y_{in}) = \frac{1 + 2\kappa + \Delta^2}{(1 + 2\kappa)^2 + \Delta^2} \tag{10.51}$$

$$\mathrm{Im}(y_{in}) = \frac{2\kappa\Delta}{(1 + 2\kappa)^2 + \Delta^2} \tag{10.52}$$

The value of the last is plotted in Fig. 10.12(b). The stabilized frequency range δ_s over which dB/df is positive can be calculated by differentiating and equating (10.52) to zero as:

$$\frac{dB}{df} = \frac{4Q_u\kappa[(1 + 2\kappa)^2 - \Delta^2]}{f_0[(1 + 2\kappa)^2 + \Delta^2]} = 0 \tag{10.53}$$

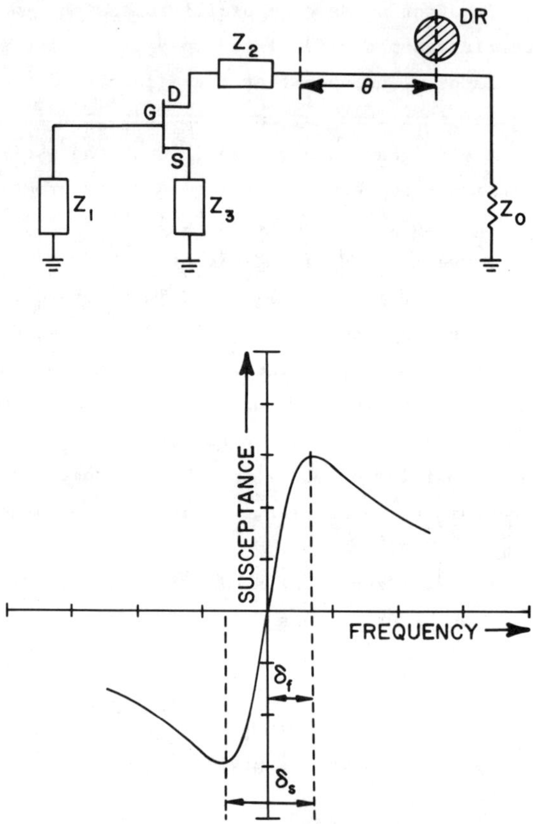

Fig. 10.12 Passive DR stabilization of a transistor oscillator: (a) configuration and (b) stabilization range (reference [7], ©1983 Artech House)

This gives us:

$$\Delta = \pm\ (1 + 2\kappa) \tag{10.54}$$

or

$$\delta_f = \pm\ \frac{f_0}{2Q}\ (1 + 2\kappa) \tag{10.55}$$

The stabilization bandwidth $\delta_S = 2\delta_f$, in terms of reflection coefficient S_{110}, can be given by

$$\delta_S = \frac{f_o}{Q_u}\left(\frac{1 + S_{110}}{S_{210}}\right) \tag{10.56}$$

This stabilization method reduces the useful RF output power in two ways. First, the free-running oscillator power is a function of the load admittance Y_L and is generally maximum for $Re(Y_L) = 1$. From (10.51), at the resonance frequency, the load admittance presented by the stabilization circuit (Fig. 10.2) is given by

$$Re(Y_L) = \frac{1}{1 + 2\kappa} = \frac{S_{210}}{1 + S_{110}} \tag{10.57}$$

This results in the decrease of power output given by

$$L_1 = P_0\left(\frac{\left(\frac{Y_0 S_{210}}{1 + S_{110}}\right)}{P_0(Y_0)}\right) \tag{10.58}$$

Second, a part of the oscillator output is also lost as the insertion loss (S_{21}) of the dielectric resonator coupled to the microstrip line as discussed in Sec. 10.2. This loss L_2 is given by

$$L_2 = \frac{1}{(1 + \kappa)^2} = \frac{1}{S_{210}^2} \tag{10.59}$$

The total insertion loss of this stabilization system is the sum of the above two losses. It may be noted that in Fig. 10.12(a), the load resistance Z_0 plays the role of a damping resistor as well, because at frequencies away from the resonance, the DR becomes transparent to the oscillator output signal which looks directly into the characteristic impedance Z_0.

The stabilized TDRO presented above, however, has certain limitations. The presence of two resonant circuits, the free-running oscillator tuned circuit, and the DR creates the problem of mode jumping and frequency hysteresis over the pulling range (stabilization bandwidth) [16]. The insertion loss introduced by this stabilization system, being a function of the performance desired, can also sometimes be excessive.

10.6 Stable Transistor DRO Design

A stable TDRO uses the dielectric resonator directly as an oscillator circuit element like the frequency determining circuit, feedback circuit, or matching circuit element. This approach, in addition to eliminating the undesirable features of mode jumping and hysteresis, reduces the size as well as the cost. A stable TDRO also has a high efficiency circuit, of more simple construction, making it an obvious choice over the stabilized TDRO.

In general, an oscillator circuit can be represented as either a series or a parallel circuit as shown in Fig. 10.13. The DR can be used to realize one or more of the immittances shown in this figure. The commonly used stable TDRO's can be divided into two types: one using the DR as a series feedback element, and the other using the DR as a parallel feedback element. We will now discuss the analysis and design of both types.

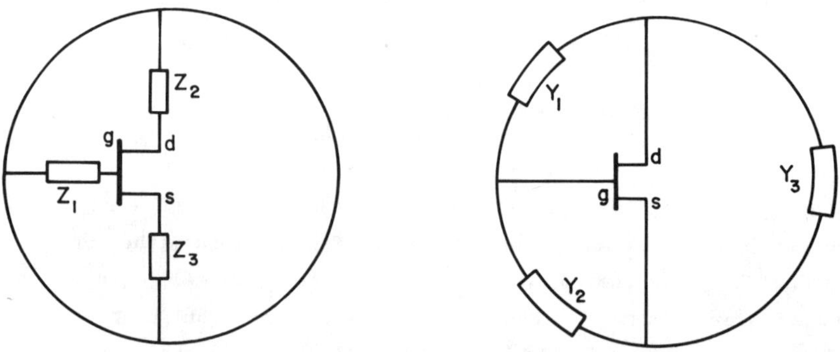

Fig. 10.13 General transistor oscillator configuration: (a) Series type and (b) Parallel type (reference [7], ©1983 Artech House)

Figure 10.14 presents the different configurations using the dielectric resonator as a series feedback element. Figures 10.14 (a) to (c) use the resonator at one terminal pair while the configuration (d) uses the DR as series feedback element at two terminal pairs of the transistors. As an example, we will now discuss the step by step procedure for the realization of the configuration shown in Fig. 10.14(a) using the 3-port S-parameters of NEC70000 described in Sec. 10.3. Comments on the design of other configurations are made where necessary.

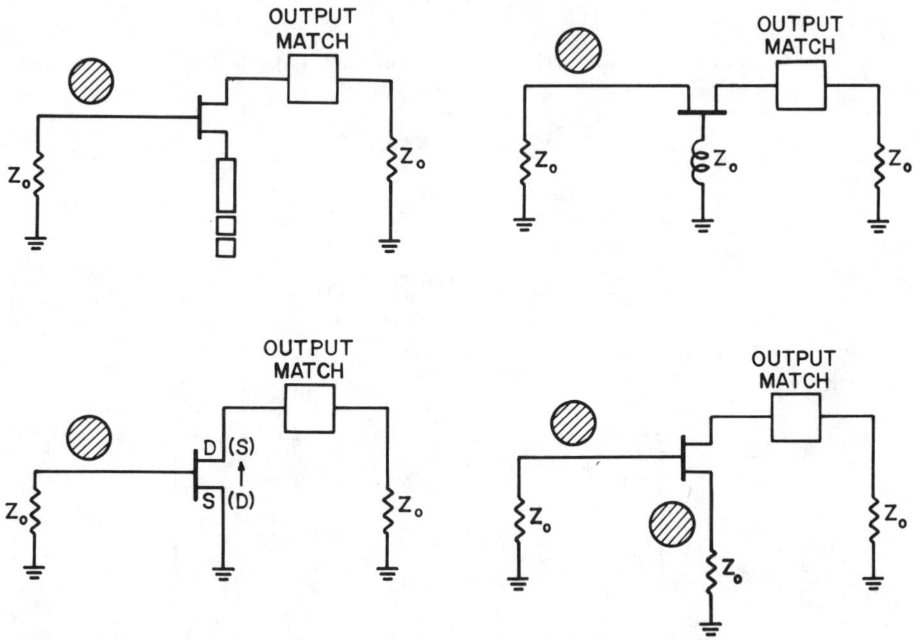

Fig. 10.14 Different configurations for series feedback transistor DRO's (reference [7], ©1983, Artech House)

As a first step in the design procedure, we have to determine the impedance Z_3 in Fig. 10.15. With the impedance Z_3 connected to the source terminal, the reflection coefficient Γ_3 becomes

$$\Gamma_3 = \frac{Z_3 - Z_0}{Z_3 + Z_0} \tag{10.60}$$

Substituting into (10.20), we obtain

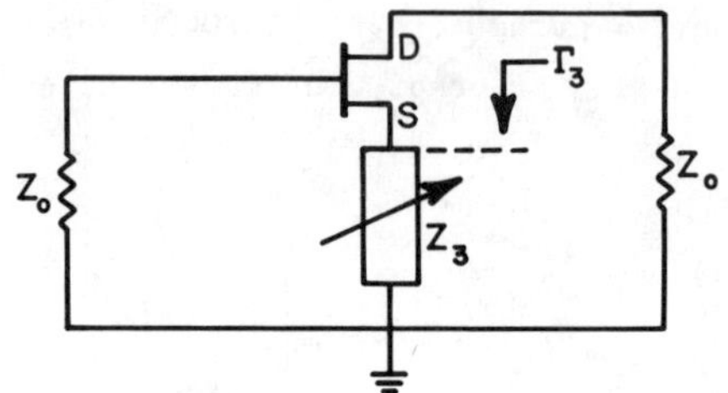

Fig. 10.15 Determination of Z_3 (reference [7], ©1983 Artech House)

$$b_1 = S_{11}\, a_1 + S_{12}\, a_2 + S_{13}\, b_3\, \Gamma_3 \tag{10.61}$$

$$b_2 = S_{21}\, a_1 + S_{22}\, a_2 + S_{23}\, b_3\, \Gamma_3 \tag{10.62}$$

$$b_3 = S_{31}\, a_1 + S_{32}\, a_2 + S_{33}\, b_3\, \Gamma_3 \tag{10.63}$$

Eliminating b_3 from (10.61) to (10.63), the reduced 2-port S-matrix is given by

$$\underset{\sim}{S}^T = \begin{pmatrix} S_{11} + \dfrac{S_{31}S_{13}\Gamma_3}{1 - S_{33}\Gamma_3} & S_{12} + \dfrac{S_{13}S_{32}\Gamma_3}{1 - S_{33}\Gamma_3} \\ & \\ S_{21} + \dfrac{S_{31}S_{23}\Gamma_3}{1 - S_{33}\Gamma_3} & S_{22} + \dfrac{S_{23}S_{32}\Gamma_3}{1 - S_{33}\Gamma_3} \end{pmatrix} \tag{10.64}$$

The aim is to determine the value of series feedback impedance Z_3 which will result in the modulus of S^T_{11} and S^T_{22} being greater than unity and, hence, create the desired instability in the transistor. For the example considered in Fig. 10.14(a), the open-circuited microstrip line represents a purely reactive impedance. The $|\Gamma_3| = 1$ plane determines a circle when mapped in the input and output reflection coefficient planes using well-known techniques [17]. The generalized mapping equations are given in Appendix 10.B. Figure 10.16 shows the $|\Gamma_3| = 1$ plane mapped into the S^T_{11} and S^T_{22} planes for the transistor NEC70000 at 10 GHz. The shaded area represents inductive impedance and the unshaded area represents the capacitive impedance in the source.

From Fig. 10.17 it may be noted that a negative reactance greater than -j30 ohms can be used to make both S^T_{11} and S^T_{12} greater than one. A value of -j159 ohms is selected, which can be realized by an open

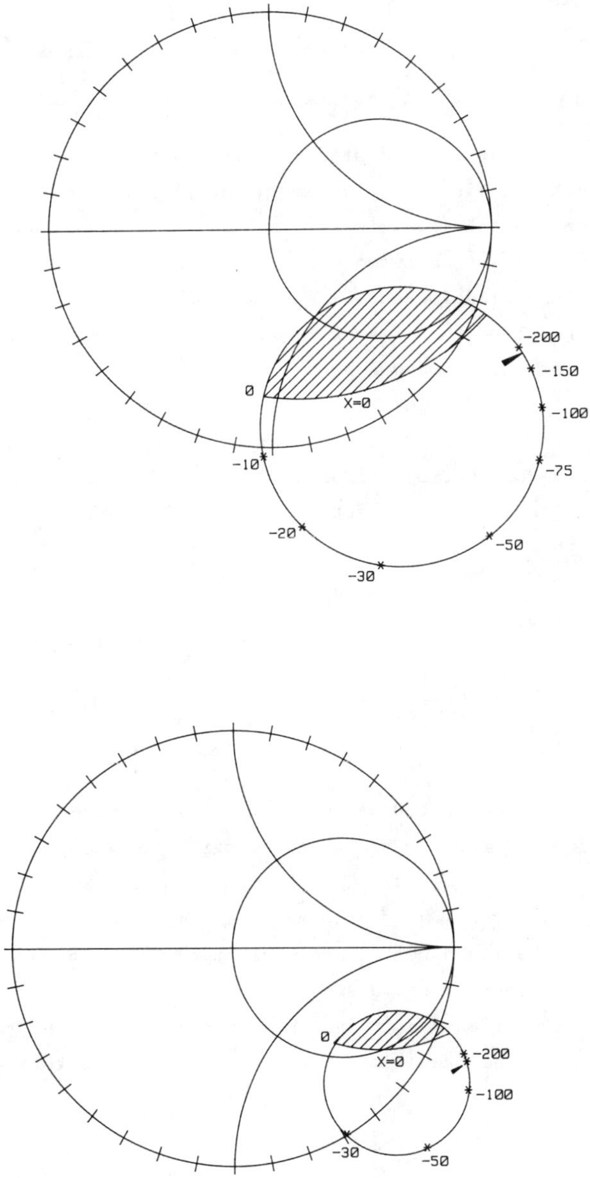

Fig. 10.16 Mapping of $|\Gamma_3| = 1$ in (a) S_{11}^T and (b) S_{22}^T

circuited stub $.048\lambda_g$ long or by a 0.1 pF capacitor. The reduced 2-port S-matrix is now given by

$$\underset{\sim}{S}^T = \begin{bmatrix} 1.34 \angle{-33.5°} & .49 \angle{70.4°} \\ .49 \angle{156°} & 1.16 \angle{-31.9°} \end{bmatrix} \quad (10.65)$$

A dielectric resonator coupled to a microstrip line can also be used as impedance Z_3 in Fig. 10.15 as shown in Fig. 10.14(d). The reflection coefficient in this case is a function of the coupling coefficient κ_3 and the distance θ_3 between the transistor plane and the resonator plane, as analyzed in Sec. 10.2. Using the matrix coefficients from (10.64), Γ_3 plane in terms of κ_3 and θ_3 can be mapped into any S-parameter of the reduced 2-port. Figure 10.17 shows, for example, mapping of the DR reflection coefficient plane into all the four S-parameters of an X-band FET at 8 GHz [12].

The non-concentric circles shown Fig. 10.17 are the constant coupling coefficient κ_3 (proportional to $|\Gamma_3|$) circles, while the radial arcs are the constant transmission electrical line length θ_3 (proportional to arg Γ_3) arcs. The relation between Γ_3, κ_3, θ_3 and frequency is given by

$$\Gamma_3 = \frac{\kappa_3}{\sqrt{(1+\kappa_3)^2 + \Delta^2}} \exp\left\{-2j\left[\theta_3 + \tan^{-1}(\Delta/(1+\kappa_3))\right]\right\} \quad (10.66)$$

with Δ as defined in (10.50). Figure 10.17 can be used to determine the DR position in order to create the desired instability in the transistor.

Continuing the TDRO design example (Fig. 10.14(a)), we have already determined the impedance Z_3; the resulting 2-port S-parameters are given in (10.65). In the second step of the procedure, we will determine the value of reflection coefficient Γ_1 which maximizes the reflection coefficient Γ_d at the drain port Fig. 10.18, using the following relation:

$$\Gamma_d = S_{22}^T + \frac{S_{12}^T S_{21}^T \Gamma_1}{1 - S_{11}^T \Gamma_1} \quad (10.67)$$

The reflection coefficient Γ_1 in this case is realized by a DR coupled to microstrip line and is characterized by the coupling coefficient κ_1 and the distance θ_1 between the transistor and the resonator plane.

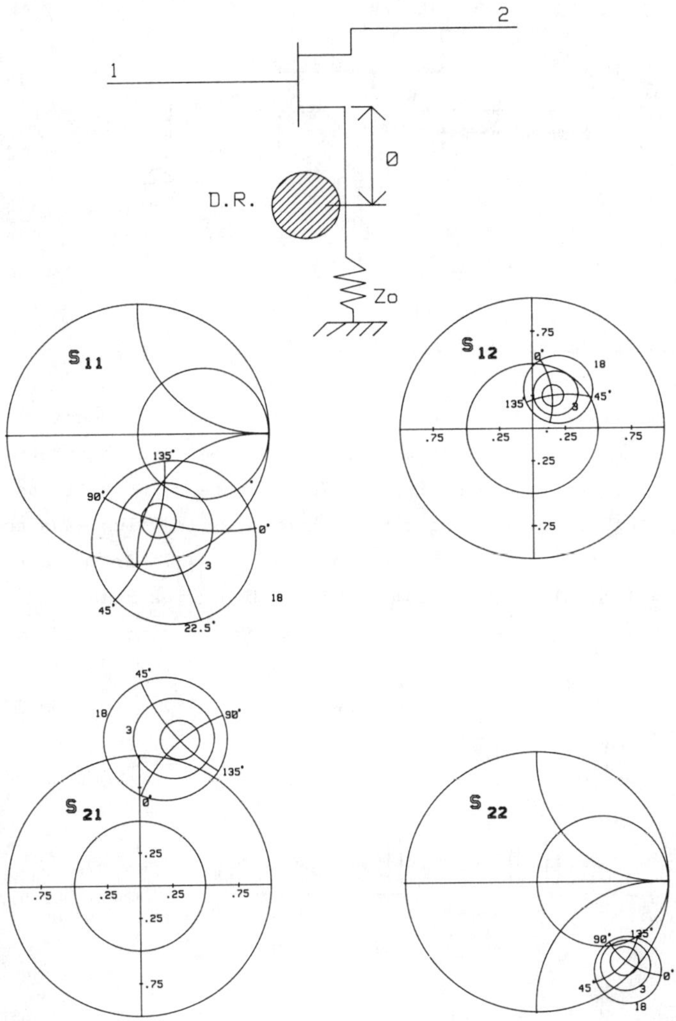

Fig. 10.17 DR as a series feedback element in source (reference [12], ©1982 IEEE)

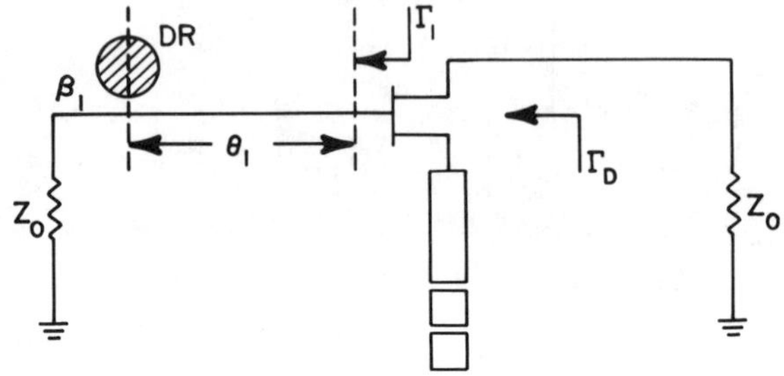

Fig. 10.18 Determination of DR position in the gate

In the present example, $|S^T_{11}|$ and $|S^T_{22}|$ being greater than unity, the mapping technique used in the first step cannot be used to determine the required Γ_1 and, hence, the position of DR. Instead, we will use the constant reflection coefficient circles approach, in which case the locus of constant reflection coefficient magnitude $|\Gamma_d|$ is drawn on the reflection coefficient plane Γ_1. From (10.67), the radius R and center Ω of the constant reflection coefficient circles can be determined to be

$$R = \frac{\left|\Gamma_d S^T_{12} S^T_{21}\right|}{\left|\Gamma_d\right|^2 \left|S^T_{11}\right|^2 - \left|\Delta\right|^2} \tag{10.68}$$

$$\Omega = \frac{S^{T^*}_{11}\left(\left|\Gamma_d\right|^2 - \left|S^T_{22}\right|\right)^2 + S^T_{22} S^{T^*}_{12} S^{T^*}_{21}}{\left|\Gamma_d\right|^2 \left|S^T_{11}\right|^2 - \left|\Delta\right|^2} \tag{10.69a}$$

where

$$\Delta = S^T_{11} S^T_{22} - S^T_{12} S^T_{21} \tag{10.69b}$$

Figure 10.19 shows various $|\Gamma_d|$ = constant circles on the $\Gamma_1(\kappa_1, \theta_1)$ plane for the example under consideration. The value of κ_1 and θ_1 can now be determined for a high value of $|\Gamma_d|$ (> 1).

The first step and the second step used the small-signal S-parameters to determine the impedances Z_3 and Z_1 to be connected at gate and source ports, respectively, in order to achieve a high value of reflection coefficient at the drain port. A number of approaches exist

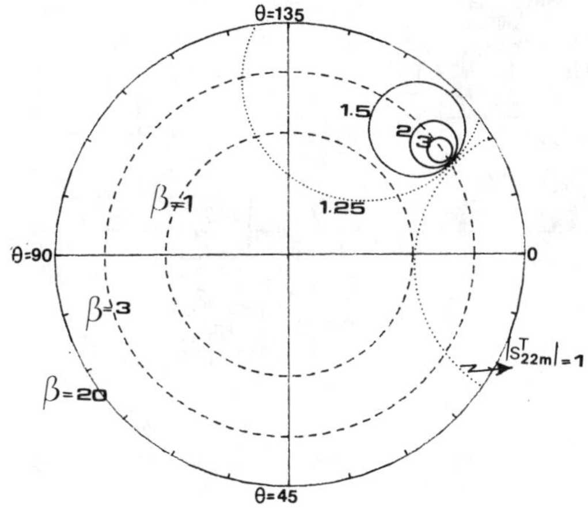

Fig. 10.19 Constant $|S^T_{22m}|$ circles in the DR reflection coefficient (κ_1, θ_1) plane

to determine the load circuit impedance Z_2 as shown in Fig. 10.13(a). The two common ones, which assume large-signal operation, are the device-line approach and the load-pull approach.

The device-line approach [17] is based on the measurement of the inverse reflection coefficient of the one-port and the added return power for different input power levels. The load impedance for maximum oscillator output power can be determined from the device line thus obtained. This approach, however, has the limitation that it requires the source resistance to be greater than the modulus of the device resistance. Otherwise, oscillation takes place and the device line cannot be measured.

The test set-up for the measurement of load-pull effects [18] is shown in Fig. 10.20. The oscillator acts as the RF power source for the system. The drain port of the transistor circuit (with optimized Z_1 and Z_3 connected) is attached to the load-pull measurement system at the input port of the reflectometer through 50 ohm line and powered up. The impedance shown on the polar display will be the impedance presented to the output of the oscillator device. Using the output tuner, contours of constant output power can then be drawn on a Smith chart using the power readings from the output power meter and an x-y recorder connected to the polar display. A typical load-pull data is shown in Fig. 10.21.

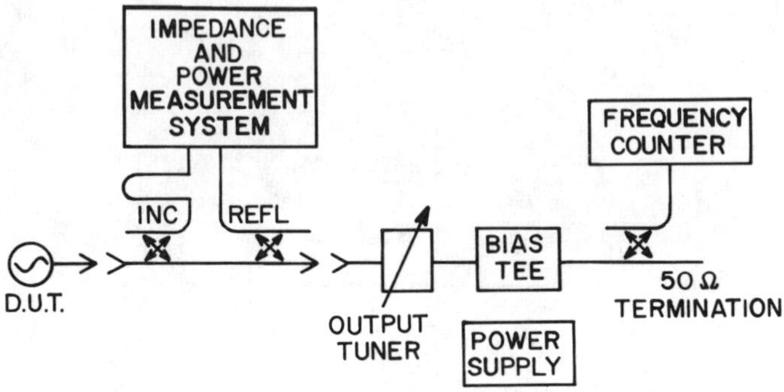

Fig. 10.20 Load-pull set-up for one-port oscillator (reference [18], ©1980 Microwaves & RF)

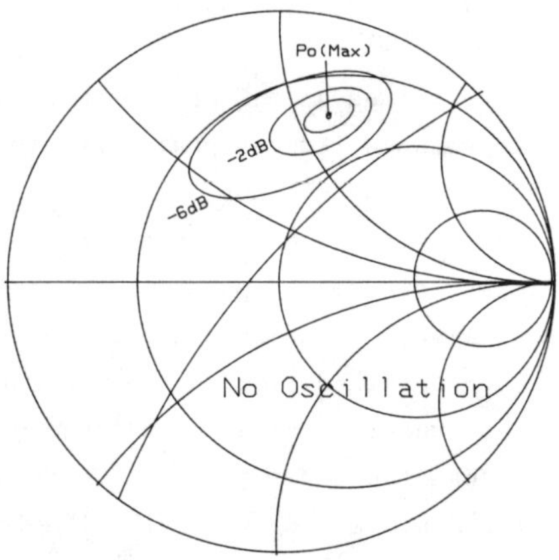

Fig. 10.21 Typical load-pull data (reference [18], ©1980 Microwaves & RF)

This load impedance chart can be used to design the output circuit for the transistor.

An alternative way of realizing a stable oscillator is using the DR simultaneously coupled to two microstrip lines as a parallel feedback element for a transistor. In this case, the transistor can be used as a two-port or a three-port device as shown in Figs. 10.22 (a) and (b), respectively.

In Fig. 10.22(a), the transistor is treated as a 2-port [19]. In this case, the input and output matching circuits for a common source transistor are designed for maximum transducer gain amplifier around the oscillator frequency f_o. Highly selective positive feedback between the input and the output can be used to create stable oscillations. This is achieved by feeding back a part of the output signal into the input through the dielectric resonator transmission filter. The lengths ℓ_1

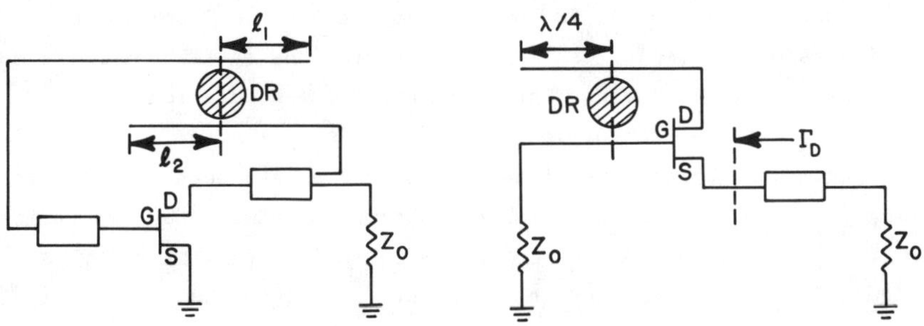

Fig. 10.22 Parallel feedback transistor DRO using transistor as (a) two-port and (b) three-port (reference [7], ©1983 Artech House)

and ℓ_2 are adjusted to achieve the phase shift around the loop, consisting of the amplifier and feedback circuit, equal to an integer multiple of 2π radians at f_o, i.e.,

$$\phi_A + \phi_R + \phi_C = 2\pi k, \quad k = 0,1,2 \tag{10.70}$$

where ϕ_A, ϕ_R, and ϕ_C are respective insertion phases of the amplifier, resonator, and the remaining part of the feedback circuit at f_o. The other condition for the oscillations to take place is that the open-loop small signal gain must exceed unity at f_o, i.e.,

$$G_A - L_R - L_C > 0 \text{ dB} \tag{10.71}$$

where G_A, L_R, and L_C are the amplifier gain, resonator filter loss, and loss in the other feedback components in dB, respectively. The necessary relations for determining the resonator insertion phase ϕ_R and insertion loss L_R can be determined from (10.16):

$$\phi_R = \tan^{-1}\left(\frac{-j2Q_u\delta}{1+\kappa_1+\kappa_2}\right). \tag{10.72}$$

and

$$L_R = 10\log\frac{(1+\kappa_1+\kappa_2)^2}{4\kappa_1\kappa_2} \tag{10.73}$$

The extent of the inequality given in (10.71) and, hence, the amplifier gain compression in the steady state oscillation condition affects the output power as well as the FM noise performance of the oscillator. Excessive gain compression can adversely affect the oscillator noise due to increased amplifier noise figure and amplitude modulation to phase modulation (AM-to-PM) conversion.

In Fig. 10.22(b) the transistor is treated as a 3-port [15,5]. In this case, the DR transmission filter is coupled between the two terminals of the transistor and the output is taken from the third. This configuration can be analyzed as a two-port containing DR as a parallel feedback network to a three-port device, the transistor (Fig. 10.23). The aim is to calculate the reflection coefficient at the output port as a function of the S-parameters of the device and those of the feedback network. Referring to Fig. 10.23, when ports 1 and 2 of the 2-port are connected to ports 1 and 2 of the 3-port, we have

$$a_{1A} = b_{1B}, \quad a_{2A} = b_{2B}, \quad a_{1B} = b_{1A}, \quad \text{and} \quad a_{2B} = b_{2A} \tag{10.74}$$

Using the above equalities, the input reflection coefficient at port 3 can be determined as follows:

$$\Gamma_3 = S_{33A} + \frac{S_{11B}S_{13A}S_{31A}+S_{12B}S_{23A}S_{31A}+S_{21B}S_{13A}S_{32A}+S_{22B}S_{23A}S_{32A}+\Delta_B\Delta'}{\Delta_B\Delta_{33A}-(S_{11B}S_{11A}+S_{12B}S_{21A}+S_{21B}S_{12A}+S_{22B}S_{22A})+1} \tag{10.75}$$

where

$$\Delta' = S_{32A}\Delta_{23A} + S_{13A}\Delta_{13A}$$

$$\Delta_{33A} = S_{11A}S_{22A} - S_{12A}S_{21A}$$

$$\Delta_{23A} = S_{12A}S_{31A} - S_{11A}S_{32A}$$

$$\Delta_{13A} = S_{21A}S_{32A} - S_{22A}S_{31A}$$

$$\Delta_B = S_{11B}S_{22B} - S_{12B}S_{21B}$$

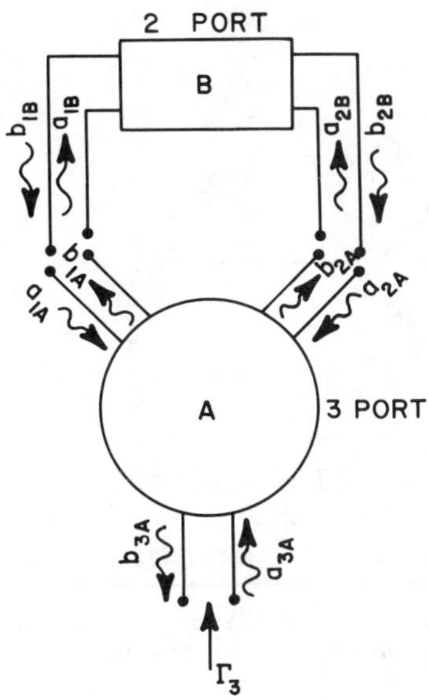

Fig. 10.23 Two-port as a parallel feedback element to a three-port (reference [5], ©1984 IEEE)

The S-parameters of the 2-port B containing the DR in terms of the coupling coefficient κ and the line length θ, were determined in Sec. 10.2. Using the relations (10.15) and (10.75), the reflection coefficient at the drain can be plotted as a function of the feedback parameters κ and θ of the network B connected between source and gate. Figure 10.24 shows such a plot for a half-micron FET at 9 GHz, which can be used to determine the feedback parameters to maximize the reflection gain (> 1). The output matching circuit can now be determined using the load pull approach described above.

The 3-port approach can, in fact, be used to determine the maximum reflection gain at any of the three terminals of the transistor as a function of parameters θ and κ of the feedback network between the other 2-ports. The preferred output port which offers maximum reflection gain can thus be determined. A damping resistance, shown in Fig. 10.22(b), is generally required at one of the FET ports in order to avoid spurious oscillations. This addition only affects the S-parameters of the feedback network (the 2-port DR transmission filter).

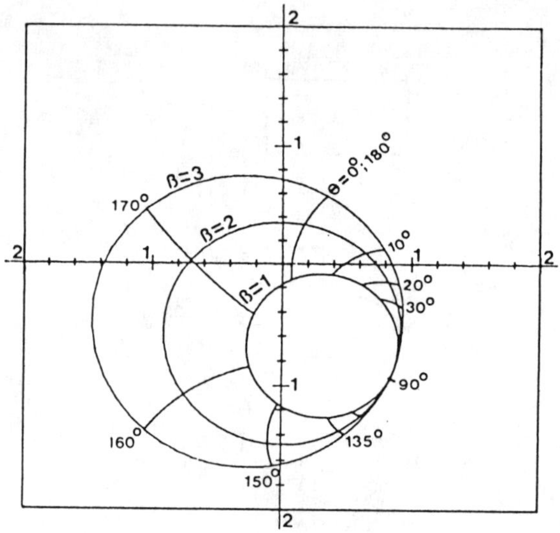

Fig. 10.24 Drain reflection coefficient as a function of DR feedback between gate and series (reference [5], ©1984 IEEE)

10.7 Temperature Stability of DRO's

Microwave sources with high temperature stabilities have been realized in the past using Invar cavities or by phase locking the source to a frequency-multiplied very high frequency (VHF) crystal oscillator operating in a temperature-stabilized environment ("ovenized"). These approaches are cumbersome as well as expensive. With the advent of high performance dielectric resonators, TDRO's now present a miniature, elegant, and inexpensive way of achieving high temperature stabilities.

A free-running transistor oscillator is known to have a negative temperature coefficient. The DR having a positive temperature coefficient is thus required to compensate for the frequency drift with temperature. At present, the temperature coefficient τ_f of the DR can be controlled by varying its composition. The DR's with τ_f equal to anywhere between +9 to -9 ppm/°C are commercially available.

The temperature stability of a TDRO can be analytically determined in terms of the coupling coefficient κ, the Q factor, and the rate of change of transistor reflection phase with temperature. The detailed temperature stability analysis of the TDRO configuration given in Fig. 10.14(a) has been treated in Ch. 7. Using this approach and a composite material DR, the temperature stabilities of 0.1 ppm/°C have been reported from -20 to +80°C [20].

A digital compensation technique for realizing high temperature stability TDRO has recently been reported [21]. In this case, a temperature sensor is mounted in the oscillator to detect the temperature changes. The data are digitized and fed to an electrically programmable read-only memory (EPROM), pre-programmed with temperature characteristics of the oscillator. The look-up table in the read-only memory (ROM) provides the necessary digital temperature correction word, which is converted to an analog signal (with proper synchronization between A/D and D/A) and applied to the DRO for frequency correction, as shown in the block diagram of Fig. 10.25. The correction signal can be applied to the varactor in case of varactor-tuned DRO [21] or to control the phase shift in the feedback loop in case of parallel feedback DRO. Using this digital compensation technique, temperature stability better than ± 0.2 ppm/°C has been achieved.

Temperature stability of a DRO can also be enhanced significantly by inserting the oscillator package in a temperature-stabilized oven. Using a heater element, a quick-response thermistor and the associated

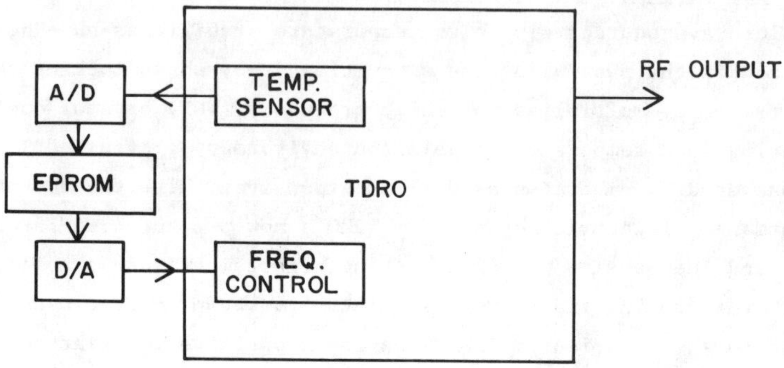

Fig. 10.25 Temperature stabilization of a TDRO using digital techniques

control circuit, the package temperature can be maintained within 5°C. The oscillator package itself is isolated from the external environment by using an outer package. The two packages are thermally connected through stand-offs which are determined taking into account the dc dissipation in the oscillator and the desired temperature of the package. The package temperature is generally controlled at 5 to 10°C above the maximum temperature for which the ovenized DRO is specified. A total frequency stability of better than ± 5 ppm can be obtained from - 40 to 70°C. This oscillator is suitable for use as a local oscillator in radio communication and test equipment, in place of certain phase-locked oscillators.

10.8 Tuning of Transistor DRO

TDRO is basically a fixed frequency oscillator with its frequency determined by the resonator material permittivity, resonator dimensions, and the shielding conditions, as discussed in earlier chapters. The oscillation frequency can, however, be tuned over a narrow frequency range using different approaches depending on the requirements. The frequency tuning of the TDRO can be accomplished mechanically or electrically, as discussed below.

Figure 10.26 shows the configuration used for mechanical frequency tuning of the TDRO. Use is made of the fact that the resonant frequency of the DR is highly sensitive to the shielding, i.e., to the proximity of the ground plane. A tuning screw is inserted from the top cover of

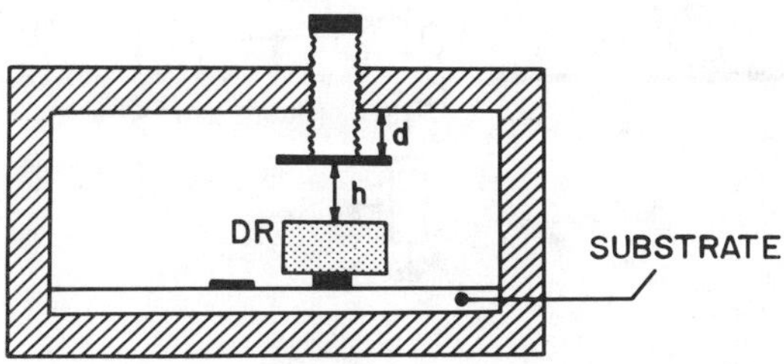

Fig. 10.26 Mechanical tuning of a dielectric resonator

the package, right above the DR. The increase in the tuning screw depth d increases the resonant frequency of the DR in the commonly used $TE_{01\delta}$ mode. Care should be taken to keep the distance h between the resonator and the tuning screw at least 0.5 times the resonator height so as not to degrade the DR quality factor. A mechanical frequency tuning range of ± 1% can be obtained in a TDRO without noticeably affecting the FM noise and output power (see also Sec. 4.9).

The frequency of the TDRO can be electrically tuned by using a number of different approaches like varactor tuning, ferrite tuning, bias tuning, and optical tuning. Electrical tuning can be accomplished over a very small bandwidth without significantly affecting the oscillator performance. This tuning can be used for different applications like digital temperature compensation, low deviation frequency modulated sources, injection locking, etc. A brief description of the various approaches for electrically tuning the TDRO follows below.

In the varactor-tuned TDRO, a varactor in association with a microstrip line is made to resonate around the DR frequency. This resonant circuit is electromagnetically coupled to the dielectric resonator, forming a pair of mutually coupled resonant circuits. By varying the varactor capacitance with the bias voltage, the resonant frequency of the DR, coupled to varactor-microstrip on one side and a 50 ohm microstrip line on the other, can now be tuned. Figure 10.27 shows a typical configuration for coupling the varactor and DR. Tighter coupling between the DR and varactor will result in greater frequency control at the cost of decreased DR Q factor and, hence, increased FM noise.

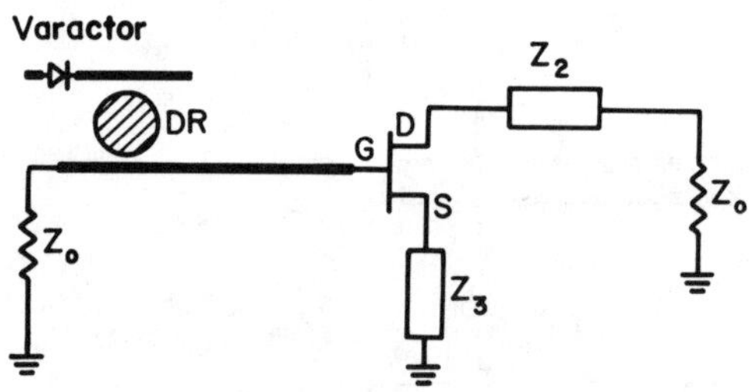

Fig. 10.27 Varactor tuning of a dielectric resonator

Figure 10.28 shows the effect of varactor tuning on a low-noise TDRO FM noise for the varactor coupling adjusted for 0.1 % frequency control at 11 GHz [22]. Using another configuration with two varactors on a quartz spacer placed directly above the resonator, a tuning bandwidth of 0.75 % has been reported [23].

The DR can also be tuned by attaching a microwave ferrite on the resonator and applying a magnetic field to it. The magnetic field controls the magnetic properties of the ferrite and, hence, the field distributions in and around the DR resulting in a shift in the resonant frequency. Tuning bandwidths on the order of 0.5 % [24] and 1 % [23] have been reported. This method, however, cannot be used effectively in practice, due to serious limitations given below:

- It is a current-driven tuning method and has a slow tuning speed like that of YIG-tuned oscillators.
- The large size and bulk of the electromagnet necessary to provide the variable magnetic field.

• Excessive power consumption of the driving circuit.

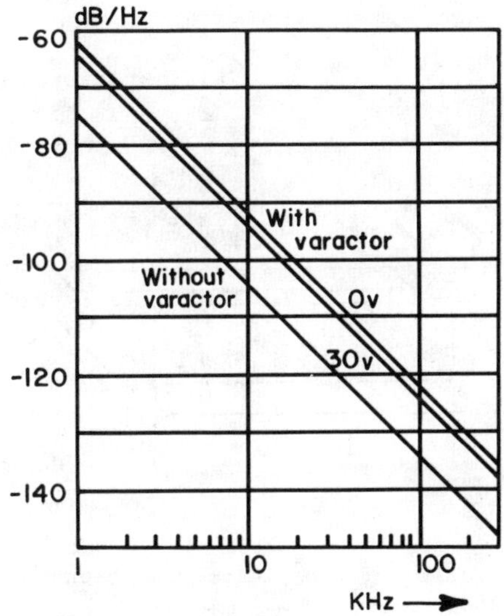

Fig. 10.28 Effect of varactor tuning on DR oscillator FM noise (reference [22], ©1983 Eur. Microwave Conf.)

Another possibility for electronic tuning is to use the bias voltage for this purpose. The frequency of any oscillator is known to be sensitive to the bias voltage. This dependence is generally described by the pushing figure. Unfortunately, the change in bias voltage also affects the output power, thus, making it difficult to use for frequency modulation purposes. The bias circuit can, however, be designed in such a way as to minimize the output power variation with bias voltage. Using such a technique, Christ and Horowitz have reported 4.5 MHz frequency tuning range at 10 GHz with less than a 1 dB variation of oscillator power output [25].

A TDRO inherently has a high Q factor and, hence, a low pushing figure. This limits the bias-tuned frequency range, which can only be increased by intentionally reducing the oscillator Q at the cost of degradation of other oscillator characteristics like FM noise, temperature stability, etc.

Optical control of microwave devices and subsystems is a rapidly growing field of research. The resonant frequency of the DR used in the TDRO can be optically modulated and tuned as shown in Fig. 10.29. A

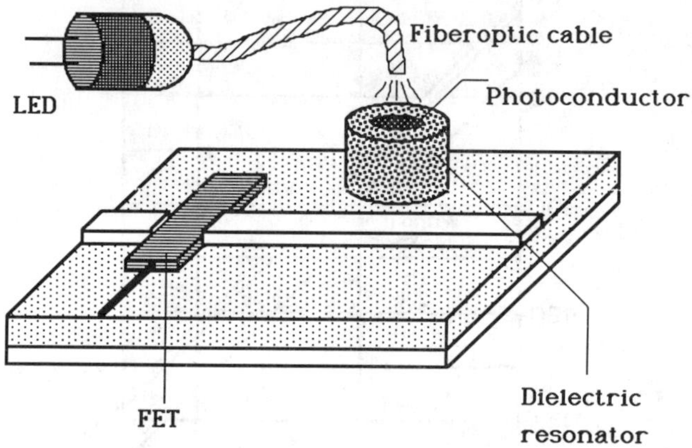

Fig. 10.29 Optical tuning of the dielectric resonator (reference [30], ©1985 IEEE)

photosensitive material like high resistivity silicon is placed directly on the dielectric resonator. Light from a laser or light emitting diode (LED) is brought through an optical fiber to illuminate the photosensitive material, changing its conductivity and perturbing the electromagnetic field in and around the resonator. This perturbation results in a shift in the center frequency of the TDRO. Using this technique, tuning bandwidths of better than 0.1 % have been achieved at X-band [26,30].

10.9 TDRO Measurements Using Network Analyzer

The measurement of oscillator characteristics such as power, frequency, harmonics, external Q factor, and FM noise are well known. These measurements, however, require a bench loaded with expensive laboratory equipment including power meters, frequency counters, spectrum analyzers, signal generators, circulators, variable attenuators, sliding shorts, etc. In this section we will discuss the measurement of major oscillator characteristics using a commonly available network analyzer and the signal generator [27,28].

The principle of measurements is based on injection locking the TDRO under test with the signal available at the "unknown" port of the network analyzer. An injection locking polar diagram (ILPD) is thus obtained and displayed on the polar display. The network analyzer is used to measure the magnitude and phase of the injection gain under locked conditions. The resulting ILPD resembles ordinary impedance measurements and can be used to measure the important characteristics of the oscillator. Moreover, the ILPD can be used to optimize key oscillator operating parameters in a manner analogous to using the network analyzer in its conventional role involving filters and amplifiers.

The Hewlett-Packard HP8410 network analyzer was used for the system described here. For low-power oscillators (up to about +10 dBm), the configuration shown in Fig. 10.30 was used. Figure 10.31 shows a higher power configuration that is useful up to about a 30 dBm output. To increase the gain measurement dynamic range and to protect the sensitive harmonic converter from inadvertent overload, attenuators of 30 dB and 3 dB are inserted in the test channel and reference channel, respectively. With a range of RF power from -16 to -43 dBm at the reference channel input of the harmonic frequency converters, the use of a signal generator with a power output of from -10 to +17 dBm allows complete coverage of the range of possible power variations encountered. For example, using an oscillator having a +10 dBm output, one can measure an injection gain range of from 13 to 40 dB (Fig. 10.30).

The equipment can be calibrated for either a transmission or a reflection coefficient mode using a through-line and a short circuit. The phase calibration can be accomplished by injection-locking the circuit with very high injection gain (> 30 dB). In this case the phase difference between the locked signal and locking source at the center frequency can be assumed to be zero at the parallel resonant circuit oscillator

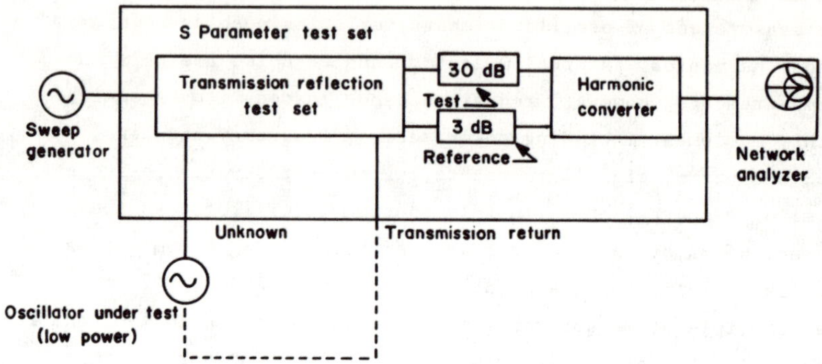

Fig. 10.30 Low-power oscillator measurement set-up (reference [28], ©1983 IEEE)

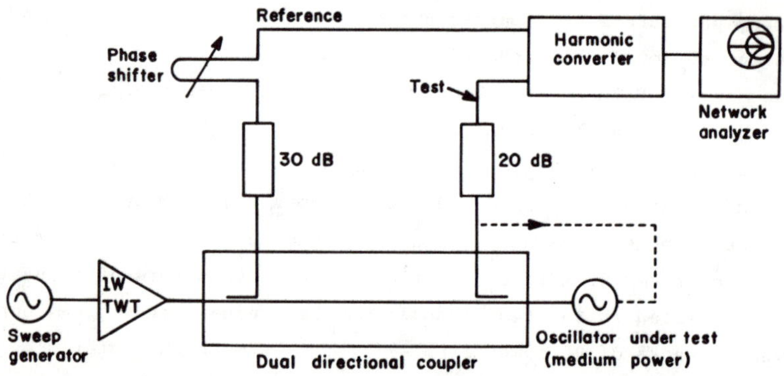

Fig. 10.31 Medium power oscillator measurement set-up (reference [28], ©1983 IEEE)

plane. The reference plane extension control is adjusted to bring the polar injection locking diagram into the desired plane. The free-running oscillator frequency can be measured using a counter to locate the θ_0 (when $\Delta\omega = 0$) point on the injection-locking diagram.

With the oscillator connected to the S-parameter test set's unknown port and the sweep generator set to sweep across about 10 % of the known oscillator frequency, noise can be observed on both the polar and rectangular display. This noise is in addition to the desired signal peak on the rectangular display and the ILPD on the polar display.

Next, the sweep band is gradually reduced, and the sweep generator power and the test channel gain are adjusted to obtain the desired ILPD on the polar display for various values of injection gain. The injection gain is thus directly measurable.

Reducing the sweep bandwidth to a usable minimum helps eliminate the noise and improve the ILPD. Figure 10.32 shows the ILPD for different values of sweep generator power output and, hence, the injection gain. Figure 10.33 is an ILPD of a FET DRO, operating at 9.5 GHz.

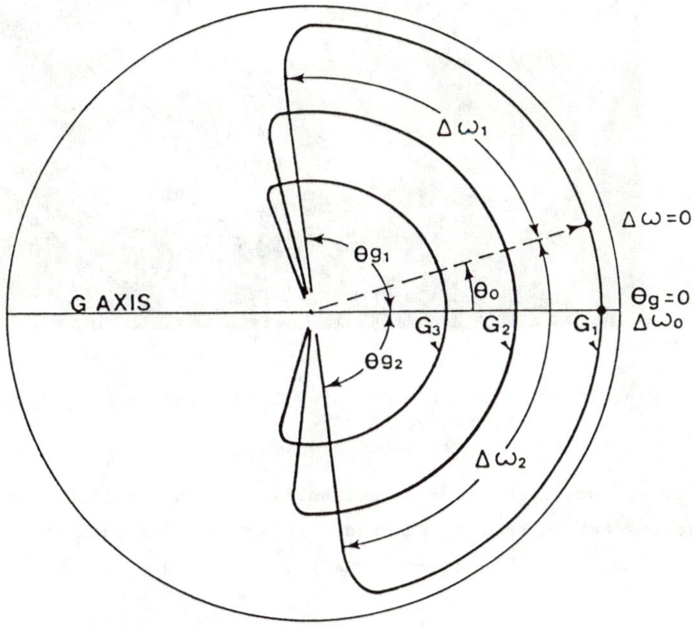

Fig. 10.32 Typical ILPD and definition of various parameters (reference [27], ©1984 Microwaves & RF)

For RF power measurements, the test channel gain is set to correspond to an injection gain of 20 dB. Using the signal generator output power level control, the G = 0 point on the ILPD is moved to the polar display's outer edge. The oscillator power can then be determined to the accuracy of the signal generator output controls without using a power meter. For the low-power test configuration of Fig. 10.30, the oscillator output power is equal to the sweep generator output power.

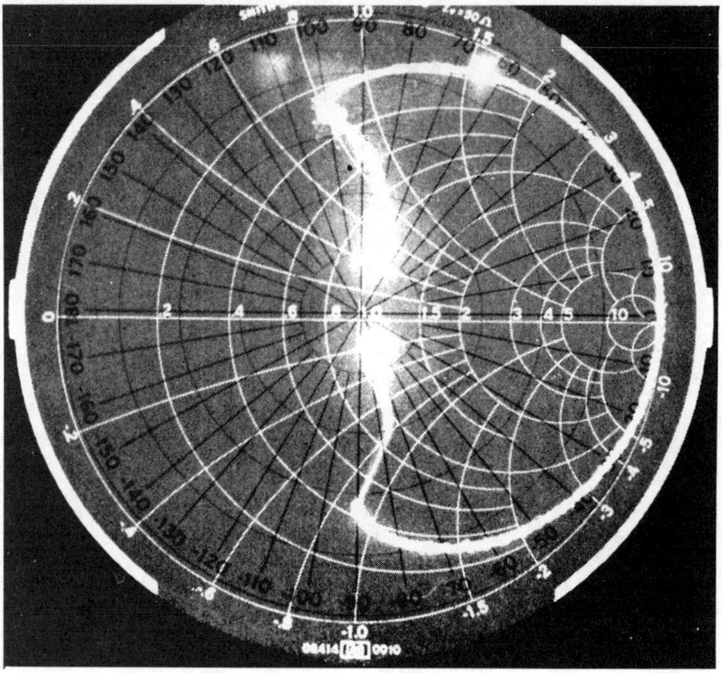

Fig. 10.33 ILPD of a DRO at 9.5 GHz (reference [27], ©1984 Microwaves & RF)

Having established the injection gain, G, on the ILPD, the remaining parameters can be easily measured. Injection bandwidth, $\Delta\omega$, is given by $|\Delta\omega_1| + |\Delta\omega_2|$ and the external quality factor, Q_{ex}, can be determined from

$$Q_{ex} = \frac{2\omega_0}{\Delta\omega} \cdot \frac{1}{\sqrt{G}} \tag{10.76}$$

Significantly, the measurement is made without a spectrum analyzer, circulator, directional coupler, and power meter, which are usually required. Moreover, the display presented by the network analyzer is generally more informative than that obtained from a spectrum analyzer because it also gives the injection gain phase information, which is not available from a spectrum analyzer.

The nonlinear constants α and K [29] of the oscillator can be determined from the measured values of $\Delta\omega_0$, θ_0, and G using the following relations [28]:

$$\alpha = \frac{G \sin \theta_0}{1 - G \cos \theta_0} \tag{6.77}$$

and

$$K = \frac{\Delta\omega_0 (G - 1)}{2\alpha} \tag{10.78}$$

This approach provides a rapid means for determining α and K and graphically depicting the effects of parameters such as biasing voltage on these nonlinear constants. For low values of injection gain, the elliptical power variation in the injection-locked frequency range as well as in the locking range asymmetry (which is a function of α) can be read directly from the ILPD.

The ILPD yields valuable insight into the oscillator output-matching circuit as well. Frequency and phase jumps within the locking range are readily displayed and recognized (Fig. 10.34). The network analyzer, thus, provides a valuable tool for testing and aligning the output load circuit. Since the injection gain represents oscillator power output (for constant oscillator signal input), the oscillator can easily be tuned for the desired power output. The method employed is comparable to using a network analyzer for tuning a filter or amplifier. This results from the comprehensive display that immediately shows the effect of changes to load or circuit impedances.

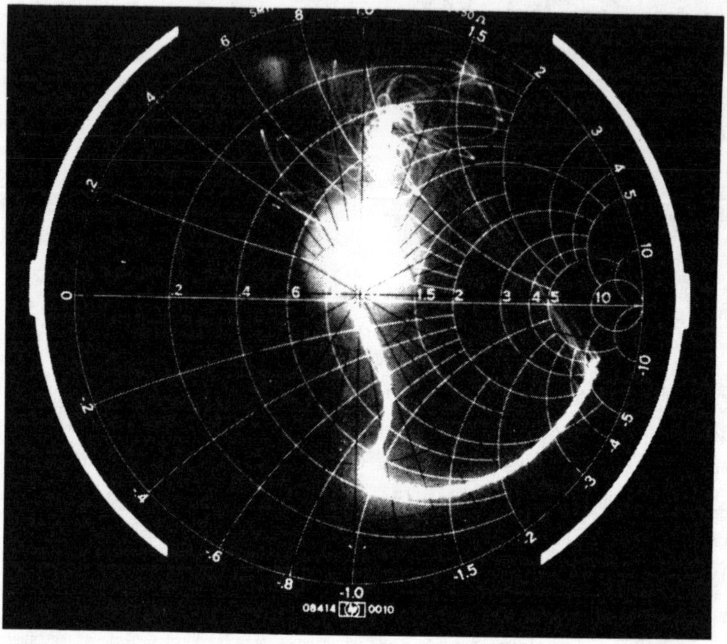

Fig. 10.34 Phase jumping of TDRO (reference [27], ©1984 Microwaves & RF)

Appendix 10.A CONVERSION OF TWO-PORT S-PARAMETERS TO THREE-PORT S-PARAMETERS

Equations (10.21a) and (10.21b) represent six equations in terms of nine 3-port S-parameters to be determined. Other equations necessary to determine all the nine S-parameters of the 3 port can be found as follows.

From (10.20) we can write:

$$b_1 = S_{11}a_1 + S_{12}a_2 + S_{13}a_3 \quad (10.A1)$$

$$b_2 = S_{21}a_1 + S_{22}a_2 + S_{23}a_3 \quad (10.A2)$$

$$b_3 = S_{31}a_1 + S_{32}a_2 + S_{33}a_3 \quad (10.A3)$$

If port 3 of the transistor shown in Fig. 10.15 is connected to ground, making it a common source configuration, we have

$$\Gamma_3 = a_3/b_3 = -1 \quad (10.A4)$$

Using (10.A4) in (10.A1) to (10.A3) and eliminating a_3 and b_3, we have

$$\begin{bmatrix} b_1 \\ b_2 \end{bmatrix} = \begin{bmatrix} S_{11} - \dfrac{S_{13}S_{31}}{1 + S_{33}} & S_{12} - \dfrac{S_{13}S_{32}}{1 + S_{33}} \\ S_{21} - \dfrac{S_{31}S_{23}}{1 + S_{33}} & S_{22} - \dfrac{S_{23}S_{32}}{1 + S_{33}} \end{bmatrix} \begin{bmatrix} a_1 \\ a_2 \end{bmatrix} \quad (10.A5)$$

or

$$|b\rangle = \underset{\sim}{S}^T\, |a\rangle \quad (10.A6)$$

where $\underset{\sim}{S}^T$ represents the reduced 2-port S-matrix of the transistor with terminal pair 3 connected to ground. This relation represents four equations, which combined with six equations from (10.21a) and (10.21b), represent the necessary ten equations to be solved to determine all the nine 3-port S-parameters. The final relations thus obtained are as given in simplified form below [31]:

$$S_{33} = \frac{\sum\limits_{\substack{i=1,2\\ j=1,2}} S_{ij}^{T}}{4 - \sum\limits_{\substack{i=1,2\\ j=1,2}} S_{ij}^{T}} \tag{10.A7}$$

$$S_{32} = \frac{1 + S_{33}}{2} (1 - S_{12}^{T} - S_{22}^{T}) \tag{10.A8}$$

$$S_{23} = \frac{1 + S_{33}}{2} (1 - S_{21}^{T} - S_{22}^{T}) \tag{10.A9}$$

$$S_{22} = S_{22}^{T} + \frac{S_{23}S_{32}}{1 + S_{33}} \tag{10.A10}$$

$$S_{11} = 1 - S_{21} - S_{31} \tag{10.A11}$$

$$S_{13} = 1 - S_{23} - S_{33} \tag{10.A12}$$

$$S_{31} = 1 - S_{33} - S_{32} \tag{10.A13}$$

$$S_{12} = 1 - S_{22} - S_{32} \tag{10.A14}$$

$$S_{21} = 1 - S_{22} - S_{23} \tag{10.A15}$$

Appendix 10.B GENERALIZED MAPPING EQUATIONS

Equation (10.64) can be written in a general from:

$$S = A + \frac{BC}{\frac{1}{\Gamma} - D} \tag{10.B1}$$

where

$$\Gamma = \frac{Z - Z_0}{Z + Z_0} \tag{10.B2}$$

The mapping equation for (10.B1) can be written as

$$S = M + R\,\frac{Z - NZ_0}{Z + N^* Z_0} \tag{10.B3}$$

where

$$M = A + \frac{BC}{1 - |D|^2} \cdot D^* \tag{10.B4}$$

$$R = \frac{BC}{1 - |D|^2} \cdot \frac{1 - D^*}{1 - D} \tag{10.B5}$$

and

$$N = \frac{1 + D^*}{1 - D^*} \tag{10.B6}$$

Equations (10.B4) and (10.B6) can be used to map the Γ plane into S plane as shown in Fig. 10.B1.

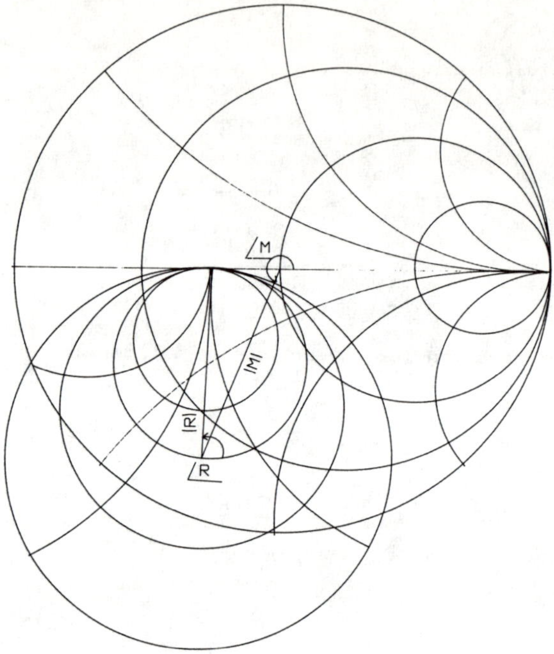

Fig. 10.B1 Mapping of Γ plane into S plane

References

[1] J.K. Plourde and C.L. Ren, "Application of dielectric resonators in microwave components," IEEE Trans. Microwave Theory Tech., vol. MTT-29, pp. 754-769, Aug. 1981.

[2] P. Guillon et al., "Microstrip bandstop filter using dielectric resonator," Proc. IEE part H., vol. 128, pp. 151-154, June 1981.

[3] A.P.S. Khanna and Y. Garault, "Determination of loaded unloaded and external quality factors of a dielectric resonator coupled to a microstrip line," IEEE Trans. Microwave Theory Tech., vol. MTT-31, pp. 261-264, March 1983.

[4] A.P.S. Khanna, "Q measurements of microstrip-coupled dielectric resonators," Microwaves and R.F., vol. 23, pp. 81-86, Jan. 1984.

[5] A.P.S. Khanna, "Parallel feedback FETDRO using 3-port S-parameters," IEEE MTT-S Int. Microwave Symp. Dig., San Francisco, pp. 181-183, May 30 - June 1, 1984.

[6] G.E. Bodway, "Circuit design and characterization of transistor by means of three-port scattering parameters," Microwave J., vol. 11, no. 5, May 1968.

[7] R. Soares, J. Graffeuil, and J. Obregon, Applications of GaAs MESFETS. Dedham, MA: Artech House, 1983.

[8] A.P.S. Khanna and J. Obregon, "Microwave oscillator analysis," IEEE Trans. Microwave Theory Tech., vol. MTT-29, pp. 606-607, June 1981.

[9] G.R. Basawapatna and R.B. Stancliff, "An unified approach to the design of wide-band microwave solid state oscillators," IEEE Trans. Microwave Theory Tech., vol. MTT-27, pp. 379-385, May 1979.

[10] A.P.S. Khanna, "Oscillateurs microondes stables integres a transistors et resonateurs dielectriques," Thesis Docteur-Ingenieur, University of Limoges, France, Sept. 1981.

[11] H. Abe et al., "A highly stabilized low-noise GaAsFET integrated oscillator with a dielectric resonator in the C-band," IEEE Trans. Microwave Theory Tech., vol. MTT-26, pp. 156-162, March 1978.

[12] A.P.S. Khanna, J. Obregon, and Y. Garault, "Efficient low noise three port x-band FET oscillator using two dielectric resonators," IEEE MTT-S Int. Microwave Symp. Dig., Dallas, pp. 277-279, 1982.

[13] A. Podcameni and L. Bermudez, "Large signal design of GaAsFET oscillators using input dielectric resonators," IEEE Trans. Microwave Theory Tech., vol. MTT-31, pp. 358-361, April 1983.

[14] A.P.S. Khanna et al., "New very high Q microwave transistor oscillator using dielectric resonators," Proc. Eur. Microwave Conf., Amsterdam, pp. 457-461, 1981.

[15] O. Ishihara et al., "A highly stabilized GaAsFET oscillator using a dielectric resonator feedback circuit in 9-14 GHz," IEEE Trans. Microwave Theory Tech., vol. MTT-28, pp. 817-824, Aug. 1980.

[16] K. Shirahata, "Stabilization of solid state microwave oscillator by loading BRF," Proc. Eur. Microwave Conf., London, 1969.

[17] W. Wagner, "Oscillator design by device line measurement," Microwave J., vol. 22, pp. 43-48, Feb. 1979.

[18] D. Poulin, "Load-pull measurements help you meet your match," Microwaves, vol. 19, pp. 61-65, Nov. 1980.

[19] Z. Galani et al., "Analysis and design of a single-resonator GaAsFET oscillator with noise degeneration," IEEE Trans. Microwave Theory Tech., vol. MTT-32, pp. 1556-1565, Dec. 1984.

[20] C. Tsironis and P. Lesartre, "Temperature stabilization of GaAsFET oscillators with dielectric resonators," Proc. Eur. Microwave Conf., Helsinki, pp. 181-186, 1982.

[21] J. Lee, et al., "Digital and analog frequency-temperature compensation of dielectric resonator oscillators," IEEE MTT-S Int. Microwave Symp. Dig., pp. 277-279, San Francisco, 1984.

[22] M. Camiade, et al., "Low noise design of dielectric resonator FET oscillators," Proc. 13th Eur. Microwave Conf., Germany, pp. 297-302, 1983.

[23] A.N. Farr, et al., "Novel technique for electronic tuning of dielectric resonators," Proc. 13th Eur. Microwave Conf., Germany, pp. 791-796, 1983.

[24] S. Watanabe, et al., "Very high Q dielectric resonators voltage controlled oscillator," Proc. 13th Eur. Microwave Conf., Paris, pp. 269-273, 1978.

[25] S. Christ and P. Horowitz, "Miniature X-band transmitter uses nine-volt battery," Microwave System News, pp. 84, Dec. 1982.

[26] P.R. Hercsfeld et al, "Optically tuned and FM modulated X-band dielectric resonator oscillator," Proc. 14th Eur. Microwave Conf., Liege, pp. 268-273, 1984.

[27] A.P.S. Khanna and J. Obregon, "Network analyzer doubles as oscillator diagnostician," Microwaves & RF, vol. 23, pp. 106-112, July 1984.

[28] A.P.S. Khanna and J. Obregon, "Direct Measurement of the non-linear MIC oscillator characteristics using injection locking polar diagram," IEEE MTT-S Int. Microwave Symp. Dig., Boston, pp. 501-503, 1983.

[29] J. Obregon and A.P.S. Khanna, "Exact derivation of the non-linear negative resistance oscillator pulling figure," IEEE Trans. Microwave Theory Tech., vol. MTT-30, pp. 1109-1111, July 1982.

[30] P.R. Herczfeld, A.S. Daryoush, V.M. Contarino, A. Rosen, Z. Turski, A.P.S. Khanna, "Optically controlled microwave devices and circuits," IEEE MTT-S Int. Microwave Symp. Dig., St. Louis, pp. 211-214, 1985.

[31] A.P.S. Khanna, "Three port S-parameters ease GaAs FET designing," Microwaves & RF, vol. 24, pp. 81-84, Nov. 1985.

INDEX